Effects of Climate Change on Birds

Effects of Climate Change on Birds

Second edition

EDITED BY

Peter O. Dunn

Department of Biological Sciences, University of Wisconsin-Milwaukee, USA

Anders Pape Møller

Ecologie Systematique Evolution, Université Paris-Sud, France

Great Clarendon Street, Oxford, OX2 6DP,
United Kingdom

Oxford University Press is a department of the University of Oxford.
It furthers the University's objective of excellence in research, scholarship,
and education by publishing worldwide. Oxford is a registered trade mark of
Oxford University Press in the UK and in certain other countries

First Edition published in 2010
Second Edition published in 2019

Impression: 1

Published in the United States of America by Oxford University Press
198 Madison Avenue, New York, NY 10016, United States of America

British Library Cataloguing in Publication Data
Data available

Library of Congress Control Number: 2019930335

ISBN 978–0–19–882426–8 (hbk.)
ISBN 978–0–19–882427–5 (pbk.)

DOI: 10.1093/oso/9780198824268.001.0001

Printed and bound by
CPI Group (UK) Ltd, Croydon, CR0 4YY

Acknowledgements

We would like to thank all authors for their efforts and timeliness. Ian Stewart was enthusiastic from the very beginning about bringing this project to fruition. Bethany Kershaw always helped resolve any problems. Julian Thomas promptly did the copy-editing bringing this to an end in hardly any time. Finally, Anna Pape Møller had the idea for the cover that reveals humans and birds in the environment. Independent of what we do, there is always the carbon footprint left behind as illustrated so aptly by our shadow in a pristine Arctic environment.

Contents

List of contributors ix

Section 1: Introduction **1**

1 **Introduction** **3**
Peter O. Dunn and Anders Pape Møller

2 **Climate change** **5**
Kevin E. Trenberth and James W. Hurrell

Section 2: Methods for studying climate change effects **27**

3 **Finding and analysing long-term climate data** **29**
Mark D. Schwartz and Liang Liang

4 **Long-term time series of ornithological data** **37**
Anders Pape Møller and Wesley M. Hochachka

5 **Quantifying the climatic sensitivity of individuals, populations, and species** **44**
Martijn van de Pol and Liam D. Bailey

6 **Ecological niche modelling** **60**
Damaris Zurell and Jan O. Engler

7 **Predicting the effects of climate change on bird population dynamics** **74**
Bernt-Erik Sæther, Steinar Engen, Marlène Gamelon, and Vidar Grøtan

Section 3: Population consequences of climate change **91**

8 **Changes in migration, carry-over effects, and migratory connectivity** **93**
Roberto Ambrosini, Andrea Romano, and Nicola Saino

9 **Changes in timing of breeding and reproductive success in birds** **108**
Peter O. Dunn

10 Physiological and morphological effects of climate change **120**
Andrew E. McKechnie

11 Evolutionary consequences of climate change in birds **134**
Céline Teplitsky and Anne Charmantier

12 Projected population consequences of climate change **147**
David Iles and Stéphanie Jenouvrier

13 Consequences of climatic change for distributions **165**
Brian Huntley

Section 4: Interspecific effects of climate change **185**

14 Host–parasite interactions and climate change **187**
Santiago Merino

15 Predator–prey interactions and climate change **199**
Vincent Bretagnolle and Julien Terraube

16 Bird communities and climate change **221**
Lluís Brotons, Sergi Herrando, Frédéric Jiguet, and Aleksi Lehikoinen

17 Fitting the lens of climate change on bird conservation in the twenty-first century **236**
Peter P. Marra, Benjamin Zuckerberg, and Christiaan Both

18 Climate change in other taxa and links to bird studies **257**
David W. Inouye

19 Conclusions **265**
Anders Pape Møller and Peter O. Dunn

Index 269

List of contributors

Ambrosini, Roberto Department of Environmental Science and Policy, University of Milano, Via Celoria 26, I 20133 Milano, Italy.
roberto.amobrosini@unimi.it

Bailey, Liam D Leibniz Institute for Zoo and Wildlife Research (IZW), Berlin, Germany.
liam.bailey@liamdbailey.com

Both, Christiaan Faculty of Science and Engineering, Groningen Institute for Evolutionary Life Sciences, 9747 AG Groningen, Netherlands.
c.both@rug.nl

Bretagnolle, Vincent Centre d'Etudes Biologiques de Chizé, UMR 7372, CNRS and Université de La Rochelle, Beauvoir sur Niort, 79360 France.
Vincent.BRETAGNOLLE@cebc.cnrs.fr

Brotons, Lluís CSIC at InForest JRU (CTFC-CREAF), E-25280 Solsona, Catalonia, Spain.
lluis.brotons@ctfc.cat

Charmantier, Anne Centre d'Ecologie Fonctionnelle et Evolutive, CEFE UMR 5175, Campus CNRS, 1919 route de Mende, F-34293 Montpellier Cedex 5, France.
anne.charmantier@cefe.cnrs.fr

Dunn, Peter O Department of Biological Sciences, University of Wisconsin-Milwaukee, P.O. Box 413, Milwaukee, WI 53201, USA.
pdunn@uwm.edu

Engler, Jan Terrestrial Ecology Unit, Department of Biology, Ghent University, K.L. Ledeganckstraat 35, B-9000 Ghent, Belgium.
JanOliver.Engler@ugent.be

Engen, Steinar Centre for Conservation Biology, Department of Biology, Norwegian University of Science and Technology, NO-7491 Trondheim, Norway.
steinar.engen@ntnu.no

Gamelon, Marlène Centre for Biodiversity Dynamics, Norwegian University of Science and Technology, Realfagsbygget, NO-7491 Trondheim, Norway.
marlene.gamelon@ntnu.no

Grøtan, Vidar Centre for Conservation Biology, Department of Biology, Norwegian University of Science and Technology, NO-7491 Trondheim, Norway.
vidar.grotan@ntnu.no

Herrando, Sergi Institut Català d'Ornitologia, Nat-Museu de Ciències Naturals de Barcelona, Plaça Leonardo da Vinci 4–5, 08019 Barcelona, Spain.
ornitologia@ornitologia.org

Hochachka, Wesley M Cornell Laboratory of Ornithology, 159 Sapsucker Woods Road, Ithaca, New York, 14850 USA.
wmh6@cornell.edu

Huntley, Brian Department of Biosciences, Durham University, Stockton Road, Durham, DH1 3LE, UK.
brian.huntley@durham.ac.uk

Hurrell, James W Department of Atmospheric Science, Colorado State University, 200 West Lake Street, 371 Campus Delivery, Fort Collins, CO 80523-1371, USA.
jhurrell@rams.colostate.edu

Iles, David Woods Hole Oceanographic Institution. Woods Hole, MA 02543-1050 U.S.A.
david.thomas.iles@gmail.com

Inouye, David W Department of Biology, University of Maryland, College Park, Maryland 20742-4415 USA.
inouye@umd.edu

Jenouvier, Stéphanie Woods Hole Oceanographic Institution. Woods Hole, MA 02543-1050 U.S.A.
sjenouvrier@whoi.edu

Jiguet, Frédéric Museum National d'Histoire Naturel, CNRS-SU UMR 7204, CESCO 43 Rue Buffon, CP135 75005 Paris, France.
fjiguet@mnhn.fr

Lehikoinen, Aleksi Finnish Museum of National History, University of Helsinki, PO Box 17, 00014, Helsinki, Finland.
aleksi.lehikoinen@helsinki.fi

Liang, Liang Department of Geography, University of Kentucky, Lexington, Kentucky 40506-0027 USA.
liang.liang@uky.edu

Marra, Peter P Migratory Bird Center, Smithsonian Conservation Biology Institute, National Zoological Park, PO Box 37012 MRC 5503, Washington, DC 20013 USA.
marrap@si.edu

McKechnie, Andrew E South African Research Chair in Conservation Physiology, National Zoological Garden, South African National Biodiversity Institute, Pretoria, South Africa.
DST-NRF Centre of Excellence at the FitzPatrick Institute, Department of Zoology and Entomology, University of Pretoria, Private Bag X20, Hatfield 0028, South Africa.
aemckechnie@zoology.up.ac.za

Møller, Anders Pape Ecologie Systematique Evolution, UMR 8079 CNRS-Université Paris-Sud XI-AgroParisTech, Batiment 362 Université Paris-Sud XI, F-91405 Orsay Cedex, France.
anders.moller@u-psud.fr

Merino, Santiago Museo Nacional de Ciencias Naturales, Consejo Superior de Investigaciones Científicas, José Gutiérrez Abascal 2, E-28006 Madrid, Spain.
mcnsm508@mncn.csic.es

Romano, Andrea Department of Ecology and Evolution, University of Lausanne, Building Biophore, 1015, Lausanne, Switzerland.
andrea.romano@unil.ch

Sæther, Bernt-Erik Center for Conservation Biology, Department of Biology, Norwegian University of Science and Technology, NO-7491 Trondheim, Norway.
bernt-erik.sather@bio.ntnu.no

Saino, Nicola Department of Environmental Science and Policy, University of Milan, Via Celoria 26, I-20133 Milano, Italy.
nicola.saino@unimi.it

Schwartz, Mark D Department of Geography, University of Wisconsin-Milwaukee, Milwaukee, WI 53211, USA.
mds@uwm.edu

Teplitsky, Céline Centre d'Ecologie Fonctionnelle et Evolutive, CEFE UMR 5175, Campus CNRS, 1919 route de Mende, F-34293 Montpellier Cedex 5, France.
teplitsky@mnhn.fr

Terraube, Julien Faculty of Environmental and Biological Sciences, University of Helsinki, Helsinki, Finland.
julien.terraube-monich@helsinki.fi

Trenberth, Kevin E National Center for Atmospheric Research, Climate Analysis Section, P.O. Box 3000, Boulder, CO 80307-3000, USA.
trenbert@ucar.edu

van de Pol, Martijn Netherlands Institute of Ecology (NIOO-KNAW), Department of Animal Ecology, Wageningen, the Netherlands.
M.vandePol@nioo.knaw.nl

Zuckerberg, Benjamin Department of Forest and Wildlife Ecology, University of Wisconsin-Madison, Madison, WI 53706-1598, USA.
bzuckerberg@wisc.edu

Zurell, Damaris Swiss Federal Research Institute WSL, Zuercherstrasse 111, CH-8903 Birmensdorf, Switzerland.
damaris@zurell.de

SECTION 1

Introduction

CHAPTER 1

Introduction

Peter O. Dunn and Anders Pape Møller

Climate change is considered the largest environmental problem of this century, and it is likely to have severe consequences for our environment (IPCC 2014). The latest special report from the Intergovernmental Panel on Climate Change (IPCC) predicts that global warming will likely surpass 1.5°C above pre-industrial levels by 2040, and it is increasing at 0.2°C per decade (IPCC 2018). Birds have been a bellwether of the impacts of climate change on animals because their behaviour and population changes have been documented for decades and even centuries in some cases. The increase in studies of the effects of climate change on birds has been exponential, and it has continued since the first edition of this book was published in 2010. One of the principal reasons for updating the previous volume is that there has been a tremendous increase in the number (7574) and complexity of studies on climate change and birds since 2010, and it is difficult for most researchers to keep track of the expanding literature. The number of papers specifically dealing with climate change and birds is now more than 11 400 and the total number of papers in the field of climate change exceeds 364 400. In the face of this complexity, we have opted for an edited volume that brings together a group of world experts to review the current state of knowledge, while simultaneously addressing alternative hypotheses and weak points in current research. Another justification for a new edition is that some new topics have become more prominent since the first edition, such as the increasing use of citizen science data, particularly eBird (over half a billion sightings from around the world), as well as new advances in theory, genomics, and ecological and demographic modelling that are providing new insights into the causes of climate change and its consequences on birds.

The book is aimed at a wide audience including undergraduate, graduate, and postgraduate students, scientists, administrators, and conservationists. Climate change issues are attracting rapidly increasing interest from many different kinds of biologists because of the widespread effects of climate change on animals and plants throughout the world. There is an enormous interest among students and post-docs for studies on this subject, and many universities are launching programmes on climate change. Furthermore, there is increasing demand for biologists trained in assessing and managing the impact of climate change on wildlife. To address this latter point, we have added a number of short methodological chapters that address specific issues of analysis with examples and key references as an entryway for students and new researchers in these areas. To facilitate researchers entering the field, several of the chapters have online supplements with R code and examples to provide help in getting started. (www.oup.co.uk/companion/dunn&moller)

The book consists of four sections. In the first section, Kevin Trenberth and James Hurrell provide a general introduction to climate and climate change (Chapter 2). In the second section, five chapters provide an introduction to methods and

Dunn, P.O., and Møller, A.P., *Introduction*. In: *Effects of Climate Change on Birds*. Second Edition. Edited by Peter O. Dunn and Anders Pape Møller: Oxford University Press (2019). © Oxford University Press.
DOI: 10.1093/oso/9780198824268.003.0001

data sources for studying climate change and its effects. In Chapter 3 Mark Schwartz and Liang Liang outline sources available for long-term climate data and some of the analytical issues that need to be addressed. In Chapter 4 Anders Møller and Wesley Hochachka review the databases on birds that are available for study, including long-term surveys like the Christmas Bird Count, Atlases, and citizen science projects like eBird. In Chapter 5 Martijn van de Pol and Liam Bailey discuss methods of indentifying the climate variables that best predict the responses of individuals and populations, and how we can compare these predictors when they differ (e.g., temperature versus precipitation). In Chapter 6 Damaris Zurell and Jan Engler explain the concepts and assumptions of ecological niche modelling, and they provide a real-world example with R code in the online supplement. In Chapter 7 Bernt-Erik Sæther, Steinar Engen, Marlène Gamelon, and Vidar Grøtan outline the steps for predicting the effects of climate change on populations using life history variables and stochastic population models.

In the third section, we have chapters that focus on the individual and population consequences of climate change, ranging from changes in physiology and behaviour to shifts in distribution and abundance and long-term evolutionary changes. The section begins with changes in migration patterns and their carry over effects by Roberto Ambrosini, Andrea Romano, and Nicola Saino in Chapter 8. Peter Dunn follows with a review of changes in the timing of breeding and its links to population trends in Chapter 9. In Chapter 10 Andrew McKechnie reviews the physiological and body size effects of climate change. The evolutionary consequences of climate change on birds are assessed by Céline Teplitsky and Anne Charmantier in Chapter 11. David Iles and Stephanie Jenouvrier expand the discussion of population models in Chapter 12 by reviewing models that incorporate changes in climate predicted by global circulation models and addressing the uncertainties in the model projections. In Chapter 13 Brian Huntley reviews changes in distributions in response to climate, starting with the Quaternary period and moving to the present.

In the fourth and last section, the chapters focus on interspecific effects of climate change, some of the conservation challenges we face, and a review of how the effects on birds are linked to other taxa. We start this section with a review by Santiago Merino of how climate changes the interactions between birds and their parasites in Chapter 14. Predator and prey interactions are reviewed by Vincent Bretagnolle in Chapter 15. Lluis Brotons, Sergi Herando, Frederic Jiguet, and Aleksi Lehikoinen review the effects of climate change, as well as the contributing role of other anthropogenic changes, on bird communities in Chapter 16. In Chapter 17 Pete Marra, Ben Zuckerberg, and Christiaan Both review the conservation challenges and some of the mitigation strategies we can use to minimize the negative impacts of climate change. Finally, in Chapter 18 David Inouye provides a broader ecological perspective in a review of climate change effects on entire food webs, including birds. In the concluding Chapter 19, Anders Møller and Peter Dunn provide an overview of advances since the previous edition of the book, as well as a summary of remaining major questions and research recommendations.

This new edition attempts to synthesize what is known about the effects of climate change on birds, as well as point out new methods and areas for future research. Each chapter attempts to provide a comprehensive review of the topic. Although some of the previous gaps in our knowledge have been filled since the first edition, there are still some notable gaps, which we discuss in the concluding chapter. We hope that readers will find some new perspectives and questions in this book that will inspire them to better understand and conserve bird populations.

References

IPCC (2014) *Climate Change 2014: Synthesis Report. Contribution of Working Groups I, II and III to the Fifth Assessment Report of the Intergovernmental Panel on Climate Change.* Core Writing Team, R.K. Pachauri, and L.A. Meyer (eds). IPCC, Geneva, Switzerland. http://www.ipcc.ch/report/ar5/syr/

IPCC (2018) *Global warming of 1.5°C.* IPCC Special Report 15. http://www.ipcc.ch/report/sr15/

CHAPTER 2

Climate change

Kevin E. Trenberth* and James W. Hurrell

2.1 Introduction

Global climate change is altering many ecosystems, leading to changes in the abundance and distribution of many populations (e.g., Stenseth et al. 2005; Rosenzweig et al. 2008). Advances in the scientific understanding of climate make it clear that there has been a change in climate that goes well beyond the range of natural variability. As stated by the Intergovernmental Panel on Climate Change (IPCC) (2007a, 2013), the warming of the climate system, which includes the atmosphere, ocean, land, and cryosphere (regions of ice and frozen ground), is 'unequivocal' and is 'very likely due to human activities'. The culprit is the astonishing rate at which heat-trapping carbon dioxide and other greenhouse gas concentrations are increasing in the atmosphere, mostly through the burning of fossil fuels and changes in land use, such as those associated with agriculture and deforestation. Greenhouse gases are relatively transparent to incoming solar radiation while they absorb and re-emit outgoing infrared radiation. The result is that more energy stays in the global climate system, most of which (over 90 per cent) goes into the oceans as heat. The ocean heat content is increasing along with sea level rise, through both expansion of the ocean and melting of land ice, and these provide a memory of the past climate change. The extra energy also further raises surface temperature and produces many other direct and indirect changes in the climate system.

The indisputable evidence of anthropogenic climate change, and the knowledge that it will continue well into the future under any plausible emission scenario, is now a factor in the planning of many organizations and governments, encapsulated by the Paris Agreement of December 2015, ratified in October 2016. Global warming does not imply, however, that future changes in weather and climate will be uniform around the globe. The land, for instance, is warming faster than the oceans, consistent with its smaller heat capacity. Moreover, uncertainties remain regarding how climate will change at regional and local scales where the signal of natural variability is large, especially over the next several decades (Hawkins and Sutton 2009). Regional differences in land and ocean temperatures arise, for instance, from natural variability such as El Niño-Southern Oscillation (ENSO) events. Natural variability can result from purely internal atmospheric processes, as well as from interactions among the different components of the climate system, such as those between the atmosphere and ocean or the atmosphere and land.

El Niño events, such as the major event in 2015–16, produce very strong warming of the central and eastern tropical Pacific Ocean while the ocean cools over portions of the subtropics and the tropical

* The National Center for Atmospheric Research is sponsored by the National Science Foundation. Any opinions, findings, and conclusions or recommendations expressed in this publication are those of the authors and do not necessarily reflect the views of the National Science Foundation.

Trenberth, K.E., and Hurrell, J.W., *Climate change*. In: *Effects of Climate Change on Birds*. Second Edition. Edited by Peter O. Dunn and Anders Pape Møller: Oxford University Press (2019). © Oxford University Press.
DOI: 10.1093/oso/9780198824268.003.0002

western Pacific. Over the Atlantic, average basin-wide warming is imposed on top of strong, natural variability on multi-decadal timescales, called the Atlantic Multi-decadal Oscillation (AMO). The level of natural variability, in contrast, is relatively small over the tropical Indian Ocean, where surface warming has been steady and large over recent decades.

Importantly, these differences in regional rates of sea surface temperature (SST) change perturb the atmospheric circulation and shift storm tracks, so that some land regions become warmer and drier, while other regions cool as they become wetter. On the regional scales on which most planning decisions are made and impacts felt, therefore, future warming will not be smooth. Instead, it will be strongly modulated by natural climate variations, and especially those driven by the slowly varying oceans on a timescale of decades. Moreover, regions that warm from both natural variability and global warming are likely to experience amplified impacts, and can endure broken records as well as sometimes disastrous outcomes for societies and ecosystems. This non-uniformity of change highlights the challenges of regional climate change that has considerable spatial structure and temporal variability.

Table 2.1 Acronyms used in the text.

AMO: Atlantic Multi-decadal Oscillation
RCP: Representative Concentration Pathway
ENSO: El Niño-Southern Oscillation
EOF: Empirical Orthogonal Function
GMST: Global Mean Surface Temperature
IPCC: Intergovernmental Panel on Climate Change
IPO: Inter-decadal Pacific Oscillation
NPI: North Pacific Index
NAM: Northern Annular Mode
NAO: North Atlantic Oscillation
PDO: Pacific Decadal Oscillation
PNA: Pacific-North American pattern
ppb: parts per billion
ppm: parts per million by volume
SAM: Southern Annular Mode
SOI: Southern Oscillation Index
SST: Sea Surface Temperature

It is the purpose of this chapter to review observed changes in climate, with a focus on changes in surface climate including variations in major patterns (modes) of climate variability. The next section describes how natural and anthropogenic drivers of climate change are assessed using climate models. The chapter concludes with a brief summary of future projected changes in climate. The physical evidence and the impacts on the environment and society, as documented by IPCC (2007a, b; 2013) and updated in the annual *State of the Climate* reports (such as for 2016, Blunden and Arndt 2017), provide

Table 2.2 Indices of circulation variability.

Southern Oscillation Index (SOI). The Tahiti minus Darwin sea level pressure anomalies, normalized by the long-term mean and standard deviation of the mean sea level pressure difference, or alternatively by the negative of the Darwin sea level pressure record (http://www.cgd.ucar.edu/cas/catalog/climind/soi.html).
Pacific-North American pattern (PNA) Index. The mean of normalized 500 hPa height anomalies at 20°N, 160°W and 55°N,115°W minus those at 45°N, 165°W and 30°N, 85°W (http://www.cpc.noaa.gov/products/precip/CWlink/pna/month_pna_index2.shtml).
North Pacific Index (NPI). The average sea level pressure anomaly over the Gulf of Alaska (30°N–65°N, 160°E–140°W; https://climatedataguide.ucar.edu/climate-data/north-pacific-np-index-trenberth-and-hurrell-monthly-and-winter).
Pacific Decadal Oscillation (PDO) index. The amplitude of the pattern defined by the leading EOF of annual mean SST in the Pacific basin north of 20°N (http://jisao.washington.edu/pdo/PDO.latest).
Atlantic Multi-decadal Oscillation (AMO) Index. The time series of annual mean SST anomalies averaged over the North Atlantic (0–60°N, 0–80°W) as departures from the global mean SST; (http://www.cgd.ucar.edu/cas/catalog/climind/AMO.html).
North Atlantic Oscillation (NAO) Index. The difference of normalized winter (December–March) sea level pressure anomalies between Lisbon, Portugal and Stykkisholmur, Iceland, or alternatively the amplitude of the leading EOF of mean sea level pressure over the North Atlantic (20°–80°N, 90°W–40°E; https://climatedataguide.ucar.edu/climate-data/hurrell-north-atlantic-oscillation-nao-index-pc-based).
Northern Annular Mode (NAM) Index. The amplitude of the pattern defined by the leading EOF of winter monthly mean Northern Hemisphere sea level pressure anomalies poleward of 20°N (https://climatedataguide.ucar.edu/climate-data/hurrell-wintertime-slp-based-northern-annular-mode-nam-index).
Southern Annular Mode (SAM) Index. The difference in average sea level pressure between southern middle and high latitudes (usually 45°S and 65°S) from gridded or station data (http://www.antarctica.ac.uk/met/gjma/sam.html), or alternatively the amplitude of the leading EOF of monthly mean Southern Hemisphere 850 hPa height poleward of 20°S.

the main basis and references for the chapter. There are numerous acronyms used in climate science, especially for names of patterns or modes of variability, and these are listed in Tables 2.1 and 2.2.

2.2 Human and natural drivers of climate change

The IPCC (2007a, 2013) concluded that most of the observed global mean surface temperature (GMST) increase of the past 50 years (Figure 2.1) is 'very likely'[1] due to human activity, while anthropogenic influences have 'likely' contributed to changes in wind patterns, affecting extratropical storm tracks and regional temperature patterns in both the Northern and Southern Hemispheres. These conclusions are based on studies that assess the causes of climate change, taking into account all possible agents of climate change (forcings), both natural and from human activities.

Forcings are factors external to the climate system and may arise naturally, such as from changes in the Sun or from changes in atmospheric composition associated with explosive volcanic eruptions, or from human activities that generate heat or which change the atmospheric composition. Feedbacks occur through interactions among the components of the climate system: the atmosphere, ocean, land, and cryosphere. Some amplify the original changes producing a positive feedback (such as warming melting snow and ice and reducing the reflection of the Sun's rays), while others diminish them: a negative feedback (such as warming causing higher temperature that radiates more heat to space). The physical processes involved are depicted in climate models. Radiative forcing is a measure of the influence that a factor has in altering the balance of incoming and outgoing energy in the Earth–atmosphere system and is an index of the importance of the factor as a potential climate change mechanism. Positive forcing tends to warm the surface while negative forcing tends to cool it.

The capability of climate models to simulate the past climate is comprehensively assessed by IPCC. Given good replications of the past, the forcings can be inserted one by one to disassemble their effects and allow attribution of the observed climate

[1] The IPCC defines the term 'very likely' as the likelihood of a result exceeding 90%, and the term 'likely' as exceeding 66%.

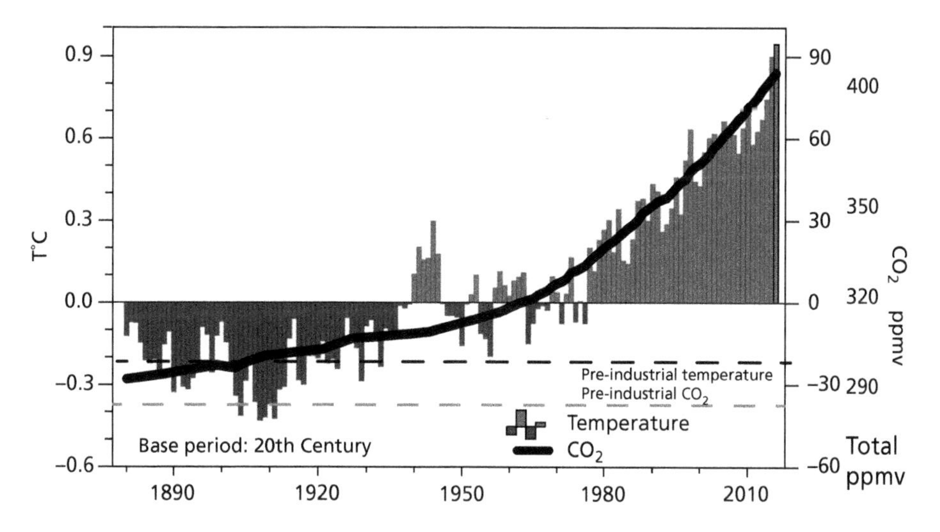

Figure 2.1 Estimated changes in annual global mean surface temperatures (°C, bars) and CO_2 concentrations (thick black line) over the past 149 years relative to 1901–2000 average values. Carbon dioxide concentrations since 1957 are from direct measurements at Mauna Loa, Hawaii, while earlier estimates are derived from ice core records. The scale for CO_2 concentrations is in parts per million (ppm) by volume, relative to the twentieth century mean of 333.7 ppm, while the temperature anomalies are relative to a mean of 14°C. Also given as dashed values are the preindustrial estimated values, with the scale at right for carbon dioxide, where the value is 280 ppm. Updated from Trenberth (1997) and from Trenberth (2016) "The hottest year on record signals that global warming is alive and well", *The Conversation*, January 20, 2016. https://theconversation.com/the-hottest-year-on-record-signals-that-global-warming-is-alive-and-well-53480

change. Therefore, climate models are a key tool to evaluate the role of various forcings in producing the observed changes in temperature and other climate variables.

The best climate models encapsulate the current understanding of the physical processes involved in the climate system, the interactions, and the performance of the system as a whole. Uncertainties arise, however, from shortcomings in the understanding and how to best represent complex processes in models. Yet, in spite of these uncertainties, today's best climate models are able to reproduce the climate of the past century, and simulations of the evolution of GMST over the past millennium are consistent with paleoclimate reconstructions.

Human activities increase greenhouse gases, such as carbon dioxide (CO_2), methane (CH_4), nitrous oxide (N_2O), and other trace gases. They also increase aerosol concentrations in the atmosphere, mainly through the injection of sulfur dioxide (SO_2) from power stations and industry, and through biomass burning. A direct effect of sulfate aerosols is the reflection of a fraction of solar radiation back to space, which tends to cool the Earth's surface. Other aerosols (like soot) directly absorb solar radiation leading to local heating of the atmosphere, and some absorb and emit infrared radiation. A further influence of aerosols is that many act as nuclei on which cloud droplets condense, affecting the number and size of droplets in a cloud and hence altering the reflection and the absorption of solar radiation by the cloud and the lifetime of the cloud (Stevens and Feingold 2009). The precise nature of aerosol/cloud interactions and how they interact with the water cycle remains a major uncertainty in our understanding of climate processes. Because human-made aerosols are mostly introduced near the Earth's surface, they are washed out of the atmosphere by rain in, typically, a few days. Thus, they remain mostly concentrated near their sources and affect climate with a very strong regional pattern, usually producing cooling.

In contrast, greenhouse gases such as CO_2 and CH_4 have lifetimes much longer; of the order of a decade for CH_4 but centuries for CO_2. Both are globally mixed and concentrations build up over time. Greenhouse gas concentrations in the atmosphere have increased markedly as a result of human activities since 1750, and they are now higher than at any time in at least the last 650 000 years. It took at least 10 000 years from the end of the last ice age (18 000 years ago) for levels of CO_2 to increase 100 parts per million (ppm) by volume to 280 ppm, but a greater increase has occurred over only the past 150 years to current values in excess of 405 ppm (Figure 2.1). About half of that increase has occurred over the last 35 years, owing mainly to combustion of fossil fuels and changes in land use. The CO_2 concentration growth-rate has been larger during the last decade than since the beginning of continuous direct measurements in the late 1950s. In the absence of controls, future projections are that the rate of increase in CO_2 amount may accelerate, and concentrations could double from pre-industrial values within the next 40 to 100 years.

Methane is the second most important anthropogenic greenhouse gas. Owing predominantly to agriculture, wetlands, and fossil fuel use, the global atmospheric concentration of CH_4 has increased 250 per cent from a pre-industrial value of 715 parts per billion (ppb) by volume to 1843 ppb in 2016. Global N_2O concentrations have increased significantly from pre-industrial values as well. The total net anthropogenic forcing includes contributions from aerosols (a negative forcing) and several other sources, such as tropospheric ozone and halocarbons.

The latest IPCC report (2013), rather than estimating future human behaviour and the resulting forcings, developed a number of 'Representative Concentration Pathways' (RCPs) tied to possible radiative forcings in 2100. The RCPs are to be used for policy planning purposes and they enable projections of possible future climates. The values chosen for the RCPs range over 2.6, 4.5, 6.0, and 8.5 Watts per square metre in 2100, relative to 1750. These are transient, not equilibrium values, but may be compared with the 3.7 W m^{-2} that corresponds to the equilibrium forcing for a doubling of pre-industrial carbon dioxide concentrations (from 280 to 560 parts per million by volume (ppm)). Including also the prescribed concentrations of methane and nitrous oxide, the combined CO_2-equivalent concentrations in 2100 are 475 ppm (RCP2.6), 630 ppm (RCP4.5), 800 ppm (RCP6.0), and 1313 ppm (RCP8.5). Current CO_2-equivalent forcing from greenhouse gases is about 3.0 W m^{-2}, although accounting also for aerosols, the net is about 2.3 W m^{-2} (IPCC 2013). Hence RCP2.6 is for very low emissions and very unlikely to be realized; while RCP8.5 is closer to business as usual.

Climate model simulations that account for such changes in forcings have now reliably shown that global surface warming of recent decades is a response to the increased concentrations of greenhouse gases and sulfate aerosols in the atmosphere. When the models are run without these forcing changes, the remaining natural forcings and intrinsic natural variability fail to capture the almost linear increase in GMSTs over the past 40 years or so. But when the anthropogenic forcings are included, the models simulate the observed GMST record with impressive fidelity (Figure 2.2). Changes in solar irradiance since 1750 are estimated to have caused a radiative forcing of +0.1 W m^{-2}, mainly in the first part of the twentieth century. Prior to 1979, when direct observations of the Sun from space began, changes in solar irradiance are more uncertain, but direct measurements show that the Sun has not caused warming since 1979. Moreover, the models indicate that volcanic and anthropogenic aerosols have offset some of the additional warming that would have resulted from observed increases in greenhouse gas concentrations alone. For instance, from 2000 to 2010 the sunspot cycle went from a maximum to a minimum and a very quiet Sun, decreasing total solar irradiance by 0.1 per cent. This perhaps offset about 10 to 15 per cent of the warming, but the solar irradiance has gone through another somewhat weak cycle since then.

The patterns of warming over each continent except Antarctica and each ocean basin over the past 50 years are only simulated by models that include anthropogenic forcing (Figure 2.2). Attribution studies have also demonstrated that many of the observed changes in indicators of climate extremes consistent with warming, including the annual number of frost days, warm and cold days, and warm and cold nights, have likely occurred as a result of increased anthropogenic forcing. In other words, many of the recently observed changes in climate are now being simulated in models.

The ability of coupled climate models to simulate the temperature evolution on continental scales, and the detection of anthropogenic effects on each continent except Antarctica, has also increased. No climate model that has used natural forcing only has reproduced either the observed global mean warming trend or the continental mean warming trends. Attribution of temperature change on smaller than continental scales and over timescales of less than 50 years or so is more difficult because of the much larger natural variability on smaller space and timescales (Hawkins and Sutton 2009).

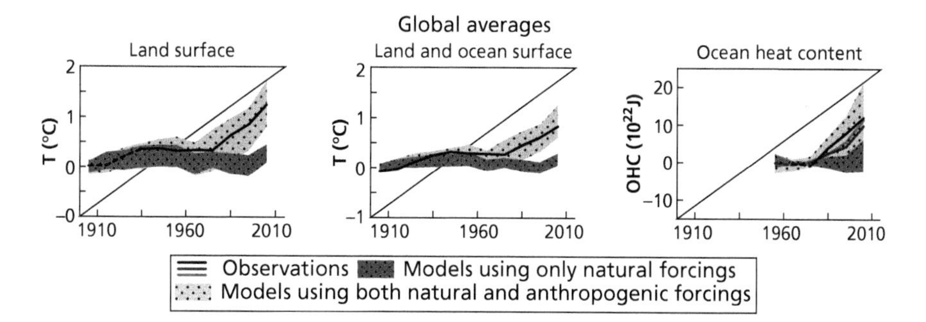

Figure 2.2 Comparison of observed global-scale changes in surface temperature with results simulated by climate models using natural and anthropogenic forcings. Decadal averages of observations are shown for 1906–2005 (black line) plotted against the centre of the decade and relative to the corresponding average for 1901–1950. Dark grey shaded bands show the 5–95% range for simulations from climate models using only the natural forcings due to solar activity and volcanoes. Light grey dotted shaded bands show the 5–95% range for simulations from climate models using both natural and anthropogenic forcings. The figure is adapted from the IPCC (2013).

2.3 Observed changes in surface climate

2.3.1 Temperature

The globe is warming dramatically compared with natural historical rates of change. GMSTs today are more than 0.9°C warmer than at the beginning of the twentieth century, and rates of GMST rise are greatest in recent decades (Figure 2.1). The average rate of increase in the GMST since 1901 is 0.78°–0.90°C century^{-1}. The warmest 16 years are the most recent, except for 1998, and the three consecutive years (2014, 2015, and 2016) each set a new GMST record, which is extremely unusual. Global land regions have warmed the most (about 0.5°C more than the oceans), with the greatest warming in the northern winter and spring months over the Northern Hemisphere continents.

There is a very high degree of confidence in the GMST estimates and their changes (Figure 2.1). The maximum difference, for instance, among three independent estimates of GMST change since 1979 is 0.01°C decade^{-1}. Spatial coverage has improved, and daily temperature data for an increasing number of land stations have also become available, allowing more detailed assessments of extremes, as well as potential urban influences on both large-scale temperature averages and microclimate. It is well documented, for instance, that urban heat island effects are real, but very local, and they have been accounted for in the analyses: the urban heat island influence on continental, hemispheric, and global average trends is at least an order of magnitude smaller than decadal and longer timescale trends, as cities make up less than 0.5 per cent of global land areas (Schneider et al. 2009).

There is no urban heat bias in the SST record and the warming is strongly evident at all latitudes over each of the ocean basins. Moreover, the warming is manifest at depth as well, indicating that the ocean is absorbing most of the heat being added to the climate system (Figure 2.3) (Cheng et al. 2017a). The upper ocean has warmed, especially since 1970, but the penetration of heat into the ocean takes time, and the deeper ocean has mainly warmed only after about 1990. Indeed,

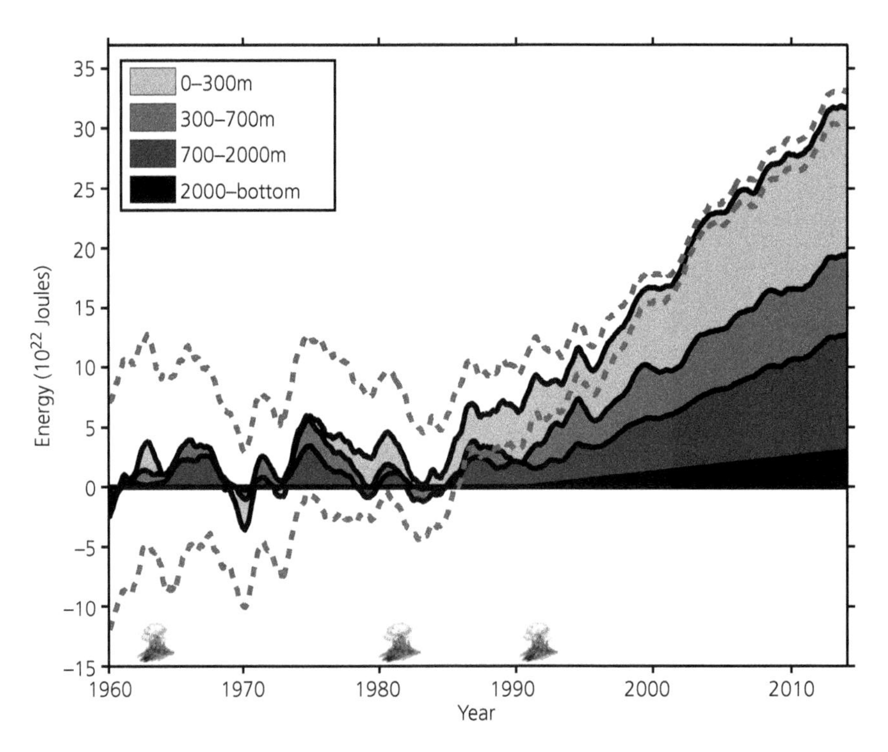

Figure 2.3 Time series of the vertically integrated ocean heat content for various layers since 1960. The total also has 95% error bars indicated as dashed grey curves to show how the uncertainty increases further into the past. From Cheng et al. (2017a).

most of the energy imbalance created by the increasing greenhouse gases (over 90 per cent) goes into the oceans, which therefore serve as the memory of past climate change and further provide an unequivocal view of the warming planet.

The largest short-term fluctuations in GMSTs come from El Niño and La Niña events. Some heat stored in the ocean is released during an El Niño, and this contributes to increases in GMSTs. From late 2007 to the first part of 2009, lower temperatures occurred in association with the large 2007–2008 La Niña event, followed by a weaker La Niña in 2008–2009. The major El Niño of 2015–16 led to these being the warmest years on record as some heat came out of the ocean (see the levelling off or slight decline at the end of Figure 2.3). The reason 1998 stands out as the warmest year last century is because of the major 1997–98 El Niño event.

2.3.2 Sea level

The ocean warming causes seawater to expand and, thus contributes to sea level rise (Figure 2.4). Melting of glaciers on land as well as ice caps and ice sheets also contribute. Instrumental measurements of sea level indicate that the global average increased approximately 17 cm over the last century. The rate has been even faster recently (about 0.31 cm per year from 1993 through 2017, see Figure 2.4, Nerem et al. 2010), when truly global values have been measured from altimeters in space. Prior to 2004, about 60 per cent of global sea level rise was from ocean warming and expansion, while 40 per cent was from melting land ice adding to the ocean volume. Since 2004 melting ice sheets have contributed more than half. The observations of consistent global sea level rise over several decades, and also an increasing rate of sea level rise in the last decade or so, along with the increasing ocean heat content, are probably the single best metrics of the cumulative global warming experienced to date (Cheng et al. 2017b). Consequences include increasing risk of coral bleaching and coastal storm surge flooding.

2.3.3 Snow cover, sea, and land ice

The observed increases in GMST are consistent with nearly worldwide reductions in glacier and small ice cap mass and extent in the twentieth century. In addition, flow speed has increased for some Greenland and Antarctic outlet glaciers, which drain ice from the interior, and melting of Greenland and West Antarctica has increased after about 2000. Critical changes (not well measured) are occurring

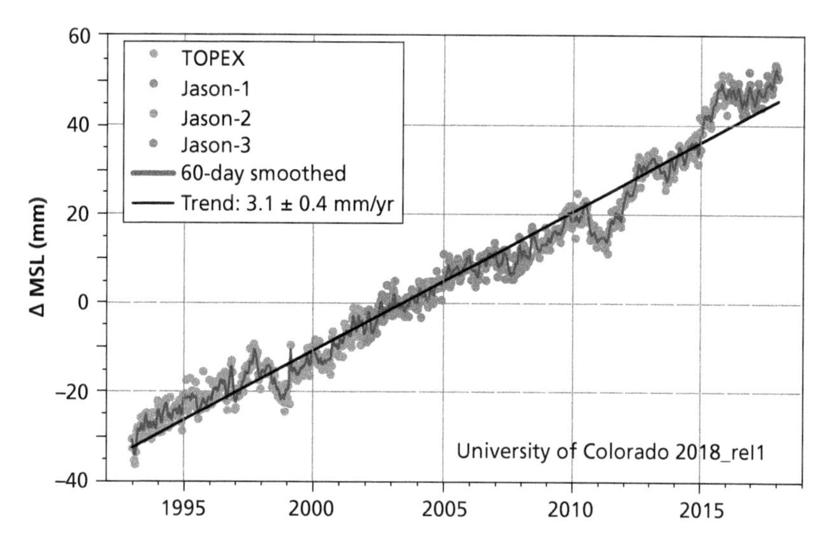

Figure 2.4 Global mean sea level with the mean annual cycle removed. The 60-day mean is indicated along with the linear trend. Updated from Nerem et al. (2010).

in the ocean and ice shelves that buttress the flow of glaciers into the ocean. Glaciers and ice caps respond not only to temperature but also to changes in precipitation, and both winter accumulation and summer melting have increased over the last half century in association with temperature increases. In some regions, moderately increased accumulation observed in recent decades is consistent with changes in atmospheric circulation and associated increases in winter precipitation (e.g., southwestern Norway, parts of coastal Alaska, Patagonia, and the South Island of New Zealand) even though increased ablation has led to marked declines in mass balances in Alaska and Patagonia. Tropical glacier changes are synchronous with those at higher latitudes and all have shown declines in recent decades. Decreases in glaciers and ice caps contributed to sea level rise by 0.05 cm per year from 1961 to 2003, and 0.08 cm per year from 1993 to 2003. Taken together, shrinkage of the ice sheets of Greenland and Antarctica contributed 0.04 cm per year to sea level rise over 1993 to 2003 and about 0.1 cm per year to sea level rise since then.

Snow cover has decreased in many regions of the Northern Hemisphere, particularly in spring, consistent with greater increases in spring than autumn surface temperatures in middle latitudes. Sea-ice extents have decreased in the Arctic, particularly in the spring and summer seasons (13.3 per cent decade^{-1} decrease from 1978 through 2016 in September), and this is consistent with the fact that the average annual Arctic temperature has increased at twice the global average rate, although changes in winds are also a major factor. The lowest sea ice cover to date was in 2012. There have also been decreases in sea-ice thickness and an unprecedented increase in amount of first year ice in the Arctic that is very vulnerable to melting. Temperatures at the top of the permafrost layer in the Arctic have increased since the 1980s (up to 3°C locally), and the maximum area covered by seasonally frozen ground has decreased by about 7 per cent in the Northern Hemisphere since 1900, with an even greater decrease (15 per cent) in the northern spring. There has been a reduction of about two weeks in the annual duration of northern lake and river ice cover.

In contrast to the Arctic, Antarctic sea ice did not exhibit any significant trend from the end of the 1970s through 2015, which is consistent with the lack of trend in surface temperature averaged south of 65°S over that period. However, in spring 2016 into autumn 2017, an exceptional drop in sea ice extent occurred in all sectors (Turner et al. 2017) in association with a strongly negative Southern Annular Mode (SAM) (see section 2.5.5). Moreover, along the Antarctic Peninsula, where significant warming has been observed, progressive break-up of ice shelves occurred beginning in the late 1980s, culminating in the break-up of the Larsen-B ice shelf in 2002 and Larsen-C in 2017. The latter created a huge iceberg the size of Delaware. Antarctic conditions are uniquely influenced greatly by the ozone hole, which alters the atmospheric circulation over the southern regions.

2.3.4 Extremes

For changes in temperature, there is likely to be an amplified change in extremes. Extreme events, such as heat waves, are exceedingly important to both natural systems and human systems and infrastructure. People and ecosystems are adapted to a range of natural weather variations, but it is the extremes of weather and climate that exceed tolerances. Widespread changes in temperature extremes have been observed over the last 50 years. In particular, the number of heat waves globally has increased, and there have been widespread increases in the numbers of warm nights. Cold days, cold nights, and days with frost have become rarer. Such changes greatly affect the range of animals, including birds.

Satellite records suggest a global trend towards more intense and longer lasting tropical cyclones (including hurricanes and typhoons) since about 1970, correlated with observed warming of tropical SSTs, and consistent with expectations of more activity with global warming. There is no clear trend in the annual number of tropical cyclones globally although a substantial increase has occurred in the North Atlantic after 1994 and the most active month (in terms of hurricane days) ever on record is September 2017. There are concerns about the quality of tropical cyclone data, particularly before the satellite era. Further, strong multi-decadal variability is observed and complicates detection of long-term trends in tropical cyclone activity. It has been estimated that heavy rains in tropical storms and hurricanes have increased by 10 to 15 per cent as a result of higher SSTs and more water vapour in the atmosphere (Trenberth 2007).

2.3.5 Precipitation and drought

Changes are occurring in the amount, intensity, frequency, and type of precipitation in ways that are also consistent with a warming planet. These aspects of precipitation generally exhibit large natural variability compared to temperature, making it harder to detect trends in the observational record. A key ingredient in changes in character of precipitation is the observed increase in water vapour and thus the supply of atmospheric moisture to all storms, increasing the intensity of precipitation events. This is consistent with the expectation that the water-holding capacity of the atmosphere increases by about 7 per cent per degree Celsius. Widespread increases in heavy precipitation events and risk of flooding have been observed, even in places where total amounts have decreased. Hence the frequency of heavy rain events has increased in most places and so too have episodic heavy snowfall events which are therefore associated with warming.

Long-term (1900–2015) trends have been observed in total precipitation amounts over some large regions. Significantly increased precipitation has been observed in eastern parts of North and South America, northern Europe, and northern Asia. Drying has been observed in the Sahel, the Mediterranean, southern Africa, and parts of eastern Asia. Precipitation is highly variable spatially and temporally. Robust long-term trends have not been observed for other large regions. The pattern of precipitation change is one of increases generally at higher northern latitudes (because as the atmosphere warms it holds more moisture) and drying in parts of the tropics and subtropics over land. Basin-scale changes in ocean salinity provide further evidence of changes in Earth's water cycle, with freshening at high latitudes and increased salinity in the subtropics.

More intense and longer droughts have been observed over wider areas since the 1970s, particularly in the tropics and subtropics. Increased drying due to higher temperatures and decreased precipitation have contributed to these changes, with the latter the dominant factor. The regions where droughts have occurred are determined largely by changes in SST, especially in the tropics (such as during El Niño), through changes in the atmospheric circulation and precipitation. In the western United States, diminishing snow pack and subsequent summer soil moisture reductions have also been a factor. In Australia and Europe, direct links to warming have been inferred through the extreme nature of high temperatures and heat waves accompanying drought.

In summary, there are an increasing number of many independent surface observations that give a consistent picture of a warming world.

2.4 Observed changes in atmospheric circulation

2.4.1 Sea level pressure

Much of the warming that has contributed to the GMST increases of recent decades (Figure 2.1) has occurred during northern winter and spring over the continents of the Northern Hemisphere. This pattern of warming is strongly related to decade-long changes in natural patterns of the atmospheric and oceanic circulation. The changes in northern winter circulation are reflected by lower-than-average sea level pressure over the middle and high latitudes of the North Pacific and North Atlantic Oceans, as well as over much of the Arctic, and higher-than-average sea level pressure over the subtropical Atlantic (Figure 2.5).

Over the North Pacific, the changes in sea level pressure correspond to an intensification of the Aleutian low-pressure system, while over the North Atlantic the changes correspond to intensified low- and high-pressure centres near Iceland and the Azores, respectively. These northern oceanic pressure systems are semi-permanent features of the winter atmospheric circulation (e.g., Hurrell and Deser 2009). Over the Southern Hemisphere, similar changes have been observed during the austral summer, with surface pressures lowering over the Antarctic and rising over middle latitudes since the late 1970s. The long-term significance of the southern sea level pressure

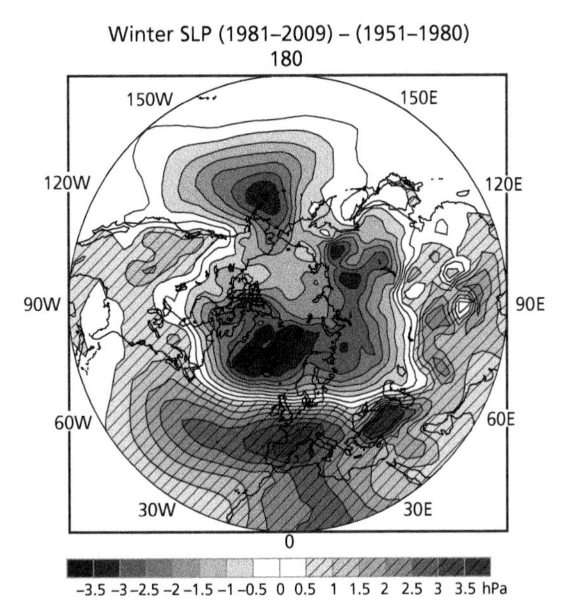

Figure 2.5 Northern winter (December–March) average Northern Hemisphere sea level pressure anomalies (hPa) 1981–2009 expressed as departures from the 1951–1980 values. Positive values are hatched. The sea level pressure data are from Trenberth and Paolino (1980).

change is more difficult to establish, however, given the greater paucity of historical data over the Southern Ocean and Antarctica.

2.4.2 Winds and storm tracks

Changes in winds naturally accompany changes in sea level pressure because of the geostrophic relationship whereby the pressure gradients are largely balanced by the Coriolis force associated with the rotation of the Earth. Accordingly, winds rotate counterclockwise around a low-pressure system in the Northern Hemisphere and clockwise in the Southern Hemisphere. Cyclones are low pressure systems or depressions associated with unsettled stormy weather, as opposed to anticyclones which are high pressure systems and are dominated by fine, settled weather. In low latitudes, 'tropical cyclone' usually refers to a low-pressure system of a certain intensity (e.g., winds above gale force) and above another threshold they become hurricanes in the Western Hemisphere, typhoons in the northwest Pacific, or 'cyclones' in the Indian Ocean. Extratropical cyclones typically have cold and warm fronts attached to them.

Westerly flow across the middle latitudes of the Atlantic and Pacific sectors occurs throughout the year. As the vigour of the flow is related to the north-south (meridional) pressure gradient, the surface winds are strongest during winter when they average more than 5 m s^{-1} from the eastern United States across the Atlantic onto northern Europe as well as across the entire Pacific. These middle latitude westerly winds extend throughout the troposphere and reach their maximum (up to more than 40 m s^{-1} in the mean) at a height of about 12 km. This 'jet stream' roughly coincides with the path of storms travelling across the northern oceans onto the continents. These storm tracks play a critical role in both weather and climate, as they are associated with much of the precipitation and severe weather in middle latitudes, and they transport large amounts of heat, moisture, and momentum toward the poles.

Several studies indicate that there has been a poleward shift in the mean latitude of extratropical cyclones, and that cyclones have become fewer and more intense over the last fifty years. For instance, the change towards a deeper polar vortex and Icelandic Low in northern winter (Figure 2.5) has been accompanied by intensification and poleward displacement of the Atlantic jet and associated enhancement of Atlantic storm track activity. Analogous changes have also been found over the North Pacific and in the Southern Hemisphere.

There are, however, significant uncertainties, with some studies suggesting that storm track activity during the last part of the twentieth century may not be more intense than the activity prior to the 1950s. Station pressure data over the Atlantic-European sector (where records are long and consistent) show a decline of storminess from high levels during the late nineteenth century to a minimum around 1960 and then a quite rapid increase to a maximum around 1990, followed again by a slight decline. Changes in storm tracks, however, are complex and are related to spatial shifts and strength changes in leading patterns of climate variability (next section).

There are a few studies of changes in surface winds themselves, but they are confounded by the nature of instrumentation, which has moving parts (that can rust or clog up) and which can be easily sheltered by growth of nearby trees (Vautard et al.

2010). Over oceans, the growing size of ships gives an apparent increase in wind that is likely spurious (Cardone et al. 1990). To the extent that they reveal real changes, they are associated mostly with decadal variations in teleconnections (next section).

2.5 Observed changes in patterns of circulation variability

2.5.1 Teleconnections

A consequence of the transient behaviour of atmospheric planetary-scale waves is that anomalies in climate on seasonal timescales typically occur over large geographic regions. Some regions may be cooler than average, while at the same time, thousands of kilometres away, warmer conditions prevail. These simultaneous variations in climate, often of opposite sign, over distant parts of the globe are commonly referred to as 'teleconnections' in the meteorological literature. Though their precise nature and shape vary to some extent according to the statistical methodology and the dataset employed in the analysis, consistent regional characteristics that identify the most conspicuous patterns emerge. Understanding the nature of teleconnections and changes in their behaviour is central to understanding regional climate variability and change, as well as impacts on humans and ecosystems.

The analysis of teleconnections has typically employed a linear perspective, which assumes a basic spatial pattern with varying amplitude and mirror image positive and negative polarities. In contrast, nonlinear interpretations identify preferred climate anomalies as recurrent states of a specific polarity. Climate change may result as a preference for one polarity of a pattern, or through a change in the nature or number of states.

Arguably the most prominent teleconnections over the Northern Hemisphere extra-tropics are the North Atlantic Oscillation (NAO) and the Pacific-North American (PNA) patterns, and their spatial structures are revealed most simply through one-point correlation maps (Figure 2.6). A positive PNA teleconnection pattern in the middle troposphere coincides with the warm-phase ENSO pattern, and is typically associated with higher-than-normal pressure near Hawaii and over the northwestern United States and western Canada, while pressures are typically lower-than-normal over the central North Pacific and the southeast United States. The difference of normalized height anomalies from these four centres forms the most commonly used time-varying index of the PNA (Table 2.2). Variations in the PNA pattern represent changes in the north-south migration of the large-scale Pacific and North

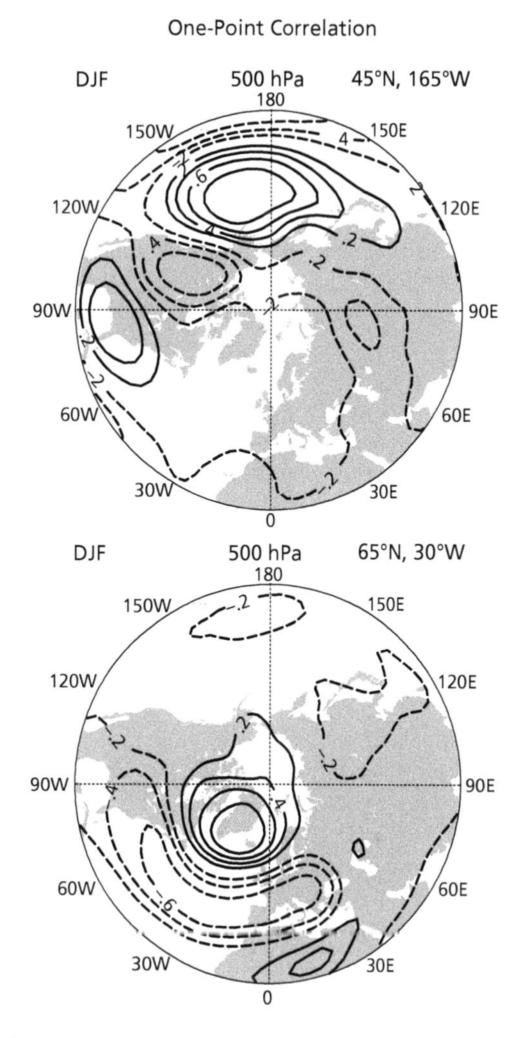

Figure 2.6 One-point correlation maps of 500 hPa geopotential heights for northern winter (December–February) over 1958–2006. In the top panel, the reference point is 45°N, 165°W, corresponding to the primary centre of action of the PNA pattern. In the lower panel, the NAO pattern is illustrated based on a reference point of 65°N, 30°W. Negative correlation coefficients are dashed, the contour increment is 0.2, and the zero contour has been excluded. Adapted from Hurrell and Deser (2009).

American air masses, storm tracks and their associated weather, affecting precipitation in western North America and the frequency of Alaskan blocking events and associated cold air outbreaks over the western United States in winter.

In the Southern Hemisphere wave structures do not emerge as readily owing to the dominance of more zonally symmetric variability (the so-called SAM). Although teleconnections are best defined over a grid, simple indices based on a few key station locations remain attractive, as the series can often be carried back in time long before complete gridded fields were available. The disadvantage of such station-based indices is increased noise from the reduced spatial sampling (Hurrell et al. 2003).

Many teleconnections have been identified, but combinations of only a small number of patterns can account for much of the interannual variability in the circulation and surface climate (Quadrelli and Wallace 2004). Trenberth et al. (2005) analysed global atmospheric mass and found four key patterns: the two annular modes (SAM and the Northern Annular Mode, or NAM), a global ENSO-related pattern, and a fourth closely related to the North Pacific Index (NPI) and the Pacific Decadal Oscillation (PDO), which in turn is closely related to ENSO and the PNA pattern.

Teleconnection patterns tend to be most prominent in winter (especially in the Northern Hemisphere), when the mean circulation is strongest. The strength of teleconnections and the way they influence surface climate also vary over long timescales, and these aspects are exceedingly important for understanding regional climate change. In the following only a few predominant teleconnection patterns are documented.

2.5.2 ENSO and the PDO

Fluctuations in tropical Pacific SSTs are related to the occurrence of El Niño, during which the equatorial surface waters warm considerably from the International Date Line to the west coast of South America. The atmospheric phenomenon tied to El Niño is termed the Southern Oscillation, which is a global-scale standing wave in atmospheric mass (thus evident in sea level pressure), involving exchanges of air between Eastern and Western Hemispheres centred in tropical and subtropical latitudes (Figure 2.7), and changes in the west-east overturning Walker Circulation near the Equator. The oscillation is characterized by the inverse variations in sea level pressure at Darwin (12.4°S, 130.9°E) in northern Australia and Tahiti (17.5°S, 149.6°W) in the south Pacific: annual mean pressures at these two stations are correlated at –0.8. A simple index of the SO is, therefore, often defined by the Tahiti minus Darwin sea level pressure anomalies, normalized by the long-term mean and standard deviation of the mean sea level pressure difference, or simply by the negative of the Darwin record (Figure 2.7 and Table 2.2). During an El Niño event, the sea level pressure tends to be higher than usual at Darwin and lower than usual at Tahiti. Negative values of the SO index (SOI), therefore, are typically associated with warmer-than-average SSTs in the near equatorial Pacific, while positive values of the index are typically associated with colder-than-average SSTs. While changes in near equatorial Pacific SSTs can occur without a swing in the SO, El Niño (EN) and the SO are linked so closely that the term ENSO is used to describe the atmosphere–ocean interactions over the tropical Pacific. Warm ENSO events, therefore, are those in which both a negative SO extreme and an El Niño occur together.

During the warm phase of ENSO, the warming of the waters in the central and eastern tropical Pacific shifts the location of the heaviest tropical rainfall eastward toward or beyond the Date Line from its climatological position centered over Indonesia and the far western Pacific, weakening the Walker Circulation. This shift in rainfall also alters the heating patterns that force large-scale waves in the atmosphere. The waves in the airflow determine the preferred location of the extratropical storm tracks. Consequently, changes from one phase of the SO to another have a profound impact on regional temperatures (Figure 2.7). Most warm phase ENSO winters, for example, are mild over the western United States, although the regional details vary considerably from one event to another.

Although the SO has a typical period of 2–7 years, the strength of the oscillation has varied considerably. There were strong variations from the 1880s to the 1920s and after about 1950, but weaker variations

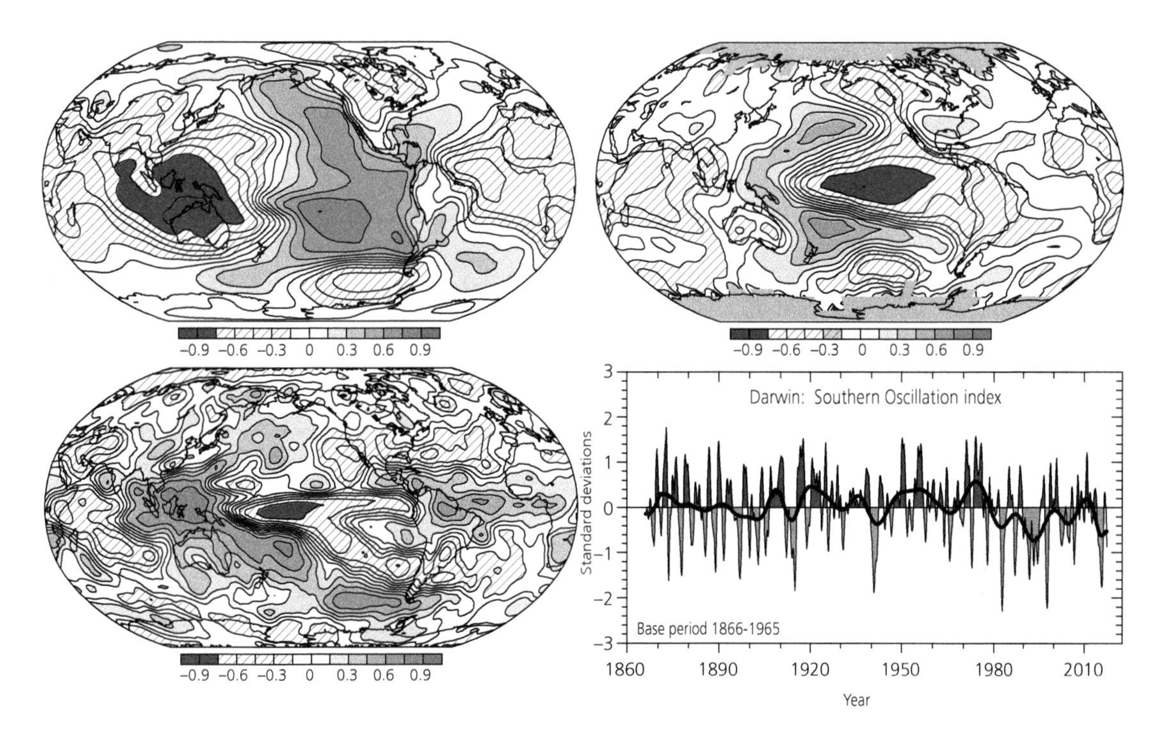

Figure 2.7 Correlations with the SOI (Table 1) for annual (May to April) means for sea level pressure (top left) and surface temperature (top right) for 1958 to 2004, and estimates of global precipitation for 1979 to 2003 (bottom left), updated from Trenberth and Caron (2000) and IPCC (2007a). The Darwin-based SOI, in normalized units of standard deviation, from 1866 to 2009 (lower right) features monthly values with an 11-point low-pass filter, which effectively removes fluctuations with periods of less than eight months. The smooth black curve shows decadal variations. Light grey values indicate positive sea level pressure anomalies at Darwin and thus El Niño conditions.

in between (with the exception of the major 1939–41 event). A remarkable feature of the SOI is the decadal and longer-term variations in recent years, which is lacking from earlier periods.

Decadal to inter-decadal variability in the atmospheric circulation is especially prominent in the North Pacific (e.g., Trenberth and Hurrell 1994) where fluctuations in the strength of the wintertime Aleutian Low pressure system, indicated by the North Pacific index (NPI; Table 2.2), co-vary with North Pacific SST in what has been termed the 'Pacific Decadal Oscillation' (PDO) or, its close cousin, the Inter-decadal Pacific Oscillation (the IPO) (Figure 2.8). The PDO/IPO has been described as a long-lived El Niño-like pattern of Indo-Pacific climate variability or as a low-frequency residual of ENSO variability on multi-decadal timescales. Phase changes of the PDO/IPO are associated with pronounced changes in temperature and rainfall patterns across North and South America, Asia, and Australia. Furthermore, ENSO teleconnections on interannual timescales around the Pacific basin are significantly modified by the PDO/IPO.

Both the PDO (Figure 2.8) and the NPI (not shown) reveal extended periods of persistently anomalous values. Low PDO goes with high NPI values, indicative of a weakened circulation over the North Pacific (1900–1924, 1945–1976, 1999–2013) and predominantly high PDO values indicate a strengthened circulation (low NPI) (1925–1944, 1977–1998, and since 2014). The well-known decrease in Aleutian Low pressure from 1976 to 1977 is analogous to transitions that occurred from 1946 to 1947 and from 1924 to 1925, and these earlier changes were also associated with SST fluctuations in the tropical Indian and Pacific Oceans (e.g., Deser et al. 2004).

The high PDO values relate to times of increases in the GMST (Figure 2.8) while the GMST no longer increases much for negative PDO values. From 1999 to 2013 this pause in the rise of GMST has also become known as a 'hiatus' in warming (Trenberth and Fasullo 2013; Trenberth et al. 2014; Trenberth

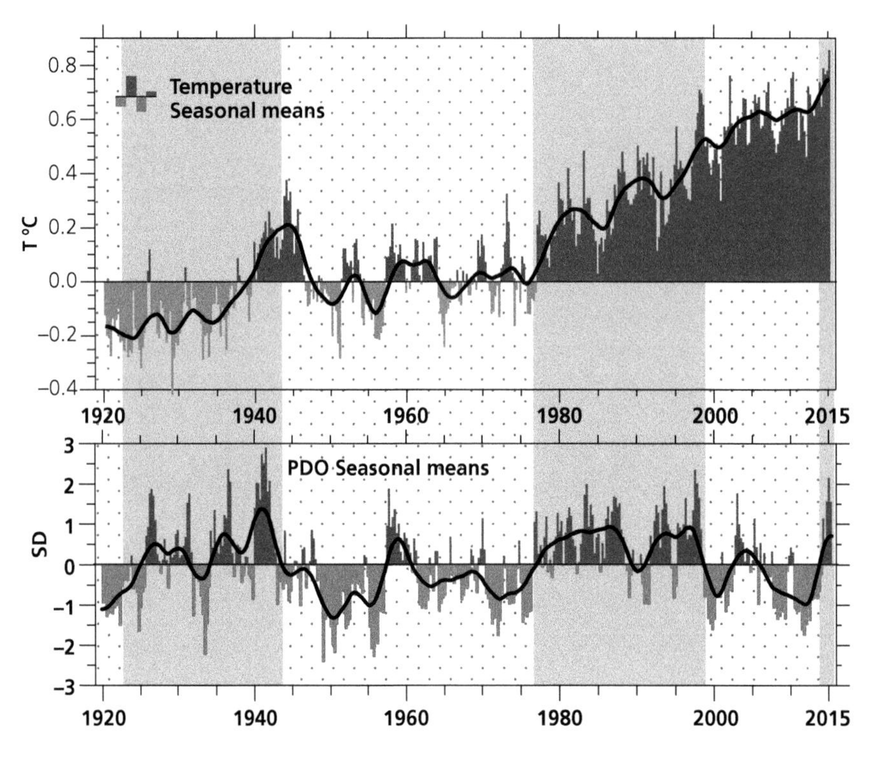

Figure 2.8 Seasonal (December–January–February; etc.) global mean surface temperatures since 1920 (relative to the twentieth-century mean) vary considerably on interannual and decadal time scales. (B) Seasonal mean PDO anomalies show decadal regimes (positive in light grey; negative in dotted) as well as short-term variability. A 20-term Gaussian filter is used in both to show decadal variations, with anomalies reflected about the end point of March to May 2015 (heavy black curves). Adapted from Trenberth (2015).

2015; Fyfe et al. 2016). Although increases in GMST stall, the ocean heat content and sea level continue to rise, showing that the heat from global warming is being redistributed within the ocean, both with depth and regionally in the West Pacific. The main pacemaker of variability in rates of GMST increase appears to be the PDO, with aerosols likely playing a role in the earlier big hiatus in 1947–76.

There have been three 'super' El Niños, where the main index has made it into the 'very strong' category: 1982–83, 1997–98 and 2015–16. The 1997–1998 event was the largest on record in terms of SST anomalies, and the GMST in 1998 was the highest on record last century. There are no 'very strong' La Niña events, which also highlights some aspects of the asymmetry between the two phases: La Niña events tend to last longer or be double-phased more often. Worldwide climate anomalies lasting several seasons have been identified with all of these events. The effects of the 2015–16 event were not as great in coastal South America where the term 'El Niño' originated. Nevertheless, a very unusual coastal El Niño occurred in the first few months of 2017, causing devastating stormy weather over northern Chile, Peru, and Colombia.

Because of the enhanced activity in the Pacific and the changes in atmospheric circulation throughout the tropics, there is a decrease in the number of tropical storms and hurricanes in the tropical Atlantic during El Niño. Good examples are 1997 and 2015 which are the most active years globally; yet 1997 was one of the quietest Atlantic hurricane seasons on record. In contrast, the El Niño events of 1990–95, 1997–98, and 2015–16 terminated before the 1995, 1998, and 2017 hurricane seasons, which unleashed storms and placed those seasons among the most active on record in the Atlantic. In 2015, super typhoon Pam ripped through Vanuatu in March causing enormous damage, enabled by warm waters from the El Niño. Less than a year later, the strongest hurricane on record in the Southern Hemisphere (Winston) severely damaged Fiji. The 2015 northern hurricane season featured by far the greatest number of category 4 and 5 hurricanes/typhoons on record (25 vs previous record 18). Strong drought and wildfires occurred in Indonesia, affecting air quality.

ENSO events involve large exchanges of heat between the ocean and atmosphere and affect GMST. Extremes of the hydrological cycle such as floods and droughts are common with ENSO and are apt to be enhanced with global warming. For example, the modest 2002–2003 El Niño was associated with a drought in Australia, made much worse by record-breaking heat. A strong La Niña event took place 2007–08, contributing to 2008 being the coolest year since the turn of the twenty-first century. 2016 is by far the warmest year on record, followed by 2015, in part because of the El Niño event, and 2014 is third (Figure 2.1). All of the impacts of El Niño are exacerbated by global warming.

2.5.3 Atlantic Multi-decadal Oscillation

Over the Atlantic sector decadal variability has large amplitude relative to interannual variability, especially over the North Atlantic. The Atlantic decadal variability has been termed the 'Atlantic Multi-decadal Oscillation' or AMO (Figure 2.9; Table 2.2) (Trenberth and Shea 2006). North Atlantic SSTs show a 65- to 75-year variation (±0.2°C range), with a warm phase 1930 to 1960, and after 1995, and cool phases during 1905 to 1925 and 1970 to 1995. Instrumental records are not long enough to determine whether AMO variability has a well-defined period rather than a simpler character, such as 'red noise'. The robustness of the signal has therefore been addressed using paleoclimate records, and similar fluctuations have been documented through the last four centuries (e.g., Delworth and Mann 2000).

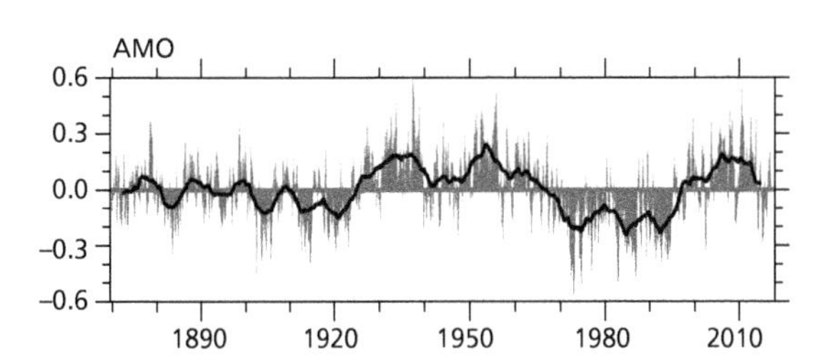

Figure 2.9 Monthly Atlantic Multidecadal Oscillation (AMO) as defined in Table 1, along with a low-pass filtered version to show decadal variability. Updated from Trenberth and Shea (2006).

The slow changes in Atlantic SSTs have affected regional climate trends over parts of North America and Europe, hemispheric temperature anomalies, sea ice concentration in the Greenland Sea, and hurricane activity in the tropical Atlantic and Caribbean (e.g., Webster et al. 2005; Trenberth and Shea 2006). In addition, tropical Atlantic SST anomalies have contributed to rainfall anomalies over the Caribbean and the Nordeste region of Brazil, and severe multi-year droughts over parts of Africa including the Sahel (e.g., Hoerling et al. 2006). Tropical Atlantic SST variations are also a factor in producing drought conditions over portions of North America, although tropical Pacific SST variations appear to play a more dominant role (e.g., Seager et al. 2008). The tropical Atlantic SSTs were at record high levels in 2005, fuelling the very active hurricane season, but have also been exceptionally high in 2010 and 2017.

2.5.4 North Atlantic Oscillation

One of the most prominent teleconnection patterns is the NAO (Hurrell 1995), which refers to changes in the atmospheric sea level pressure difference between the Arctic and the subtropical Atlantic (Figures 2.9 and 2.10). Although it is the only teleconnection pattern evident throughout the year in the Northern Hemisphere, the climate anomalies associated with the NAO are largest during the northern winter months when the atmosphere is dynamically the most active.

A time series since 1900 of wintertime NAO variability, the spatial pattern of the oscillation, and NAO impacts on winter surface temperature and precipitation are shown in Figure 2.10. Most modern NAO indices are derived either from the simple difference in surface pressure anomalies between various northern and southern locations, or from the principal component time series of the leading (usually regional) mode[2] of sea level pressure

[2] The analysis to determine the dominant modes of variability is a principal component analysis which produces Empirical Orthogonal Functions (EOFs) as the eigenvectors of the covariance matrix.

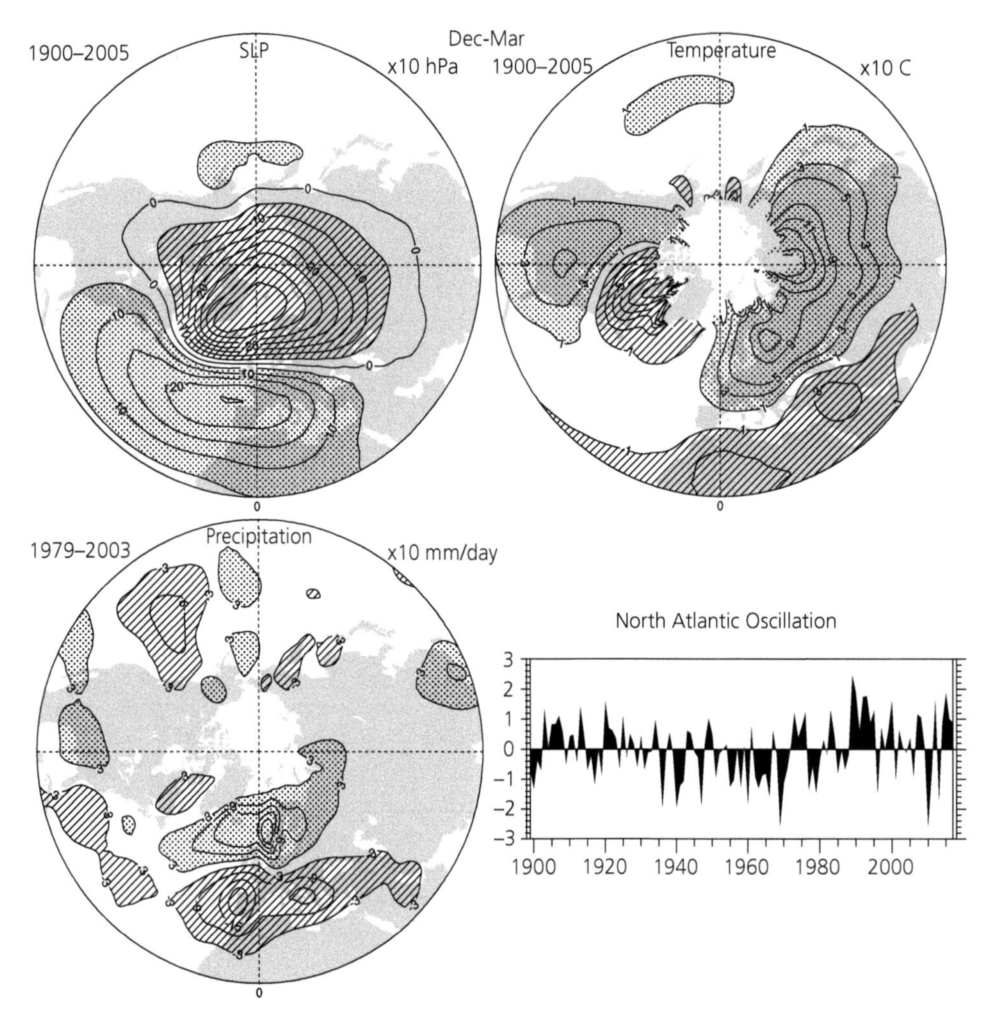

Figure 2.10 Changes in northern winter (December–March) surface pressure, temperature, and precipitation corresponding to a unit deviation of the NAO index over 1900 to 2009. (Top left) Mean sea level pressure (0.1 hPa). (Top right) Land-surface air and SST (0.1 °C; contour increment 0.2 °C): regions of insufficient data (e.g., over much of the Arctic) are not contoured. (Bottom left) Precipitation for 1979 to 2009 based on global estimates (0.1 mm day^{-1}; contour interval 0.6 mm day^{-1}). (Bottom right) Station-based index of winter NAO (Table 1). The heavy solid line represents the index smoothed to remove fluctuations with periods less than four years. The indicated year corresponds to the January of the winter season (e.g., 1990 is the winter of 1989/1990). Adapted and updated from Hurrell et al. (2003) and IPCC (2007a).

(Hurrell and Deser 2009). A commonly used index (Figure 2.10; Table 2.2) is based on the differences in normalized sea level pressure anomalies between Lisbon, Portugal and Stykkisholmur, Iceland. This NAO index correlates with the NAM index (Table 2.2) at 0.85, which emphasizes the NAO and NAM reflect essentially the same mode of tropospheric variability.

The NAO exerts a dominant influence on winter surface temperatures across much of the Northern Hemisphere, and on storminess and precipitation over Europe and North Africa (Figure 2.10) (Hurrell et al. 2003). When the NAO index is positive, enhanced westerly flow across the North Atlantic in winter moves warm moist maritime air over much of Europe and far downstream, while stronger northerly winds over Greenland and northeastern Canada carry cold air southward and decrease land temperatures and SST over the northwest Atlantic. Temperature variations over North Africa and the

Middle East (cooling) and the southeastern United States (warming), associated with the stronger clockwise flow around the subtropical Atlantic high-pressure centre, are also notable.

Positive NAO index winters are also associated with a northeastward shift in Atlantic storm activity, with enhanced activity from Newfoundland into northern Europe and a modest decrease to the south. Positive NAO index winters are also typified by more intense and frequent storms in the vicinity of Iceland and the Norwegian Sea. The correlation between the NAO index and cyclone activity is highly negative in eastern Canada and positive in western Canada. The upward trend towards more positive NAO index winters from the mid-1960s to the mid-1990s was associated with increased wave heights over the northeast Atlantic and decreased wave heights south of 40°N.

The NAO modulates the transport and convergence of atmospheric moisture and the distribution of precipitation. More precipitation than normal falls from Iceland through Scandinavia during high NAO index winters, while the reverse occurs over much of central and southern Europe, the Mediterranean, parts of the Middle East, the Canadian Arctic, and much of Greenland (Figure 2.10). As far eastward as Turkey, river runoff is significantly correlated with NAO variability. There are also significant NAO effects on ocean heat content, sea ice, ocean currents, and ocean heat transport, as well as very significant impacts on many aspects of the north Atlantic/European biosphere (e.g., IPCC 2007b).

2.5.5 Southern Annular Mode

The principal mode of variability of the atmospheric circulation in the Southern Hemisphere is known as the SAM. It is essentially a zonally symmetric structure associated with synchronous pressure or height anomalies of opposite sign in middle and high-latitudes, and therefore reflects changes in the main belt of subpolar westerly winds. Enhanced Southern Ocean westerlies occur in the positive phase of the SAM. The SAM contributes a significant proportion of southern mid-latitude circulation variability on many timescales. Trenberth et al. (2005) showed that the SAM is the leading mode in an analysis of monthly mean global atmospheric mass, accounting for around 10 per cent of total global variance.

As with the NAO/NAM, the structure and variability of the SAM results mainly from the internal dynamics of the atmosphere, and the SAM is an expression of storm track and jet stream variability. The SAM index (not shown; Table 2.2) reveals a general increase beginning in the 1960s consistent with a strengthening of the circumpolar vortex and intensification of the circumpolar westerlies that has been associated with the development of the ozone hole, especially in the southern summer (Swart et al. 2015). The trend in the SAM has contributed to Antarctic temperature trends, specifically a strong summer warming in the Peninsula region and little change or cooling over much of the rest of the continent. Only in spring 2016 did the SAM turn abruptly negative in association with large decreases in Antarctic sea ice in all sectors (Turner et al. 2017).

The earlier positive SAM was also associated with low pressure west of the Peninsula leading to increased poleward flow, warming, and reduced sea ice in the region. The positive trend in the SAM led to more cyclones in the circumpolar trough and hence a greater contribution to Antarctic precipitation from these near-coastal systems. The SAM also affects spatial patterns of precipitation variability in Antarctica and southern South America and southern Australia.

2.6 Projected future climate change

The ability of climate models to simulate the past climate record gives us confidence in their ability to simulate the future. We can look back at projections from earlier IPCC assessments and see that the observed rate of global warming since 1990 (about 0.18°C decade^{-1}) is within the projected range (0.15°C– 0.30°C decade^{-1}). Moreover, the attribution of the recent climate change to increased concentrations of greenhouse gases in the atmosphere has direct implications for the future. Because of the long lifetime of CO_2 and the slow equilibration of the oceans, there is a substantial future commitment to further global climate change even in the absence of further emissions of greenhouse gas into the atmosphere. Several coupled model experiments have explored the concept of climate change commitment. For instance, if concentrations of greenhouse gas were held constant at year 2000 levels (implying a very large reduction in emissions), a further warming trend would occur over the next 20 years at a rate of about 0.1°C decade^{-1}, with a smaller warming rate continuing after that. Such committed climate change is due to (1) the long lifetime of CO_2 and other greenhouse gases; and (2) the long time it takes for warmth to penetrate into the oceans. Under the aforementioned scenario, the associated sea level rise commitment is much longer term, due to the effects of thermal expansion on sea level. Water has the physical property of expanding as it heats up; therefore, as the warming penetrates deeper into the ocean, an ever-increasing volume of water expands and contributes to ongoing sea level rise. Since it would take centuries for the entire volume of the ocean to warm in response to the effects of greenhouse gases already in the air, sea level rise would continue for centuries. Further glacial melt is also likely.

Some of the major IPCC results include:

- Over the next two decades, all models produce similar warming trends in global surface temperatures, regardless of the emission scenario. The rate of the projected warming is near 0.2°C decade^{-1}, or about twice that of the 'commitment' runs.
- Decadal-average warming over each inhabited continent by 2030 is insensitive to the emission scenario; moreover, the temperature change is very likely to exceed the model generated natural temperature variability by at least a factor of two.
- By the middle of the twenty-first century the choice of scenario becomes more important for the magnitude of surface warming, and by the end of the twenty-first century there are clear consequences for which scenario is followed. The best estimate of the GMST change from 1986–2005 to the end of the century depends on the emissions scenario and is 0.3°C to 1.7°C (RCP2.6), 1.1°C to 2.6°C (RCP4.5), 1.4°C to 3.1°C (RCP6.0), and 2.6°C to 4.8°C (RCP8.5).
- Geographical patterns of warming show the greatest temperature increases at high northern latitudes and over land, with less warming over the southern oceans and North Atlantic, as has been observed in recent decades. In spite of a slowdown of the meridional overturning circulation and changes in the Gulf Stream in the ocean across models, there is still warming over the North Atlantic and Europe due to the overwhelming effects of the increased concentrations of greenhouse gases.
- Snow cover is projected to continue to contract in spring and summer. Widespread increases in thaw depth are projected over most permafrost regions.
- Sea ice coverage is projected to shrink. Large parts of the Arctic Ocean are expected to no longer have year-round ice cover by the middle of the twenty-first century.
- It is very likely that hot extremes, heat waves, and heavy precipitation events will continue to become more frequent. Models also project a 50 to 100 per cent decline in the frequency of cold air outbreaks in most regions of the winter Northern Hemisphere. Related decreases in frost days contribute to longer growing seasons.
- Projections of sea level rise by 2081–2100 range from 26 to 55 cm for RCP2.6, to 45 to 82 cm for RCP8.5, and RCP8.5 has values of 52 to 98 cm by 2100, with rates of 8 to 16 mm year^{-1} after 2081. These ranges are derived from climate projections in combination with process-based models and literature assessment of glacier and ice sheet contributions by the IPCC (2013).
- About half of the projected sea level rise is due to thermal expansion of sea water. There is less certainty of the future contributions from other sources. For instance, the projections include a contribution due to increased ice flow from Greenland and Antarctica, but how these flow rates might change in the future is not known, although there is a risk of much faster melt and sea level rise.
- Increases in the amount of precipitation are very likely in high latitudes, while decreases are likely in most subtropical land regions, continuing recent trends.
- Sea level pressure is projected to increase over the subtropics and middle latitudes, and decrease over high latitudes associated with annular mode changes in the Northern Hemisphere (NAM/NAO). Consequently, storm tracks are projected

to move poleward, with consequent changes in wind, precipitation, and temperature patterns outside the tropics, continuing the pattern of observed trends over the last few decades. In the Southern Hemisphere, the healing of the ozone hole counters effects from increasing greenhouse gases.

- Most models warm the central and eastern equatorial Pacific more than the western equatorial Pacific, with a corresponding mean eastward shift in precipitation. ENSO interannual variability is projected to continue in all models, but there are large inter-model differences. Regardless, the impacts of droughts and floods are apt to increase.

As our knowledge of the different components of the climate system and their interactions increases, so does the complexity of climate models. Historical changes in land use and effects of dams and irrigation, for instance, are now beginning to be considered. Future projected land cover changes due to human activities are also likely to significantly affect climate, especially locally, and these effects are now being included in climate models.

Climate change is expected to influence the capacities of the land and oceans to act as repositories for anthropogenic CO_2, and hence provide a feedback to climate change. Increasingly, global coupled climate models include the complex processes involved with modelling the carbon cycle, and suggest that this feedback is positive (adding to more warming) in all models so far considered, thereby giving higher values on the warm end of the uncertainty ranges.

2.7 Conclusions

Consequences of the physical changes in climate are addressed extensively in IPCC (2013). Considerable evidence suggests that recent warming is strongly affecting terrestrial biological systems, including earlier timing of spring events, such as leafing, bird migration and egg-laying, and poleward and upward altitudinal shifts in ranges in plant and animal species. Moreover, the resilience of many ecosystems is likely to be exceeded this century by an unprecedented combination of climate change, associated disturbances (e.g., flooding, drought, wildfire, insects, ocean acidification), and other human effects such as land use and change, pollution, and over-exploitation of resources.

Global warming promotes increases in both drought through drying (evapotranspiration) and temperature. With atmospheric temperature increases the water holding capacity goes up at 7 per cent per degree C, and has the effect of drawing moisture out of plants and soils. In many places, even as rains have become heavier (more intense), so too have dry spells become longer. A consequence of more intense but less frequent precipitation events is that what were once 500-year flood events are now more like 30- or 50-year events. After a certain point where the ground is dry and plants have reached wilting point, all of the heat goes into raising temperature and creating heat waves, and then wild fire risk goes up substantially. 'Dry lightning' can then be disastrous, especially in areas where trees are damaged, such as by bark beetles. The risk of wild fire does not necessarily translate into a fire if care has been taken in managing the risk by building wild fire breaks, cutting down on litter, and removing diseased and dead trees and vegetation near buildings.

For humans, autonomous adaptation occurs to changing conditions to some degree. Climate change effects occur amidst increases in life expectancy in most places, and are thus hard to sort out. Direct effects are nonetheless evident from changes in heat, cold, storms (including hurricanes and tornadoes), drought, and wild fires. The drought-related heat waves in Europe in 2003 and Russia in 2010 caused, respectively, almost 70 000 and 55 000 deaths. On the other hand, fewer cold waves reduce mortality. Safe drinking water is jeopardized by more intense rains and runoff which can lead to contamination and increased microbial loading. Hence, waterborne diseases have been observed to increase. Also, drought and observed earlier snow melt and runoff jeopardize water supplies, especially in summer. Changes in temperatures, humidity, and precipitation also affect the environment for pests and disease, and have increased risk of certain problems in plants, animals, and humans (Chapter 14).

The reality of anthropogenic climate change is no longer debated in the scientific community. The imperative for policymakers is to act aggressively to reduce carbon emissions and dependency on fossil fuels, creating instead a sustainable and clean energy future. Mitigation actions taken now mainly have benefits 50 years and beyond because of the huge inertia in the climate system. Therefore, society will have to adapt to climate change, including its many adverse effects on human health and ecosystems, even if actions are taken to reduce the magnitude and rate of climate change. The projected rate of change far exceeds anything seen in nature in the past 10 000 years and is therefore apt to be disruptive in many ways.

References

Blunden, J., and Arndt, D. S. (eds) (2017). State of the Climate in 2016. *Bulletin of the American Meteorological Society*, 98 (8), Si–S277, doi:10.1175/2017BAMSStateoftheClimate.1.

Cardone, J.S., Greenwood J. G., and Cane, M.A. (1990). On trends in historical marine wind data. *Journal of Climate*, 3, 113–27.

Cheng, L., Trenberth, K., Fasullo, J., Boyer, T., Abraham, J., and Zhu, J. (2017a). Improved estimates of ocean heat content from 1960 to 2015. *Science Advances*, 3, 3, e1601545, doi:10.1126/sciadv.1601545.

Cheng, L., Trenberth, K.E., Fasullo, J., Abraham, J., Boyer, T., von Schuckmann, K., and Zhu J. (2017b). Taking the pulse of the planet. *Eos*, https://doi.org/10.1029/2017EO081839.

Delworth, T. L., and Mann, M. E. (2000). Observed and simulated multidecadal variability in the Northern Hemisphere. *Climate Dynamics*, 16, 661–76.

Deser, C., Phillips, A. S., and Hurrell, J. W. (2004). Pacific interdecadal climate variability: Linkages between the tropics and the north Pacific during boreal winter since 1900. *Journal of Climate*, 17, 3109–24.

Fyfe J.C., Meehl, G.A., England, M.H., Mann, M.E., Santer, B.D., Flato, G.M., Hawkins, E., Gillett, N.P., Xie, S.-P., Kosaka, Y., and Swart, N.C. (2016). Making sense of the early-2000s warming slowdown. *Nature Climate Change*, 6, 224–8.

Hawkins, E., and Sutton, R. (2009). The potential to reduce uncertainty in regional climate predictions. *Bulletin of the American Meteorological Society*, 90, 1095, doi:10.1175/2009BAMS2607.1.

Hoerling, M.P., Hurrell, J.W., Eischeid, J., and Phillips, A. (2006). Detection and attribution of 20th century Northern and Southern African rainfall change. *Journal of Climate*, 19, 3989–4008.

Hurrell, J.W. (1995). Decadal trends in the North Atlantic Oscillation and relationships to regional temperature and precipitation. *Science*, 269, 676–9.

Hurrell, J.W., Kushnir, Y., Ottersen, G., and Visbeck, M. (2003). An overview of the North Atlantic Oscillation. In: J.W. Hurrell, Y. Kushnir, G. Ottersen, and M. Visbeck (eds), *The North Atlantic Oscillation: Climatic significance and environmental impact. Geophysical Monographs*, 134, pp. 1–35. Amer. Geophys. U., Washington, DC.

Hurrell, J.W., and Deser, C. (2009). Atlantic climate variability. *Journal of Marine Systems*, 78, 28–41, doi:10.1016/j.jmarsys.2008.11.026.

IPCC (2007a). *Climate change 2007. The physical science basis.* S. Solomon, et al. (eds). Cambridge University Press, Cambridge, UK.

IPCC (2007b). *Climate change 2007: Impacts, adaptation and vulnerability.* M.L. Parry, et al. (eds). Cambridge University Press, Cambridge, UK.

IPCC (2013). *Climate Change 2013: The Physical Science Basis.* Fifth Assessment Report of the Intergovernmental Panel on Climate Change. T.F. Stocker, et al. (eds). Cambridge University Press, Cambridge, UK, 1535 pp.

Nerem, R.S., Chambers, D., Choe, C., and Mitchum, G. T. (2010). Estimating mean sea level change from the TOPEX and Jason altimeter missions. *Marine Geodesy*, 33, supp 1, 435.

Quadrelli, R., and Wallace, J. M. (2004). A simplified linear framework for interpreting patterns of Northern Hemisphere wintertime climate variability. *Journal of Climate*, 17, 3728–44.

Rosenzweig, C., Karoly D., Vicarelli, M., Neofotis, P., Wu, Q., Casassa, G., Menzel, A., Root, T.L., Estrella, N., Seguin, B., Tryjanowski, P., Liu, C., Rawlins, S., and Imeson, A. (2008). Attributing physical and biological impacts to anthropogenic climate change. *Nature*, 453, 353–8.

Schneider, A., Friedl, M.A., and Potere, D. (2009). A new map of global urban extent from MODIS satellite data. *Environmental Research Letters*, 4, doi:10.1088/1748-9326/4/4/044003.

Seager, R., Kushnir, Y., Ting, M., Cane, M., Naik, N., and Miller, J. (2008). Would advance knowledge of 1930s SSTs have allowed prediction of the dust bowl drought? *Journal of Climate*, 21, 3261–81.

Stenseth, N.C., Ottersen, G., Hurrell, J.W., and Belgrano, A. (2005). *Marine ecosystems and climate variation*. Oxford University Press, Oxford, UK.

Stevens, B., and Feingold, G. (2009). Untangling aerosol effects on clouds and precipitation in a buffered system. *Nature*, 461, 607–13.

Swart, N.C., Fyfe, J.C., Gillett, N., and Marshall, G.J. (2015). Comparing trends in the Southern Annular Mode and surface westerly jet. *Journal of Climate*, 28, 8840–55, doi:10.1175/JCLI-D-15-0334.1.

Trenberth, K.E. (2007). Warmer oceans, stronger hurricanes. *Scientific American*, July, 2007, 45–51.

Trenberth, K.E. (2015). Has there been a hiatus? *Science*, 349, 691–2, doi:10.1126/science.aac9225.

Trenberth, K.E., and Fasullo J.T. (2013). An apparent hiatus in global warming? *Earth's Future*, 1, 19–32, doi: 10.002/2013EF000165.

Trenberth, K.E., and Hurrell, J.W. (1994). Decadal atmosphere–ocean variations in the Pacific. *Climate Dynamics*, 9, 303–19.

Trenberth, K.E., and Shea, D.J. (2006). Atlantic hurricanes and natural variability in 2005. *Geophysical Research Letters*, 33, L12704.

Trenberth, K.E., Stepaniak, D.P., and Smith, L. (2005). Interannual variability of patterns of atmospheric mass distribution. *Journal of Climate*, 18, 2812–25.

Trenberth, K.E., Fasullo, J.T., Branstator, G., and Phillips, A.S. (2014). Seasonal aspects of the recent pause in surface warming. *Nature Climate Change*, 4, 911–16, doi:10.1038/NCLIMATE2341.

Turner, J., Phillips, T., Marshall, G.J., Hosking, J.S., Pope, J.O., Bracegirdle, T.J., and Deb, P. (2017). Unprecedented springtime retreat of Antarctic sea ice in 2016, *Geophysical Research Letters*, 44, 6868–75, doi:10.1002/2017GL073656.

Vautard, R., Cattiaux, J., Yiou, P., Thepaut, J-N., and Ciais, P. (2010). Northern Hemisphere atmospheric stilling partly attributed to an increase in surface roughness. *Nature Geoscience*, 3, 756–61, doi:10.1038/ngeo979.

Webster, P.J., Holland, G.J., Curry, J.A., and Chang, H.-R. (2005). Changes in tropical cyclone number, duration and intensity in a warming environment. *Science*, 309, 1844–6.

SECTION 2

Methods for studying climate change effects

CHAPTER 3

Finding and analysing long-term climate data

Mark D. Schwartz and Liang Liang

3.1 Introduction

There is generally much more long-term climate and related data available to support ecological and biological research now than in previous decades. However, while more data is almost always better, it does underscore the need for every researcher to be better informed of: 1) the wide range and types of climate data available; 2) the different forms that these data typically take, in collection, pre-processing, and output attributes; 3) the optimal research strategies to best utilize these data; and 4) the many types of related supporting data. All this information is necessary to developing the most efficacious research designs.

Thus, in this chapter we will give an overview of many major sources of long-term climate and related data, while also providing some general guidance and recommendations to help in selecting the best data for specific ecological/biological research projects.

3.2 Major data sources

3.2.1 Global historical climatology network

The National Oceanic and Atmospheric Administration (NOAA's) National Centers for Environmental Information (formerly the National Climatic Data Center) maintains the Global Historical Climatology Network (GHCN), which is the official U.S. archived dataset, as well as one of the most comprehensive collections of station-based land surface climate data gathered from around the world (https://www.ncdc.noaa.gov/data-access/land-based-station-data/land-based-datasets/global-historical-climatology-network-ghcn).

These data are compiled from over twenty sources, and include information older than 175 years, up to as recently as within the last hour. The data can be accessed by FTP for researchers who know their exact needs, or guided through a Climate Data Online system. There are two primary subsets: 1) the GHCN-Monthly which provides monthly mean temperature data for over 7000 stations and over 200 countries and territories, with ongoing updates of more than 2000 stations. Non-climatic influences that can bias the observed temperature records are removed using homogeneity adjustments (Lawrimore et al. 2011); and 2) the GHCN-Daily, which includes records from at least 80 000 stations in more than 150 countries and territories. Many of these are also updated daily. While some stations include daily maximum and minimum temperature, total precipitation, snowfall, and snow depth, the clear majority report only precipitation (roughly two-thirds). Basic quality assurance corrections are applied to the full dataset, but the data are not corrected for systematic biases (Menne et al. 2012).

Schwartz, M.D., and Liang, L., *Finding and analysing long-term climate data*. In: *Effects of Climate Change on Birds*. Second Edition. Edited by Peter O. Dunn and Anders Pape Møller: Oxford University Press (2019). © Oxford University Press.
DOI: 10.1093/oso/9780198824268.003.0003

3.2.2 WorldClim

WorldClim is a very high spatial resolution (approx. 1 km^2) spatially interpolated (gridded) monthly climate dataset for global land areas, which assimilated station data for the current climate, as well as providing downscaled Global Climate Model (GCM) scenario output for representation of past and future climates (http://www.worldclim.org/).

Using WorldClim Version 1 (see Hijmans et al. 2005), monthly average maximum and minimum temperatures, total precipitation, and 'bioclimatic' ('more biologically meaningful' variables, derived from the monthly rainfall and temperature data) can be obtained for a variety of GCM past climate (mid-Holocene, Last Glacial Maximum, and Last Interglacial) and future climate scenarios, as well as the current climate (1960–1990, from station-derived data). The latest version (WorldClim 2) is so far only available for the current climate (1970–2000, from station-derived data), but the past and future climate scenarios are expected to eventually also be available. Besides monthly temperatures (minimum, maximum, and average), and precipitation, the newest version also includes solar radiation, vapour pressure, and wind speed (Fick and Hijmans 2017).

3.2.3 Climatic Research Unit

The Climatic Research Unit of the University of East Anglia (Norwich, UK) is one of the world's leading institutions dedicated to the study of climate change. The unit has compiled global land air temperature data from various sources around the world since 1978, and incorporated sea surface temperature data from the Hadley Centre, UK Met Office starting in 1986. The current temperature dataset (HadCRUT4; Morice et al. 2012) provides global coverage on a 5° by 5° grid, with its land component known as the CRUTEM4 (Jones et al. 2012) and the oceanic portion as HadSST3 (Kennedy et al. 2011), respectively. The land near-surface air temperature data are based on station records from national meteorological services around the world; and the sea surface temperatures are primarily from measurements taken by ships and buoys. This instrumental climate dataset includes monthly and annual temperature anomalies worldwide dating back to 1850, and is updated regularly to the most recent month.

In addition, the unit has produced a global land climatic data time series (CRU TS) at a higher spatial resolution (i.e., 0.5° by 0.5°) that dates back to 1901 (Harris et al. 2014). This dataset includes additional monthly variables such as mean temperature, maximum and minimum temperatures, diurnal temperature range, precipitation total, vapour pressure, cloud cover, rain-day counts, and potential evapotranspiration. The CRU TS dataset is not updated as frequently as HadCRUT4, and, thus, the current version (v. 4.01) only covers the period up to 2016. The unit also hosts some paleoclimate data from dendrochronology research, as well as regularly updated data on air pressure and atmospheric circulation indices such as the North Atlantic Oscillation, among others. These data are available free of charge to the public under the Open Database License (http://www.cru.uea.ac.uk/data.).

3.2.4 Vegetation data from satellites

Long-term changes in terrestrial vegetation are monitored in real time by satellites using Normalized Differential Vegetation Index (NDVI; Goward et al. 1991; Tarpley 1991). NDVI is derived from the difference between near infrared and red band reflectances divided by their sum (Rouse Jr et al. 1974). As a part of NOAA's Climate Data Records (CDR), high quality daily NDVI data are available at a 0.05° by 0.05° grid collected by Advanced Very High Resolution Radiometer (AVHRR) sensors on NOAA satellites from 1981 to present for all land regions of the world. This dataset is regularly updated (up to 10 days from the present) to reflect ongoing changes in vegetation conditions across the world. The global Leaf Area Index (LAI) and Fraction of Absorbed Photosynthetically Active Radiation (FAPAR) are also available from the same satellites at similar temporal and spatial resolutions. In comparison to NDVI, these biophysical variables are more directly related to plant photosynthesis and their quantities are more reflective of primary production and carbon assimilation. These datasets, along with additional CDRs, are open to public access via NOAA's National Centers for Environmental Information (http://www.ncdc.noaa.gov).

Vegetation indices (VI) are also available from newer satellite sensors, especially the Moderate Resolution Imaging Spectroradiometer (MODIS) on board National Aeronautics and Space Administration (NASA) Terra and Aqua satellites (Huete et al. 2002). MODIS provides finer spatial resolution than AVHRR and supports two primary vegetation indices. The first is NDVI which maintains continuity with that of AVHRR. The second is the newer Enhanced Vegetation Index, which reduces soil background noise and has improved sensitivity over high biomass regions (Huete and Justice 1999). Current MODIS vegetation indices (Version 6) are available at several spatial resolutions: 250m, 500m, 1km, and 0.05° according to the Climate Modelling Grid. To remove cloud contamination, MODIS VI products are composited to monthly or 16-day windows from which the best-quality values were chosen. The Terra and Aqua VI composite windows are phased apart, so it is possible to achieve a higher temporal resolution (i.e., 8-day) by using the two sources in tandem. Bioclimatic variables such as LAI and FAPAR are also available from MODIS observations. These data are freely distributed by the Land Processes Distributed Active Archive Center (https://lpdaac.usgs.gov), a partnership between the U.S. Geological Survey and NASA. The Terra and Aqua satellites are more recent, so MODIS data are only available since February 2000.

3.2.5 Berkeley Earth

Berkeley Earth (http://berkeleyearth.org/), a non-profit organization, developed an open dataset of global temperature that extends back to 1750 for land regions. It combines their land data with HadSST ocean temperature data (see section 3.2.3) to cover the entire globe back to 1850. Members of the group were motivated to address concerns brought up by climate change sceptics in detecting long-term climatic trends. The Berkeley Earth gridded data include monthly temperature data for land and ocean dating back to the 1700s or 1800s, depending on specific datasets. These products are available in two grid systems/spatial resolutions. The 'Equal Area' divides the Earth's surface into 15 984 equal-area grid cells. The other grid follows 1° by 1° latitude-longitude cells. In comparison with other long-term gridded global products, the Berkeley dataset offers generally better spatial resolution, and employed a somewhat different interpolation method that results in lower data uncertainties (Rohde et al. 2013). Two useful products that are in the experimental stage include higher resolution (1/4° by 1/4°) land temperature data for the contiguous U.S. and Europe, and a global daily land temperature dataset.

3.2.6 PRISM and DAYMET

In comparison to global datasets, gridded climate products for regions with more weather information, such as the United States or North America, typically offer higher spatial and temporal resolutions. The PRISM (http://prism.oregonstate.edu/) dataset, for example, is available for the conterminous United States at 4km (2.5 arcmins) or 800m (30 arc-seconds) grids. The dataset was developed by the PRISM Climate Group at Oregon State University using a Parameter-elevation Relationships on Independent Slopes Model (i.e., PRISM) interpolation method (Daly et al. 2008). The dataset provides estimates of precipitation, minimum and maximum temperatures, mean dew point, and minimum and maximum vapour pressure deficits. The dataset includes a 30-year normal (1981–2010) of monthly and annual conditions and historical monthly conditions from 1895 to 1980. Both monthly and daily products are additionally provided for recent years (1981-present). DAYMET (https://daymet.ornl.gov/) is another climate data source for North America, which is unique in providing daily meteorological estimates at 1 km spatial resolution. Data are available for 1980 through the latest full calendar year. The variables cover temperature, precipitation, humidity, shortwave radiation, snow water equivalent, and day length, complementing those of other data sources.

3.2.7 Regional fluctuations: ENSO, NAO, and PDO

There are a number of large-scale anomalies that influence the spatial and temporal variability of atmospheric circulations in different regions around the world. Among the major ones are El Niño Southern Oscillation (ENSO), the North Atlantic

Oscillation (NAO), and the Pacific Decadal Oscillation (PDO). More detailed information about these phenomena is provided in Chapter 2. ENSO can be tracked in time and space by sea surface temperature (SST) anomalies in the equatorial Pacific region, or by time series of specific ENSO indices (without geographic details). An often-used index to characterize the oceanic component of global ENSO status is the Oceanic Niño Index (ONI, officially adopted by NOAA; Barnston et al. 1997; Trenberth 1997). The North Atlantic Oscillation features primarily an atmospheric component. Its teleconnection pattern maps and index time series are derived from mean standardized 500-mb height anomalies over the North Atlantic (20°~80°N) using a Rotated Principal Component Analysis (Barnston and Livezey 1987). Similarly, the Pacific Decadal Oscillation index is derived from the leading principal component of North Pacific (poleward from 20°N) sea surface temperature anomalies (Mantua and Hare 2002). Detailed information about the development and use of these indices are available from NOAA's Climate Prediction Center (http://www.cpc.ncep.noaa.gov/) and National Centers for Environmental Information's Teleconnections web pages (https://www.ncdc.noaa.gov/teleconnections/). Longer-term time series of these climatic anomaly indices (that date back to the nineteenth century) are available from the Global Climate Observing System Working Group on Surface Pressure (https://www.esrl.noaa.gov/psd/gcos_wgsp/).

3.3 Analysing and integrating data

3.3.1 RNCEP software

Tools have been developed to facilitate access to and utilization of climate data. For instance, RNCEP is a package of functions in the open source R language that allows convenient retrieval and use of long-term climate data for ecological research (Kemp et al. 2012). RNCEP supports access to the National Centers for Environmental Prediction (NCEP)/National Center for Atmospheric Research (NCAR) Reanalysis I dataset (Kalnay et al. 1996), as well as the NCEP/Department of Energy (DOE) Reanalysis II dataset (Kanamitsu et al. 2002). The NCEP/NCAR Reanalysis I provides a comprehensive set of meteorological variables, including global 4-times daily, daily, and monthly air temperatures, air pressures, and precipitation since 1948 at a 2.5° by 2.5° grid. As an improved version, the NCEP/DOE Reanalysis II dataset corrected known data assimilation errors, and updated process parameterizations. It provides global grids of similar variables with the same spatial and temporal resolutions back to 1979.

3.3.2 Spring indices

A notable contribution to long-term climate monitoring is the integration of weather data with biospheric processes such as onset of the spring season. Given the closely coupled relationship between atmosphere and biosphere, and the sensitive response of plant life cycle timing to changing weather and climate, weather-based phenological models offer a unique perspective on climate variability and its impact on ecosystems. Spring Indices (SI) are widely tested models that integrate seasonal temperature changes, matching the needs of temperate plants for initiating growth in spring (Schwartz 1990, 1997). The SI models were developed using decades-long spring phenological observations of a cloned lilac cultivar *Syringa chinensis* 'Red Rothomagensis', and two honeysuckle cloned cultivar varieties (*Lonicera tatarica* 'Arnold Red' and *L. korolkowii* 'Zabeli').

Two primary SI predictions are 'first leaf date' and 'first bloom date', which can be used to mark the start and ending of the spring onset period. As biological observations are often not as available as weather measurements in time and space, the SI models expand our ability to relate long-term climate change to a crucial biological process over broad geographic regions (Schwartz et al. 2006). The original SI accounted for chilling (required by many temperate trees for dormancy release), which limited the use of SI models in subtropical climates of temperate land regions. Recent studies suggested that this chilling requirement can be lifted without compromising the prediction quality of SI models, allowing the use of more broadly applicable extended Spring Indices (SI-x; Schwartz et al. 2013).

The greatest advantages of using SI models for evaluating basic phenological regional patterns and trends in the context of biological research are: 1) SI models can be generated at any location that has daily maximum–minimum air temperature (1.5–2.0 m level) time series data, so they can be produced and evaluated over much larger geographic areas than any currently available conventional phenological data; 2) they are process-based in terms of atmospheric phenomena (effectively capturing and translating nonlinearities inherent in large-scale air temperature variations into forms relevant to plant growth) and spatially independent, which allows them to be fully scalable from individual sites upward to continental and larger areas; 3) SI model output is consistent over all areas, which may not be true for conventional phenological data due to different species and event definitions; and 4) the code has been published and is available for use by anyone (Ault et al. 2015a).

While SI models do not reproduce all the details of multi-species phenological data at any site, or the specific phenology of some types of plants, these models process weather data into indices directly related to growth and development of many plant species. As such, they provide baseline assessment of each location's general phenological response (with the above noted limitations) over a standard period, supplying a needed context for evaluating and comparing regional or local-scale studies, similar to what the Palmer Drought Index provides for evaluating moisture stress. Further, SI is expected to be robust under future conditions where climate departs significantly from the historical mean, as the models are optimized for continental-scale applications (Schwartz 1997). The latest (SI-x) models have been widely adopted as standardized measures of the start of the spring growing season. They are employed as one of the 'Climate Change Indicators in the United States' by the U.S. Environmental Protection Agency (https://www.epa.gov/climate-indicators/climate-change-indicators-leaf-and-bloom-dates), and as one of the 'USGCRP Indicators' developed by the National Climate Assessment (NCA) within the U.S. Global Change Research Program (https://www.globalchange.gov/browse/indicator-details/3661). The SI-x models can be run on long-term gridded climate data to produce continuous, 'wall-to-wall' predictions of growing season onset over most temperate land regions (Ault et al. 2015b). The gridded first leaf and first bloom dates since 1981 as well as the current year near-real time maps for the continental U.S. and Alaska are available from the USA-National Phenology Network (https://www.usanpn.org/data/spring). Experimental forecasts for the start of the spring season (months in advance, termed 'Springcasting') are also now being produced by the Emergent Climate Risk Lab (ECRL) at Cornell University, using the SI-x models (http://ecrl.eas.cornell.edu/Misc/SpringOnset/SpringCasting/).

3.4 Other long-term data sources

The *National Ecological Observatory Network* (NEON, http://www.neonscience.org/) is a continental-scale facility being developed under the sponsorship of the National Science Foundation. NEON is designed to collect and make freely available high-quality, standardized data that characterize and quantify complex, rapidly changing ecological processes across the United States (including Alaska, Hawaii, and Puerto Rico) from 81 field sites (47 terrestrial and 34 aquatic). Data collection methods include *in situ* instrument measurements, and field sampling and airborne remote sensing that are standardized across all sites. The strategically selected field sites represent different regions of vegetation, landforms, climate, and ecosystem performance across the continental United States and additional areas. As the full observatory is still being developed and coming online, the *NEON Data Portal* (http://www.neonscience.org/data/neon-data-portal) may be the best way to explore what data are currently available. Publications providing both descriptions of the datasets and examples of studies using NEON data are also available online (http://www.neonscience.org/community/papers-publications). Individual Biological Field Stations are another potential source of long-term climate data, which can be reached by using the directory of contact information on the Organization of Biological Field Stations (http://www.obfs.org/) web page to determine what local information they may be collecting.

3.5 Choice of climate time window for analyses

A crucial consideration in any analysis of the relationships among climate data and biological activity is the period of influence. Time periods closer to the start of an activity are often thought to be better predictors than those further back in time, but this may vary considerably, depending on the organism, and typically must be determined empirically for each study. Another important factor, however, is the ways in which climate data are commonly collected. Typically, data are either daily or monthly averages or accumulations. Monthly data are more commonly available and offer considerable advantages in terms of volume and ease of use. A note of caution, though, about monthly data is the artificial nature of the averaging categories. The influence of climate is unlikely to be constrained in any way by essentially arbitrary monthly boundaries. Thus, monthly averages may tend to obscure the details of biological linkages to climatic drivers. Perhaps the best approach is to start with monthly data to identify general relationships, and then shift to daily data (if available) to refine the understanding (see also Chapter 5).

3.6 Downscaling

Most gridded products of climate information (including both past climate and future climate predictions produced by Global Circulation Models (GCMs)) are developed at continental to global scales. The spatial resolutions of these products are often too coarse (e.g., >100 km) to support local to regional scale applications, such as evaluating the impact of climate change on specific land surface systems. Therefore, spatial downscaling is necessary to make these data useful for addressing finer scale (e.g., <100 km) problems (Wilby and Fowler 2010; Wilby and Wigley 1997). For example, the WorldClim dataset (see section 3.2.2) includes downscaled climate variables at several relatively high spatial resolutions, as fine as 1km (Hijmans et al. 2005). The same approach may be used to temporally downscale climate data (e.g., from monthly to daily).

Climate data downscaling assumes that the finer-scale climate conditions are a function of the combined effect of coarser-scale climate conditions along with local characteristics such as the presence of water bodies and topography. Two general approaches have been used to execute the downscaling. The first, dynamical downscaling, employs Regional Climate Models that incorporate local variables with outputs from GCMs as boundary conditions. The other approach, statistical downscaling, establishes empirical relationships between local/regional scale climate variables with the corresponding continental/global scale climate variables. The latter is less computationally intensive but requires sufficient observational data to establish the cross-scalar relationships. These general approaches are employed in different downscaling projects using many specific Regional Climate Models and statistical methods (reviewed by Trzaska and Schnarr 2014). Additional uncertainties are expected from these downscaling processes. Therefore, users should be aware of the uncertainties associated with specific downscaling techniques/climate datasets to avoid potential pitfalls in data processing and interpretation of results.

3.7 Discussion and recommendations

When pondering the best climate data to use in a scientific study, it is important to consider measurement issues and the impact of means or extremes, as well as variations in scale and time. For many local studies, where often only a single nearby collection site is available, such considerations may be unnecessary. However, we would argue that even in these cases it is well worth the effort to review one's assumptions. Climate data are almost exclusively available as records of changes in air, soil, or water properties (typically temperature), at standardized levels above or below the ground surface—not direct measurements of organisms—thus the timing and methods of collection (even if they cannot be changed) need to be acknowledged and their impacts considered. Another important issue is whether to use mean data or extremes. With air temperatures, for example, will the studied biological activity best correspond to mean temperatures, or will it more likely be driven by maximum or minimum values? Such considerations may be more important for other (less common) variables, such as wind speed or wind direction. Also, the choice of a time window (see section 3.5) is another important issue.

Larger-scale studies will need to make a critical decision regarding whether to use station-based or gridded data (presuming both are viable options for the area and time of interest). Each type of data offers advantages and concerns, which are often reciprocal. For example, station-based observations are discrete, but may not be representative of conditions in the larger region. Conversely, gridded data may provide a workable regional average (especially if topography is considered in the gridding process), but can homogenize extreme values in ways that obscure vital environmental variations. Further, when working with gridded data outside the range of instrumental records (anytime in the future, or before the late nineteenth century) one must contend with the implications of the downscaling process (see section 3.6) on climate variables. Finally, when considering using related information, such as satellite sensor-derived indices (see section 3.2.4) or regional atmospheric fluctuation indices (see section 3.2.7 and Chapter 2), their spatial scale and compatibility with the chosen climate data must be evaluated.

References

Ault, T.R., Zurita-Milla, R., and Schwartz, M.D. (2015a). A Matlab© toolbox for calculating spring indices from daily meteorological data. *Computers & Geosciences*, 83, 46–53.

Ault, T.R., Schwartz, M.D., Zurita-Milla, R., Weltzin, J.F., and Betancourt, J.L. (2015b). Trends and natural variability of spring onset in the coterminous United States as evaluated by a new gridded dataset of spring indices. *Journal of Climate*, 28, 8363–78.

Barnston, A.G., and Livezey, R.E. (1987). Classification, seasonality and persistence of low-frequency atmospheric circulation patterns. *Monthly Weather Review*, 115, 1083–1126.

Barnston, A.G., Chelliah, M., and Goldenberg, S.B. (1997). Documentation of a highly ENSO-related SST region in the equatorial Pacific: Research note. *Atmosphere-Ocean*, 35, 367–83.

Daly, C., Halbleib, M., Smith, J.I., Gibson, W.P., Doggett, M.K., Taylor, G.H., Curtis, J., and Pasteris, P.P. (2008). Physiographically sensitive mapping of climatological temperature and precipitation across the conterminous United States. *International Journal of Climatology*, 28, 2031–64.

Fick, S.E., and Hijmans, R.J. (2017). WorldClim 2: New 1-km spatial resolution climate surfaces for global land areas. *International Journal of Climatology*, (published online). doi: 10.1002/joc.5086.

Goward, S.N., Markham, B., Dye, D.G., Dulaney, W., and Yang J.L. (1991). Normalized difference vegetation index measurements from the Advanced Very High-Resolution Radiometer. *Remote Sensing of Environment*, 35, 257–77.

Harris, I., Jones, P.D., Osborn, T.J., and Lister, D.H. (2014). Updated high-resolution grids of monthly climatic observations—the CRU TS3.10 Dataset. *International Journal of Climatology*, 34, 623–42.

Hijmans, R.J., Cameron, S.E., Parra, J.L., Jones, P.G., and Jarvis, A. (2005). Very high resolution interpolated climate surfaces for global land areas. *International Journal of Climatology*, 25, 1965–78.

Huete, A., Didan, K., Miura, T., Rodriguez, E.P., Gao, X., and Ferreira, L.G. (2002). Overview of the radiometric and biophysical performance of the MODIS vegetation indices. *Remote Sensing of Environment*, 83, 195–213.

Huete, A., and Justice, C. (1999). *MODIS Vegetation Index (MOD 13) Algorithm Theoretical Basis Document, Version 3*. NASA Goddard Space Flight Center, Greenbelt, MD.

Jones, P.D., Lister, D.H., Osborn, T.J., Harpham, C., Salmon, M., and Morice, C.P. (2012). Hemispheric and large-scale land-surface air temperature variations: An extensive revision and an update to 2010. *Journal of Geophysical Research: Atmospheres* 117, D05127.

Kalnay, E., Kanamitsu, M., Kistler, R., Collins, W., Deaven, D., Gandin, L., Iredell, M., Saha, S., White, G., Woollen, J., Zhu, Y., Chelliah, M., Ebisuzaki, W., Higgins, W., Janowiak, J., Mo, K.C., Ropeliewski, C., Wang, J., Leetmaa, A., Reynolds, R., Jenne, R., and Joseph, D. (1996). The NCEP/NCAR 40-Year Reanalysis Project. *Bulletin of the American Meteorological Society*, 77, 437–71.

Kanamitsu, M., Ebisuzaki, W., Woollen, J., Yang, S-K., Hnilo, J.J., Fiorino, M., and Potter, G.L. (2002). NCEP–DOE AMIP-II Reanalysis (R-2). *Bulletin of the American Meteorological Society*, 83, 1631–43.

Kemp, M.U., van Loon, E., Shamoun-Baranes, J., and Bouten, W. (2012). RNCEP: Global weather and climate data at your fingertips. *Methods in Ecology and Evolution*, 3, 65–70.

Kennedy, J., Rayner, N., Smith, R., Parker, D., and Saunby, M. (2011). Reassessing biases and other uncertainties in sea surface temperature observations measured in situ since 1850: 2. Biases and homogenization. *Journal of Geophysical Research: Atmospheres*, 116, D14104.

Lawrimore, J.H., Menne, M.J., Gleason, B.E., Williams, C.N., Wuertz, D.B., Vose, R.S., and Rennie, J. (2011). An overview of the Global Historical Climatology Network monthly mean temperature data set, version 3. *Journal of Geophysical Research: Atmospheres* 116, D19121.

Mantua, N.J., and Hare, S.R. (2002). The Pacific Decadal Oscillation. *Journal of Oceanography* 58, 35–44.

Menne, M.J., Durre, I., Vose, R.S., Gleason, B.E., and Houston, T.G. (2012). An overview of the Global Historical Climatology Network daily database. *Journal of Atmospheric and Oceanic Technology*, 29, 897–910.

Morice, C.P., Kennedy, J.J., Rayner, N.A., and Jones, P.D. (2012). Quantifying uncertainties in global and regional temperature change using an ensemble of observational estimates: The HadCRUT4 Data Set. *Journal of Geophysical Research: Atmospheres*, 117, D8101.

Rohde, R., Muller, R., Jacobsen, R., Perlmutter, S., Rosenfeld, A., Wurtele, J., Curry, J., Wickham, C., and Mosher, S. (2013). Berkeley Earth temperature averaging process. *Geoinformatics & Geostatistics: An Overview*, 1, 2.

Rouse Jr., J.W., Haas, R.H., Schell, J.A., and Deering, D.W. (1974). Monitoring vegetation systems in the Great Plains with ERTS. In: *Third Earth Resources Technology Satellite-1 Symposium, Volume I: Technical Presentations*, pp. 310–317, NASA SP-351, Greenbelt, MD.

Schwartz, M.D. (1990). Detecting the onset of spring: A possible application of phenological models. *Climate Research*, 1, 23–9.

Schwartz, M.D. (1997). Spring Index Models: An approach to connecting satellite and surface phenology. In: H. Lieth and M.D. Schwartz (eds), *Phenology in Seasonal Climates I*, pp. 23–38, Backhuys Publishers, Leiden, Netherlands.

Schwartz, M.D., Ahas, R., and Aasa, A. (2006). Onset of spring starting earlier across the Northern Hemisphere. *Global Change Biology*, 12, 343–51.

Schwartz, M.D., Ault, T.R., and Betancourt, J.L. (2013). Spring onset variations and trends in the continental United States: Past and regional assessment using temperature-based indices. *International Journal of Climatology*, 33, 2917–22.

Tarpley, J. (1991). The NOAA Global Vegetation Index Product: A review. *Global and Planetary Change*, 4, 189–94.

Trenberth, K.E. (1997). The definition of El Niño. *Bulletin of the American Meteorological Society*, 78, 2771–7.

Trzaska, S., and Schnarr, E. (2014). *A Review of Downscaling Methods for Climate Change Projections*. United States Agency for International Development by Tetra Tech ARD, Washington, D.C.

Wilby, R.L., and Fowler, H.J. (2010). Regional climate downscaling. In: C.F. Fung, A. Lopez, and M. New (eds), *Modelling the Impact of Climate Change on Water Resources*. Wiley-Blackwell Publishing, Chichester, UK.

Wilby, R.L., and Wigley, T.M.L. (1997). Downscaling general circulation model output: A review of methods and limitations. *Progress in Physical Geography: Earth and Environment*, 21, 530–48.

CHAPTER 4

Long-term time series of ornithological data

Anders Pape Møller and Wesley M. Hochachka

4.1 Introduction

We have estimated that there may be more than 200 000 persons with a significant knowledge of birds worldwide, making this specialty an unparalleled source of knowledge about nature. This opens up possibilities for citizen science and a significant role of amateurs in research and conservation projects. This is even more the case for research on climate change, where we can observe changes in front of our eyes.

Here, we briefly outline a diversity of potential sources of information on phenology, demography, life history, and distribution that potentially can be exploited for long-term studies of effects of climate change on birds. This list of potential sources of suitable data is not exhaustive, although it includes many types of data that are novel in the context of research on climate change.

4.2 Data

We provide brief overviews, both noting potential uses of these data and (in section 4.3) discussing potential sources of error or bias for which uses of these data will need to account. An overview of these sources of data and their strengths and weaknesses is given in Table 4.1.

4.2.1 Long-term population studies

Evidence of long-term changes in demographic parameters—survival, fecundity, phenology, recruitment, morphology, and dispersal—can be found in data from long-term studies of populations of individually marked birds. These studies are typically facilitated by studying birds nesting in artificial nest boxes, which leads to a taxonomic over-representation of species such as tits, *Ficedula* flycatchers, and *Tachycineta* swallows (Møller et al. 2014). We believe that academic researchers are running more than 250 currently active long-term individual-based projects worldwide lasting more than 10 years; some studies have been ongoing for more than 50 years (nest box studies of great tits *Parus major* in UK and the Netherlands). There are many more time series collected meticulously by amateur bird ringers perhaps increasing the number of long-term projects by a factor ten.

4.2.2 Nest record schemes

Some of the data from long-term amateur studies have been provided to projects that compile data from the monitoring of bird nests. There is a national nest record scheme in UK run by the British Trust for Ornithology, with schemes with fewer data in other countries including the NestWatch programme

Møller, A.P., and Hochachka, W.M., *Long-term time series of ornithological data*. In: *Effects of Climate Change on Birds*. Second Edition. Edited by Peter O. Dunn and Anders Pape Møller: Oxford University Press (2019). © Oxford University Press. DOI: 10.1093/oso/9780198824268.003.0004

Table 4.1 An overview of different sources of long-term data on birds with potential sources of error and bias.

Data type	Frequency of data type	Kinds of data	Potential errors and biases
Long-term population studies	More than 300 of a duration of more than 10 years worldwide	Population size, demographic and life history data, including recruitment and dispersal	Changes in methodology Inter-observer variability
Bird survey data	National and continental databases	Population trends for breeding and wintering birds	Changes in methodology Changes in quality of observers Spatial and temporal heterogeneity in research effort
Atlases	At least four repeated atlas projects	Breeding and winter distributions	Changes in methodology Changes in quality of participants Spatial heterogeneity in effort
Nest record schemes	The only large programmes are the U K nest record scheme and the North American nest record scheme	Phenology, clutch size, reproductive success, and duration of breeding seasons	Changes in methodology Spatial and temporal variation in sampling effort
Bird ringing	National programmes in most countries in the developed world	Age distributions, recruitment rates, sex- and age-specific survival rates, dispersal rates, and migration distances	Changes in methodology Changes in quality of bird ringers Changes in spatial and temporal distribution of ringing effort
Bird migration stations	More than 200 worldwide	Phenology and sex and age distributions	It remains unknown whether observations and captures reflect migration Changes in methodology
Bird observation databases	Extensive databases in many developed countries	Phenological data, age distributions, and geographical distribution	Change in observer effort Spatial and temporal variation in observer effort Change in quality of observers
Museum collections	Extensive collections in most developed countries	Phenological, demographic, and life history data	Temporal change in inclusion criteria Spatial and temporal variation in sampling effort Little collection performed the last 50 years
DNA collections	Many long-term population studies	Blood samples including samples of Haematozoa, bacteria, and virus in blood	

in the United States (http://watch.birds.cornell.edu/nest/home/index).

4.2.3 Bird surveys

Data from long-term surveys of all bird species, conducted across countries and even continents, can be used to provide evidence of the effects of climates on birds (e.g., La Sorte et al. 2009). Such surveys have been conducted in many countries for decades, with breeding-season surveys dating back to 1962 in the UK (http://www.bto.org/birdtrends2004/bbs.htm) and 1996 in North America (data available at: https://www.pwrc.usgs.gov/bbs/RawData/). Continent-wide European bird monitoring has begun more recently across all of Europe (http://www.ebcc.info/pecbm.html). The National Audubon Society provides access to over a century of data, mostly from North America, that have been gathered in Christmastime surveys (http://netapp.audubon.org/CBCObservation/).

Surveys also exist for more restricted taxonomic groups in some regions, although these data are not necessarily stored in easily accessible repositories. For example, populations of swans, geese, ducks, seabirds, and raptors are surveyed annually in UK, Germany, the Netherlands, and many other countries. Waterbird surveys in the old world have been running since 1962 (www.wetlands.org). North American surveys of breeding and wintering waterfowl date back decades, with some of these data being freely available (http://www.fws.gov/migratorybirds/).

4.2.4 Breeding and winter bird atlases

One specialized form of bird survey is the bird atlas, with repeated atlases being created in an increasing number of countries, states, and provinces. Changes in bird distributions between atlases have been used to infer effects of climate change (e.g., Thomas and Lennon 1999; Zuckerberg 2009). The European breeding bird atlas (Hagemeijer and Blair 1997) was the basis for projections of potential future distributions of birds across the continent (Huntley et al. 2007), and these projections can be compared with observed changes in the new European breeding bird atlas under way. These European atlas data are freely available through the Global Biodiversity Information Facility (https://www.gbif.org/dataset/c779b049-28f3-4daf-bbf4-0a40830819b6). A map of atlas projects in North America can be found at http://www.pwrc.usgs.gov/bba/. There is also a worldwide list at http://avibase.bsc-eoc.org/links/links.jsp?page=g_atlas.

4.2.5 Bird ringing information

There are national bird ringing schemes in most developed countries, whose data can be used to investigate diverse potential impacts of climate change. Ringing schemes have been running continuously since 1899 in Denmark, 1904 in Germany, and 1909 in the UK; Europe's ringing data are well managed and documented by an umbrella organization, EURING (http://www.euring.org). The North American ringing programme is also centralized. Constant effort ringing programmes are running in most European countries, and elsewhere. These programmes record information on temporal and climate-related change in (1) age distribution of captured birds (and hence indirectly mortality rates (Snow 1954)), (2) recruitment rate (because ringing information for young can be compared with later recovery information (e.g. Møller 1983; Albert et al. 2016)), (3) sex- and age-specific survival rate, (4) natal and breeding dispersal rates, and (4) changes in migration distances and winter quarters (e.g. Fiedler et al. 2004).

4.2.6 Bird migration stations

Highly standardized ringing and other migration-monitoring activities are ongoing at more than 200 bird observatories worldwide. Bird migration stations have existed since 1853 on Helgoland, Germany, established by Heinrich Gätke. The information recorded at these observatories range from visual observations of migration to bird ringing. Rubolini et al. (2007) have recently assembled 672 time series of first arrival dates and 289 time series of mean/median arrival dates mainly from bird migration stations from Europe, and a large European data base is currently being developed.

A major assumption at bird observatories is that the number and composition of birds observed and captured reflects the number and composition of birds migrating at a site. In other words, studies at bird observatories assume that data constitute a random sampling of a constant proportion of migratory birds passing through a site. This assumption should be viewed with caution (e.g. Hjort and Lindholm 1978; Winstanley et al. 1974; Hochachka and Fiedler 2008).

4.2.7 Bird observation depositories

The popularity of bird watching has resulted in the creation of many regional, national, and international databases of observations made by bird watchers. The largest of these is eBird, which has more than 440 million observations worldwide. An example of a national extensive database is DOFbasen in Denmark, which has 20 million observations (https://dofbasen.dk/). Often the observation periods over which observations were recorded are variable, and so ancillary information such as the duration of the observation period and distance travelled are an important component of these datasets; these ancillary data allow for statistical correction for variation in the probability that birds will be detected. Another consideration with these data is whether they are collected in the form of presence-only or presence-absence data; Guillera-Arroita et al. (2015) describes the limits of interpretation of data of these two forms.

These databases contain information that can be used for studies of changes in phenology, and changes in geographic distribution, and perhaps even age distribution linked to climate change.

4.2.8 Museum collections

The concept of what constitutes a museum specimen is broadening (e.g. Webster 2017), as are the potential effects of climate change that can be examined with data from museum collections. Physical specimens of skins, clutches of eggs, and nests in museum collections date back more than 150 years. Examination of these complete specimens can be used for a diversity of lines of research into the effects of climate change, as varied as examinations of changes in morphology (Salewski et al. 2010), and climate-dependent changes in nest morphology (Møller 2006; Møller & Nielsen 2015).

Examination of specific aspects of physical specimens also can shed light on effects of climate change. Many feather characters can be assessed in museum specimens. Fault bars (Michener & Michener 1938), daily growth bars during moult (Grubb 2006; Møller et al. 2018), and feather wear (Møller et al. 2018) can be assessed from historical museum specimens, and these records can be compared with recent samples obtained from the same sites in the field during recent climate change. Fault bars, daily growth bars, and feather wear are more common or more severe under adverse weather conditions (Møller et al. 2018; A. P. Møller unpubl. data). Feathers can be used for studies of change in composition of stable isotopes (and hence locations, habitats, and diets), and potentially also for long-term studies of coloration, pigments, hormones, and communities of microorganisms.

Technical advances are allowing the DNA of very old physical specimens to be sequenced, allowing examination of changes through time down to a molecular genetic level. Purpose-designed collections of DNA are becoming an important part of the physical specimens that museums are collecting, such as recent efforts in German collections to permanently preserve DNA that has been used for taxonomic purposes and keep it available (http://www.dnabank-network.org).

Time series of behaviours exist in the virtual, or electronic, specimens being curated by museums. Most notably, recordings of vocal activity sometimes date back decades. Archives of recordings of bird song can be found on XenoCanto (https://www.xeno-canto.org/) and Macauley Library, Cornell University (www.birds.cornell.edu/page.aspx?pid=1676).

The impetus to collect museum specimens has changed through time, and therefore spatial and temporal variation in sampling effort is likely to affect conclusions. For example, many museum collections in Europe have hardly increased their collections with new specimens during the last 50 years, which is the period with the most pronounced change in climate.

4.3 Methodological considerations

The types of long-term data described in this chapter were never collected for the purpose of studying effects of climate change, and as a result potentially sources of random noise and bias exist in the data that could otherwise have been eliminated in the design of data collection. Thus, a necessary aspect of analysing these data is the testing and correction for these sources of statistical noise in the process of fitting statistical models.

Data from individual study sites, for example long-term demographic studies, may be particularly sensitive to changes in a project's methods or study site over time: even a single new researcher who begins to work on a project after it has begun may use slightly different field practices that can unwittingly produce a systematic change in bias in a long-term time series of data. There are few explicit tests for inter-observer variation in measurements and their influence on conclusions (Gosler et al. 1998; Møller et al. 2006b). Long-term time series can have noise introduced deliberately by changes made to aid a study's proximate goals. For example, the Wytham Wood study of great tits introduced predator-proof nest boxes in 1976 (McCleery et al. 1996) to eliminate predation by weasels *Mustela nivalis*, but this change in design also profoundly affected the behaviour and life history of the birds. Use of data from one or a small number of local population studies for studies of climate change also assumes that the effects of climate are not confounded by changes in habitat or the extent of dispersal, although such potential effects are rarely considered (Møller et al. 2006a).

Data from broad-scale, country- or continent-wide time series are likely buffered from many systematic biases that vary through time, because the larger number of participants in data collection will have their individual idiosyncrasies average out, resulting in random noise rather than systematic biases. Similarly, it is less likely that environmental conditions at all locations could simultaneously change in ways that are unrelated to changes in climate. Nevertheless, broad-scale studies are not immune from systematic changes in bias through time. Changes could be introduced by increases in the quality of field guides and the knowledge of species identification, ageing, and sexing criteria. For data from ringing sites and large bird observatories, changes in capture practice such as the introduction of mist nets affect the species and numbers of birds ringed. Spatial and temporal changes in ringing effort may affect conclusions as well as spatial and temporal changes in recovery and recovery reporting probabilities (Sauter et al. 2009). The methodology used in large citizen science projects can also change through time, as occurred with the British Trust for Ornithology's breeding season population monitoring, for which the entire protocol was changed. This specific transition, however, was carefully planned (Freeman et al. 2007), with multiple years of overlap of these two monitoring schemes so that the systematic differences between the data collection methods could be measured and calibration performed during data analysis. In this case, the primary data cannot be directly compared across the entire time series, but the 'secondary data'—the output from analysis and calibration—is a robust measure of changes in population sizes through time.

A final way in which projects can change through time is the collection of additional forms of data through time. There have also been improvements in the quality of ringing data, with greater proportions of individuals being sexed and aged in recent years. Atlas projects have also typically collected new types of data when repeated, for example often gathering supplemental information from which densities of breeding birds can be estimated (for example, this happened for the second Atlas project in Denmark). Thus, with some data sources, only a 'least common denominator' of data is available for use in retrospective studies of long-term changes.

In summary, an important early step of working with any set of long-term data is learning about the process of data collection, in order to identify potential factors—changes in personnel, protocols, and study areas as examples—that could have changed through the course of a long-term project's history. With citizen science projects for which participants select the locations at which data are collected, researchers will need to consider potential biases resulting from the non-random locations selected by participants (Tulloch and Szabo 2012), and the potential for the distribution of locations to change through time, which can affect conclusions (e.g. Sparks et al. 2008). Exploratory analyses should be conducted in order to try to detect evidence of systematically changing biases, prior to building the final statistical models that are to be used to extract biological patterns from the raw data.

4.4 Discussion

There are vast amounts of data relevant for studies of the effects of climate change on birds, although quality and availability vary. However, applications for access to data have sometimes lasted many years for no obvious reasons without a resolution in sight, although fortunately most scientists and amateurs in the field are open for collaborations.

A major difficulty is cross-validation and critical tests of data quality. Surprisingly, there are few published attempts to cross-validate phenological, demographic, or life history estimates by relying on different estimators (e.g. Møller et al. 2004).

Large databases by necessity contain a lot of data. While recent progress in computer technology has allowed vast increases in the size of computerized databases, the quality of data remains unknown. There are numerous opportunities for errors entering such databases. Surprisingly, there are no published estimates of error rates from any of the databases listed here. One of us checked information on more than 40 000 ringed birds in a national bird ringing database against the original notes made by the bird ringers. More than 3% of the observations in the national database contained errors in dates, localities, or even species.

We are aware of unpublished estimates of error rates in databases for long-term studies that indicate similarly high rates. Clearly, many more methodological studies are required for estimating error rates and for setting acceptable limits to error rates. We also suggest that all scientists in the field consistently attempt to address this issue to allow others to assess the quality of data. Publication of such assessments will give credibility to individual studies, but also to the field.

4.5 Conclusions and recommendations

Numerous data allow studies of the effects of climate change on birds. There have been few attempts to coordinate databases. Some depositories of data require written applications for data access, payment for access, and co-authorship. Others provide free access on the internet such as the US and several European breeding bird surveys databases. Clearly, in a world where collection of many data has been funded by national or international funding agencies, access for scientific reasons should not be an issue. We may also question the ethics of limiting restriction to data with potentially significant conservation implications.

Free and anonymous access through the internet to large datasets and sometimes complex data is not always the optimal solution. Inclusion of curators of data or the scientists who initially collected all the data as co-authors may actually improve the quality of the analyses (Mills et al. 2015). However, we recommend that straightforward procedures for gaining access to data should be offered to all potential users. Unfortunately, even today access to some databases may imply delays of months or even years.

References

Albert, S.K., DeSante, D.F., Kaschube, D.R., and Saracco, J.F. (2016). MAPS (Monitoring Avian Productivity and Survivorship) data provide inferences on demographic drivers of population trends for 158 species of North American landbirds. *North American Bird Bander*, 41, 133–40.

Fiedler, W., Bairlein, F., and Köppen, U. (2004). Using large scale data from ringed birds for the investigation of effects of climate change on migrating birds: pitfalls and prospects. In: A.P. Møller, W. Fiedler, and P. Berthold (eds), *Effects of climatic change on birds*, pp. 49–67. Elsevier, Amsterdam, Netherlands.

Freeman, S.N., Noble, D.G., Newson, S.E., and Baillie, S.R. (2007). Modelling population changes using data from different surveys: The common birds census and the breeding bird survey. *Bird Study*, 54, 61–72.

Gosler, A.G., Greenwood, J.J.D., Baker, J.K., and Davidson, N.C. (1998). The field determination of body size and condition in passerines: A report to the British Ringing Committee. *Bird Study*, 45, 92–103.

Grubb, T.C., Jr. (2006). *Ptilochronology: Feather time and the biology of birds*. Oxford University Press, Oxford, UK.

Guillera-Arroita, G., Lahoz-Monfort, J.J., Elith, J., Gordon, A., Kujala, H., Lentini, P.E., McCarthy, M.A., Tingley, R., and Wintle, B.A. (2015). Is my species distribution model fit for purpose? Matching data and models to applications. *Global Ecology and Biogeography*, 24, 276–92.

Hagemeijer, W.J.M., and Blair, M.J. (eds) (1997). *The EBCC atlas of European breeding birds*. T. and A.D. Poyser, London, UK.

Hjort, C., and Lindholm, C.-G. (1978). Annual ringing totals and population fluctuations. *Oikos*, 30, 387–92.

Hochachka, W.M., and Fiedler, W. (2008). Trends in trappability and stop-over duration can confound interpretations of population trajectories from long-term migration ringing studies. *Journal of Ornithology*, 149, 375–91.

Huntley, B., Green, R.E., Collingham, Y.C., and Willis, S.G. (2007). *A climatic atlas of European breeding birds*. Durham University, the RSPB, and Lynx Edicions, Barcelona.

La Sorte, F.A., Lee, T.M., Wilman, H., and Jetz, W. (2009). Disparities between observed and predicted impacts of climate change on winter bird assemblages. *Proceedings of the Royal Society of London Series B: Biological Sciences*, 276, 3167–74.

McCleery, R.H., Clobert, J., Julliard, R., and Perrins, C.M. (1996). Nest predation and delayed cost of reproduction in the great tit. *Journal of Animal Ecology*, 65, 96–104.

Michener, H., and Michener, J.R. (1938). Bars in flight feathers. *Condor*, 40, 149–60.

Mills, J.A., Teplitsky, C., Arroyo, B., et al. (2015). Archiving primary data: solutions for long-term studies. *Trends in Ecology & Evolution*, 30, 581–89.

Møller, A.P. (1983). Time of breeding, causes of recovery and survival of European sandwich terns. *Vogelwarte*, 32, 123–41.

Møller, A.P. (2006). Rapid change in nest size of a bird related to change in a secondary sexual character. *Behavioural Ecology*, 17, 108–16.

Møller, A.P., and Nielsen, J.T. (2015). Large increase in nest size linked to climate change: An indicator of life history, senescence and condition. *Oecologia*, 179, 913–21.

Møller, A.P., Berthold, P., and Fiedler, W. (2004). The challenge of future research on climate change and avian biology. In: A.P. Møller, W. Fiedler, and P. Berthold (eds), *Effects of climatic change on birds*, pp. 237–245. Elsevier, Amsterdam, Netherlands.

Møller, A.P., Flensted-Jensen, E., and Mardal, W. (2006b). Rapidly advancing laying date in a seabird and the changing advantage of early reproduction. *Journal of Animal Ecology*, 75, 657–65.

Møller, A.P., Hobson, K.A., and Laursen, K. (2018). Retrospectively analysing body condition and nitrogen fertilizer content in historical samples of birds. *Journal of Zoology*, 305, 188–95.

Møller, A.P., Adriaensen, F., Artemyev, A., et al. (2014). Clutch-size variation in Western Palaearctic secondary hole-nesting passerine birds in relation to nest box design. *Methods in Ecology and Evolution*, 5, 353–62.

Møller, A.P., Chabi, Y., Cuervo, J.J., de Lope, F., Kilpimaa, J., Kose, M., Matyjasiak, P., Pap, P.L., Saino, N., Sakraoui, R., Schifferli, L., and von Hirschheydt, J. (2006a). An analysis of continent-wide patterns of sexual selection in a passerine bird. *Evolution*, 60, 856–68.

Rubolini, D., Møller, A.P., Rainio, K., and Lehikoinen, E. (2007). Assessing intraspecific consistency and geographic variability in temporal trends of spring migration phenology among European bird species. *Climate Research*, 35, 135–46.

Salewski, V., Hochachka, W.M., and Fiedler, W. (2010). Global warming and Bergmann's rule: Do central European passerines adjust their body size to rising temperatures? *Oecologia*, 162, 247–60.

Sauter, A., Korner-Nievergelt, F., Atkinson, P., Guélat, J., Kania, W., Kéry, M., Köppen, U., Robinson, R., Schaub, M., Thorup, K., van der Jeugd, H., and van Noordwijk, A.J. (2009). Improving the analysis of movement data from marked individuals through explicit estimation of observer heterogeneity. *Journal of Avian Biology*, 41, 8–17.

Snow, D.W. (1954). The annual mortality of the blue tit in different parts of its range. *Ibis*, 49, 174–7.

Sparks, T., Huber, K., and Tryjanowski, P. (2008). Something for the weekend? Examining the bias in avian phenological recording. *International Journal of Biometeorology*, 52, 505–10.

Thomas, C.D., and Lennon, J.J. (1999). Birds extend their ranges northwards. *Nature*, 399, 213.

Tulloch, A.I.T., and Szabo, J.K. (2012). A behavioural ecology approach to understand volunteer surveying for citizen science datasets. *Emu*, 112, 313–25.

Webster, M. (ed.) (2017). *The extended specimen: Emerging frontiers in collections-based ornithological research*. CRC Press, Boca Raton, FL.

Winstanley, D., Spencer, R., and Williamson, K. (1974). Where have all the whitethroats gone? *Bird Study*, 21, 1–14.

Zuckerberg, B. (2009). Poleward shifts in breeding bird distributions in New York State. *Global Change Biology*, 15, 1866–83.

CHAPTER 5

Quantifying the climatic sensitivity of individuals, populations, and species

Martijn van de Pol and Liam D. Bailey

5.1 Introduction

To understand the effect of weather variables on ecosystems, biologists relate variation in biological variables to temporal or spatial variation in weather variables. This type of analysis has become increasingly common as researchers seek to quantify and predict the impacts of global climate change. Biological response variables may range from behavioural traits, phenotypic traits, demographic or population parameters, or even the biodiversity of a community. Although the exact biological question may depend on the field of research, from an analytical perspective these studies generally boil down to the same simple questions: 'How much does my biological variable of interest Y change with weather variable X (e.g. body mass increases with β = 0.1 g/°C)?', or 'How much of the variation in Y is explained by X (e.g. partial-R^2)?'. The strength of these variables (e.g. β or R^2) is what we term 'climatic sensitivity'.

Although it may appear straightforward to quantify climatic sensitivities, there are several challenges. Here we focus on two in particular. First, how do we identify the weather signal to which individuals, populations, or species respond? Or, in other words: Which weather variables should we study? Over which period and in what way does this weather variable affect organisms? For example, is it the mean spring temperature or total rainfall over the past fortnight that is the most relevant driver? Answering these questions is far from trivial as we often have limited a priori knowledge about the underlying mechanisms that mediate how weather affects organisms, while many different analytical methods exist.

Second, how do we compare the climatic sensitivity of multiple individuals, populations, or species and explain differences among them? This challenge is important across a range of research fields. For example, evolutionary biologists are interested in individual variation in phenotypically plastic responses to weather to study genotype by environment interactions. Population biologists are interested in variation among populations in responses to weather, as spatial variation may lead to portfolio effects that buffer impacts on the entire (meta-)population. Conservation biologists are interested in identifying ecological or life history characteristics that predict variation among species in responses to weather, to prioritize conservation efforts. In all such comparative analyses different individuals, populations, or species may be affected by the same weather variables in different ways, be impacted by completely different weather variables, or even be insensitive to changes in weather altogether. How do we make sure that comparative studies are comparing real biological differences and not differences in climatic sensitivities caused by the methods or implicit assumptions used?

Van de Pol, M., and Bailey, L.D., *Quantifying the climatic sensitivity of individuals, populations, and species*. In: *Effects of Climate Change on Birds*. Second Edition. Edited by Peter O. Dunn and Anders Pape Møller: Oxford University Press (2019).
 DOI: 10.1093/oso/9780198824268.003.0005

5.2 Identifying the critical climatic predictors

5.2.1 The need for systematic exploratory approaches

Organisms are likely to be affected by a mixture of different weather variables acting at different periods of time. Consequently, the number of possible candidate weather signals can be substantial, making analyses difficult. A confirmatory approach uses pre-existing biological knowledge to limit the number of potential variables to a few testable hypotheses (Frederiksen et al. 2014). Such an approach can be useful in study systems where the ecophysiology or behaviour of the focal organism provides clues to identify appropriate weather variables. For example, strong winds are known to cause flooding of shorebird nests leading to widespread reproductive failure (Bailey et al. 2017), and high temperatures that approach avian body temperature are known to trigger thermoregulatory behaviour and limit foraging (McKechnie and Wolf 2010; Smit et al. 2016). However, in many study systems we have limited a priori knowledge about potentially important weather variables. Effects of weather variables may be lagged or may act through indirect pathways (e.g. via effects on food, predators, or parasites), making the observation of direct weather effects difficult.

Sometimes weather signals may be identified using experimental manipulation. For example, work in climate-controlled aviaries has shown that the rate of seasonal temperature increase, rather than the mean or daily temperature range, best determines the onset of avian egg-laying (Schaper et al. 2011). However, meaningful experimental manipulation is very difficult, particularly for mobile species. Furthermore, responses to experimental manipulation may not reflect the impacts of climate change in the wild (Wolkovich et al. 2012). Therefore, in many study systems a more exploratory approach based on observational data is required.

Exploratory approaches do not require much a priori knowledge on a study system, but they do require systematic thought about the variables to be tested. In particular, it will be necessary to decide which weather variables will be considered, over which time period, and in what way they might affect the biological response. For example, the commonly used variable 'mean annual temperature' includes three—typically implicit—assumptions: (i) temperature is the key weather variable affecting the biological response, (ii) temperature during the entire year affects the biological respone, and (iii) mean temperature over this period best explains variation in the biological response. We call these the 'which-when-how' assumptions that together define the 'weather signal' (Table 5.1). Unfortunately, most studies have failed to consider these different aspects of weather signals in a systematic way, with the choice of weather signal often being arbitrarily based on a range of untested assumptions (reviewed by van de Pol et al. 2016).

The use of an unsystematic approach is problematic, as overly simplistic, suboptimal, or wrongly identified weather signals can lead to erroneous conclusions which will limit our ability to make reliable projections of climate change impacts. If a trait displays no response to an arbitrarily selected weather signal, it is difficult to determine whether this is evidence of climatic insensitivity or simply a reflection of a flawed choice of signal. Even when we find a relationship between weather and the biological response, we cannot be sure that we have selected the most relevant weather signal (e.g., one that explains most of the variation in the biological response, or best predicts new observations). These problems not only hamper projections for single populations, but also cloud whether reported intra- or interspecific variation in climatic sensitivities reflect true biological differences or methodological flaws (section 5.3).

An exploratory approach aims to systematically validate the assumptions underlying a weather signal using a 'climate window analysis' that considers different climate variables (precipitation, humidity, etc.), periods of different length (spring, summer, the past week, etc.), and different aggregate statistics (minimum, or number of days with freezing temperatures, etc.). In fact, such an exploratory approach allows testing of many more detailed hypotheses, as summarized in Table 5.1 (reviewed by van de Pol et al. 2016). For example, for migratory species an additional challenge is to identify whether it is the weather at the overwintering grounds, en route, or at the breeding grounds that is most important for, e.g., reproduction and survival.

Table 5.1 Aspects of weather signals that can be considered.

Climate aspect	Challenge	Possible solution
Which weather variable?		
Identity of weather variables	Most studies focus on temperature, some on precipitation, while other variables are rarely considered (e.g., wind, humidity).	Consider a larger set of weather variables. E.g., strong winds can decrease insect abundance and reduce performance of insectivorous birds (Møller 2013).
Number of weather variables	Few studies consider more than one variable simultaneously.	Allow for the possibility that weather variables (rain, humidity, temperature) act concurrently (McLean et al. 2018). Be aware of collinearity.
Interactions between weather variables	Few studies consider the possibility of interactions between weather variables (e.g., impact of hot temperatures may be most noticeable during dry spells).	Test for interactions or use compound variables (e.g., drought index).
When (or where) is weather important?		
Window duration	Studies mostly consider a few long (annual or seasonal) time periods.	Allow for time windows to vary in length, including short windows (weeks, days) for more recent weather (Kruuk et al. 2015).
Critical period	Often studies consider weather to act only in one specific season (e.g., winter for annual survival).	Explore weather influences in different parts of the year. Consider lags beyond a year if weather acts indirectly via food or predators (Mesquita et al. 2015).
Multiple windows of same variable	Few studies consider that a single weather variable can have multiple signals, even in opposing directions. E.g., warm temperatures during winter have positive effects, while more recent hot days are detrimental (Kruuk et al. 2015).	Employ methods that allow testing for multiple signals (Table 5.2).
Variation in windows among subjects	Individuals, populations, or species may not be affected by weather at the same time. E.g., the mass of early offspring (born in May) is likely affected by weather in a different period than the mass of late offspring (born in July).	Allow windows to vary among subjects, e.g., using relative time windows (weather during the month before offspring were born) instead of absolute time windows (April weather) (van de Pol et al. 2012).
Spatial scale over which weather acts	Species moving over large areas are likely to be affected by weather in different locations (e.g., migratory species).	Use weather from different areas. Or find the spatial region over which weather best explains variation in the biological variable (Mesquita et al. 2015; Suggitt et al. 2017; Haest et al. 2018a).
How does weather act?		
Summarizing how weather acts (aggregate statistic)	Most studies consider that weather effects are best described by mean weather, but it may also act via changes in the variability in weather, or extreme values that exceed a certain threshold.	Consider a larger set of aggregate statistics (e.g., mean, variance, growing degree days, days with values above X). Test where thresholds lie (Phillimore et al. 2013), e.g., do effects of heat waves occur at >30 or >35°C?
Shape of the biological response	Responses to weather may be nonlinear, e.g., due to threshold effects or organism having a thermal range beyond which it can't perform properly.	Consider both linear and nonlinear functions of weather variables.

The question then is over what spatial scale weather acts (Mesquita et al. 2015; Suggitt et al. 2017; Haest et al. 2018a), and how likely carry-over effects occur of weather much earlier in the year.

5.2.2 What methods are available?

Exploratory approaches have become more relevant with increases in computational power. A variety of exploratory methods are now available to identify weather signals for a species and quantify its corresponding climatic sensitivity. However, so far there has been no review comparing these different methodological approaches. Many of the methods have been developed for specific problems in disparate fields of biology, such as analyses of phenological events (Gienapp et al. 2005; Schleip et al. 2008; van de Pol and Cockburn 2011; Holloway et al. 2018) or abundance and presence/absence data of species (Elith et al. 2011; Elston et al. 2017). Consequently, there has been limited cross-referencing and methodological synthesis, and the application of each method is often restricted to a specific field. However, most approaches are generally applicable to a wide variety of data-types (quantitative and phenological trait data, demographic data, time series abundance data, presence-absence data, etc.) and can be implemented using a broad range of statistical models (generalized linear (mixed) models, time-to-event models, autoregressive models, species distribution models, etc.). Most methods are thus more broadly relevant than they originally appear, and this section categorizes them and discusses their pros and cons (Table 5.2).

The most widely used approach is the 'multiple regression' approach (e.g. Sparks and Carey 1995), in which multiple weather covariates are entered (e.g. stepwise) into a regression model. While this approach is simple and easy to implement, multiple regression methods cannot deal with a large set of predictor variables nor with highly correlated predictors. Because of these weaknesses, this method typically uses only a few seasonal or monthly periods as predictor variables (e.g. mean temperature of each month of the year). The limitation on the number of predictor variables makes the method rather limited in how well it can explore hypotheses about the duration and length of windows and other 'which-when-how' assumptions (Table 5.1).

An increasingly frequently used approach is the 'sliding window' approach (e.g. Husby et al. 2010; Kruuk et al. 2015; formalized by van de Pol et al. 2016; Bailey and van de Pol 2016), which instead of

Table 5.2 Overview of different exploratory approaches for climate window analysis.

Exploratory approach	Pro	Con	Performance tested?	Implementation
Multiple regression, many studies	Simplicity	Cannot handle many different weather variables, time windows and aggregate statistics.	No	dredge-function in R-package MuMIn, stepwise model selection
Sliding window, many studies, formalized by van de Pol et al. 2016 Bailey and van de Pol 2016	Can directly compare many different weather variables, time and spatial windows and aggregate statistics.	No weighted mean. Can only sequentially detect multiple signals of the same weather variable.	Yes	R-package *climwin* (Bailey and van de Pol 2015; spatial extension by Haest et al. 2018a)
Weighted mean, Gienapp et al. 2005, Schleip et al. 2008, van de Pol & Cockburn 2011	Quantifies how the importance of weather gradually changes throughout year.	Only works with mean aggregate statistic. Can only sequentially detect multiple signals of the same weather variable.	No	R-package *climwin* (excl. Schleip method)
Regularization, Sims et al. 2007, Roberts 2008, Teller et al. 2016, Thackeray et al. 2016, Elston et al. 2017	Quantifies how the importance of weather gradually changes throughout year. Can detect multiple windows for the same weather variable.	Does not directly output a single measure of climatic sensitivity. No choice of aggregate statistics.	Only Sims et al. 2007 and Teller et al. 2016	Code provided in Sims et al., Teller et al., and Elston et al.
Machine learning, Teller et al. 2016, Holloway et al. 2018	Maximizes predictive value. Can compare many different weather variables, time windows, and aggregate statistics.	Data hungry. Identified weather signal provides limited mechanistic understanding.	Only Teller et al. 2016	Code provided in papers

including many windows as separate predictor variables, only calculates a single predictor variable over a specific time period. By repeating this procedure for all possible windows within a year (or previous years if longer lags are expected), and by fitting each window using a separate regression model, one can then select the best among a large set of models (Figure 5.1a). This approach is much more flexible than the multiple regression approach. It can be highly sensitive to spurious results due to multiple testing, but recent new tools have been developed to deal with this problem (Bailey and van de Pol 2016; van de Pol et al. 2016). A sliding window approach also allows for systematic comparison of a variety of aggregate statistics in addition to the mean (e.g. variance, rate of temperature change), which sets it apart from several other methods described below. Some aggregate statistics, such as growing degree days and the number of cold days, are quite mechanistic (Bonhomme 2000) and include additional variables that can be tested (e.g. the temperature threshold value above which one starts counting growing degrees; Phillimore et al. 2013).

When considering the mean of a weather variable, a major drawback from both above methods is that all days in a window are assumed to contribute equally to the overall mean of the weather variable (Figure 5.1b-i). This assumption will often be unrealistic, and it may be more appropriate to allow the importance of weather to gradually change over time (e.g. initially increase and later decrease; Gienapp et al. 2005; van de Pol and Cockburn 2011). Such a hypothesis can be tested using a 'weighted mean' approach (Schleip et al. 2008; van de Pol and Cockburn 2011), in which flexible functions are used to model the importance (weight) of weather over time (e.g. the past year) and these weights are then used to calculate a weighted mean of a weather variable (Figure 5.1b-ii). In this way, the importance of weather in all days can be accounted for, although some days may be given a weight of zero and thus will be unimportant. The weighted mean approach can be useful on its own, but can also be used to refine hypotheses generated by the sliding window approach where mean is the most relevant aggregate statistic (van de Pol et al. 2012).

'Regularization' approaches have been developed to extend the multiple regression approach to deal with many correlated predictor variables. These methods can include daily weather data over the entire year as separate predictor variables, and then use various tools to regularize (constrain) the 365 regression coefficients with some form of penalization or shrinkage (ridge regression (Sims et al. 2007), difference penalized regression (Sims et al. 2007), penalized spline regression (Roberts 2008; Teller et al. 2016), or least absolute shrinkage and selection operator (Teller et al. 2016); for a parametric approach see Elston et al. (2017)). For example, a penalized spline regression smooths the regression coefficients of consecutive days to have similar values. We can then look at the effect size for the coefficients over the year and identify periods over which β-estimates are consistently positive or negative, which likely reflect critical periods (Figure 5.1c). The regularization approach also allows for a weather variable to differ in importance (magnitude of β-estimates) on different days, akin to the weight used in the weighted mean approach. Another advantage is that the regularization approach can identify multiple signals for a single weather variable (e.g. positive effect of warm winters, but negative effect of hot summers). However, a major drawback of this method is that it does not estimate a single measure (β) of climatic sensitivity, but instead an entire vector of βs that cannot be easily used for subsequent comparative or meta-analysis.

Finally, 'machine learning' approaches are particularly good at dealing with situations of variable selection out of a large number of correlated predictors. This approach has similar flexibility to the sliding window approach to test different weather variables, time periods, and aggregate statistics (Holloway et al. 2018; Teller et al. 2016). In contrast to other approaches, machine learning tools do not aim to select the weather signal that explains most of the observed variation in the response variable, but instead they aim to identify the weather signal that best predicts the biological response in an independent dataset. A key drawback of machine learning is that it requires large amounts of data (e.g. many time series on a species) for training the model and testing the model's predictive value, which will often be unavailable in biological studies. Furthermore, decision trees

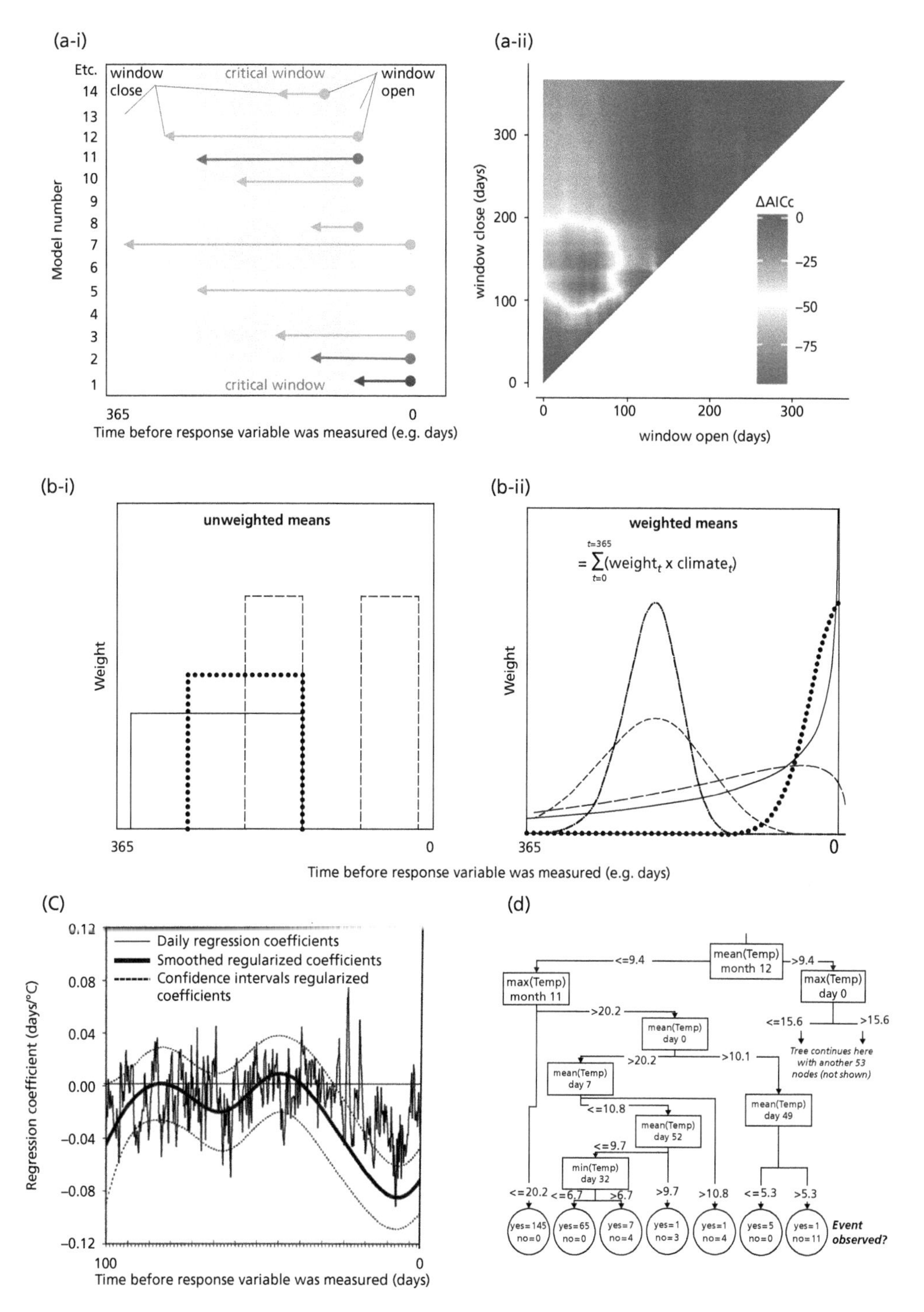

Figure 5.1 Illustrations of different exploratory approaches. (a-i) The sliding window approach tests different models that vary in the time the window opens and closes, and then compares models to determine how well each explains the observed data. One expects that the model that best describes the unknown true critical window will have the best model fit (e.g. model number 11). (a-ii) By plotting all possible time windows in a year, one can observe a clear peak during which a weather variable is important (redness). (b-i) When taking the mean of a weather variable, the sliding window approach assumes that all days within the window contribute equally to the mean (unweighted). (b-ii) The weighted mean approach relaxes this assumption, allowing the importance of weather variables to decay over time in various forms (different lines).
(c) Regularization approaches smooth the daily regression coefficients, allowing for the identification of the critical period(s) during which the weather variable is important (here ~20 days before the biological variable was measured). (d) Machine learning tools may be used to classify the response data. Decision trees can have many nodes that use signals that vary in identity, window, and aggregate statistics. Figures are adapted from Bailey and van de Pol (2016), van de Pol et al. (2016), van de Pol and Cockburn (2011), Roberts (2008), Holloway et al. (2018).

created through machine learning may involve many nodes which makes the weather signal difficult to interpret (Figure 5.1d), and less useful for mechanistic understanding and experimental manipulation.

5.2.3 Method bias and performance testing

Exploratory methods can have intrinsic biases, either because they test many different models or due to overfitting caused by high model flexibility or the inclusion of many predictor variables (allowing for overly complex models). Method bias can have important biological consequences. For example, if a method suffers from a high rate of false positives, models may incorrectly identify weather signals that are associated with the response variable simply due to chance. Furthermore, even if the results are not spurious, but the model has identified a suboptimal weather signal, climatic sensitivity may be biased (e.g. too high R^2 or low β; van de Pol et al. 2016).

Tools exist to deal with these biases, such as model comparison, cross-validation, and regularization. But how well these tools overcome biases, and how this depends on the sample size and structure of the biological data as well as on the temporal resolution of the weather data (e.g. daily, weekly, monthly, annual weather data), is largely unknown. Van de Pol et al. (2016) used simulations to show that using information theoretic model selection procedures in the sliding window approach results in too many false positives when sample sizes are moderate or small (e.g. $\leqq$ 20 time-steps). This bias can be overcome using randomization techniques. Furthermore, they found that the R^2 of the weather signal was overestimated, but this bias can be overcome using cross-validation. Performance testing of different methods is not only important to detect any intrinsic biases, but also allows one to compare the precision of estimates across methods and data structures. For example, Teller et al. (2016) used simulations to show that both their regularization methods outperformed a machine learning approach when applied to time series of 10–50 years, but that all methods performed poorly when the number of different climatic environments was below 20.

5.2.4 Which method to use?

There is no straightforward answer when deciding which of the available methods to use. Which method is most appropriate strongly depends on one's research aims. For example, some studies might be most interested in identifying weather signals that have the highest predictive value, while others might be more interested in understanding or identifying weather signals that could be used in experiments. These two aims may require different methods, as from a statistical perspective maximizing predictive and explanatory power is not equivalent (Shmueli 2010). The lack of method validation (Table 5.2) also makes it difficult to determine how well each method performs in terms of bias and precision, making comparison based on performance statistics currently infeasible. Finally, some methods have specific requirements, limiting their suitability for some studies. Machine learning approaches require large amounts of data, limiting their usefulness to large multi-species/population studies. Similarly, weighted mean and regularization methods will be inappropriate for those studies where one is interested in aggregate statistics other than the mean, while the sliding window and machine learning approaches offer the flexibility to use other aggregate statistics.

Notwithstanding, the following considerations may provide useful guidelines. For studies interested in prediction, machine learning approaches are likely to be the most powerful. Studies that lack the required sample size for machine learning (e.g. only a single time series available), can consider applying cross-validation on other methods (Bailey and van de Pol 2016). For studies interested in explanation, the sliding window approach is flexible and easy to implement using the R-package *climwin* (the package includes a vignette with worked examples). The sliding window approach has the added benefit that it is the only method for which the performance is extensively tested on simulated datasets, such that biases are known and tools have been developed to deal with them (van de Pol et al. 2016). However, in cases where the climate signal is likely to be best represented by the mean of a weather variable, both the weighted mean and regularization approaches will probably provide climate signals that have the highest explanatory power (e.g. van de Pol et al. 2012), as these two methods remove the assumption that all days within a window are equally important. The regularization approach has the limitation that it does not directly produce a single estimate of climatic sensitivity (β, though R^2 is available), while the weighted mean approach is limited by the fact that it can only detect one climate window per year (although multiple windows can be detected in a sequential manner; van de Pol et al. 2016). Of the regularization methods available the Teller method (2016) is most easily implemented (only a few lines of R code) and has been performance tested. Of the weighted mean approaches, the method by van de Pol and Cockburn (2011) generalizes the method of Gienapp et al. (2005), and is included in the R-package *climwin*. Finally, the multiple regression approach offers few benefits compared to other approaches, as it has rather limited exploratory possibilities.

5.2.5 What biological data are needed?

We already highlighted that long time series are needed to reliably detect weather signals (e.g. >20 years; Teller et al. 2016; van de Pol et al. 2016). The main determinant of sample size is the number of different climatic environments experienced. If the biological response is only measured once a year (e.g. reproductive success), and all individuals experience similar climatic conditions, then the effective sample size will be the number of study years (assuming a sufficient number of individuals are sampled to accurately estimate, e.g., the mean annual reproductive success). If the biological response is measured multiple times a year (e.g., body condition), then effective sample size will be determined by the total number of sampling periods (e.g., months).

Effective sample size can be increased not only by collecting more data over time but also by utilizing spatial variation in weather. The number of different climatic environments experienced can be increased by collecting time series in different populations that are sufficiently distant for weather patterns to differ. In situations where individuals of the same population experience micro-climatic variation (e.g., climate is buffered by nest-burrows, and nest temperatures are used), such information will increase the number of sampled environments and the power to detect weather signals. Similarly, if some individuals reproduce much earlier than others, they may be affected by weather at different times of the year (Table 5.1).

Augmenting sample size using spatial or intra-annual variation in weather conditions requires careful consideration, as associations between a weather signal and the biological variable may come about by different mechanisms. The observed association may reflect not only inter-annual covariation between weather and the biological response (e.g., warmer years lead to higher reproduction), but also spatial (e.g., populations in warmer areas produce more offspring) or intra-annual associations (e.g., early reproducers experience cold weather and have high reproduction). The observed overall covariation is a mixture of all these processes, and the associations at each level of variation need not have the same sign, strength, or causality. Mixed modelling techniques exist to estimate the weather signal at the appropriate level of variation (Phillimore et al. 2010; van de Pol and Wright 2009).

5.2.6 Non-climatic causes of temporal variation

There is no requirement for a temporal trend in climate (e.g., global warming) to be present in the data to identify a weather signal, as statistical methods can use the variability among (or within) years to estimate patterns of covariation between a weather and biological variable. However, nowadays often a systematic (e.g., linear) trend exists in weather over time, in which case it is important to consider whether an association between weather and a biological variable may exist solely due to both the biological and climate variable changing directionally over time (Sparks and Tryjanowski 2005). For example, it has been suggested that most of the studies reporting a relationships between migrant arrival date and the North Atlantic Oscillation index have been caused by shared trends over time (Haest et al. 2018b). If the shared temporal trend is not due to weather but due to another unknown factor changing over time (e.g., habitat fragmentation), then one would expect no covariance between the annual deviations of temperature from the trend over time and the biological response variable. De-trending the data, or including a linear year covariate in climate window analyses, can thus be used to determine whether climate sensitivities are caused by shared trends or real weather impacts (Haest et al. 2018b). A similar argument can be made for seasonal de-trending of weather data in studies that use intra-annual variation in weather (van de Pol et al. 2016), as biological variables can also show seasonal patterns due to non-climatic influences (e.g., change in day length).

Furthermore, in most climate window analyses it will be useful to account for specific confounding factors that fluctuate or change over time, such that it becomes easier to identify the weather signal, and to obtain more precise estimates of climatic sensitivities. Confounding variables may be abiotic or biotic factors, such as changing levels of human disturbance or predator abundance over time. In addition, it can be important to consider whether changes in methods over time might affect biological data, and if so statistically account for them by including methodology as a covariate into the climate window model. Finally, if observations are prone to observer bias (e.g., biometric or behavioural measurements) one could account for that source of heterogeneity by including observer identity as a random effect.

5.2.7 How do we interpret results?

Although an exploratory approach avoids the need for untested assumptions, care should be taken in interpreting the results. An exploratory approach on its own provides no evidence that the identified weather signal is the causal driver; this can only be corroborated by other studies or experimental manipulation. Even if such corroboration exists, a detected weather signal may only be a proxy for another ecological driver, as weather effects may often act indirectly (e.g., via food abundance; Møller 2013). Weather signals may also be highly collinear (e.g., when it's warm, it is usually not raining), which may cloud the interpretation of results, although simulations have suggested that exploratory methods can handle such collinearity (Sims et al. 2007; Teller et al. 2016).

Interpretation of results will be greatly aided by complementary behavioural or physiological studies to identify the mechanisms by which weather acts. For example, a behavioural assay was used to understand the relationship between high temperatures and thermoregulatory behaviour in arid bird species, which provides insights into the temperature thresholds that impact body mass and survival (Smit et al. 2016). Greater understanding of organisms' physiological constraints (upper and lower thermal thresholds) has already been invaluable in improving species niche modelling (Kearney and Porter 2009).

5.2.8 Future challenges for climate window analysis

Many methods have been developed, but most are not validated. Therefore, a pressing future challenge is assessing how well each method performs. Various studies have compared how different methods perform on real datasets (van de Pol and Cockburn 2011; Phillimore et al. 2013; Roberts et al. 2015). Although this provides information about robustness of results to the methods used, performance testing should be done on simulated datasets with known effect sizes, such that bias and precision can be quantified. Performance testing on simulated datasets should focus on identifying potential bias due to multiple testing and overfitting, and such bias should be quantified in all statistics that directly influence our biological interpretation (β, R^2, P-value of weather signal). Furthermore, by performance testing with the same test data much needed benchmarks can be generated that allow comparison of the strengths and weaknesses of each method. We do not expect there to be one single method that performs best in all situations and on all key metrics, but it is important to know which method performs best in different situations.

We also do not understand well how the bias and precision of methods depends on the temporal resolution of weather data (Holloway et al. 2018). Many methods can deal with daily weather data, but possibly coarser resolutions (e.g., weekly) may provide a better compromise between biological relevance and the statistical challenges of dealing with many predictors. Furthermore, we know little about how well different methods can detect multiple signals of the same weather variable (but see Teller et al. 2016). Finally, while methods produce a 'best' weather signal, there is inevitable model uncertainty, and currently studies ignore how model uncertainty in the identity of the weather signal propagates into uncertainty in the climatic sensitivity and subsequent (comparative) analyses. Model averaging procedures may be a logical way to deal with multi-model inference (Bailey and van de Pol 2016), but this requires further investigation. In addition, climate data are typically collected at a coarser scale than biological data, and while spatially interpolated weather data are often available, these interpolations contain uncertainty (Suggitt et al. 2017) that if ignored will lead to regression dilution when estimating climatic sensitivities.

The main practical challenge is to determine how to most efficiently improve our power to detect true weather signals. Where should we invest our research effort? Spatial replication and measuring microclimatic variation are two ways to circumvent the slower process of collecting more years of data. However, we have no insights yet in how efficient these strategies are. How many different areas need to be sampled and how geographically and climatically distinct would these areas have to be to make estimation of climatic sensitivities reliable? Simulation studies could provide practical guidelines.

Finally, greater biological information about the mechanism by which weather affects biological processes will be an important step towards making our analyses less exploratory. A meta-analysis of what type of weather signals are currently reported in the literature, and how this depends on the biological response variable and taxa of the model system, would provide a useful first step towards achieving this goal. More confirmatory climate window analysis could be supported by supplementary observations that pinpoint the behavioural or physiological mechanism, particularly for threshold effects driven by extremes (e.g., by observing at which temperatures organisms stop their normal activities to avoid allostatic overload; Wingfield et al. 2017). Mechanistic knowledge and models may be particularly crucial for estimating climatic sensitivities of extremes, as detection of their signals is statistically by far the most challenging due to the inherent rareness of extremes (Solow 2017; van de Pol et al. 2017).

5.3 Variation among individuals, populations, and species

Thus far we have discussed methods to estimate a single measure of climatic sensitivity. Next, we focus on comparing climatic sensitivities among subjects, which will be relevant for biologists interested in understanding variation in climatic sensitivity among individuals, populations, and species. Understanding how each method and its implicit assumptions can affect the variation we see in climatic sensitivity is an important step towards understanding the biological patterns of interest.

Once we start comparing across subjects there is the additional complexity that subjects may differ not only in their climatic sensitivity (β or R^2 of weather signal), but also in the identity of their weather signal. For example, early migrating individuals may be affected by weather earlier in the year than late migrating individuals. Populations across a species range may be more rainfall-dependent in arid zones, while populations in wet zones may be more temperature-dependent. Variation in weather signals may be even more likely when comparing among species, as species typically differ in ecological and life history characteristics that may mediate weather impacts (e.g., diet, migratory status). In common European birds there is no evidence for a phylogenetic signal in the identity of the weather signal for body mass (McLean et al. 2018), suggesting that weather signals even vary across closely related species.

The key methodological challenge in comparative studies is that it is generally hard to identify differences in weather signals across subjects. This is problematic as a suboptimal choice of weather signal for a given subject will also affect the estimate of climatic sensitivity and consequently affect our interpretation of across-subject variation in climatic sensitivity. For example, fairywrens (Maluridae) in Australia follow the common pattern that climatic sensitivity is highest in populations at higher latitudes, if we assume that they are all affected by the same weather signal. However, climatic sensitivities vary little across populations if we allow each population to be affected by rainfall in a different part of the year (which was found to be the best weather signal; van de Pol et al. 2013). Thus, the challenges of identifying variation in weather signals and in climatic sensitivity are intricately linked.

5.3.1 Modelling variation among subjects in weather signal and climatic sensitivity

Ideally, one should aim to identify the best weather signal(s) for each individual, population, or species separately. However, often this is practically impossible, as for a single subject data across at least 20 different environments should be collected to reliably identify the weather signal and climatic sensitivity (Teller et al. 2016; van de Pol et al. 2016). What to do if such data is unavailable for each subject?

First, one should consider how realistic it is that all subjects are affected by the same weather signal. This may not only depend on the diversity of subjects compared (e.g., related or phylogenetically distant species) and where they live (different weather variables may be limiting in different habitats), but also on the biological response variable. For example, across European birds there is little variation in weather signals for the timing of egg-laying, such that assuming a common weather signal reflecting spring temperatures would give similar climatic sensitivity patterns to an analysis where weather signals are allowed to vary across species. However, for these same species there is huge variation in weather signals for offspring productivity and so assuming a common weather signal across species is inappropriate in this context (McLean et al. 2016).

Instead of identifying weather signals for each subject separately, another solution may be to assume that inter-subject variation in weather signals can be described by a specific factor. For example, Phillimore et al. (2016) analysed how the timing of egg-laying depends on temperature by allowing the starting point and duration of sliding temperature windows to vary with latitude across populations. They showed that a weather signal where northern populations are affected by temperatures earlier in the season was best supported. Strikingly, the decision to allow subjects to vary in their weather signal directly affected their conclusions about climatic sensitivity: their latitudinal gradient model suggested phenotypic plasticity did not track the temperature sensitivity of selection on egg-laying date, while a model assuming a shared weather signal suggested tracking did occur. Although Phillimore et al. (2016) is an elegant example, it is unclear whether the latitudinal gradient can account for most of the among-subject variation in weather signal. More generally, at the moment we have very little general knowledge about how and why weather signals vary across individuals, populations, and species. While phenology often exhibits latitudinal gradients allowing for such an approach, it remains to be seen whether their approach is feasible for other questions and model systems.

5.3.2 Comparing climate sensitivities

Allowing for subjects to vary in their weather signal can make their comparison more difficult, as it can cause climatic sensitivities to have different units. Similar issues will occur in meta-analyses on climate sensitivities derived from different studies. For example, body mass response to climate may be in gram/°C in some species while for others it may be in gram/mm rainfall. How do we compare these sensitivities such that we are not comparing apples and oranges, or at least make it more like comparing mandarins and oranges?

Various possibilities exist due to the different ways to standardize effect sizes and define climatic sensitivities (β or R^2). One often used option is to standardize all β-estimates by dividing it by the standard deviation (e.g., across years) in the weather signal, such that the units become 'change with one standard deviation of the weather signal'. When the effect of the weather signal is nonlinear (e.g., polynomial, splines), the strength of the relationship varies with the value of the weather signal. In such cases, one could 'linearize' the climatic sensitivity, by estimating the linear slope of the weather signal at the mean climatic conditions (the average of the weather signal across years; e.g., McLean et al. 2018).

An alternative way to make units more comparable is to use the (partial) correlation instead of the standardized (partial) regression coefficient as a measure of climatic sensitivity, although this does not work well for nonlinear relationships. The partial-R^2 is another climatic sensitivity metric that is dimensionless and can be used for nonlinear relationships. However, it does not distinguish responses that differ in sign.

If there are multiple weather signals affecting a single subject, the above approaches become problematic (except when using partial-R^2). How do we compare the climatic sensitivity of species A and B, when the former is rainfall and temperature sensitive and the latter only rainfall sensitive? A solution could be to instead compare climate vulnerabilities (McLean et al. 2018). Vulnerabilities are determined by the product of climatic sensitivity and climate exposure. For example, if the climatic sensitivity of body mass is 0.5 g/°C and temperature is projected by climatologists to change by 0.02 °C/year (the exposure), then the projected change in body mass due to temperature change is 0.01 g/year. Similarly, if the climatic sensitivity is 0.05 g/mm and rainfall exposure is 0.2 mm/year, then the projected change in mass due to rainfall change is also 0.01 g/year. The total vulnerability of species A to climate change will then be 0.02 g/year (sum of the vulnerability to temperature and rainfall), which can be directly compared to the vulnerability of the rainfall-sensitive species B.

Which of the above options is most appropriate will depend on the situation: are there nonlinear effects, do effects differ in sign, are there multiple signals acting simultaneously, are you interested in strength of relationship (β) or association (R^2, r)? In cases where multiple options are possible, it seems prudent to investigate whether the conclusions are robust to the choice of climatic sensitivity metric.

5.3.3 How to deal with a lack of climatic sensitivity?

When performing climate window analysis on many subjects, it is likely that for some subjects no weather signal will be identified (i.e., could not be distinguished from occurring due to chance). This may be evidence of climatic insensitivity. However, the failure to detect a weather signal may also result from limited statistical power caused by small sample size, or from the effect of the signal being weak or being estimated over a narrow range of climatic conditions, making a signal hard to detect. In such situations it is important to consider how confident one can be that statistical power was sufficient to detect a true weather signal. Did the probability of detecting a weather signal depend on sample size and the range of conditions over which they were observed? Numerical simulations can provide guidance on what effect size would be detectable for a given sample size (Bailey and van de Pol 2016; Teller et al. 2016).

Often analysis should be restricted to those subjects for which statistical power is likely to be sufficient, for example by only including subjects that were measured over 20 years (Thackeray et al. 2016). In such situations it is important to think about how this restricts the generalization of results. For example, in birds most studies over 20 years will be biased towards Northern Hemisphere hole-breeding passerines. In situations of ample statistical power, the failure to detect a weather signal likely suggests true climatic insensitivity. This raises issues for comparative analyses, because when no weather signal is detected, we do not know the value of climatic sensitivity, its precision, or its units. How do we then deal with apparently climatically insensitive subjects in comparative analyses?

An intuitive option may be to use the estimate of climatic sensitivity from the 'best' weather signal, even if we have no evidence it is a true positive signal. However, this may not be advisable when using highly exploratory methods. The rationale behind this option is that non-significant weather signals on average should have a climatic sensitivity that is close to zero with large standard errors. However, simulation studies on datasets that do not contain a weather signal have shown that the 'best' weather signal in sliding window approaches can have estimates of climatic sensitivity that are far from zero and have little uncertainty due to multiple testing (van de Pol et al. 2016). Thus climatic sensitivities that are known to be false positives should not be used in comparative analysis if derived using sliding window analyses, and it remains to be determined whether this also holds for other methods.

Alternatively, one could assume that subjects for which no weather signal was detected have a climatic sensitivity of zero. Imputing with zero in situations where the true climatic sensitivities are likely to be small has the drawback that this underestimates the variance across species, which may mean that some test-statistics in the comparative analysis are anti-conservative. If many subjects are climatically insensitive then imputing with zero may also cause the climatic sensitivity response variable to become zero-inflated. In such case one could instead perform a two-step analysis: First, one can analyse what characteristics differentiate subjects that are climatically sensitive from those where no weather signal was detected. Next, one can analyse what characteristics explain variation in the strength of climatic sensitivity among subjects for which a weather signal could be detected. This two-step approach allows us to gain biological knowledge from climatically insensitive subjects without the potential issues associated with imputing with zero.

5.3.4 Explaining variation in climatic sensitivities

Explaining the biological mechanism driving among subject variation in climatic sensitivity will be the most interesting parts of any comparative study. A logical first step will be to quantify the among-subject variation in climatic sensitivities. For example, evolutionary biologists studying phenotypic plasticity may be interested in quantifying the evidence for individual by environment interactions. By including a weather signal as a random slope term that can vary across subjects, its variance provides an estimate of the amount of heterogeneity present. However, reliable estimation of random slope variance requires large sample sizes (Martin et al. 2011), and underestimation of the amount of variation among subjects in climatic sensitivity occurs when few subjects are sampled (< 10; van de Pol 2012).

If there is substantial variation among subjects in climatic sensitivity, then one could investigate which factors might explain this variation. Comparative multi-species studies could use ecological and life history characteristics of species (e.g., dispersal ability, diet, and habitat specialization) to explain interspecific variation in climatic sensitivity (Buckley and Kingsolver 2012). Studies that compare multiple populations of the same species could consider population characteristics, such as habitat type or resource diversity, to explain intraspecific variation (Iles et al. 2018; McLean et al. 2018). If data are available on multiple populations of many species, then comparison of how much of the variation in climatic sensitivity is due to differences between species and population can help tease apart the effects of environment from phylogeny. For example, if the climatic sensitivity of two populations of the same species are on average as different as the climatic sensitivity of two populations of different species, then species characteristics are less likely than population characteristics to explain the observed heterogeneity (McLean et al. 2018).

5.3.5 Future challenges for comparing climate sensitivities

The main future challenge for comparing climatic sensitivities is partly methodological and partly biological. Biologists need to become more aware that assuming a shared weather signal across subjects is a rather specific assumption that can have large consequences for the interpretation of variation in climatic sensitivities. More studies are now needed to explore how often and in what situations individuals, populations, and species will differ in their weather signal, and what may explain weather signal variation. Until we have a good understanding of this, it seems prudent to specifically allow each subject to have its own weather signal. As this will require long time series for each subject, it will mean that only particularly long-term studies will be suitable for comparative analyses.

5.4 Conclusions

Understanding the relationship between weather and biological data has often required biologists to make simplifying but rather critical assumptions. However, with the growing availability of long-term datasets, increasing computational power, and many new analytical methods, there is now ample possibility for exploratory approaches. Exploratory approaches allow us to systematically test a wide range of assumptions when limited a priori knowledge is available. Birds provide a particularly useful focal group for these approaches, as there are many individual-based long-term studies and large multi-species databases of time series (e.g., Thackeray et al. 2016). Many analytical methods have recently been developed in disparate biological fields that make better use of this arduously collected biological data. We have summarized these methods in this chapter to make them more easily comparable and accessible, to allow biologists to better quantify climatic sensitivities at different levels of organization and understand why they vary. Understanding variation in climatic sensitivities is crucial for making long-term predictions about global change. However, careful consideration of assumptions and methodology is needed to make sure results are not biased.

It is still unclear exactly how much bias is introduced by relying on simplifying assumptions when estimating climatic sensitivity. However, the few studies that have explicitly tested these assumptions have overwhelmingly found they can have important impacts. The degree to which one explores the identity of weather signals (McLean et al. 2018), the critical time window (Pearce-Higgins et al. 2014), spatial resolution (Gillingham et al. 2012; Trivedi et al. 2008), aggregate statistics (Husby et al. 2010), and the possibility of multiple signals (Kruuk et al. 2015) can directly affect the strength of climatic sensitivities as well as the interpretation of biological outcomes. Similarly, the few comparative studies that allow weather signals to vary among subjects suggest that the assumption of shared weather signals will not generally hold, and that this assumption can strongly affect biological conclusions (van de Pol et al. 2013; Phillimore et al. 2016; McLean et al. 2018; Haest et al. 2018b). It therefore appears important to explicitly consider how robust results are to the methods used and assumptions made before making biological inferences. Ultimately, if we are able to gain a deeper understanding of the mechanisms by which weather affects the biological systems we study, we can reduce our need to explore these many assumptions in the first place.

References

Bailey, L., and van de Pol, M. (2015). climwin: Climate window Analysis. URL https://cran.r-project.org/web/packages/climwin/

Bailey, L.D., Ens, B.J., Both, C., Heg, D., Oosterbeek, K., and van de Pol, M. (2017). No phenotypic plasticity in nest-site selection in response to extreme flooding events. *Philosophical Transactions of the Royal Society of London Series B: Biological Sciences*, 372, 20160139.

Bailey, L.D., and van de Pol, M. (2016). climwin: an R toolbox for climate window analysis. *PLoS One*, 11, e0167980.

Bonhomme, R. (2000). Bases and limits to using 'degree day' units. *European Journal of Agronomy*, 13, 1–10.

Buckley, L.B., and Kingsolver, J.G. (2012). Functional and phylogenetic approaches to forecasting species' responses to climate change. *Annual Review of Ecology, Evolution and Systematics*, 43, 205–26.

Elith, J., Phillips, S.J., Hastie, T., Dudík, M., Chee, Y.E., and Yates, C.J. (2011). A statistical explanation of MaxEnt for ecologists. *Diversity and Distributions*, 17, 43–57.

Elston, D.A., Brewer, M.J., Martay, B., Johnston, A., Henrys, P.A., Bell, J.R., Harrington, R., Monteith, D., Brereton, T.M., Boughey, K.L., and Pearce-Higgins, J.W. (2017). A new approach to modelling the relationship between annual population abundance indices and weather data. *Journal of Agricultural, Biological and Environmental Statistics*, 22, 427–45.

Frederiksen, M., Lebreton, J.-D., Pradel, R., Choquet, R., Gimenez, O. (2014). Identifying links between vital rates and environment: a toolbox for the applied ecologist. *Journal of Applied Ecology*, 51, 71–81.

Gienapp, P., Hemerik, L., and Visser, M.E. (2005). A new statistical tool to predict phenology under climate change scenarios. *Global Change Biology*, 11, 600–6.

Gillingham, P., Huntley, B., Kunin, W., and Thomas, C. (2012). The effect of spatial resolution on projected responses to climate warming. *Diversity and Distributions*, 18, 990–1000.

Haest, B., Hüppop, O., and Bairlein, F. (2018a). The influence of weather on avian spring migration phenology: What, where and when? *Global Change Biology*, 24(12), 5769–88.

Haest, B., Hüppop, O., and Bairlein, F. (2018b). Challenging a 15-year-old claim: The North Atlantic Oscillation index as a predictor of spring migration phenology of birds. *Global Change Biology*, 24, 1523–37.

Holloway, P., Kudenko, D., and Bell, J.R. (2018). Dynamic selection of environmental variables to improve the prediction of aphid phenology: A machine learning approach. *Ecological Indicators*, 88, 512–21.

Husby, A., Nussey, D.H., Visser, M.E., Wilson, A.J., Sheldon, B.C., and Kruuk, L.E. (2010). Contrasting patterns of phenotypic plasticity in reproductive traits in two great tit (*Parus major*) populations. *Evolution*, 64, 2221–37.

Iles, D.T., Rockwell, R.F., and Koons, D.N. (2018). Reproductive success of a keystone herbivore is more variable and responsive to climate in habitats with lower resource diversity. *Journal of Animal Ecology*, 87, 1182–91.

Kearney, M., and Porter, W. (2009). Mechanistic niche modelling: combining physiological and spatial data to predict species' ranges. *Ecology Letters*, 12, 334–50.

Kruuk, L.E., Osmond, H.L., and Cockburn, A. (2015). Contrasting effects of climate on juvenile body size in a Southern Hemisphere passerine bird. *Global Change Biology*, 21, 2929–41.

Martin, J.G., Nussey, D.H., Wilson, A.J., and Reale, D. (2011). Measuring individual differences in reaction norms in field and experimental studies: a power analysis of random regression models. *Methods in Ecology and Evolution*, 2, 362–74.

McKechnie, A.E., and Wolf, B.O. (2010). Climate change increases the likelihood of catastrophic avian mortality events during extreme heat waves. *Biology Letters*, 6, 253.

McLean, N., Lawson, C.R., Leech, D.I., and van de Pol, M. (2016). Predicting when climate-driven phenotypic change affects population dynamics. *Ecology Letters*, 19, 595–608.

McLean, N., van der Jeugd, H.P., and van de Pol, M. (2018). High intra-specific variation in avian body condition responses to climate limits generalisation across species. *PLoS One*, 13, e0192401.

Mesquita, M.d.S., Erikstad, K.E., Sandvik, H., Barrett, R.T., Reiertsen, T.K., Anker-Nilssen, T., Hodges, K.I., and Bader, J. (2015). There is more to climate than the North Atlantic Oscillation: a new perspective from climate dynamics to explain the variability in population growth rates of a long-lived seabird. *Frontiers in Ecology and Evolution*, 3, 10.3389/fevo.2015.00043

Møller, A.P. (2013). Long-term trends in wind speed, insect abundance and ecology of an insectivorous bird. *Ecosphere*, 4, art6.

Pearce-Higgins, J.W., Green, R.E., and Green, R. (2014). *Birds and climate change: impacts and conservation responses*. Cambridge University Press, Cambridge, UK.

Phillimore, A.B., Hadfield, J.D., Jones, O.R., and Smithers, R.J. (2010). Differences in spawning date between populations of common frog reveal local adaptation. *Proceedings of the National Academy of Sciences of the United States of America*, 107, 8292–7.

Phillimore, A.B., Leech, D.I., Pearce-Higgins, J.W., and Hadfield, J.D. (2016). Passerines may be sufficiently plastic to track temperature-mediated shifts in optimum lay date. *Global Change Biology*, 22, 3259–72.

Phillimore, A.B., Proios, K., O'Mahony, N., Bernard, R., Lord, A.M., Atkinson, S., and Smithers, R.J. (2013). Inferring local processes from macro-scale phenological pattern: a comparison of two methods. *Journal of Ecology*, 101, 774–83.

Roberts, A.M.I. (2008). Exploring relationships between phenological and weather data using smoothing. *International Journal of Biometeorology*, 52, 463–70.

Roberts, A.M., Tansey, C., Smithers, R.J., and Phillimore, A.B. (2015). Predicting a change in the order of spring phenology in temperate forests. *Global Change Biology*, 21, 2603–11.

Schaper, S.V., Dawson, A., Sharp, P.J., Gienapp, P., Caro, S.P., and Visser, M.E. (2011). Increasing temperature, not mean temperature, is a cue for avian timing of reproduction. *American Naturalist*, 179, E55–E69.

Schleip, C., Menzel, A., and Dose, V. (2008). Norway spruce (*Picea abies*): Bayesian analysis of the relationship between temperature and bud burst. *Agricultural and Forest Meteorology*, 148, 631–43.

Shmueli, G. (2010). To explain or to predict? *Statistical Science*, 25, 289–310.

Sims, M., Elston, D.A., Larkham, A., Nussey, D.H., and Albon, S.D. (2007). Identifying when weather influences life-history traits of grazing herbivores. *Journal of Animal Ecology*, 76, 761–70.

Smit, B., Zietsman, G., Martin, R.O., Cunningham, S.J., McKechnie, A.E., and Hockey, P.A.R. (2016). Behavioural responses to heat in desert birds: implications for predicting vulnerability to climate warming. *Climate Change Responses*, 3, 9.

Solow, A.R. (2017). On detecting ecological impacts of extreme climate events and why it matters. *Philosophical Transactions of the Royal Society of London Series B: Biological Sciences*, 372, 20160136.

Sparks, T.H., and Carey, P.D. (1995). The responses of species to climate over two centuries: An analysis of the Marsham phenological record, 1736–1947. *Journal of Ecology*, 83, 321–9.

Sparks, T.H., and Tryjanowski, P. (2005). The detection of climate impacts: some methodological considerations. *International Journal of Climatology*, 25, 271–7.

Suggitt, A.J., Platts, P.J., Barata, I.M., Bennie, J.J., Burgess, M.D., Bystriakova, N., Duffield, S., Ewing, S.R., Gillingham, P.K., and Harper, A.B. (2017). Conducting robust ecological analyses with climate data. *Oikos*, 126, 1533–41.

Teller, B.J., Adler, P.B., Edwards, C.B., Hooker, G., and Ellner, S.P. (2016). Linking demography with drivers: climate and competition. *Methods in Ecology and Evolution*, 7, 171–83.

Thackeray, S.J., Henrys, P.A., Hemming, D., Bell, J.R., Botham, M.S., Burthe, S., Helaouet, P., Johns, D.G., Jones, I.D., and Leech, D.I. (2016). Phenological sensitivity to climate across taxa and trophic levels. *Nature*, 535, 241.

Trivedi, M.R., Berry, P.M., Morecroft, M.D., and Dawson, T.P. (2008). Spatial scale affects bioclimate model projections of climate change impacts on mountain plants. *Global Change Biology*, 14, 1089–1103.

van de Pol, M. (2012). Quantifying individual variation in reaction norms: how study design affects the accuracy,

precision and power of random regression models. *Methods in Ecology and Evolution*, 3, 268–80.
van de Pol, M., Bailey, L.D., McLean, N., Rijsdijk, L., Lawson, C.R., and Brouwer, L. (2016). Identifying the best climatic predictors in ecology and evolution. *Methods in Ecology and Evolution*, 7, 1246–57.
van de Pol, M., Brouwer, L., Brooker, L.C., Brooker, M.G., Colombelli-Négrel, D., Hall, M.L., Langmore, N.E., Peters, A., Pruett-Jones, S., Russell, E.M., Webster, M.S., and Cockburn, A. (2013). Problems with using large-scale oceanic climate indices to compare climatic sensitivities across populations and species. *Ecography*, 36, 249–55.
van de Pol, M., and Cockburn, A. (2011). Identifying the critical climatic time window that affects trait expression. *American Naturalist*, 177, 698–707.
van de Pol, M., Jenouvrier, S., Cornelissen, J.H., and Visser, M.E. (2017). Behavioural, ecological and evolutionary responses to extreme climatic events: challenges and directions. *Philosophical Transactions of the Royal Society of London Series B: Biological Sciences*, 372, 10.1098/rstb.2016.0134
van de Pol, M., Osmond, H.L., and Cockburn, A. (2012). Fluctuations in population composition dampen the impact of phenotypic plasticity on trait dynamics in superb fairy-wrens. *Journal of Animal Ecology*, 81, 411–22.
van de Pol, M., and Wright, J. (2009). A simple method for distinguishing within- versus between-subject effects using mixed models. *Animal Behaviour*, 77, 753–8.
Wingfield, J.C., Pérez, J.H., Krause, J.S., Word, K.R., González-Gómez, P.L., Lisovski, S., and Chmura, H.E. (2017). How birds cope physiologically and behaviourally with extreme climatic events. *Philosophical Transactions of the Royal Society of London Series B: Biological Sciences*, 372, 20160140.
Wolkovich, E.M., Cook, B.I., Allen, J.M., Crimmins, T.M., Betancourt, J.L., Travers, S.E., Pau, S., Regetz, J., Davies, T.J., Kraft, N.J.B., Ault, T.R., Bolmgren, K., Mazer, S.J., McCabe, G.J., McGill, B.J., Parmesan, C., Salamin, N., Schwartz, M.D., and Cleland, E.E. (2012). Warming experiments underpredict plant phenological responses to climate change. *Nature*, 485, 494.

CHAPTER 6

Ecological niche modelling

Damaris Zurell and Jan O. Engler

6.1 Context

In recent decades, ecology and biogeography have become increasingly quantitative disciplines. Spurred by technological advances, the amount of digital information is growing rapidly. Remote sensing and geographic information systems facilitate the mapping of environmental variables at high spatial and temporal resolution, and global positioning systems and citizen science platforms allow collecting biodiversity data in unprecedented amounts and precision. Many applications, however, require further processing and synthesis of these raw data. For example, conservation planning, reserve design, risk assessment, and invasive species management often require maps of species' actual or potential distributions rather than a limited number of samples of known occurrences (Franklin 2010). Ecological niche models (ENMs) provide an efficient way to interpolate (or sometimes extrapolate) species' distributions. They achieve this by relating species' observations to environmental predictor variables using statistical and machine-learning methods. If digital maps of environmental variables are available, the potential distribution of the species can then be predicted based on the estimated species–environment relationship (Franklin 2010; Guisan et al. 2017).

Applications of ENMs have exploded in recent years as more data are becoming available and the methodological tools are constantly improving. Also, in global change research, ENMs constitute the most widely applied modelling framework to project potential future range shifts of species (IPBES 2016). Among all organisms, birds have received tremendous attention in ENM applications (Engler et al. 2017). This may in part relate to the deep knowledge of birds (Newton 2003), their high taxonomic coverage (Jarvis 2016), and huge public interest leading to myriads of opportunistic and systematic observations (cf. Chapter 4).

In this chapter, we first explain the concept behind ENMs and their underlying assumptions (section 6.2). We then describe the basic modelling steps and illustrate them using a simple real-world example (section 6.3), and provide an overview of potential sources of uncertainty in underlying data and in the models (section 6.4). After this conceptual and methodological overview, we discuss potential limitations of ecological niche models in a global change context (section 6.5) and outline the latest developments and future perspectives (section 6.6). As ENMs are a vast field under constant development, we will not be able to cover all aspects in depth but aim to provide a concise overview. For more detailed insights, we refer readers to excellent reviews (Elith and Leathwick 2009; Guisan and Thuiller 2005; Guisan and Zimmermann 2000) and books (Franklin 2010; Guisan et al. 2017) that summarize ENM research over the last few decades.

6.2 Theory and concept

What limits species' distributions? ENMs help us understand why species occur in some places but

Zurell, D., and Engler, J.O., *Ecological niche modelling*. In: *Effects of Climate Change on Birds*. Second Edition.
Edited by Peter O. Dunn and Anders Pape Møller: Oxford University Press (2019). © Oxford University Press.
DOI: 10.1093/oso/9780198824268.003.0006

not in others, and where else we could expect to find the species. The central conceptual framework underlying ENMs is the niche theory as formulated by Joseph Grinnell and later enhanced by G. Evelyn Hutchinson (see Soberón 2007, for an overview). Hutchinson distinguished two main factors that limit species' ranges: (i) the abiotic environment that determines the fundamental niche of a species, and (ii) the biotic environment that comprises all interactions with other species. First, Hutchinson described a species' fundamental niche as an n-dimensional hyperspace comprising all environmental conditions where a species has a positive population growth rate and can persist indefinitely. Second, as species rarely live in solitude but interact with other species, the biotic environment determines if a species can prevail in the presence of other species. The intersection between abiotic and biotic environment is usually referred to as the realized niche of a species.

Later, Pulliam (1988) and Hanski (1999) added source-sink and metapopulation dynamics as important determinants of species' distributions. According to these theories, a species could also be present in unsuitable 'sink' habitats that do not contribute to population growth or be absent from suitable habitats because of dispersal limitations. For example, stochastic events could lead to local extinction of species. Dispersal constraints will then determine whether the species is able to recolonize those areas or not. Figure 6.1 illustrates this heuristic in geographic space (Soberón 2007). A species can only persist in habitats where both the abiotic (A) and biotic environmental conditions (B) allow positive population growth. The intersection between A and B thus marks the potential distribution of the species. The movement or dispersal capacity of the species (M) will then determine which source habitats (allowing positive population growth) are occupied and which sink habitats (having negative population growth) are within reach of the species.

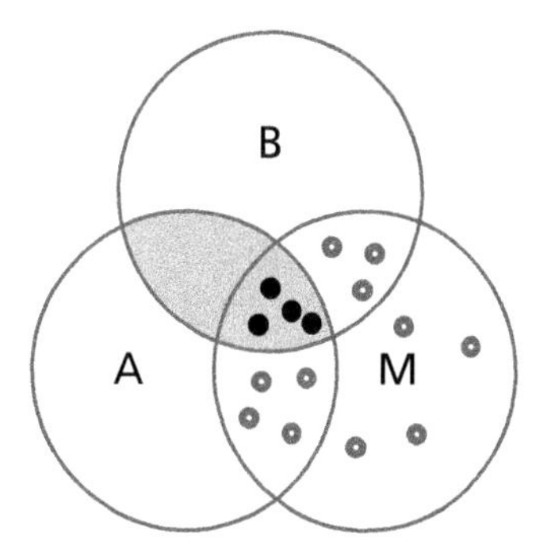

Figure 6.1 Heuristic representation of the factors limiting species' distributions in geographic space (adapted from Soberón 2007). *A* represents the geographic area where the species has positive population growth. The environmental conditions found within this area correspond to the fundamental niche. *B* represents the geographic area where a species can coexist with competitors. *M* represents the geographic area that is accessible to the species within the time period of interest and limited by the species' dispersal abilities. The intersection of *A*, *B*, and *M* is the occupied area that contains source populations (closed circles). The grey shaded area represents the region where abiotic and biotic environments are potentially suitable for the species and thus correspond to the classic view of the realized niche (Soberón 2007). Open circles represent sink populations with negative population growth rates due to suboptimal abiotic conditions or due to competitive exclusion.

How to quantify species' ecological niches? ENMs provide a framework for quantifying the species–environment relationship and predicting a species' distribution in space and time. They are comparably easy to build because many software packages are available and they often have low data requirements (Franklin 2010; Guisan et al. 2017). As input, ENMs use data on the geographic distribution of species' observations (occurrence, presence–absence, or abundance; the response or dependent variable) and environmental variables (the predictor or independent variables), which are now widely available in digital format (Chapters 3–4). Using adequate statistical algorithms, species' observations are then related to prevailing environmental conditions at the same locations. Apart from some purely spatial approaches such as Kriging, the statistical model is fitted in environmental rather than geographic space (Figure 6.2). Last, the species' potential distribution is predicted by calculating the occurrence probability of the species for any location based on the estimated species–environment relationship and the local environmental conditions as provided by digital maps of the current environment. If digital environmental layers for future (or past) climate scenarios are available, the potential distribution of the species can be forecasted (or hindcasted).

There is an ongoing debate whether the fitted species–environment relationship approximates the fundamental niche (area *A* in Figure 6.1), the realized niche (intersection of *A* and *B* in Figure 6.1), or

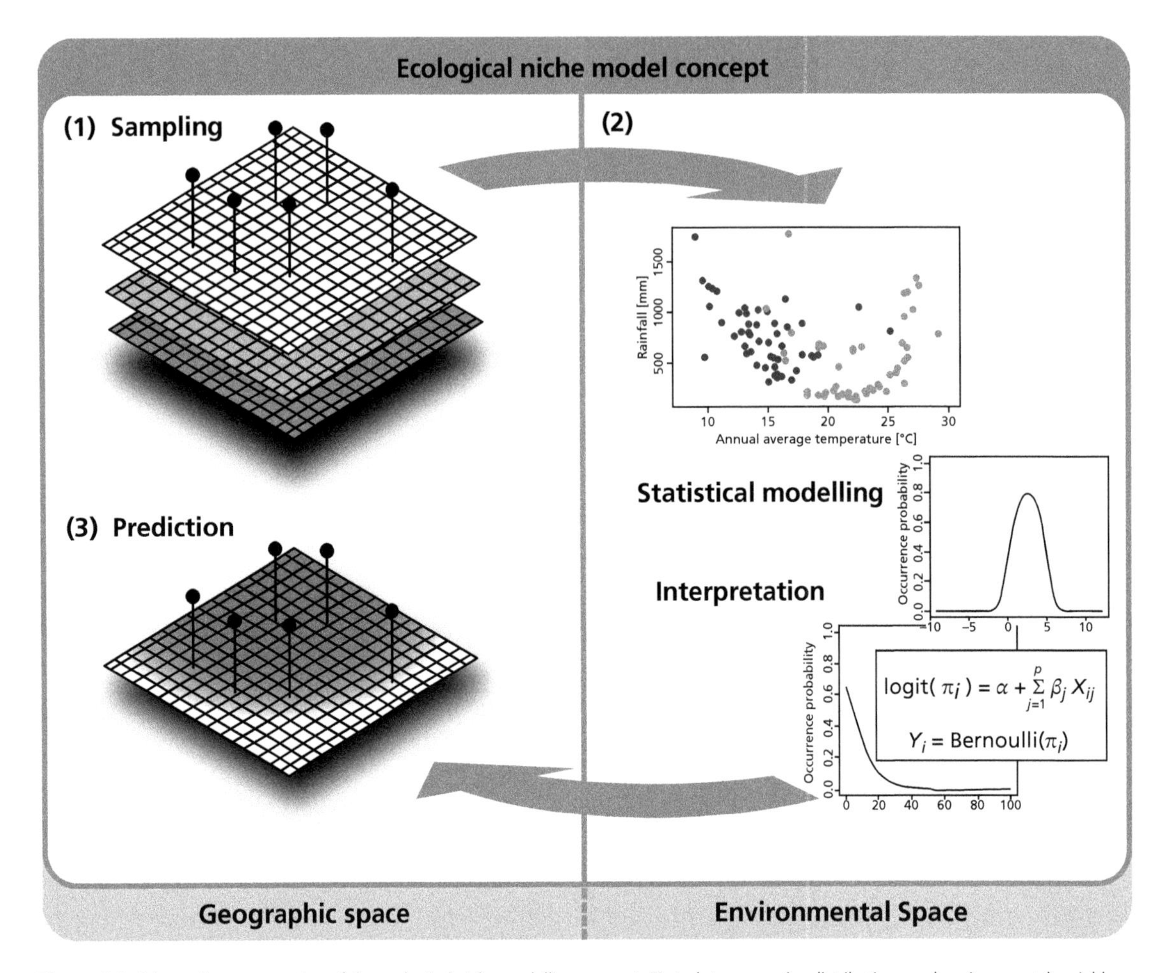

Figure 6.2 Schematic representation of the ecological niche modelling concept. First, data on species distributions and environmental variables are sampled in geographic space. Second, a statistical model (e.g. generalized linear model) is used to relate species observations to prevailing environmental conditions. Third, the potential distribution of the species is mapped by projecting the estimated species–environment relationship back onto geographic space, either to current environmental conditions or to selected climate change scenarios.

the occupied niche (intersection of *A*, *B*, and *M* in Figure 6.1). For example, it has been argued that ENMs based on coarse-scale climatic data describe the fundamental niche, models based on presence-only data the realized niche, and models based on presence–absence data the occupied niche (Franklin 2010). This debate is also reflected in the multitude of names that are used as synonyms for ENMs, for example climate envelopes, habitat model, resource selection functions, species distribution models, to name only a few. Here, we want to take a rather neutral view (similar to Elith and Leathwick 2009) and emphasize that which parts of a species' niche are modelled will depend on the underlying data, the scale of the study, and its methods. It is important to carefully consider these factors in the model building process (sections 6.3 and 6.4).

Ideally, only ecologically relevant environmental variables should be used to infer species' niches. Thereby, it is useful to distinguish between proximal (resource and direct) and distal (indirect) gradients (Austin 1980; Engler et al. 2017). In the case of birds, resource gradients refer to all materials and energy that can be consumed by the species and are essential for their metabolism such as water, seeds, and insects. Direct gradients are environmental variables, such as temperature, that do not act as resources, but exert a causal (proximal) effect on physiology and demography of species. By contrast, indirect gradients do not directly affect distributions of species, but are correlated with ecologically more relevant (proximal) predictors. Typical indirect gradients are topographic variables such as elevation or slope, and land cover variables such as habitat types. Indirect variables often provide simple surrogates of more complex combinations of resources and direct gradients (Guisan and Zimmermann 2000). For example, elevation may correlate strongly with temperature, moisture, and solar radiation. We want to stress that when ENMs are used for extrapolation as is the case in climate change research, direct and resource variables are preferable over indirect variables because correlation structures vary across the globe and may change under climate change. For deciding which variables are likely proximal or distal, prior knowledge on species' ecology and ecophysiology is certainly helpful (Guisan et al. 2017).

Ecological processes are highly scale-dependent and, thus, the spatial resolution of environmental and species data plays an important role for understanding and modelling species' niches (Pearson and Dawson 2003). Climate is generally believed to limit ranges at macro-scales, whereas habitat is thought to affect species' distributions at intermediate spatial scales. It is almost impossible to define one single best scale of study. Birds select resources and habitats in a hierarchical manner (Franklin 2010). Also, competition with heterospecifics may happen at a range of scales; for example, birds may compete for food over a large area (e.g., foraging seabirds), but compete for nest sites at a much finer scale (e.g., on a cliff face for seabirds).

6.3 Modelling steps

The main modelling steps in ENMs are (i) conceptualization, (ii) data preparation, (iii) model fitting, (iv) model evaluation, and (v) spatiotemporal predictions, which we briefly explain below and which are detailed in Franklin (2010) and Guisan et al. (2017). Additionally, we develop a simple, real-world example to illustrate these modelling steps (Box 6.1) using the silvereye (*Zosterops lateralis*) as study species (Figure 6.3). Accompanying code and data are provided as the electronic supplementary material ESM 6.1. (www.oup.co.uk/companion/dunn&moller)

(i) Conceptualization: Before modelling, it is important to carefully consider the conceptual setup of the study. Foremost, this involves formulating clear research objectives and summarizing the available ecological knowledge on the species and study system. Based on these considerations, the main underlying assumptions are checked (cf. section 6.5), appropriate environmental predictors are identified and checked for availability, and, if necessary, a sampling strategy is designed for obtaining species' observations. Also, decisions on appropriate modelling algorithms should be taken at this stage and hypotheses about the expected shape of the species–environment relationship (linear or unimodal) should be formulated.

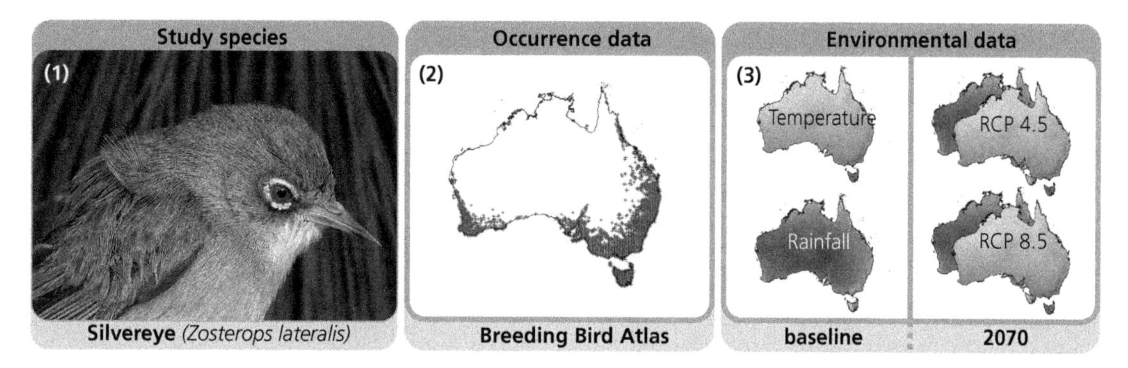

Figure 6.3 (1) The silvereye (*Zosterops lateralis*) is a small passerine bird of the south-west Pacific. (2) In Australia, the species occurs in more humid coastal areas in the south-west and east. (3) Bioclimatic variables such as annual average temperature and annual rainfall sums are available for current climate (baseline) and for climate scenarios under two Representative Concentration Pathways (RCP 4.5 and 8.5) for the year 2070. Picture credit: Rob Davis.

Box 6.1 Case study

Using publicly available data, we describe the main modelling steps for successfully fitting an ENM and projecting species' response to climate change. All data and source codes are available in the Electronic Supplementary Material ESM 6.1. We exemplify all modelling steps using the statistical software environment R (www.r-project.org) that offers different functionalities for ENMs. For simplicity, we use mainly functions contained in the R package *dismo* that provides various functions for data preparation and visualization and integrates different ENM algorithms for modelling and predictions.

(i) Conceptualization

We aim at assessing potential climate change effects on the silvereye (*Zosterops lateralis*, Figure 6.3), a common Australian bird species with stable population trends. We expect its distribution to be at equilibrium with the environment and its range edges to be predominantly determined by climatic conditions. Specifically, the silvereye is widespread along more humid coastal regions, while avoiding the dry and hot interior as well as the northern coast. We, thus, hypothesize that the species is limited mainly by dry and hot conditions on the Australian continent and expect a linearly increasing response to precipitation and a linearly decreasing response to temperature.

(ii) Data preparation

Occurrence information for the period 1977–1981 were available from the first Atlas of Australian Birds (Blakers et al. 1984) at a 10-minute resolution (accessed through www.gbif.org on February 6th 2018) (Figure 6.3). We use these atlas data rather than newer citizen science based occurrence records here to reduce any sampling biases. For subsequent modelling, we randomly selected 500 occurrence records and sampled 5000 random pseudo-absences outside known silvereye occurrences. As predictors, two bioclimatic variables, annual mean temperature (bio1) and sum of annual rainfall (bio12), were obtained from worldclim (v. 1.4; Hijmans et al. 2005) for the period 1960–1990 at the same 10-minute resolution and cropped to the Australian mainland (Figure 6.3). Additionally, we obtained CMIP5 climate change scenarios for the period 2061–2080 from one climate model (ACCESS1-0 GCM) and two representative concentration pathways (RCP 4.5 and 8.5).

(iii) Model fitting

The species–environment relationship was estimated using a generalized linear model (GLM) with a logit link function (Figure 6.4). GLMs constitute a comparably simple, parametric ENM approach and we chose it mainly for its simplicity. In practice, we recommend comparing multiple ENM algorithms (Franklin 2010; Guisan et al. 2017). As described above, and based on the species' distribution in Australia, we only included linear terms of annual temperature and precipitation sums. To avoid overfitting, we weighted pseudo-absences such that the sum of their weights equals the sum of presences.

(iv) Model evaluation

Model behaviour was assessed using inflated response curves (Figure 6.4). As expected, species' occurrence probability decreased with temperature and increased with precipitation. Other response shapes (e.g. unimodal) did not produce a better model fit. The resulting model was evaluated using an 80-20 split-sample approach. For simplicity, we only used a single iteration while, ideally, split samples should be repeated (e.g. 100 times) to reduce possible effects of outliers. Then, we calculated two evaluation statistics for the hold-out test data. Both the AUC (area under the receiver operating characteristic curve) value of 0.915 and the maximum kappa value of 0.58 indicate high predictive performance (Figure 6.4). For transforming continuous ENM output to binary predictions, we chose a threshold that maximizes the sum of sensitivity (true positive rate) and specificity (true negative rate).

(v) Spatiotemporal predictions

Maps of predicted silvereye distribution for current climate closely resemble the observed distribution (Figures 6.3 and 6.5). This strong niche filling confirms our equilibrium assumption and allows a solid climate change impact assessment. Our future projections suggest potential range contractions towards coastal areas in response to climate change with severe range losses especially in Southwest Australia. This could put the subspecies (ssp. *chloronothus*) that is endemic to the Southwest at high risk from climate change. Our uncertainty analysis showed that novel climatic conditions can be mainly expected in Northern Australian regions well outside the species' potential distribution, and will thus not affect our climate impact assessment.

(ii) Data preparation: Next, the data are prepared at the appropriate scale and preliminary analyses are performed. If new observations of species need to be obtained, this step will also involve consideration of the sampling design. GIS layers of the environmental predictor variables are prepared at the desired spatial and temporal grain (size of map unit) and extent (number of maps units). Potentially, some scale mismatches occur between different data layers that need to be dealt with by upscaling or downscaling grain and extent. ENMs are a classic application of spatial ecology. Nevertheless, we deliberately also mention the temporal scale here, because it is important to also match the temporal grain and extent of the environmental data with that of the species data and the assumptions underlying the study system. For example, we need to ensure that the considered climatic, land cover, and remote sensing data match the timeframe of the species data. If the species distribution data contain only presence records, then adequate background data (pseudo-absences) need to be selected for the ENM. The best strategy for deriving pseudo-absences will depend on the research question and on the ENM algorithm (Guisan et al. 2017).

(iii) Model fitting: A number of different ENM algorithms are available (cf. section 6.4.2), from which one or several have been selected during the conceptualization phase. Important aspects to consider during model fitting are how to deal with multicollinearity in environmental predictor variables, with spatial autocorrelation, how many predictors can be included in the model (without overfitting), how to select relevant predictors, and how to select from or average different candidate models.

(iv) Model evaluation: The fitted model should be analysed in depth (Figure 6.4). First, the realism of the fitted response curves and the residuals need to be checked. Second, to assess predictive performance, the model should ideally be validated against independent data. Truly independent data are rarely available, in which case resampling methods can be useful. The predictive accuracy is then assessed based on different measures that evaluate discrimination or classification ability as well as goodness-of-fit. Most ENM algorithms will yield continuous output, the probability of species' occurrence. For some validation measures, this continuous output needs to be transformed into binary predictions, for which an appropriate thresholding approach needs to be selected.

(v) Spatiotemporal predictions: Once the ENM has been fitted, the model can be used for making spatial predictions. In most cases, these will be continuous maps of predicted occurrence probability or binary maps of potential species' distribution (Figure 6.5). Predictions to new times and places, for example under different climate change scenarios, are usually referred to as projections to make explicit that uncertainty is involved. When attempting projections, it is imperative to check for novel environmental conditions in the extrapolation data. If the future climatic range has no contemporary analogue and the ENM is extrapolating beyond the environmental range it was calibrated on, these projections should be clearly marked as uncertain (Zurell et al. 2012a). Also, different ENM algorithms will likely show different extrapolation behaviours and, thus, the fitted response curves (cf. model evaluation above) might need re-investigation to assess potential consequences. Projections also require paying special attention to the assumptions underlying ENMs, foremost the equilibrium assumption that is likely to be violated under climate change (cf. section 6.5).

Model building is an iterative process. Each step in the model building process may yield new insights on the species and the study system. Subsequently, single steps could be improved, for example, by adjusting the sampling design, the selection of environmental data or the model fitting, which will re-initiate the modelling cycle. Although many software packages are available, successfully fitting and evaluating predictive ENMs is far from being trivial and we highly recommend studying advanced books on ENMs (Franklin 2010; Guisan et al. 2017).

6.4 Sources of uncertainty

Different sources of uncertainty can affect the quality of ENMs and their projections for climate change scenarios. Here, we provide a brief overview of potential sources of uncertainty in underlying data, in the modelling approaches and in future scenarios. Thus, this section deals with uncertainty related to rather technical issues while we will provide more

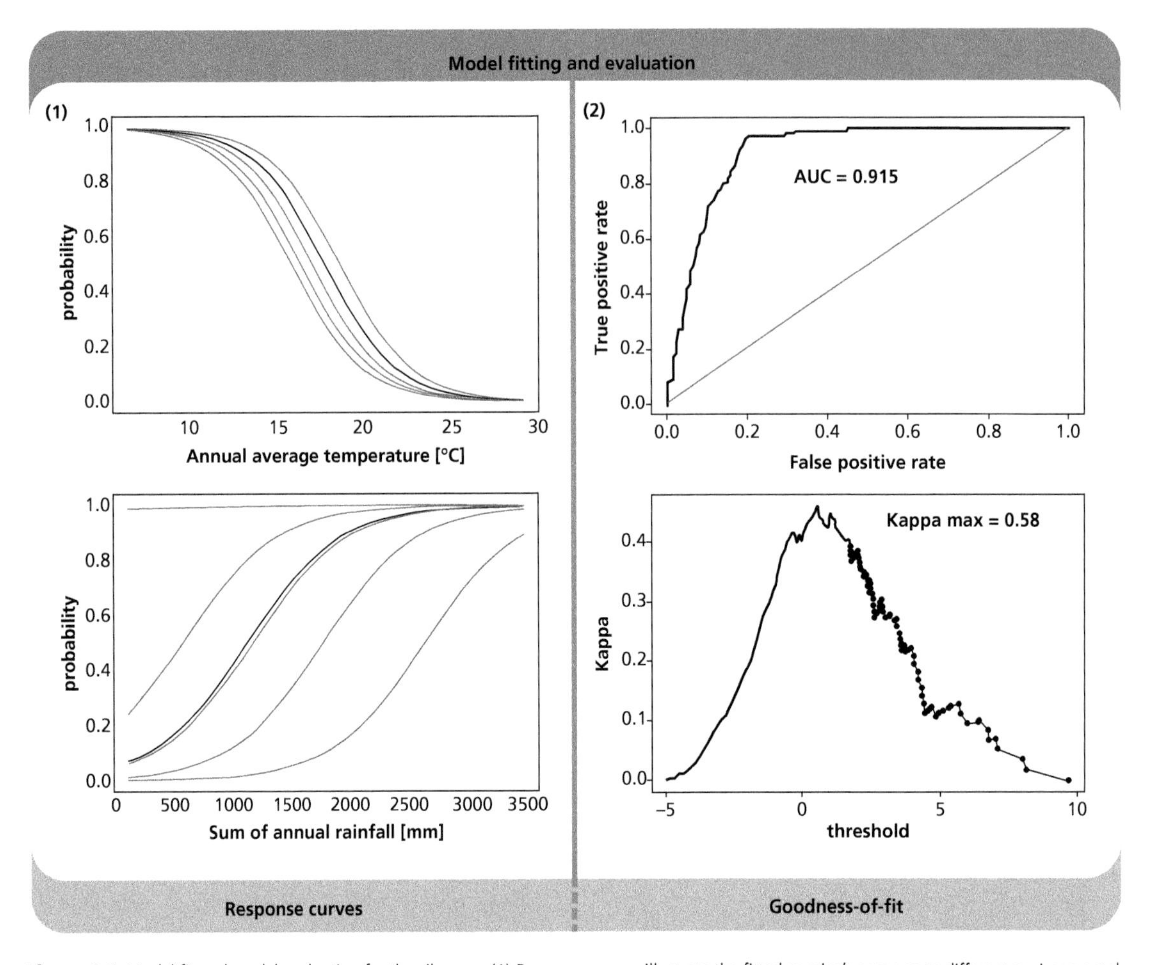

Figure 6.4 Model fit and model evaluation for the silvereye. (1) Response curves illustrate the fitted species' response to different environmental gradients and facilitate interpretation of ENMs as well as comparison to initial hypotheses about species' biology. (2) Evaluation statistics allow judging the model's predictive accuracy. The silvereye model shows overall high predictive performance.

critical thoughts on underlying assumptions and bird biology in section 6.5.

6.4.1 Data sources

Species data are available from many different sources including range maps, national monitoring programmes, atlas projects, and citizen science data (Chapter 4). In the cause of an ENM study, also one's own species data might be sampled. Generally, any source of species data needs to be checked for reliability and potential biases as well as its adequacy for answering the question at hand. This is especially true when downloading data from large databases such as GBIF (www.gbif.org) that contain records from different sources with different reliability. Typical challenges of species data for ENMs include uncertainty in spatial and temporal resolution, positional uncertainty, biased sampling design or uneven spatial coverage, low sample sizes, and imperfect detection.

The adequate spatial resolution of the species data depends on the research question and on the error, or positional uncertainty, the user is willing to accept in the model (Graham et al. 2008). If point observations are available, ENMs could, in principle, be applied at comparably fine resolution. However, in mobile animals such as birds, it is crucial to think about the meaning of a sampled point location. Is the observation of an individual at any place representative of a foraging location, a chance passage, or a home range, and which spatial resolution would best represent this? At the very least, the spatial resolution used for the ENM must be at least twice as coarse as the positional error, e.g. the potential error around a GPS location. At the other extreme are expert-drawn range maps. In order to use these as input for an ENM, the range maps need to be rasterized to an arbitrary spatial resolution. As species do not occur at all locations throughout their range, these maps represent species' occurrences only at very coarse resolution while they contain a lot of false positives at fine resolutions (Hurlbert and Jetz 2007).

A good sampling design should systematically cover all major environmental gradients of the study area and should, ideally, encompass the entire niche of the species in order to yield robust ENMs. Most monitoring and atlas programmes follow this principle while museum and citizen science data often suffer from spatial sampling bias, as a consequence of uneven coverage of presence records throughout the species' range. Often, museum and citizen science data are biased towards better-surveyed areas (e.g. popular bird watching areas, in national parks and hiking areas, or near urban areas) while having low sampling density or even data voids in remote areas. Strategies to deal with such spatial sampling bias include spatial filtering of the occurrence records or manipulation of the background (pseudo-absence) data such that the latter contain the same spatial bias as the presence records (Fourcade et al. 2014). Spatial filtering will be especially difficult in data-poor situations. As a rule of thumb, ten presence observations should be available per environmental predictor variable used in the ENM. In general, the minimum amount of presence records is advised to be between 20 and 50 (Guisan et al. 2017). Many species occurring in Asia, Africa, or South America suffer from few available samples (Meyer et al. 2015) as well as spatial sampling bias, making reliable ENM predictions for these regions difficult.

Additional uncertainty in the data may arise from imperfect detection. Low detectability can lead to false absences, meaning the species is not recorded although present. Indirect observations such as aural detection or environmental DNA can easily lead to false presences, meaning the species is recorded although absent. If data from repeated surveys or additional information on detection distances or detection time are available, then hierarchical frameworks can be used to model detection probability (additional to the species' environmental niche) by explicitly taking into account the observation process (Guillera-Arroita 2017).

Environmental data from remote sensing products (He et al. 2015) and climate databases (Hijmans et al. 2005; see also Chapter 3) are becoming increasingly available at fine spatial resolution for large regions, often the globe. Yet, precision as well as spatial and temporal coverage of the data may vary across regions and across datasets, and it is important to understand underlying assumptions and potential biases in these data.

Climate data are now available at resolution of 30 seconds for the entire globe, for example from the WorldClim database (Hijmans et al. 2005). Here, it is important to note that the climate data were not sampled at a 30-second resolution but were interpolated from available weather stations. The precision of the data thus relies on the density of weather stations, which varies considerably across the globe. Additionally, these datasets differ in their baseline (e.g. the time period used to average climate baseline values) and the downscaling methods used for interpolation. These uncertainties from different data sources could be accounted for by using ensemble approaches (Araújo and New 2007). Ensembles combine multiple alternative model predictions, which may arise from differences in the input data, the modelling algorithms, or the future scenarios. Instead of choosing one best model setup, ensembles provide information on the expected variability in model predictions additional to the main trend.

Choosing appropriate environmental variables to describe species' niches is non-trivial and could also constitute a considerable source of uncertainty. As we have discussed in section 6.2, it is generally advisable to use direct environmental variables that are limiting the species' distribution. For example, temperature could have a direct, physiological effect on the species' niche. However, climate data provide a range of different temperature variables such as mean annual temperature, temperature seasonality, or minimum and maximum temperatures of the coldest and warmest months, respectively. Previous knowledge of the species' ecology can help guide the variable choice. Another possibility is to separately test the explanatory power of each variable and choose those that best explain the distribution of the species (and are not too highly correlated). Also, uncertainty in model predictions due to the choice of predictor variables could be accounted for in an ensemble framework.

Differences in spatial and temporal coverage of environmental variables may lead to scaling issues as mentioned in section 6.3. For successfully building ENMs, all environmental data layers must fulfil the underlying assumptions and fit the spatial and temporal grain and extent of the species data. For example, serious temporal scale mismatches could arise if the species' presence records cover a longer time period, including records from before and after a major disturbance such as a hurricane, while the land cover data only cover the time period after the hurricane. Climate data are usually averaged over longer time periods. Still it is important to check for potential temporal scale mismatches between the species distribution and climatic data.

6.4.2 Different modelling approaches

A multitude of models have been developed for constructing ENMs, ranging from parametric to machine learning approaches (Franklin 2010; Guisan et al. 2017). All of these approaches make specific assumptions about the species–environment relationship that can lead to inconsistent projections under climate change. Many non-parametric or machine-learning approaches such as MaxEnt, random forests, and boosted regression trees can fit very complex response surfaces and potentially overfit data, while (semi-)parametric methods fit simpler response surfaces and potentially underfit data (at least when fitted with lower order polynomials and derivatives). The necessary complexity should be determined by the research question and previous knowledge on the study system. If the species is not at equilibrium with its environment, then overly complex responses may lead to spurious predictions (Elith et al. 2010). At the same time, overly simplistic responses may lead to models that are not much better than the null expectation. The use of MaxEnt has been increasing rapidly over the last years. However, we want to stress that there is no evidence that MaxEnt is generally performing better than other approaches. Rather, its popularity may be related to the fact that a stand-alone user interface is available for MaxEnt, which makes it easy to use even without extensive statistical and programming experience. Generally, it is advisable to compare multiple model techniques to account for uncertainty in algorithmic choices (IPBES 2016).

For climate change projections, it is indispensable to check for novel environmental predictions in the climate change scenario (Figure 6.5) and to check

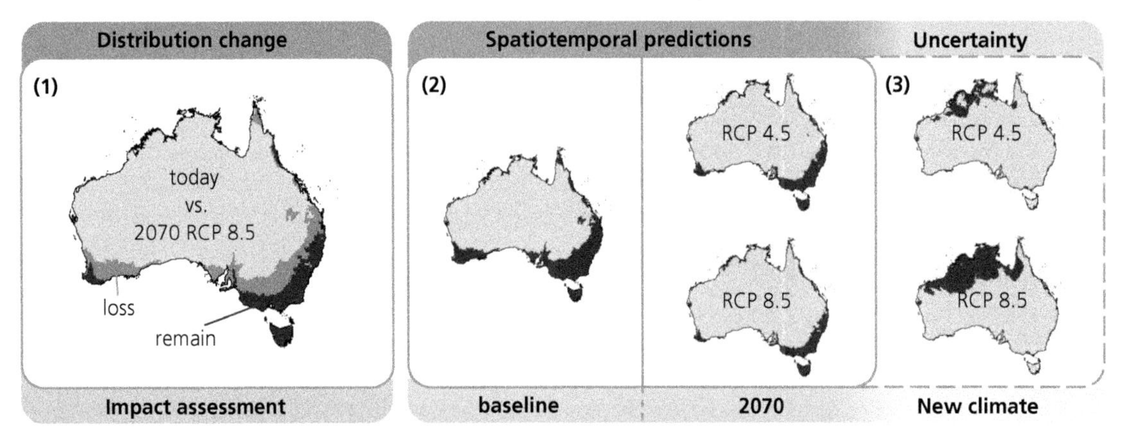

Figure 6.5 Maps of potential future climate impact on the silvereye. (1) The map of projected range changes indicates substantial loss in occupied area until 2070. (2) Distribution change can be easily derived by comparing potential distribution under current (baseline) climate and projected distribution under future climate scenarios (Representative Concentration Pathways, [RCP]). (3) One source of uncertainty in future projections of species distribution could be the emergence of novel climatic conditions. In Australia, novel climatic conditions will mainly occur in the north where the silvereye does not occur.

the extrapolation behaviour of the fitted ENM (cf. section 6.3) (Elith et al. 2010). For example, response surfaces that imply that a species could survive for any range of future temperatures should be judged as implausible (Elith et al. 2010; Zurell et al. 2012a). It is important to note that uncertainty through different modelling approaches cannot be studied in isolation from uncertainty in data sources. Problems with extrapolation will mainly occur in cases where the distribution data do not cover the entire range of environments experienced by the species, which could lead to truncated niches in the data (Zurell et al. 2012a).

Uncertainty due to different modelling approaches can be accounted for by combining these in an ensemble framework (Araújo and New 2007). However, we want to stress that even in ensembles it is of utmost importance to understand the peculiarities of the different ENM algorithms and check their extrapolation behaviour.

6.4.3 Different climate models and scenarios

A prerequisite for projecting species' distributions into the future are information about future climates that are provided by climate scenarios. Climate models simulate the complex interactions between atmosphere, ocean, land surface, snow and ice, ecosphere, and several chemical and biological processes. Naturally, many assumptions are being made in the modelling processes, and currently many different models exist ranging from very simple to more comprehensive climate models to Earth System Models (ESMs) that include an interactive carbon cycle. No single best climate model can be identified but rather the available models differ in their ability to simulate certain climate attributes (IPCC 2013). For simulating future climate, these models additionally rely on socio-economic scenarios about future anthropogenic forcings, including emissions of greenhouse gases and aerosols, which depend on socioeconomic factors and geopolitical agreements. The newest climate simulations currently available have been carried out within the Coupled Model Intercomparison Project Phase 5 (CMIP5) of the World Climate Research Programme. Under this framework, a large number of comprehensive climate models and ESMs have been compared. Climate simulations from CMIP5 are based on a new set of scenarios of anthropogenic forcings, called the Representative Concentration Pathways (RCPs). Four different RCPs represent different mitigation scenarios and different targets of radiative forcing by 2100.

The IPCC (2013) provides details about the key uncertainties in these climate scenarios. In summary, uncertainty in climate models and scenarios may arise through uncertainty in process representation in the different climate models, spatial and temporal scales, uncertainty in observational data used for model evaluation, and uncertainty in other factors such as anthropogenic forcings. When projecting species' distributions into the future, uncertainty in climate change scenarios should be taken into account, for example by exploring a number of plausible climate models and socio-economic scenarios.

6.5 Robustness and validation against observed distribution and population changes

In the previous sections, we have learned that building and applying ecological niche models requires critical consideration of the necessary modelling steps and of the data used to build and project models. Yet, careful construction and evaluation of the models does not guarantee high transferability in space and time. As ecological niche models are phenomenological, they are strictly not valid beyond the range of variables they are calibrated on. The emergence of novel environments could thus seriously hamper prediction accuracy. Also, several of the assumptions underlying ecological niche models may limit their use in a global change context. Specifically, ENMs typically assume that: 1) species are at equilibrium with the environment, 2) interspecific interactions do not affect the distribution of the species, 3) the selected environmental variables are adequate for describing the niche of species, and 4) the species–environment relationship will stay constant under future climates.

The equilibrium assumption could be violated in two ways. First, if the distribution of a species and, thus, the observed data are affected by transient (time-delayed) dynamics such as dispersal limitations and adaptation, it may lead to biased niche estimates in ENMs. This could be the case if the species is currently expanding its range, for example during invasion or due to competitor release. Second, when making projections under climate change ENMs assume an instantaneous realization of a new equilibrium situation. This would mean that newly suitable habitat will be colonized immediately without any time lags due to dispersal limitations or due to competition. Projections of historical ranges paired with DNA analyses in willow and rock ptarmigan suggested that these cold-adapted species were able to track changes in their suitable habitat over the last 20 millennia (Lagerholm et al. 2017). However, the authors caution that increasing current-day fragmentation may hamper future range shifts.

The question whether the species' niche is adequately described by the selected environmental variables is extremely difficult to evaluate. Distinguishing between proximal and distal effects on birds is often non-trivial. Birds are highly mobile organisms that often show seasonal variations in their range boundaries, or even spatially disjunct breeding and overwintering ranges as is the case in many migrants. These seasonal variations are likely driven by variations in seasonal resource availability closely intertwined with high climate seasonality (Eyres et al. 2017). Additionally, birds are endothermic and their distributions may be less tightly bounded by climatic conditions than in ectotherms (for an overview in birds see Engler et al. 2017, and references therein). Recent analyses indicate that climate variables may both have a direct effect on species' ranges, indicative of thermal tolerances (Khaliq et al. 2017), and an indirect effect due to their correlation with resource availability (Buckley et al. 2012). One impressive example is the massive northward shift in winter distributions of many North American bird species through supplemental feeding (Newton 2003). This example indicates that the winter climate is not physiologically (i.e. proximally) limiting northern range edges in these species but is correlated with resource limitations in winter. Similarly, range edges along upper thermal limits can be relaxed through artificial water supplements or behavioural adaptations in arid environments, allowing birds to compensate for overheating and thermal stress (Pattinson and Smit 2017). However, increased desertification as a consequence of climate change is expected to exacerbate upper thermal limitations in birds, even in desert-adapted species (Albright et al. 2017).

Numerous studies have projected bird species' response to climate change (cf. Engler et al. 2017), but few have confronted projections with observations to test if projections were successful. In a cross-taxon analysis forecasting historic range changes of tetrapods over the last 60 years in Australia, Morán-Ordóñez et al. (2017) found ENMs to produce useful predictions. Similarly, Fordham et al. (2018) predicted range changes of British birds over the last 40 years and found that ENMs can provide reasonable first approximations of the magnitude of potential range shifts. Still more complex models, taking into account information on dispersal and demography, were better at predicting range changes at the grid-cell level. Although not explicitly focusing on climate change, Oliver et al. (2012) showed that ENM predictions were more strongly correlated with occupancy and population density than with population stability, suggesting that ENMs might be poor at predicting long-term population persistence.

6.6 Latest developments and perspectives in a global change context

ENMs have greatly enhanced our understanding of avian species–environment relationships and potential climate change impacts on birds. Nevertheless, because they do not explicitly take into account the effect of biotic interactions, demography, and adaptation, ENMs are of limited use for understanding the interplay of different niche components. Overall, we are still in the infancy of having a functional and holistic understanding of species–environment relationships. Therefore, it is not surprising that this is an active field with many exciting results and much debate.

In recent years, several approaches have been discussed to establish ENM frameworks that integrate demographic processes and interspecific interactions (Zurell 2017). Such a framework would help in disentangling the different niche constraints illustrated in Figure 6.1. Demographic models help to better understand the complex dependencies between species' life history, their distribution, and environmental processes. To date, few studies have integrated demographic information for projecting bird species' response to climate change (Fordham et al. 2018; Zurell et al. 2012b). In part, this may be due to the higher data requirements needed for parameterizing the demographic model. At the same time, technical challenges remain. Mostly, hybrid models that have been used still rely on ENM predictions and couple these with a population dynamic model. Although these models are able to predict transient dynamics into the future, the hybrid approach could lead to circularity problems. If the current distribution of the species is affected by any transient dynamics such as dispersal constraints, then this is implicitly included in the underlying ENM and the predictions would be biased that way. So-called dynamic range models provide an alternative approach by directly relating demographic rates to environmental gradients and simultaneously fitting all model components (demography, dispersal, and environmental response) to data (Pagel and Schurr 2012). However, these models have not been applied to real-world species yet, and the underlying demographic model is as yet too inflexible to incorporate the complex life histories of birds.

Joint species distribution models (JSDMs) have been introduced as extensions to single-species ENMs. These models simultaneously model the species–environment relationships of multiple species as well as the residual correlation between those species. The residual correlations can be indicative of interspecific interactions but also of missing environmental information. Thus, JSDMs may help in deriving hypotheses about potential range-limiting interactions between species, but they are also highly scale-dependent (Zurell 2017).

Another branch of progress towards a more functional understanding of species–environment relationships is the integration of molecular genetic information below the species level. While ENMs have frequently been combined with phylogenetic data to understand niche evolution, integrating population genetic or genomic information into ENMs is a more recent innovation (see Engler et al. 2017 for an overview and references). Incorporating the information on population genetic structure into ENMs would allow capture of local adaptations that may be particularly important for anticipating species' response to climate change (Valladares et al. 2014). As genomic tools are getting more sophisticated and allow studying gene regulation and expression for many bird species (Kraus and Wink 2015), it will soon be possible to apply ENMs to study genotype–environment relationships in birds. In combination with enhanced ENM frameworks incorporating demography and biotic interactions, this will open up new perspectives for studying climate change impacts on birds at a functional-molecular level and for understanding adaptive responses to sudden environmental changes.

References

Albright, T.P., Mutiibwa, D., Gerson, A.R., Smith, E.K., Talbot, W.A., O'Neill, J.J., McKechnie, A.E., et al. (2017). Mapping evaporative water loss in desert passerines reveals an expanding threat of lethal dehydration. *Proceedings of the National Academy of Sciences of the United States of America*, 114, 2283–8.

Araújo, M.B., and New, M. (2007). Ensemble forecasting of species distributions. *Trends in Ecology & Evolution*, 22, 42–7.

Austin, M.P. (1980). Searching for a model for use in vegetation analysis. *Vegetatio*, 42, 11–21.

Blakers, M., Reilly, P.N., and Davies, S. (1984). *The atlas of Australian birds*. Melbourne Univ. Press, Melbourne, Australia.

Buckley, L.B., Hurlbert, A.H., and Jetz, W. (2012). Broad-scale ecological implications of ectothermy and endothermy in changing environments. *Global Ecology and Biogeography*, 21, 873–85.

Elith, J., Kearney, M., and Phillips, S. (2010). The art of modelling range-shifting species. *Methods in Ecology and Evolution*, 1, 330–42.

Elith, J., and Leathwick, J.R. (2009). Species distribution models: Ecological explanation and prediction across space and time. *Annual Review of Ecology, Evolution, and Systematics*, 40, 677–97.

Engler, J.O., Stiels, D., Schidelko, K., Strubbe, D., Quillfeldt, P., and Brambilla, M. (2017). Avian SDMs: current state, challenges, and opportunities. *Journal of Avian Biology*, 48, 1483–1504.

Eyres, A., Böhning-Gaese, K., and Fritz, S.A. (2017). Quantification of climatic niches in birds: adding the temporal dimension. *Journal of Avian Biology*, 48, 1517–31.

Fordham, D.A., Bertelsmeier, C., Brook, B.W., Early, R., Neto, D., Brown, S.C., Ollier, S., et al. (2018). How complex should models be? Comparing correlative and mechanistic range dynamics models. *Global Change Biology*, 24, 1357–70.

Fourcade, Y., Engler, J.O., Rödder, D., and Secondi, J. (2014). Mapping species distributions with MAXENT using a geographically biased sample of presence data: a performance assessment of methods for correcting sampling bias. *PLoS One*, 9, e97122.

Franklin, J. (2010). *Mapping species distributions: Spatial inference and prediction*. Cambridge University Press, Cambridge, UK.

Graham, C.H., Elith, J., Hijmans, R.J., Guisan, A., Townsend Peterson, A., Loiselle, B.A., and The NCEAS Predicting Species Distributions Working Group (2008). The influence of spatial errors in species occurrence data used in distribution models. *Journal of Applied Ecology*, 45, 239–47.

Guillera-Arroita, G. (2017). Modelling of species distributions, range dynamics, and communities under imperfect detection: advances, challenges, and opportunities. *Ecography*, 40, 281–95.

Guisan, A., and Thuiller, W. (2005). Predicting species distribution: offering more than simple habitat models. *Ecology Letters*, 8, 993–1009.

Guisan, A., Thuiller, W., and Zimmermann, N.E. (2017). *Habitat suitability and distribution models: with applications in R*. Cambridge University Press, Cambridge, UK.

Guisan, A., and Zimmermann, N. E. (2000). Predictive habitat distribution models in ecology. *Ecological Modelling*, 135, 147–86.

Hanski, I. (1999). *Metapopulation ecology*. Oxford University Press, Oxford, UK.

He, K.S., Bradley, B.A., Cord, A.F., Rocchini, D., Tuanmu, M.-N., Schmidtlein, S., Turner, W., et al. (2015). Will remote sensing shape the next generation of species distribution models? *Remote Sensing in Ecology and Conservation*, 1, 4–18.

Hijmans, R.J., Cameron, S.E., Parra, J.L., Jones, P.G., and Jarvis, A. (2005). Very high resolution interpolated climate surfaces for global land areas. *International Journal of Climatology*, 25, 1965–78.

Hurlbert, A.H., and Jetz, W. (2007). Species richness, hotspots, and the scale dependence of range maps in ecology and conservation. *Proceedings of the National Academy of Sciences of the United States of America*, 104, 13384–9.

IPBES (2016). *The Methodological Assessment Report on Scenarios and Models of Biodiversity and Ecosystem Services*. Secretariat of the Intergovernmental Platform for Biodiversity and Ecosystem Services, Bonn, Germany.

IPCC (2013). *Climate Change 2013: The Physical Science Basis: Working Group I Contribution to the Fifth Assessment Report of the Intergovernmental Panel on Climate Change*. Cambridge University Press, Cambridge, UK.

Jarvis, E.D. (2016). Perspectives from the Avian Phylogenomics Project: Questions that can be answered with sequencing all genomes of a vertebrate class. *Annual Review of Animal Biosciences*, 4, 45–59.

Khaliq, I., Böhning-Gaese, K., Prinzinger, R., Pfenninger, M., and Hof, C. (2017). The influence of thermal tolerances on geographical ranges of endotherms. *Global Ecology and Biogeography*, 26, 650–68.

Kraus, R.H.S., and Wink, M. (2015). Avian genomics: fledging into the wild! *Journal of Ornithology*, 156, 851–65.

Lagerholm, V.K., Sandoval-Castellanos, E., Vaniscotte, A., Potapova, O.R., Tomek, T., Bochenski, Z.M., Shepherd, P., et al. (2017). Range shifts or extinction? Ancient DNA and distribution modelling reveal past and future responses to climate warming in cold-adapted birds. *Global Change Biology*, 23, 1425–35.

Meyer, C., Kreft, H., Guralnick, R., and Jetz, W. (2015). Global priorities for an effective information basis of biodiversity distributions. *Nature Communications*, 6, 8221.

Morán-Ordóñez, A., Lahoz-Monfort, J.J., Elith, J., and Wintle, B.A. (2017). Evaluating 318 continental-scale species distribution models over a 60-year prediction horizon: what factors influence the reliability of predictions? *Global Ecology and Biogeography*, 26, 371–84.

Newton, I. (2003). *Speciation and Biogeography of Birds*. Academic Press, London, UK.

Oliver, T.H., Gillings, S., Girardello, M., Rapacciuolo, G., Brereton, T.M., Siriwardena, G. M., Roy, D.B., et al. (2012). Population density but not stability can be predicted from species distribution models. *Journal of Applied Ecology*, 49, 581–90.

Pagel, J., and Schurr, F.M. (2012). Forecasting species ranges by statistical estimation of ecological niches and spatial population dynamics. *Global Ecology and Biogeography*, 21, 293–304.

Pattinson, N.B., and Smit, B. (2017). Seasonal behavioral responses of an arid-zone passerine in a hot environment. *Physiology and Behavior*, 179, 268–75.

Pearson, R.G., and Dawson, T.P. (2003). Predicting the impacts of climate change on the distribution of species: are bioclimate envelope models useful? *Global Ecology and Biogeography*, 12, 361–71.

Pulliam, H.R. (1988). Sources, sinks, and population regulation. *American Naturalist*, 132, 652–61.

Soberón, J. (2007). Grinnellian and Eltonian niches and geographic distributions of species. *Ecology Letters*, 10, 1115–23.

Valladares, F., Matesanz, S., Guilhaumon, F., Araújo, M.B., Balaguer, L., Benito-Garzón, M., Cornwell, W., et al. (2014). The effects of phenotypic plasticity and local adaptation on forecasts of species range shifts under climate change. *Ecology Letters*, 17, 1351–64.

Zurell, D. (2017). Integrating demography, dispersal and interspecific interactions into bird distribution models. *Journal of Avian Biology*, 48, 1505–16.

Zurell, D., Elith, J., and Schröder, B. (2012a). Predicting to new environments: tools for visualizing model behaviour and impacts on mapped distributions. *Diversity and Distributions*, 18, 628–34.

Zurell, D., Grimm, V., Rossmanith, E., Zbinden, N., Zimmermann, N.E., and Schröder, B. (2012b). Uncertainty in predictions of range dynamics: black grouse climbing the Swiss Alps. *Ecography*, 35, 590–603.

CHAPTER 7

Predicting the effects of climate change on bird population dynamics

Bernt-Erik Sæther, Steinar Engen, Marlène Gamelon, and Vidar Grøtan

7.1 Introduction

We are currently living in a time with large environmental changes occurring at a global scale. In many regions, important habitat types are disappearing at an alarmingly high rate, resulting in massive landscape alterations that affect ecosystem structures and functions over large areas. Simultaneously, we are experiencing rapid changes in climate, likely to affect ecological processes almost everywhere on Earth. These changes accentuate the need for developing ecology as a predictive science, enabling us to quantitatively assess how global changes will affect ecological dynamics and future characteristics of ecosystems.

A major challenge in transforming ecology into a predictive science is that this requires quantification of processes acting at different organismic levels, from genes via populations and communities of species up to entire ecosystems. In many respects, we have already identified general patterns and processes that affect the structure of ecological systems, especially at large spatial scales. However, for most ecological processes we are able to make inferences about what should be observed before making actual observations (i.e. make predictions) only to a small extent. This forms the core content of being a predictive science (Mouquet et al. 2015). Here, we focus on how to derive expectations for future trajectories of single populations and how to include the effects of climate change in the models forming the basis for such predictions.

Developing reliable population projections is particularly difficult in a changing environment for several reasons. First, the chosen model must describe the dynamics reasonably well. Second, whatever population model is chosen, the effects of a change in the environment must be introduced into the population model in an appropriate way, thus requiring environmental effects to be specified for specific parameters in the model. For example, in a population subject to density dependence a trend in the environment may influence population growth, carrying capacity, or both (Sæther and Engen 2010a). Third, model predictions must take into account the stochastic influences on population dynamics. These can be caused by stochastic variation in the environment, which affects all individuals or groups in a similar way, and by demographic stochasticity, which is due to random variation in survival or reproduction occurring independently among individuals (Lande et al. 2003). The effects of demographic stochasticity are of particular importance to include in analyses of climate effects on small populations because they reduce the population growth rate and increase the probability of extinction (Lande et al. 2003). Fourth, uncertainties in population censuses or in parameter estimates may be large, and should be assessed when constructing population forecasts.

Sæther, B.-E., Engen, S., Gamelon, M., and Grøtan, V., *Predicting the effects of climate change on bird population dynamics*. In: *Effects of Climate Change on Birds*. Second Edition. Edited by Peter O. Dunn and Anders Pape Møller: Oxford University Press (2019). DOI: 10.1093/oso/9780198824268.003.0007

Birds are excellent candidates for developing techniques to predict how expected changes in climate will affect future population abundances. Firstly, we know that climate variation has a huge impact on the population dynamics of many bird species (e.g. Sæther et al. 2004; Jenouvrier 2013; Pearce-Higgins et al. 2015; Martay et al. 2017). Secondly, the relationship between climate-induced demographic variation and population dynamics is relatively well understood in several species covering a wide range of life histories (e.g. Reed et al. 2013b; Barbraud et al. 2012; Jenouvrier et al. 2005). Thirdly, some of the longest and most comprehensive individual-based demographic studies of any free-living species exist for birds (Clutton-Brock and Sheldon 2010) offering unique opportunities to parameterize quite complex demographic models.

The purpose of this chapter is to outline a potential general approach for predicting the effects of expected climate changes on the dynamics of single populations. We do this in three steps. First, we introduce and define useful concepts for making and interpreting ecological forecasts. Second, we review some simple population models, illustrating important processes to include for obtaining reliable projections of future changes in population size. Finally, we identify general patterns that have appeared in studies developing models for analyses of population responses to expected changes in climate.

7.2 The concept of Population Prediction Interval

Here we will argue that the concept of Population Prediction Interval (PPI) can be a useful tool for making predictions about future population sizes. A PPI can be interpreted similarly to a confidence interval, making inferences about a range of projected population sizes based on simulations. Specifically, the PPI is defined as the stochastic interval that includes a given population size with probability ($1-\alpha$), where α is the probability that the variable we want to predict is not contained in the stochastic interval. For example, in Figure 7.1 we model a population for 50 time steps (t = 50) with an initial size of 20 individuals and a carrying capacity (K) of 50. After 50 time steps, the distribution of predicted population sizes at each time step is shown in Figure 7.1. Here, PPI is graphed using quantiles, so each bar on the right side of the figure shows the proportion of population size estimates in 10 percentile groups. The PPI has the convenient property that its width (e.g., the difference between the 5 per cent and 95 per cent percentiles) is affected by deterministic processes in the population dynamics (e.g. due to density regulation), stochastic fluctuations in population size (due to demographic and environmental stochasticity), uncertainties in parameter estimates, and biases in estimates of population size (Sæther et al. 2007b).

An important aspect of the PPI is that it can be used to assess the quality of our predictions (Sæther et al. 2009). If our model makes correct predictions of future population fluctuations, then the quantiles of simulated population sizes at a given time step will be uniformly distributed across the quantiles of the PPI (Figure 7.1a): i.e., quantiles with equal probabilities will include an equal number of simulated population trajectories. If the distribution resembles a normal distribution (Figure 7.1b), then stochastic influences on the population dynamics have been overestimated, caused either by demographic or environmental stochasticity or underestimation of the strength of density regulation. In contrast, in populations subject to density regulation, we may also encounter a dome-shaped downward distribution, which indicates that we have overestimated the strength of density regulation or underestimated stochastic influences on the dynamics (Figure 7.1c). If there is a predominance of population estimates in the upper quantiles of the PPI, our predictions are *biased* because they underestimate future population sizes (Figure 7.1d). In this case, we predict too many estimates in the lower quantiles. This occurs if we have estimated too small population growth rates or if we have underestimated the carrying capacity. In contrast, skewed distributions of the population sizes towards smaller quantiles indicate a bias caused by overestimating future population sizes (Figure 7.1e).

This stochastic approach of the PPI differs from the deterministic approach typically considering a single trajectory at a future time step (Petchey et al. 2015). Because we build our predictions from stochastic simulations of an underlying model, we define a *correct* prediction as a uniform distribution of recorded

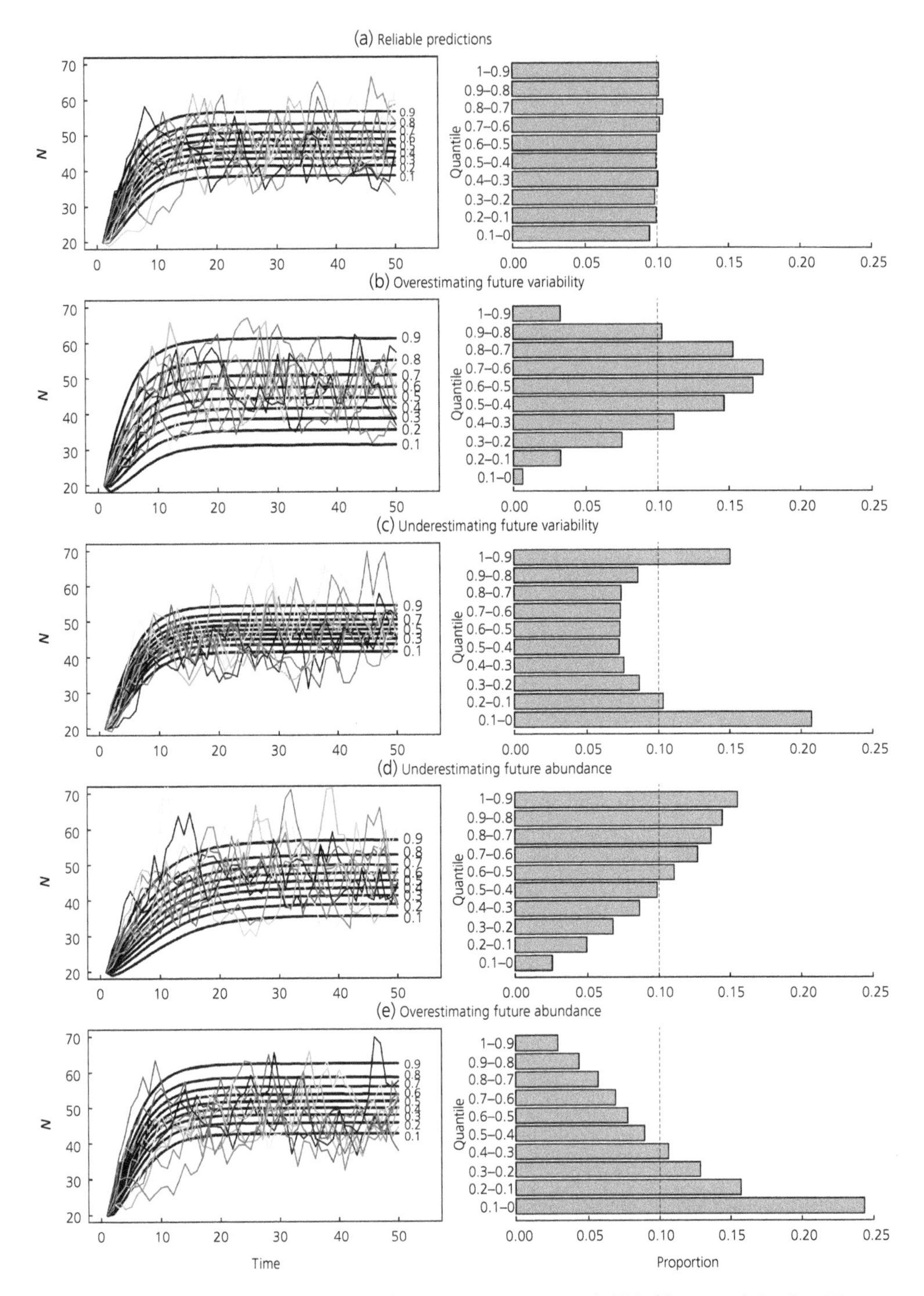

Figure 7.1 Illustration of concepts related to interpretation of Population Prediction Intervals (PPI) of future population sizes. PPIs were constructed by stochastic simulations of population sizes forward in time. Here we assume that the real dynamics of a population follows a discrete logistic form of density dependence with parameters $r = 0.3$, $K = 50$, $\sigma_e^2 = 0.01$, and $\sigma_d^2 = 1$. For a given set of estimated parameters and starting at population size $N = 20$ at time $t = 1$, we then simulate 10 000 stochastic time series from time $t = 2$ to $t = 50$. At each time step we calculate the quantiles of the population sizes across all simulated time series. In the figures (left columns) we show quantiles 0.1, 0.2, …, 0.9. Quantiles should be interpreted as the probability that the population size is lower than the given quantile at time t. From this it follows that at

population sizes when they are summarized in quantiles (Figure 7.1a). Such a prediction interval is also unbiased because half of the population sizes will be above and below the 50 per cent quantile, respectively. A *precise* prediction happens when the difference in predicted population size between the upper and lower quantiles of the PPI is small. In general, the width of the PPI and hence the precision of the projections will decrease with increasing environmental stochasticity and decreasing strength of density dependence (Figure 7.1). *Uncertainty* refers to the level of standard errors in population estimates or in estimates of model parameters. The larger the uncertainties, the less precise become our predictions of future population sizes.

7.3 The dynamical consequences of climate change

Analyses of time series of fluctuations in population size have so far been the most common approach for examining how climate variation affects avian population dynamics and has provided important general insights. However, this approach relies on several simplifying assumptions. In particular, vital rates are assumed to be equal among age classes and respond similarly to density dependence. In addition, we also now know that climate changes have contrasting effects on different stages of the life history of many bird species (Sæther and Engen 2010a; Herfindal et al. 2015). This involves timing of reproduction (Visser et al. 2004), the rate of offspring production, and survival at different seasons (Grosbois et al. 2008). Reliable predictions of how these climate-induced changes in vital rates will influence future population sizes are dependent on the way different parameters characterizing the population dynamics are affected.

In many cases, this means that necessary information required to make population projections will be difficult to obtain from statistical analyses of time series alone. To illustrate this point more precisely, assume a simple population model with a theta-logistic type of density regulation (Gilpin et al. 1976) and environmental stochasticity. Let us consider a situation in which climate change causes a proportional change either in population growth rate r or in carrying capacity. Solbu et al. (2015) generated 200 new time series by simulation of this model and estimated the parameters using the Bayesian approach outlined by Bolker et al. (2013). Some results of these analyses are summarized in Table 7.1 and illustrate two important points:

(1) In species living in a seasonal environment, climate-induced changes of the carrying capacity can strongly influence their future abundance (Sæther et al. 2000b). However, such climate-induced changes affecting the carrying capacity can be extremely difficult to detect when entirely based on time series of population estimates and hence their consequences for future population sizes are almost impossible to predict.
(2) A wide range of variation in population size is required to obtain reliable estimates of the deterministic specific growth rate r or even large climate-induced changes in r. If data on population growth at small population sizes are unavailable, crude estimates of r based on demographic data (Sæther et al. 2002) may

time t we expect that 10% of realizations will be found at population sizes between the 0.1 and 0.2 quantiles, 10% between 0.2 and 0.3 quantiles, 40% between 0.3 and 0.7 quantiles and so on. Thus, if the PPI is constructed with the correct model/parameters the realizations of the process should be uniformly distributed across the quantiles. In a) we constructed both the PPI and subsequently simulated 10 000 stochastic time series with parameters given above. Coloured lines in the leftmost figure show a few example simulations of the process. As expected, the distribution of recorded quantiles (49 × 10 000) is uniform. In b) we show the consequence of a biased parameter estimate when constructing the PPI. Here we set $\hat{\sigma}_e^2 = 0.02$ instead of $\sigma_e^2 = 0.01$ when constructing the PPI. Increased environmental variance leads to larger variability in population size and a wider PPI. When subsequently simulating 10 000 time series of the correct dynamics ($\sigma_e^2 = 0.01$) and recording the corresponding quantiles of simulated population sizes in the PPI we observe that the distribution of quantiles show that most simulated population sizes are within the central region of the PPI. This illustrates that the PPI is too wide because of being constructed based on parameters leading to a process variance in N that is too high. For comparison, in c) we constructed a PPI based on $\hat{\sigma}_e^2 = 0.005$ leading to a too narrow PPI. The variability of population sizes could also be biased low or high because of biased estimates in the strength of density regulation (not shown here). In d) we constructed the PPI using $\hat{r} = 0.2$ as a biased estimate of the growth rate at small population size. In the short term this leads to an underestimation of future population sizes and the distribution of quantiles becomes skew. Another example of skewness of the prediction interval can be found in e) where we used $\hat{K} = 45$ as our estimate of the carrying capacity.

Table 7.1 The effects of a step-wise change in population growth rate r or the carrying capacity in a theta-logistic model of density regulation on the bias in parameter estimates based on 200 simulations of the model. The density regulation is described by the parameter θ (Lande et al. 2003). The true parameter values for the model were deterministic growth rate $r = 0.2$, $\theta = 1.5$, carrying capacity $K = 1000$, and environmental variance $\sigma_e^2 = 0.01$, assuming a proportional decrease due to climate change in r of $\rho = 0.5$ (Case A or B) or in K of $\kappa = 0.5$ (Case C). In case A the initial population size is $N_0 = 20$, whereas in case B the simulations are started at the carrying capacity $K = 1000$ (from Solbu et al. 2013, 2015).

Case	r		ρ or κ		K	
	Estimate	**Credibility interval**	**Estimate**	**Credibility interval**	**Estimate**	**Credibility interval**
A	0.204	0.153, 1.942	0.370	0.021, 2.308	972.8	851.2, 7233.5
B	0.033	0.019, 4.362	0.101	0.021, 4.993	1060.5	5.2, 16 191.2
C			0.556	0.46, 0.736	951.0	798.1, 1140.1

provide a better approach than using time series data alone (Clark et al. 2010).

Another important implication of the simulation studies by Solbu et al. (2013, 2015) is that the type of density regulation will strongly affect the probability of detecting climate-induced changes in population dynamics from time series analyses alone. In particular, the effects of climate change on future population sizes will be difficult to reliably predict for populations subject to weak density regulation.

Although several approaches have recently appeared to estimate uncertainty in key parameters such as r and the environmental variance σ_e^2 (e.g. Bolker et al. 2013; Abadi et al. 2010), biases and standard errors in parameter estimates decrease only moderately with increasing length of the time series (Solbu et al. 2015). Consequently, most available time series on fluctuations in population size are too short to provide useful information about the effects of climate change on future population dynamics (Knape and de Valpine 2011).

Thus, our recommendation for predicting climate effects is to make specific assumptions about *how* key climate variables are likely to affect different parameters determining population dynamics. The complexity of these effects will depend on the life history of the species in question.

7.4 Identification of key climate predictors for avian population predictions

In many cases, the climate variable or the time window for explaining changes in population size are chosen a priori based on varying levels of ecological justification, usually by including mean values for a specific (e.g. monthly) period of time. However, identifying such critical periods or variables is not trivial (Stenseth and Mysterud 2005), and often requires in-depth knowledge about how different life history traits are affected by local weather and regional climate phenomena throughout the year (Hallett et al. 2004).

van de Pol et al. (2016; Chapter 5) have proposed a statistical approach to identify and quantify critical weather signals, implemented into the R-package *climwin* (Bailey and van de Pol 2016). They compare a null model without any climate effects to a set of models identifying all potentially relevant weather variables and range of time windows using a model selection approach, reporting the final model in a standardized way that includes potential interactions between candidate variables. A major advantage of this approach is that it assures transparency in the choice of critical weather signal. Furthermore, it also enables extrapolation of the future climate effects by providing statistical estimates of the slope of the relationship between the weather signal and the population parameter in question. However, the reliability of this approach for climate projections depends heavily on the effect sizes of the climate windows. With small effect sizes, high levels of false positives often appear (van de Pol et al. 2016). Thus, the best use of this approach probably involves choosing a reasonably small set of candidate climate variables and possible time windows based on a priori information and then subsequently applying statistical procedures to select the best variables and time windows.

7.4.1 Age structure and the effects of environmental autocorrelations

A key challenge for making population predictions is to account for the influence of temporal (or spatial) autocorrelations, which is the correlation between two variables measured at different times (or locations). Trends in climate tend to cause temporal autocorrelation in environmental noise. In a population structured by age or life history stages, this is represented by a noise matrix. However, autocorrelation in the noise must not be confused with autocorrelation in population size. Even in the absence of autocorrelations in the noise (white noise), there may still be large autocorrelations between population sizes $n_{i,t}$ and $n_{j,t+h}$ of age-classes i and j at times t and $t + h$.

Environmental autocorrelations are important to consider when making inferences about the population dynamical consequences of changes in climate because they may influence mean and variance in population size (Heino et al. 2000; Tuljapurkar et al. 2003; Tuljapurkar et al. 2009; Morris et al. 2011). In the simplest density-independent case without any age structure, environmental autocorrelation has no impact on the long-run growth rate of the population because this is the limit of $(\ln N_t - \ln N_0)/t$ as t approaches infinity, where N_t is the population size at time t (Dennis et al. 1991). From the relation $\ln N_t - \ln N_0 = \sum_{u=1}^{t}(\ln N_u - \ln N_{u-1})$, the long-run growth rate is the expected value of $\ln N_u - \ln N_{u-1}$ for a noise process even if there are autocorrelations because expectations in general are additive. Thus, only the environmental variance $\sigma_e^2 = \text{var}(N_{t+1} \mid N_t)/N_t^2$ (Lande et al. 2003) influences the long-run growth rate s, which can be approximated by

$$s \approx \ln \lambda - \sigma_e^2/(2\lambda^2), \qquad (7.1)$$

where $r = \ln \lambda$ is the deterministic growth rate defined by $\lambda = \text{E}(N_t / N_{t-1})$.

In age-structured populations, as in the case without any age-dependency, there is an effect of approximately $-\sigma_e^2/(2\lambda^2)$ of the environmental variance in the age-structured dynamics. In addition, there is an effect due to temporal autocorrelations and cross-correlations among elements in the projection matrix (Engen et al. 2005, 2007; Doak et al. 2005; Morris et al. 2011; Sæther and Engen 2010a). This term is usually negative and may be smaller or larger in magnitude than σ_e^2 depending on the mean projection matrix and the strength of temporal autocorrelations. Tuljapurkar and Haridas (2006) found that for large environmental autocorrelation it may even be substantially larger in magnitude than σ_e^2, hence reducing the long-run growth rate s.

Engen et al. (2013) proposed a method to estimate the influence of the environmental autocorrelation on the dynamics of age-structured populations, ignoring density-dependence. This approach is based on Fisher's (1930) concept of the total reproductive value of the population (Engen et al. 2007, 2009), which has the useful property that even though the total population size may be substantially autocorrelated, because of stochastic fluctuations in the environment, the total reproductive value still does not show any autocorrelation. If temporal autocorrelation in total reproductive value is present, then we know that this only results from environmental autocorrelation. We can use this to estimate the long-run growth rate under temporally correlated environmental noise and the component due to this autocorrelation, provided that data on individual contributions to future generations are available.

Let us consider the additive component τ in the long-run growth rate generated by temporal autocorrelation in the projection matrices $\mathbf{L}$ in complex ways. It can generally be described as the difference between the expected multiplicative growth rate and the multiplicative growth rate λ for the corresponding model in the average environment. It can be expressed by the covariances between mean individual reproductive values for each age class (Engen et al. 2009) with a stochastic component depending solely on the noise matrix of the actual year, and the size of the same age class, which has a stochastic component depending only on noise matrices in previous years. Accordingly, if the noise through some time has contributed to build a strong age class and this age class also strongly contributes to future population sizes relative to its own size by a large mean individual contribution to future generations, then this also provides a positive contribution to τ. In other words, if an age class that is large in a given year also tends to contribute more recruits to future generations that year than the temporal average of individuals in that age class, then it contributes positively to τ. In contrast, if this age class tends to provide smaller contributions than the average when the age class is strong, this age-specific contribution to τ is negative.

So far, we are unaware of any study that has used this approach to explore the effects of autocorrelations in the environment on avian population dynamics. In an analysis of four mammalian populations, Engen et al. (2013) showed small non-significant effects of τ, being positive in two species and negative in two others. This supports similar conclusions by Morris et al. (2011) for primates. Accordingly, using another approach, van de Pol et al. (2011) found a relatively small effect of the colour of environmental noise on the dynamics of a declining Eurasian oystercatcher (*Haematopus ostralegus*) population, mainly because demographic rates depend nonlinearly on environmental variables or on several variables that filtered out the effects of noise colour.

7.4.2 Predictions in space

A major impact of expected changes in climate is that it will alter the distribution of abundances in space of many species (Parmesan 2006) including birds (Illan et al. 2014). For many scientists and environmental managers predicting such effects is a major challenge for ecological research, due to its implications for biodiversity conservation (Sutherland et al. 2006). The most common approach to develop such predictions is so-called climate envelope models (Berry et al. 2002; Huntley et al. 2004, Chapter 13) in which the current distribution of abundances is correlated with different climate variables, and then these relationships are used to predict future abundances in space based on projected changes in key climate variables. However, the use of this type of modelling to predict the impact of climate change has two major shortcomings. First, it fails to include the effects of non-climate factors that can be important contributors to spatial variation in the abundance of species (Mustin et al. 2007; Oliver and Roy 2015). Secondly, this approach ignores the effects of local dynamics. It has been well known since the work of Moran (1953) that spatial correlation in climate may induce spatial correlations in population fluctuations, but it is less well known that the spatial distribution of abundances will also depend on local population processes. For example, if density dependence is locally strong, then the degree of spatial synchrony in population fluctuations will be less than for weakly density-regulated populations even under the same environmental conditions because of increased contribution to population synchrony from dispersal (Lande et al. 1999; Engen and Sæther 2005). Similarly, it is not necessarily the climate variables that most strongly affect the local population fluctuations that contribute the most to the degree of spatial synchrony in the population dynamics (Sæther et al. 2007a). Thus, although climate envelope models may capture some broad overall patterns (Guisan and Thuiller 2005), their inherent weaknesses make them difficult to use for projecting changes in the spatial distribution of abundances as a consequence of changes in climate.

7.5 Examples

7.5.1 The white-throated dipper in southern Norway

As an example of the use of PPI, we explored how expected changes in climate are likely to affect the future abundance of white-throated dippers (*Cinclus cinclus*) in the river system of Lyngdalselva in southern Norway (58°08′–58°40′N, 6°56′–7°20′E). The white-throated dipper is a 50–70 g small passerine bird widely distributed at northern latitudes. This short-lived species depends on open water for foraging and running water for nesting. The amount of ice during the winter thus influences the availability of feeding habitats and affects the winter survival (Loison et al. 2002). As a consequence, an increase in winter temperature is expected to result in increased carrying capacity (Sæther et al. 2000b). Our population in southern Norway has been monitored since 1978. All breeding sites were visited during the nest building period to identify breeding pairs and record occupied nests. During visits in the breeding season, fledglings were ringed, ringed mothers identified, and unringed mothers given a ring to allow future identifications (Gamelon et al. 2017).

To predict how climate change is expected to affect future population sizes, we built an age-structured model accounting for the combined influence of density dependence and climate variation. (Figure 7.2). An important aspect of this model was also to consider the influence of immigrants on population dynamics. Our estimate of population size (at year *t*) included the number of locally born females and immigrants from surrounding areas. As climate variables that can potentially affect survival, recruitment, and more generally population size, we considered the North Atlantic Oscillation index during winter months (December–February) (*NAO*) (Hurrell 1995), local mean winter temperature (*temp*), and local mean winter precipitation (*prec*) (Figure 7.2).

We found that warmer winters had a positive effect on survival and recruitment rates for all age classes. The effects of winter temperature on recruitment were particularly strong for first-year survival as winter is a critical season for survival of the white-throated dipper (Loison et al. 2002). Importantly, we found significant interactions between winter temperatures and density, indicating particularly reduced survival and recruitment rates in cold winters at high population sizes. This suggests an indirect effect of climate through resource limitation or social exclusion from access to territories during severe winters. There was also a complex interaction between mean winter temperature and the number of local females in explaining

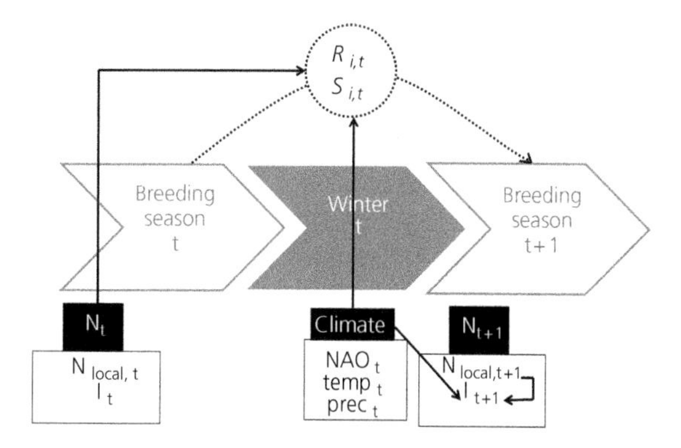

Figure 7.2 Climate-density-dependent population model for the white-throated dipper population in southern Norway (Gamelon et al. 2017). Recruitment rates $R_{i,t}$ and survival rates $S_{i,t}$ of age class *i* at year *t* may be influenced by density (N_t) (i.e., sum of local ($N_{local,t}$) and immigrant breeding females (I_t)) and by winter conditions (i.e., NAO_t, mean winter temperature $temp_t$, mean winter precipitation $prec_t$). I_{t+1} may be influenced by $N_{local,t+1}$ and by winter conditions.

the number of immigrant females entering the local population. In particular, milder winters were associated with higher number of immigrants than cold winters, but only when the number of local females was low. Mean winter temperature was consequently the main environmental factor affecting demographic rates.

These results illustrate that the dynamical impact of variation in climate depends on population density. This implies that the strength of density dependence must also be estimated and included into models developed to predict the effects of climate change. Under global warming, the probability of an increase in numbers at small population sizes (e.g. $N = 20$) over a period of 40 years is much larger because of these interactions between density and temperature, compared to the case where density dependence and environmental stochasticity are considered as independent effects. Hence, predicted future population size becomes a complex function including the effects of environmental stochasticity and deterministic processes caused by density dependence, as well as their interaction. Fortunately, our model was able to reconstruct the past fluctuations in population size very closely (Gamelon et al. 2017).

From the Large Ensemble Community Project of the Community Earth System Model (CESM), we obtained daily average surface temperatures at the closest model grid point of the study site (58.9°N, 7.5°E) for 40 ensemble members (Deser et al., submitted; for a full description of the climate projections, see Kay et al. 2015) from 1921 to 2050. For these 40 ensemble simulations, we calculated the expected mean temperature for each winter (December–February) from 2013 onwards and simulated stochastic trajectories of future temperatures taking observed variance and autocorrelation of observed past temperatures into account. Using these predicted winter temperatures for the future (2013–2050), we then projected the population size until 2050 starting with the age-specific densities in 2013 and, incorporating the interaction between population size and winter conditions.

Our results indicate that warmer winters will result in increased size of this dipper population mainly due to increased survival of local birds (Figure 7.3). In fact, in 30 years we predict that

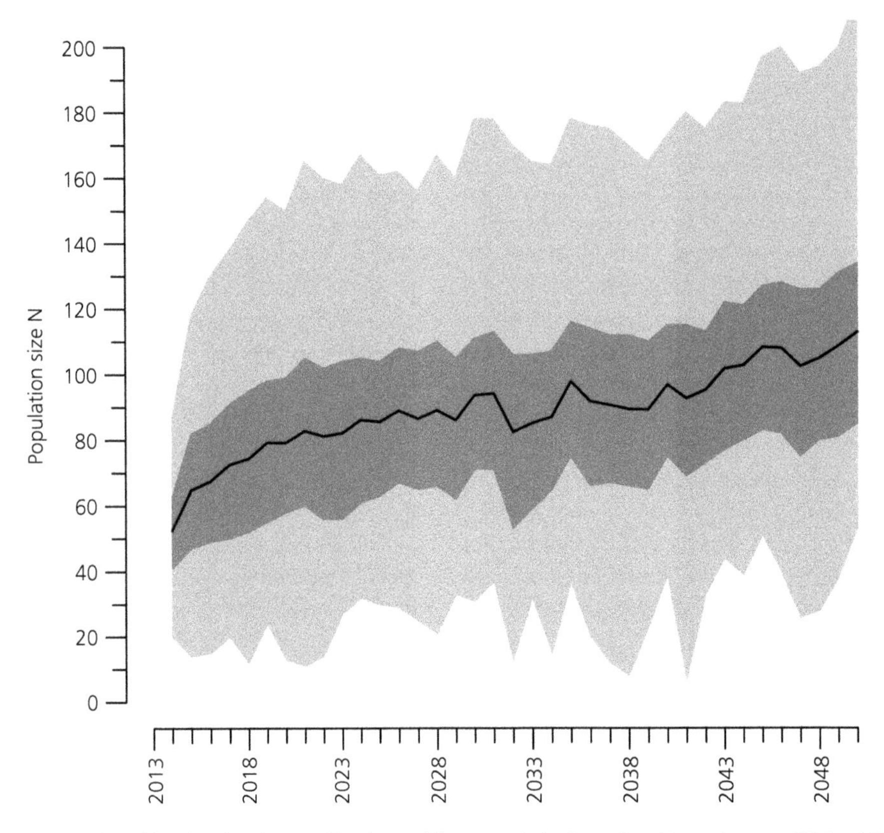

Figure 7.3 Predicted number of breeding females in a white-throated dipper population in southern Norway between 2014 and 2050 according to Gamelon et al. (2017), based on local climate scenarios. Shaded dark grey corresponds to 50% and light grey corresponds to 95% prediction intervals associated with the predicted densities.

population size will be approximately 35 per cent larger than the current population size. However, the width of the PPI was large, mainly caused by the strong influences of environmental stochasticity on the dynamics of this population. Still, the range of the interval becomes relatively constant in width after a short time, which is caused by the influence of density dependence. These density-dependent effects also reduce the number of immigrants joining the population (Gamelon et al. 2017). This illustrates that a full understanding of the demographic mechanisms under climate change may not be achieved using a more typical population model that does not include density-dependent effects of climate. Indeed, models that do not include the interaction between winter conditions and density tend to overestimate the number of immigrants joining the local population at high densities during mild winters and underestimate survival and recruitment rates of local birds. This occurs because at low densities survival and recruitment rates only slightly depend on winter temperatures, whereas at high density they become strongly dependent on winter conditions and are less buffered against poor environmental conditions (cold winters).

These findings highlight that the local population will likely be less buffered against occasional cold winters in coming years. This illustrates that predicting the putative effects of global warming on natural populations even for bird species with quite simple life histories such as the dipper require a quite detailed population model (Figure 7.2).

7.5.2 A Dutch oystercatcher population

The Eurasian oystercatcher is a long-lived monogamous shorebird that occurs year-round along the coasts in northwestern Europe. This species was studied from 1983 to 2007 on the Dutch island of Schiermonnikoog (53°29′N, 6°14′W), where the breeding population has been declining (van de Pol et al. 2010, 2011). Oystercatchers are strictly territorial. The population dynamics of oystercatchers is sensitive to climate variation because harsh, cold winters negatively affect survival, whereas fecundity is greater following cold winters, which promote egg-laying by one of the preferred prey species (*Nereis diversicolor*) (van de Pol et al. 2010). Thus, winter warming is expected to affect different demographic rates in opposite ways.

To assess the influence of climate change on the future growth of this oystercatcher population, van de Pol et al. (2010, 2011) built a stochastic stage-structured model that included demographic stochasticity, differences in habitat quality that affected the density regulation, variation in key climate variables, and residual environmental variance. The analyses of this model showed that the expected increase in winter temperatures in this area is likely to have a positive effect on growth rate of this population (Figure 7.4). An increase in mean winter temperature of 1°C is likely to stop the decline in size of this population. This occurs because the positive effects of an increase in temperature on adult survival exceed the negative effects on population growth of a decrease in fecundity due to reduced food availability for the chicks. Furthermore, in this population the influence of a change in the mean climate on the population growth rate was much larger than the effects of a change in the climate variability (van de Pol et al. 2010).

7.5.3 Seabirds in the southern Atlantic Ocean

Analyses of population dynamics and demography of seabirds in the Southern Ocean have revealed large influences of expected changes in climate on the future trajectories of local populations (Jenouvrier et al. 2012; Barbraud et al. 2012), as well as shifts in the spatial distribution of breeding pairs across large geographical areas (Ainley et al. 2010). In many cases, these effects are related to changes in the sea ice extent and coverage. Using stochastic stage-structured models parameterized from a unique set

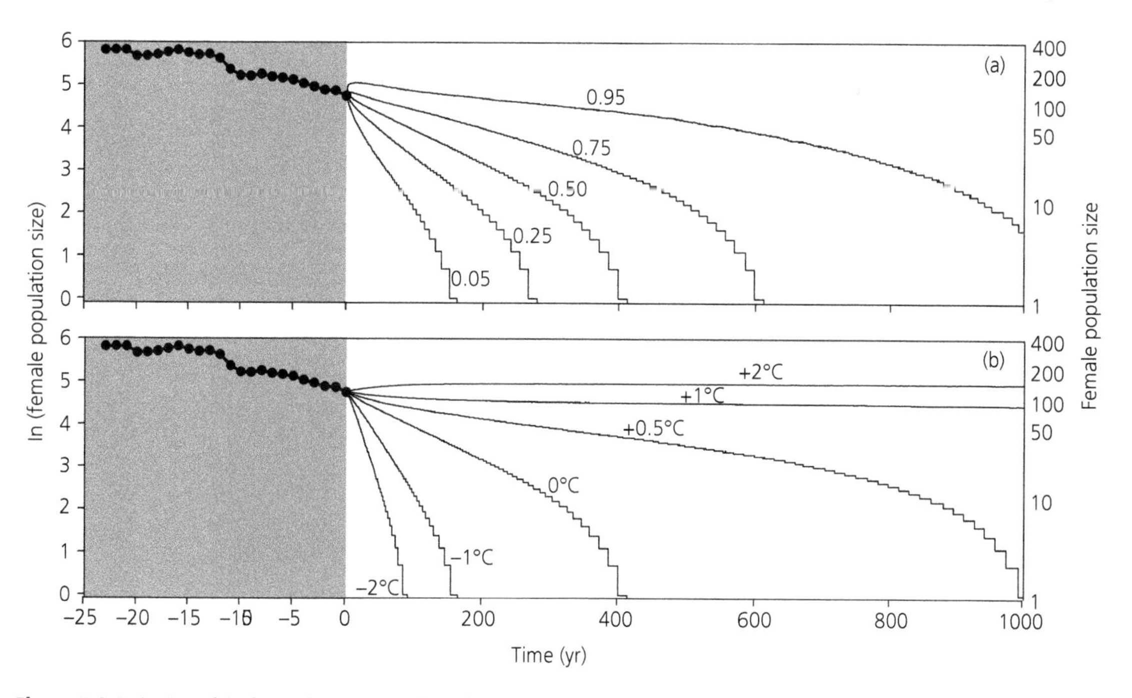

Figure 7.4 Projections of the future changes in numbers of a declining Dutch Oystercatcher population (van de Pol et al. 2010). The dark grey areas indicate the period in which time series of population counts are available. In the light grey areas, in (a) the 5%, 25%, 50%, 75%, and 90% quantile of the population size at different future time steps are shown and in (b) the 50% quantile for the projected numbers of breeding pairs is plotted for different level of change in mean winter temperature while keeping the standard deviation constant.

of long-term individual-based demographic data, the effects of variation in sea-ice conditions on the population growth rate and the consequences of expected changes in climate can be predicted even at a continental scale. In some cases, this is based on the assumption that the climate models most accurately reproducing the past sea-ice variation are also those that will provide the most reliable scenarios for the future (Jenouvrier et al. 2012).

Analyses of the population dynamics of emperor penguin (*Aptenodytes forsteri*) (Figure 7.5) and Adelie penguin (*Pygoscelis adeliae*) (Figure 7.6) breeding on the Antarctic continent have revealed that the population growth rate is strongly dependent on sea ice conditions. For example, the growth rate of Adelie penguin colonies was positively associated with an increase in the winter maximum concentration of

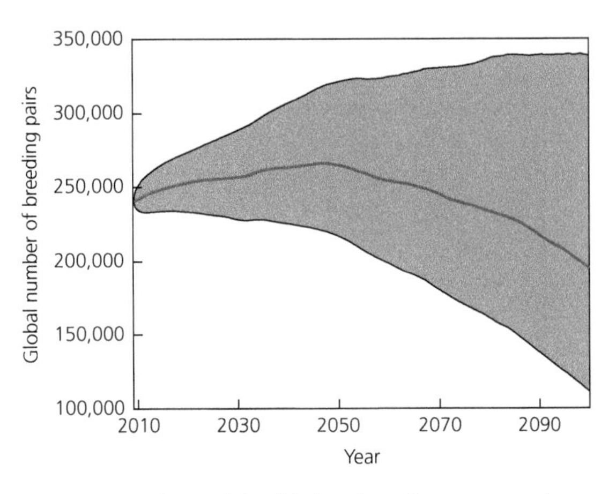

Figure 7.5 Prediction of the global numbers of emperor penguins as a consequence of projected changes in sea-ice concentration using atmosphere–ocean generalized circulation models with a local selection approach (Jenouvrier et al. 2014).

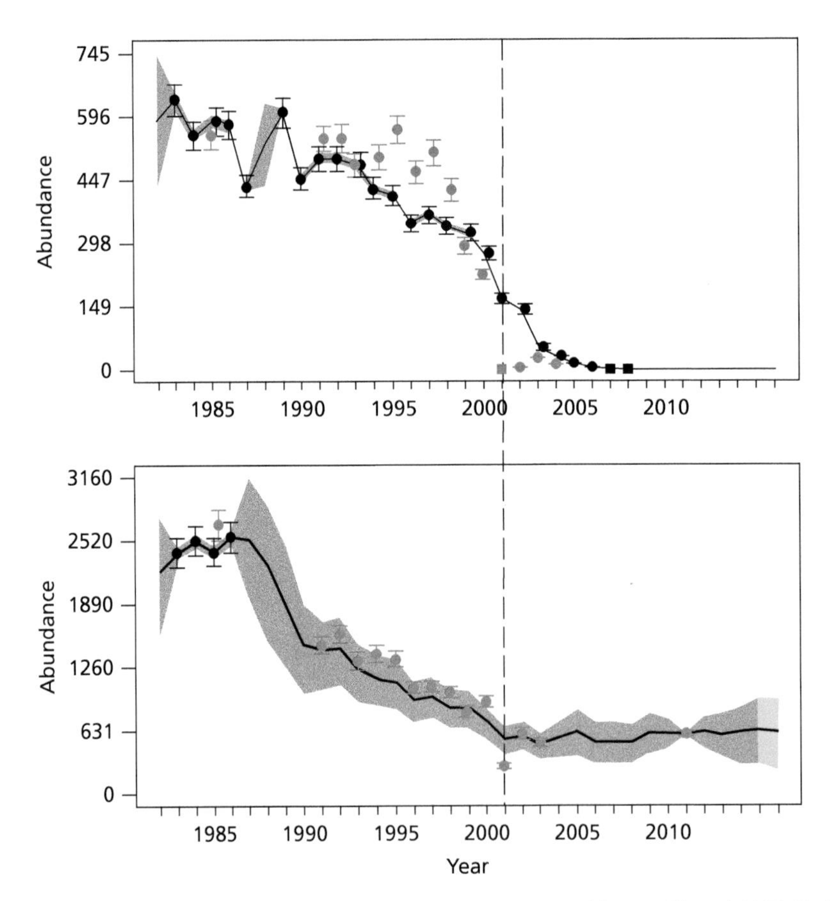

Figure 7.6 Predicted changes in the number of Adélie penguins at two localities in Antarctica (Che-Castaldo et al. 2017). The black lines represent the median of the posterior density credible interval, whereas the grey (1982–2005) and the green (2006) shaded areas indicate the 75% highest posterior credible intervals. Black and red circles show nest and chick counts, respectively, whereas squares indicate years when no nests (black) or chicks (red) were found. All abundances for 2016 represent population forecasts from the stochastic projection model.

sea ice (Che-Castaldo et al. 2017). An important contribution of this study is that it clearly demonstrates that not only the mean but also the temporal fluctuations in winter sea ice conditions provide a substantial contribution to the long-run population growth rate (Eq. 1). A similar effect of variation in sea ice conditions was also documented for the emperor penguin (Jenouvrier et al. 2014), indicating large effects of global warming on the dynamics of these two species (Jenouvrier et al. 2009, 2012; Che-Castaldo et al. 2017).

Detailed demographic modelling of effects on the demography and population dynamics of variation in sea surface temperature in two Procellariiform species revealed in a similar way that not only changes in mean climate but also the variance can have a huge impact on the population growth rate. In the black-browed albatross (*Thalassarche melanophris*) studied at Kerguelen Island in the southern Atlantic, Pardo et al. (2017) found that around the historical mean value of the sea surface temperature a change of the standard deviation of this variable had a huge impact on the stochastic growth rate of the population. Similarly, in southern fulmars (*Fulmarus glacialoides*), foraging trips of parents feeding young became longer under unfavourable sea ice conditions, resulting in less food provided to the chicks (Jenouvrier et al. 2015). Consequently, fewer chicks fledged, which reduced the stochastic population growth rate. Thus, the long-run growth of this population becomes a function of the frequency of bad years with far-away ice affecting both the deterministic population growth rate in the average environment as well as the environmental variance (Figure 7.7).

7.6 Discussion

A large body of studies, mainly based on time series analysis of population estimates, has shown that climate variation strongly influences fluctuations in size of avian populations (Martay et al. 2017;

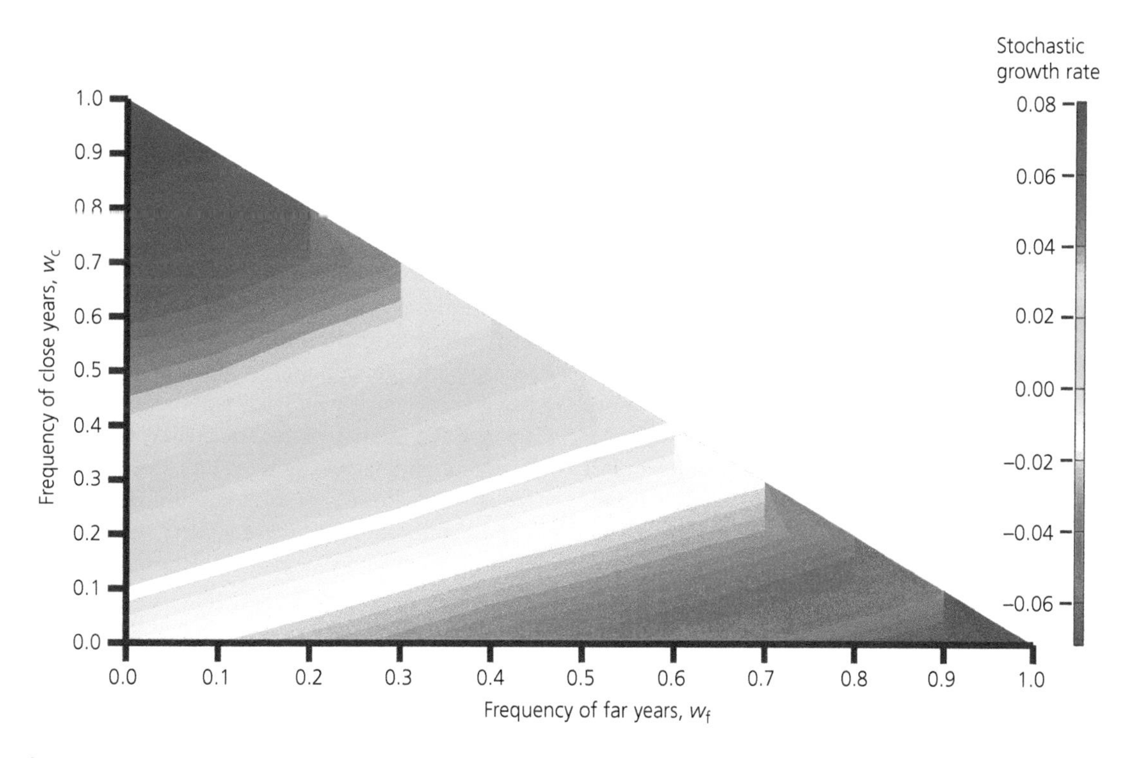

Figure 7.7 Stochastic population growth rate s of the southern fulmar as a function of the frequency of extreme unfavourable far-ice conditions w_f and favourable years with ice close to the colony w_c (Jenouvrier et al. 2015). The white contour indicates a stable population ($s = 0$). Warm colours (red, orange, yellow) show the parameter space giving a declining population whereas population increase is expected in areas with greenish and bluish colours.

Pearce-Higgins et al. 2015; Sæther et al. 2004). Here we have shown that it is difficult to use such retrospective analyses for predictions of future responses to climate change. Our survey of examples of such forecast studies show that reliable predictions necessarily contain a high level of uncertainty (Figures 7.3 and 7.4). A major reason for this is that avian population dynamics are strongly influenced by environmental stochasticity, which is for most species, irrespective of their life history, the most important driver of fluctuations in population size (Sæther et al. 2016). Furthermore, most available time series are in a statistical sense short, thus often resulting in large uncertainties in parameter estimates. Credible population predictions must therefore assess the effects of such uncertainties as well as biases in population estimates. In addition, the deterministic processes (e.g. caused by density dependence) affecting the population dynamics must also be included. For example, great tits (*Parus major*) in the Netherlands had lower reproductive success when their breeding was mistimed relative to the maximum food supply available to the nestlings; however, this had a small effect on the population growth rate (Reed et al. 2013c) because there was higher juvenile survival when density was lower (Reed et al. 2013a). These insights can only be achieved by applying stochastic stage- or age-structured models (Figure 7.2) even for bird species with quite simple life histories. Integrated Population Models (Kéry and Schaub 2012) provide a promising Bayesian framework to estimate the parameters including their uncertainties used in construction of such models.

Although predictions of climate-induced responses in population sizes at a specific time in the future often become quite uncertain, the application of PPI can still provide important insights about future changes in abundance (Figure 7.1). This is facilitated by the interpretation of PPI, which is the same as for a confidence interval, except that we draw inference about a stochastic quantity (e.g. population size) rather than a parameter. Uncertainty in the parameters affects the confidence we have in the population predictions. The lack of consideration of uncertainties and biases in population parameters has often made verification of the predictions from population viability models difficult (Ludwig 1999; Brook et al. 2000). By assuming no uncertainty in parameter estimates, this effect can be quantitatively assessed by examining the effects on the width of the PPI, which in many cases can be quite large (Sæther et al. 2000a; Engen et al. 2001; Sæther et al. 2009). Furthermore, the application of PPI can also be used to predict the probability that the population will reach a certain size (e.g., extinction at $N = 1$) within a given time period (Figure 7.4). Thus, the application of PPI enables us to characterize how patterns in the population dynamics are likely to be affected by expected changes in climate. The importance of such a process-based approach is well illustrated by the analyses of Jenouvrier et al. (2014) who explored the impact of expected climate changes on the future global abundance of emperor penguins, which first showed an increasing phase followed after a few years by a relatively steep decline (Figure 7.5).

The advantage of applying demographic models when performing ecological forecasting is that mechanisms for the impact of climate change on population growth must be explicitly specified (Frederiksen et al. 2008). Some examples presented in this chapter show that these effects can be quite complicated. For example, seabird population studies in the Southern Ocean reveal complex age-dependent effects in which a climate variable can have a positive effect on the population growth at one stage whereas it can have a negative contribution at other life history stages. For example, in the wandering albatross (*Diomedea exulans*) increasing sea surface temperature had a positive effect on recruitment age, but affected juvenile survival rate negatively, which could perhaps be explained by age-specific spatial segregation (Fay et al. 2017). A similar effect was also found for the influence of winter temperature on demography of the Dutch population of oystercatchers (van de Pol et al. 2010, 2011). Another demographic influence is that climate change at an early stage can translate into long-lasting effects throughout several life history stages, generating consistent heterogeneity among individuals in demography (Jenouvrier et al. 2015; Herfindal et al. 2015; van de Pol et al. 2010). Although these complex relationships can be challenging to parameterize, the identification of age-classes causing particularly large contributions to the population growth rate (Sæther et al. 2013) may help to identify critical life history stages that should receive high attention when constructing population models.

A general pattern that appears from forecast analyses of the influence of expected changes in climate on avian population dynamics is that not only do changes in the mean matter, but also alterations of the variance are important to include in predictive models. This occurs because environmental stochasticity tends to reduce the long-run growth rate of populations at a logarithmic scale (Eq. 7.1), assuming density-independent (Lande et al. 2003) and linear (Lawson et al. 2015) effects of the environmental variable on population growth. This influence of environmental variability on the long-run population growth rate (Eq. 7.1) represents a particular challenge for developing predictions because climate projections typically only involve deterministic projections of climate models based on given sets of parameter values. Consequently, it becomes extremely difficult to assess the variability in climate changes and thus how to make predictions about the effects of extreme climate events (van de Pol et al. 2017). An important challenge is, therefore, to improve the spatio-temporal downscaling of process-based climate models.

Another challenge for developing reliable population predictions for birds is the ability for most of them to fly. In principle, this ability should simplify relocations in space when facing more unfavourable environmental conditions. Accordingly, many recent population studies of bird have revealed that immigrants from surrounding areas constitute a relatively high proportion of the birds that breed locally (Gamelon et al. 2017; Grøtan et al. 2009). Thus, reliable population predictions require that spatial processes affecting the local population dynamics are properly modelled.

Although several challenges remain for obtaining reliable population predictions, a clear pattern appears from the studies so far: the expected changes in climate according to projections from IPCC-class climate models are likely to have a large influence on the future abundance of many bird species (Jenouvrier 2013; Figures 7.3 and 7.4). This influence can be either positive or negative. However, the projections conducted until now with all the methodological limitations strongly indicate that the magnitude of the population responses from the range of variation in future climate, independent of emissions scenario, will be large. It is therefore an urgent need for the development of robust predictive models that can be used as tool for developing population projections for a variety of species based on variables that can be easily measured such as life history characteristics. Birds will definitely be flagship taxa to reach such a goal.

Acknowledgments

This study was funded by the Research Council of Norway (SFF-III, project no. 223-257).

References

Abadi, F., Gimenez, O., Arlettaz, R., and Schaub, M. (2010). An assessment of integrated population models: bias, accuracy, and violation of the assumption of independence. *Ecology*, 91, 7–14.

Ainley, D., Russell, J., Jenouvrier, S., Woehler, E., Lyver, P.O., Fraser, W.R., and Kooyman, G.L. (2010). Antarctic penguin response to habitat change as Earth's troposphere reaches 2 degrees C above preindustrial levels. *Ecological Monographs*, 80, 49–66.

Bailey, L.D., and van de Pol, M. (2016). climwin: An R toolbox for climate window analysis. *PLoS One*, 11, e0167980.

Barbraud, C., Rolland, V., Jenouvrier, S., Nevoux, M., Delord, K., and Weimerskirch, H. (2012). Effects of climate change and fisheries bycatch on Southern Ocean seabirds: a review. *Marine Ecology Progress Series*, 454, 285–307.

Berry, P.M., Dawson, T.P., Harrison, P.A., and Pearson, R.G. (2002). Modelling potential impacts of climate change on the bioclimatic envelope of species in Britain and Ireland. *Global Ecology and Biogeography*, 11, 453–462.

Bolker, B.M., Gardner, B., Maunder, M., Berg, C.W., Brooks, M., Comita, L., Crone, E., Cubaynes, S., Davies, T., de Valpine, P., Ford, J., Gimenez, O., Kéry, M., Kim, E.J., Lennert-Cody, C., Magnusson, A., Martell, S., Nash, J., Nielsen, A., Regetz, J., Skaug, H., and Zipkin, E. (2013). Strategies for fitting nonlinear ecological models in R, AD Model Builder, and BUGS. *Methods in Ecology and Evolution*, 4, 501–12.

Brook, B.W., O'Grady, J.J., Chapman, A.P., Burgman, M.A., Akcakaya, H.R., and Frankham, R. (2000). Predictive accuracy of population viability analysis in conservation biology. *Nature*, 404, 385–7.

Che-Castaldo, C., Jenouvrier, S., Youngflesh, C., Shoemaker, K.T., Humphries, G., McDowall, P., Landrum, L., Holland, M.M., Li, Y., Ji, R.B., and Lynch, H.J. (2017). Pan-Antarctic analysis aggregating spatial estimates of Adelie penguin abundance reveals robust dynamics despite stochastic noise. *Nature Communications*, 8. doi: 10.1038/s41467-017-00890-0

Clark, F., Brook, B.W., Delean, S., Akcakaya, H.R., and Bradshaw, C.J.A. (2010). The theta-logistic is unreliable for modelling most census data. *Methods in Ecology and Evolution*, 1, 253–62.

Clutton-Brock, T., and Sheldon, B.C. (2010). Individuals and populations: the role of long-term, individual-based studies of animals in ecology and evolutionary biology. *Trends in Ecology & Evolution*, 25, 562–573.

Dennis, B., Munholland, P.L., and Scott, J. M. (1991). Estimation of growth and extinction parameters for endangered species. *Ecological Monographs*, 61, 115–43.

Doak, D.F., Morris, W.F., Pfister, C., Kendall, B.E., and Bruna, E.M. (2005). Correctly estimating how environmental stochasticity influences fitness and population growth. *American Naturalist*, 166, E14–E21.

Engen, S., Lande, R., Sæther, B.-E., and Dobson, F.S. (2009). Reproductive value and the stochastic demography of age-structured populations. *American Naturalist*, 174, 795–804.

Engen, S., Lande, R., Sæther, B.-E., and Festa-Bianchet, M. (2007). Using reproductive value to estimate key parameters in density-independent age-structured populations. *Journal of Theoretical Biology*, 244, 208–317.

Engen, S., Lande, R., Sæther, B.-E., and Weimerskirch, H. (2005). Extinction in relation to demographic and environmental stochasticity in age-structured models. *Mathematical Biosciences*, 195, 210–227.

Engen, S., and Sæther, B.-E. (2005). Generalizations of the Moran effect explaining spatial synchrony in population fluctuations. *American Naturalist*, 166, 603–12.

Engen, S., Sæther, B.-E., Armitage, K.B., Blumstein, D.T., Clutton-Brock, T.H., Dobson, F.S., Festa-Bianchet, M., Oli, M.K., and Ozgul, A. (2013). Estimating the effect of temporally autocorrelated environments on the demography of density-independent age-structured populations. *Methods in Ecology and Evolution*, 4, 573–84.

Engen, S., Sæther, B.-E., and Møller, A. P. (2001). Stochastic population dynamics and time to extinction of a declining population of barn swallows. *Journal of Animal Ecology*, 70, 789–97.

Fay, R., Barbraud, C., Delord, K., and Weimerskirch, H. (2017). Contrasting effects of climate and population density over time and life stages in a long-lived seabird. *Functional Ecology*, 31, 1275–84.

Fisher, R.A. (1930). *The genetical theory of natural selection*. Clarendon Press, Oxford, UK.

Frederiksen, M., Daunt, F., Harris, M.P., and Wanless, S. (2008). The demographic impact of extreme events: stochastic weather drives survival and population dynamics in a long-lived seabird. *Journal of Animal Ecology*, 77, 1020–9.

Gamelon, M., Grøtan, V., Nilsson, A.K., Engen, S., Hurrell, J.W., Jerstad, K., Phillips, A.S., Røstad, O.W., Slagsvold, T., Walseng, B., Stenseth, N.C., and Sæther, B.-E. (2017). Interactions between demography and environmental effects are important determinants of population dynamics. *Science Advances*, 3, e1602298.

Gilpin, M.E., Case, T.J., and Ayala, F.J. (1976). Θ-selection. *Mathematical Biosciences*, 32, 131–9.

Grosbois, V., Gimenez, O., Gaillard, J.M., Pradel, R., Barbraud, C., Clobert, J., Møller, A.P., and Weimerskirch, H. (2008). Assessing the impact of climate variation on survival in vertebrate populations. *Biological Reviews*, 83, 357–99.

Grøtan, V., Sæther, B.-E., Engen, S., Van Balen, J.H., Perdeck, A.C., and Visser, M.E. (2009). Spatial and temporal variation in the relative contribution of density dependence, climate variation, and migration to fluctuations in the size of great tit populations. *Journal of Animal Ecology*, 78, 447–59.

Guisan, A., and Thuiller, W. (2005). Predicting species distribution: offering more than simple habitat models. *Ecology Letters*, 8, 993–1009.

Hallett, T.B., Coulson, T., Pilkington, J.G., Clutton-Brock, T.H., Pemberton, J.M., and Grenfell, B.T. (2004). Why large-scale climate indices seem to predict ecological processes better than local weather. *Nature*, 430, 71–5.

Heino, M., Ripa, J., and Kaitala, V. (2000). Extinction risk under coloured environmental noise. *Ecography*, 23, 177–84.

Herfindal, I., van de Pol, M., Nielsen, J.T., Saether, B.-E., and Møller, A.P. (2015). Climatic conditions cause complex patterns of covariation between demographic traits in a long-lived raptor. *Journal of Animal Ecology*, 84, 702–11.

Huntley, B., Green, R.E., Collingham, Y.C., Hill, J.K., Willis, S.G., Bartlein, P.J., Cramer, W., Hagemeijer, W.J.M., and Thomas, C.J. (2004). The performance of models relating species' geographical distributions to climate independent of trophic level. *Ecology Letters*, 7, 417–26.

Hurrell, J.W. (1995). Decadal trends in the North Atlantic Oscillation: Regional temperatures and precipitation. *Science*, 269, 676–9.

Illan, J.G., Thomas, C.D., Jones, J.A., Wong, W.K., Shirley, S.M., and Betts, M.G. (2014). Precipitation and winter temperature predict long-term range-scale abundance changes in Western North American birds. *Global Change Biology*, 20, 3351–64.

Jenouvrier, S. (2013). Impacts of climate change on avian populations. *Global Change Biology*, 19, 2036–57.

Jenouvrier, S., Barbraud, C., and Weimerskirch, H. (2005). Long-term contrasted responses to climate of two Antarctic seabird species. *Ecology*, 86, 2889–903.

Jenouvrier, S., Caswell, H., Barbraud, C., Holland, M., Stroeve, J., and Weimerskirch, H. (2009). Demographic models and IPCC climate projections predict the decline of an emperor penguin population. *Proceedings of the National Academy of Sciences of the United States of America*, 106, 1844–7.

Jenouvrier, S., Holland, M., Stroeve, J., Barbraud, C., Weimerskirch, H., Serreze, M., and Caswell, H. (2012). Effects of climate change on an emperor penguin population: analysis of coupled demographic and climate models. *Global Change Biology*, 18, 2756–70.

Jenouvrier, S., Holland, M., Stroeve, J., Serreze, M., Barbraud, C., Weimerskirch, H., and Caswell, H. (2014). Projected continent-wide declines of the emperor penguin under climate change. *Nature Climate Change*, 4, 715–18.

Jenouvrier, S., Peron, C., and Weimerskirch, H. (2015). Extreme climate events and individual heterogeneity shape life-history traits and population dynamics. *Ecological Monographs*, 85, 605–24.

Kay, J.E., Deser, C., Phillips, A., Mai, A., Hannay, C., Strand, G., Arblaster, J.M., Bates, S.C., Danabasoglu, G., Edwards, J., Holland, M., Kushner, P., Lamarque, J.-F., Lawrence, D., Lindsay, K., Middleton, A., Munoz, E., Neale, R., Oleson, K., Polvani, L., and Vertenstein, M. (2015). The Community Earth System Model (CESM) large ensemble project: A community resource for studying climate change in the presence of internal climate variability. *American Meteorological Society*, August 2015, 1333–49. doi:10.1175/BAMS-D-13-00255.1

Kéry, M., and Schaub, M. (2012). *Bayesian population analysis using WinBUGS*. Academic Press, Amsterdam.

Knape, J., and de Valpine, P. (2011). Effects of weather and climate on the dynamics of animal population time series. *Proceedings of the Royal Society B: Biological Sciences*, 278, 985–92.

Lande, R., Engen, S., and Sæther, B.-E. (1999). Spatial scale of population synchrony: environmental correlation versus dispersal and density regulation. *American Naturalist*, 154, 271–81.

Lande, R., Engen, S., and Sæther, B.-E. (2003). *Stochastic population dynamics in ecology and conservation*. Oxford University Press, Oxford, UK.

Lawson, C.R., Vindenes, Y., Bailey, L., and van de Pol, M. (2015). Environmental variation and population responses to global change. *Ecology Letters*, 18, 724–36.

Loison, A., Sæther, B.-E., Jerstad, K., and Røstad, O.W. (2002). Disentangling the sources of variation in the survival of the European dipper. *Journal of Applied Statistics*, 29, 289–304.

Ludwig, D. (1999). Is it meaningful to estimate a probability of extinction? *Ecology*, 80, 298–310.

Martay, B., Brewer, M.J., Elston, D.A., Bell, J.R., Harrington, R., Brereton, T.M., Barlow, K.E., Botham, M.S., and Pearce-Higgins, J.W. (2017). Impacts of climate change on national biodiversity population trends. *Ecography*, 40, 1139–51.

Moran, P.A.P. (1953). The statistical analysis of the Canadian lynx cycle. II. Synchronization and meteorology. *Australian Journal of Zoology*, 1, 291–8.

Morris, W.F., Altmann, J., Brockman, D.K., Cords, M., Fedigan, L.M., Pusey, A.E., Stoinski, T.S., Bronikowski, A.M., Alberts, S.C., and Strier, K.B. (2011). Low demographic variability in wild primate populations: fitness impacts of variation, covariation, and serial correlation in vital rates. *American Naturalist*, 177, E14–E28.

Mouquet, N., Lagadeuc, Y., Devictor, V., Doyen, L., Duputie, A., Eveillard, D., Faure, D., Garnier, E., Gimenez, O., Huneman, P., Jabot, F., Jarne, P., Joly, D., Julliard, R., Kefi, S., Kergoat, G.J., Lavorel, S., Le Gall, L., Meslin, L., Morand, S., Morin, X., Morlon, H., Pinay, G., Pradel, R., Schurr, F.M., Thuiller, W., and Loreau, M. (2015). Review: Predictive ecology in a changing world. *Journal of Applied Ecology*, 52, 1293–1310.

Mustin, K., Sutherland, W.J., and Gill, J.A. (2007). The complexity of predicting climate-induced ecological impacts. *Climate Research*, 35, 165–75.

Oliver, T.H., and Roy, D.B. (2015). The pitfalls of ecological forecasting. *Biological Journal of the Linnean Society*, 115, 767–78.

Pardo, D., Jenouvrier, S., Weimerskirch, H., and Barbraud, C. (2017). Effect of extreme sea surface temperature events on the demography of an age-structured albatross population. *Philosophical Transactions of the Royal Society of London, Series B: Biological Sciences*, 372.

Parmesan, C. (2006). Ecological and evolutionary responses to recent climate change. *Annual Review of Ecology, Evolution, and Systematics*, 37, 637–69.

Pearce-Higgins, J.W., Eglington, S.M., Martay, B., and Chamberlain, D.E. (2015). Drivers of climate change impacts on bird communities. *Journal of Animal Ecology*, 84, 943–54.

Petchey, O.L., Pontarp, M., Massie, T.M., Kefi, S., Ozgul, A., Weilenmann, M., Palamara, G.M., Altermatt, F., Matthews, B., Levine, J.M., Childs, D.Z., McGill, B.J., Schaepman, M.E., Schmid, B., Spaak, P., Beckerman, A.P., Pennekamp, F., and Pearse, I.S. (2015). The ecological forecast horizon, and examples of its uses and determinants. *Ecology Letters*, 18, 597–611.

Reed, T.E., Grøtan, V., Jenouvrier, S., Sæther, B.-E., and Visser, M.E. (2013a). Population growth in a wild bird is buffered against phenological mismatch. *Science*, 340, 488–91.

Reed, T.E., Jenouvrier, S., and Visser, M.E. (2013b). Phenological mismatch strongly affects individual fitness but not population demography in a woodland passerine. *Journal of Animal Ecology*, 82, 131–44.

Reed, T.E., Gienapp, P., and Visser, M.E. (2013c). Density dependence and microevolution interactively determine effects of phenology mismatch on population dynamics. *Oikos*, 124, 81–91.

Solbu, E.B., Engen, S., and Diserud, O.H. (2013). Changing environments causing time delays in population dynamics. *Mathematical Biosciences*, 244, 213–23.

Solbu, E.B., Engen, S., and Diserud, O.H. (2015). Guidelines when estimating temporal changes in density dependent populations. *Ecological Modelling*, 313, 355–76.

Stenseth, N.C., and Mysterud, A. (2005). Weather packages: finding the right scale and composition of climate in ecology. *Journal of Animal Ecology*, 74, 1195–8.

Sutherland, W.J., Armstrong-Brown, S., Armsworth, P.R., Brereton, T., Brickland, J., Campbell, C.D., Chamberlain, D.E., Cooke, A.I., Dulvy, N.K., Dusic, N.R., Fitton, M., Freckleton, R.P., Godfray, H.C.J., Grout, N., Harvey, H.J., Hedley, C., Hopkins, J.J., Kift, N.B., Kirby, J., Kunin, W.E., Macdonald, D.W., Marker, B., Naura, M., Neale, A.R., Oliver, T., Osborn, D., Pullin, A.S., Shardlow, M.E.A., Showler, D.A., Smith, P.L., Smithers, R.J., Solandt, J.L., Spencer, J., Spray, C.J., Thomas, C.D., Thompson, J., Webb, S.E., Yalden, D.W., and Watkinson, A.R. (2006). The identification of 100 ecological questions of high policy relevance in the UK. *Journal of Applied Ecology*, 43, 617–27.

Sæther, B.-E., Coulson, T., Grøtan, V., Engen, S., Altwegg, R., Armitage, K.B., Barbraud, C., Becker, P.H., Blumstein, D.T., Dobson, F.S., Festa-Bianchet, M., Gaillard, J.M., Jenkins, A., Jones, C., Nicoll, M.A.C., Norris, K., Oli, M.K., Ozgul, A., and Weimerskirch, H. (2013). How life history influences population dynamics in fluctuating environments. *American Naturalist*, 182, 743–59.

Sæther, B.-E., and Engen, S. (2010a). Population analyses. In: A.P. Møller, W. Fiedler, and P. Berthold (eds), *Effects of Climate Change on Birds*. Oxford University Press, Oxford, UK.

Sæther, B.-E., and Engen, S. (2010b). Population consequences of climate change. In: A.P. Møller, W. Fiedler, and P. Berthold (eds), *Effects of Climate Change on Birds*. Oxford University Press, Oxford, UK.

Sæther, B.-E., Engen, S., Grøtan, V., Fiedler, W., Matthysen, E., Visser, M.E., Wright, J., Møller, A.P., Adriaensen, F., van Balen, H., Balmer, D., Mainwaring, M.C., McCleery, R.H., Pampus, M., and Winkel, W. (2007a). The extended Moran effect and large-scale synchronous fluctuations in the size of great tit and blue tit populations. *Journal of Animal Ecology*, 76, 315–25.

Sæther, B.-E., Engen, S., Lande, R., Arcese, P., and Smith, J.N.M. (2000a). Estimating the time to extinction in an island population of song sparrows. *Proceedings of the Royal Society of London Series B: Biological Sciences*, 267, 621–6.

Sæther, B.-E., Engen, S., and Matthysen, E. (2002). Demographic characteristics and population dynamical patterns of solitary birds. *Science*, 295, 2070–3.

Sæther, B.-E., Grøtan, V., Engen, S., Coulson, T., Grant, P.R., Visser, M.E., Brommer, J.E., Grant, B.R., Gustafsson, L., Hatchwell, B.J., Jerstad, K., Karell, P., Pietiäinen, H., Roulin, A., Røstad, O.W., and Weimerskirch, H. (2016). Demographic routes to variability and regulation in bird populations. *Nature Communications*, 7. doi: 10.1038/ncomms12001

Sæther, B.-E., Grøtan, V., Engen, S., Noble, D.G., and Freckleton, R.P. (2009). Critical parameters for predicting populations of some British passerines. *Journal of Animal Ecology*, 78, 1063–75.

Sæther, B.-E., Lillegård, M., Grøtan, V., Filli, F., and Engen, S. (2007b). Predicting fluctuations of re-introduced ibex populations: the importance of density-dependence, environmental stochasticity and uncertain population estimates. *Journal of Animal Ecology*, 76, 326–36.

Sæther, B.-E., Sutherland, W. J., and Engen, S. (2004). Climate influences on a population dynamics. *Advances in Ecological Research*, 35, 185–209.

Sæther, B.-E., Tufto, J., Engen, S., Jerstad, K., Røstad, O.W., and Skåtan, J.E. (2000b). Population dynamical consequences of climate change for a small temperate songbird. *Science*, 287, 854–6.

Tuljapurkar, S., Gaillard, J.M., and Coulson, T. (2009). From stochastic environments to life histories and back. *Philosophical Transactions of the Royal Society of London, Series B: Biological Sciences*, 364, 1499–1509.

Tuljapurkar, S., and Haridas, C.V. (2006). Temporal autocorrelation and stochastic population growth. *Ecology Letters*, 9, 327–37.

Tuljapurkar, S., Horvitz, C.C., and Pascarella, J.B. (2003). The many growth rates and elasticities of populations in random environments. *American Naturalist*, 162, 489–502.

van de Pol, M., Bailey, L.D., McLean, N., Rijsdijk, L., Lawson, C.R., and Brouwer, L. (2016). Identifying the best climatic predictors in ecology and evolution. *Methods in Ecology and Evolution*, 7, 1246–57.

van de Pol, M., Jenouvrier, S., Cornelissen, J.H.C., and Visser, M.E. (2017). Behavioural, ecological and evolutionary responses to extreme climatic events: challenges and directions. *Philosophical Transactions of the Royal Society of London, Series B: Biological Sciences*, 372, 10.1098/rstb.2016.0134.

van de Pol, M., Vindenes, Y., Sæther, B.-E., Engen, S., Ens, B.J., Oosterbeek, K., and Tinbergen, J.M. (2011). Poor environmental tracking can make extinction risk insensitive to the colour of environmental noise. *Proceedings of the Royal Society of London Series B: Biological Sciences*, 278, 3713–22.

van de Pol, M., Vindenes, Y., Sæther, B.-E., Engen, S., Ens, B.J., Oosterbeek, K., and Tinbergen, J.M. (2010). Effects of climate change and variability on population dynamics in a long-lived shorebird. *Ecology*, 91, 1192–1204.

Visser, M.E., Both, C., and Lambrechts, M.M. (2004). Global climate change leads to mistimed avian reproduction. *Advances in Ecological Research*, 35, 89–110.

SECTION 3

Population consequences of climate change

CHAPTER 8

Changes in migration, carry-over effects, and migratory connectivity

Roberto Ambrosini, Andrea Romano, and Nicola Saino

8.1 Introduction

Some of the first evidence for an effect of climate change on organisms came from studies of the timing (phenology) of bird migration (see, e.g., Berthold 1991). Ornithologists probably have a privileged point of view on the impacts of climate change because they have a long tradition of the study of migratory animals, which, in turn, seem to be more sensitive to changing climate than sedentary ones. This may occur because the rate of climate change is uneven across the globe, with northern latitudes experiencing faster warming trends than tropical areas. Consequently, animals moving across latitudes are subject to diverging trends of climate change in different stages of their annual life cycle and can, therefore, become mistimed with the local ecological conditions, with potentially negative effects on populations (Both and Visser 2001; Both et al. 2009; Saino et al. 2011).

The extent of such effects depends on a range of factors. First, climate change is already affecting the distribution of migratory individuals by determining shifts at the breeding and non-breeding grounds and on migration routes. Second, migratory populations show different degrees of connection between the areas where they spend different parts of their annual life cycle ('migratory connectivity'; Webster et al. 2002). Generally, if individuals from the same population 'stay together' during different phases of their annual life cycle ('strong' connectivity), then populations are considered to be at greater risk from climate change, because any negative impact of new climatic conditions in one area will immediately reverberate across the whole population. In contrast, mixing of individuals with those from other populations ('weak' connectivity) implies that only a portion of a population visits each site and, thus, the negative effects will be diluted.

Clearly, populations of migratory birds often respond to climate change in an integrated way, so it is possible to observe modifications in different traits, such as the timing of life history events, the distribution of organisms, and the direction and the speed of movements. In addition, current individual conditions can have downstream effects by influencing subsequent ones, thus determining the so called 'ecological carry-over effects', which further complicate the study of variation in flexible phenotypic traits like migration events, because variation in any life history stage at any one time cannot be fully understood while ignoring historical effects from previous ones (O'Connor et al. 2014). This chapter reviews studies addressing all these factors to elucidate how climate change is affecting the migration behaviour of individuals and the possible consequences of such changes on populations.

8.2 Patterns of changing migration

Migration is a complex phenomenon that depends on a multitude of consecutive decisions on timing,

Ambrosini, R., Romano, A., and Saino, N., *Changes in migration, carry-over effects, and migratory connectivity.* In: *Effects of Climate Change on Birds.* Second Edition. Edited by Peter O. Dunn and Anders Pape Møller: Oxford University Press (2019).
 DOI: 10.1093/oso/9780198824268.003.0008

direction, and speed taken by individuals, conditions encountered during each phase of the life cycle, as well as in previous ones (Newton 2011). For clarity, we will attempt to analyse each of these traits separately, but the reader should be aware that this is a rather artificial dissection of an integrated response to climate change.

8.2.1 Arrival date at the breeding grounds

Mean global temperature has increased by about 1°C worldwide over the last century, causing huge impacts on various ecosystems. However, global warming has not affected the entire globe at the same magnitude, as the Northern hemisphere has undergone a more marked warming during the last decades than tropical and equatorial regions. The advancement of spring events due to the increase of temperature at the breeding grounds is considered the main driver of the advanced bird phenology that has been observed during the last decades (Bussière et al. 2015). This is also the case for avian pre-nuptial ('spring') migration, which has advanced consistently in all continents during recent decades (Cotton 2003; Jonzén et al. 2006; Nakata et al. 2011; Bitterlin and Van Buskirk 2014; Chambers et al. 2014; Usui et al. 2017), with Antarctica representing the only exception (Barbraud and Weimerskirch 2006). Indeed, birds have significantly advanced their spring migration time by 2.1 (95 per cent CI: 1.4 to 2.9) days per decade in a meta-analysis of 73 studies primarily from the last 50 years (Usui et al. 2017). However, the length of the phenological time series available for these analyses seldom exceeds 35 years, and there is, therefore, limited information on responses to climate conditions starting at the beginning of large scale anthropogenic warming. One of the few bird phenological studies spanning more than one century found that, in the Czech Republic, spring arrival dates have fluctuated, and birds were arriving earlier during the cooler early part of the nineteenth century than in the recent warm period (Kolářová et al. 2017). However, this study also found a trend toward earlier arrivals in recent decades.

Variation in phenological response among species displaying different life history traits has frequently been documented. A major difference has been observed between species that migrate over short distances and trans-continental, long-distance migrants. Indeed, the former have usually greatly advanced their arrival date, while the migration schedule of the latter seems to be more constrained by carry-over effects of previous annual activities (e.g., moult at the wintering grounds), crossing large ecological barriers, and weather conditions during the migratory journey, resulting in a less pronounced advancement in their arrival date. Such a difference between birds adopting different migratory strategies was documented by observations by single authors in specific locations, but also by comparative and meta-analytical studies (e.g., Végvári et al. 2010; Bitterlin and Van Buskirk 2014; Usui et al. 2017). An extraordinary example is the temporal variation in laying date between two insectivorous cavity-nesting passerines breeding in the same areas across Europe: the great tit (*Parus major*), a resident to short-distance migratory species, and the pied flycatcher (*Ficedula hypoleuca*), a trans-Saharan migrant. In all the studied populations, the advancement in laying date in the pied flycatcher was less strong than in the great tit, because of later arrival on the breeding grounds (Samplonius et al. 2018). Therefore, residents and short-distance migrants seem more sensitive than long-distance migrants to temperature variation and better able to cope with current and, probably also future, climatic changes.

Other traits that have been suggested to affect the advancement in migration by different species are body mass (Chambers et al. 2014; Usui et al. 2017), diet (Végvári et al. 2010; Bitterlin and Van Buskirk 2014), timing of moult (Végvári et al. 2010), or preferred habitat (Chambers et al. 2014). However, the effect of these traits cannot be easily generalized across taxa and continents because they seem also to depend on the specific area where the phenological observations have been carried out and to the phenological variable considered (e.g., first arrival date vs mean arrival date). Future studies on different taxa in other biogeographic regions could help to understand the generality of these findings.

Variation in arrival date to the breeding grounds has received a great deal of attention in studies of the impacts of climate change because it is related to population trends. For instance, a study of 100

European bird species revealed that species that that have not advanced arrival dates showed population declines, while species with stable or increasing populations advanced their arrival dates considerably (Møller et al. 2008). It has been hypothesized that this has occurred because shifts in the timing of life history events are often unequal across trophic levels, causing a mismatch between the phenology of organisms and their food (Both et al. 2009). For instance, in a study on 117 European migratory bird species, Saino et al. (2011) used accumulated winter and spring temperatures ('degree-days') as a proxy for timing of spring biological events and showed that birds, despite arriving earlier at their breeding grounds, now arrive at higher degree-days than in the past. Therefore, they may have become increasingly mismatched relative to spring phenology. In addition, the species that showed the largest mismatches also showed the largest declines.

The mechanisms linking mismatched arrival and population decline are still unclear. For instance, by analysing extensive data on temporal change in arrival date, laying date, and clutch size of birds from Europe and North America, Dunn and Møller (2014) showed that although many species have advanced their arrival date, it was not related to changes in laying date or clutch size. Franks et al. (2018) used survey data on plant and insect species and the egg-laying dates of British songbirds for calculating an index of trophic asynchrony between bird and spring phenology and found that birds were more asynchronous in warmer springs, but offspring productivity was only marginally reduced. However, those species whose offspring annual productivity was more greatly reduced by asynchrony also declined more. These studies suggest that population change cannot be directly attributed to the effects of trophic asynchrony on offspring productivity. Further studies are therefore needed to elucidate the causal links between arrival dates and population trends.

8.2.2 Departure time from the breeding and the wintering grounds

Departure times have been investigated to a lesser extent than arrival times. This is an important gap in our knowledge, as the decision to leave clearly affects the timing of all the following migration steps. The few studies that have investigated departure dates from the breeding grounds found remarkable variability in phenological changes among species, which has been related to different constraints on the onset of migration (see section 8.2.5). Generally, long-distance migrants seem to have advanced post-breeding migration, while short-distance migrants seem to have delayed it. These results appear consistently in studies conducted in Europe (Jenni and Kéry 2003), North America (Bitterlin and Van Buskirk 2014; Usui et al. 2017), and Australia (Chambers et al. 2014). Thus, phenological trends appear to be consistent across continents.

Even fewer studies have addressed departure dates from the wintering grounds, but the small amount of information available suggests that climate in the breeding grounds, particularly spring phenology, is the most consistent driver. For instance, Palearctic migrants wintering in South Africa have advanced their departure from their non-breeding grounds, while intra-African migrants have not (Bussière et al. 2015). Differences among populations seem to exist, however. For instance, last departure of barn swallows (*Hirundo rustica*) from a major wintering roost in South Africa tended to be delayed during 1993–2010 (Møller et al. 2011). Departure dates from the wintering grounds can also be influenced by wintering habitat quality and conditions; several studies have found that individuals wintering in higher quality habitats are able to depart earlier for spring migration (Studds and Marra 2005; McKellar et al. 2013).

8.2.3 Range shift of the wintering area

Climate change is thought to promote the poleward movement of geographic ranges (Parmesan and Yohe 2003). Bird range shifts can be detected by comparing bird atlases and censuses compiled in different periods or from analysis of winter recoveries of marked birds. According to predictions, the wintering ranges of several bird species that winter in the Northern Hemisphere have shifted northwards toward their breeding areas, thus decreasing the length of migratory journeys. Rates of range shift have been estimated to be 0.45–1.48 km/year in North America (La Sorte and Thompson III 2007) and 1.2 km/year in the UK (Siriwardena and Wernham 2002; see also Brommer and Møller 2010).

The earlier onset of spring in the breeding grounds of the Northern Hemisphere, which should favour migrants that arrive earlier at the breeding grounds, should produce a northward shift also in the wintering range of those populations that winter in the Southern Hemisphere because this would reduce the length of the migration journey. An analysis of ring recoveries of barn swallow showed that both populations wintering in southern Africa and those wintering to the north of the Equator have shifted their wintering ranges northwards (Ambrosini et al. 2011). Importantly, this northward shift of the wintering ranges occurred in a direction opposite to that of climate regions, so migratory species may end up wintering in different, probably sub-optimal habitats. Similarly, species breeding in temperate or polar areas of the Southern Hemisphere and wintering in the tropics can be predicted to shift their winter distribution southwards. However, to the best of our knowledge, so far there is no published information on actual range shifts in these species.

Observations suggest that long-distance migrant species may retain a high degree of plasticity, which should allow them to drastically change their migration strategy. A clear example of this ability is represented by the discovery in 1980 of six pairs of barn swallows breeding in Argentina, within the species' historical wintering range (Billerman et al. 2011). Since then, the population has grown to thousands of pairs, also thanks to immigration from the ancestral population breeding in North America, with individuals shifting their annual cycles of moult, migration, and breeding by about six months and developing new wintering ranges in the north of South America (Winkler et al. 2017). Similar mechanisms of plasticity may allow long-distance migrants to avoid crossing geographical barriers, eventually turning into short-distance migrants. For instance, about 150 barn swallows spent the winter and moulted in the town of Aveiro, Portugal in winter 2015/2016 (van Nus and Neto 2017), and during the period 2007–2010 about 50 trans-Saharan species were observed each year in Spain during winter (Morganti 2014). Admittedly, such observations are still scattered in the scientific literature, and a more systematic investigation would be necessary to shed light on this potential mechanism of adaptation to climate change.

Birds may also change their migratory mode of life. The likelihood of a species to become sedentary when climate conditions ameliorate is probably larger for species that migrate short-distance and that are partially migratory than for long-distance migrants (Pulido 2011). However, the only published multispecies study that investigated variation in the proportion of resident and migratory individuals in Finland found evidence for increased winter residency in five of nine waterbird species (ducks and gulls), but in none of the 18 terrestrial passerine species (Meller et al. 2016). This result is consistent with the hypothesis that migration behaviour of partial migrators is driven by seasonal food limitation rather than by temperature per se.

The possibility of a species coping with climate change seems also related to the degree of variability in migration behaviour. For instance, Gilroy et al. (2016) analysed phenological trends of 340 European breeding birds and showed that partial migrants advanced their spring arrival date more than full migrants, suggesting that within-population variability in migratory movements and destinations may help facilitate species' responses to climate change.

8.2.4 Migration speed

After the start of migration, weather and climate can affect its duration in different ways. Indeed, the progress of migration depends on flight speed and on the duration of stopovers, key periods during which migrants recover their fat store and body condition.

Flight speed is reduced under unfavourable conditions, such as intense precipitation or headwind. Small migrants, like passerines, usually avoid resuming migration when winds are too strong, or they are in the opposite direction to the migration route (Åkesson and Hedenström 2000; Bolus et al. 2017; Smith and McWilliams 2018). Larger birds are more capable of actively selecting and exploiting tailwinds and can rely on updrafts (Illan et al. 2017). Climate change can also produce favourable weather conditions and improve progression speed, as has been documented, for instance, on the song thrush (*Turdus philomelos*) whose migration speed has increased during the last 40 years due to more frequent tailwinds (Sinelschikova et al. 2007).

Climate change can also affect frequency and duration of stopovers, because negative weather conditions en route can coerce birds to remain in a place until weather becomes suitable again (Åkesson and Hedenström 2000; Tøttrup et al. 2012b; Bolus et al. 2017; Smith and McWilliams 2018), or force them to stop even if not physiologically needed (e.g., because they still have fuel reserves), until weather conditions ameliorate. The amount of food available at stopover sites, which determines refuelling opportunities for migratory birds, is also affected by weather conditions, and a warmer climate can improve foraging opportunities, particularly for primary and secondary consumers, which can thus refuel in shorter times and speed up migration progression. A very interesting example was recently documented in two songbird species, the red-backed shrike (*Lanius collurio*) and the thrush nightingale (*Luscinia luscinia*), whose extreme delay in spring arrival recorded in Sweden in 2011 could be traced back to a significant prolongation of time spent at a stopover site located on the Horn of Africa, which was at that time impacted by an exceptional drought that limited the ability to accumulate fat stores (Tøttrup et al. 2012a).

Birds are not only passively affected by climate during migration, but also seem able to produce adaptive phenotypic responses to spatial variability in conditions en route (Tøttrup et al. 2008). Indeed, they can use a number of reliable weather variables to optimize their migration time. Variation in temperature seems to be the most important predictor of migration progression. Indeed, it has been demonstrated that distantly related species adopting different migratory strategies are able to track temperature as migration advances in order to adjust the speed of their journey accordingly. This is especially the case for populations breeding at northern latitudes (Ahola et al. 2004; Bauer et al. 2008), while migration timing of populations of the same species which breed at lower latitudes, and thus migrate earlier, seems to be less flexible in following environmental conditions (Both 2010).

Weather conditions can also affect the timing of passage across an ecological barrier. For instance, at a small island used by migratory birds as a stopover site after the crossing of the Mediterranean Sea capture frequencies decreased with increasing head and crosswinds at the departure sites in North Africa, and with tail and crosswinds at stopover sites (Saino et al. 2010). The strength of these effects varied among species, however, and was lower for species whose wing morphology is more typical of long-distance migrants, suggesting that long-distance migrants are less susceptible to weather.

8.2.5 Constraints on change in timing of migration

Large variability in change of migration timing seems linked to different ecological and life history constraints. For instance, variability in the phenological changes in post-nuptial ('autumn') migration in European and North American species have been linked to the length of the migratory journey and to the timing of post-nuptial moult (Jenni and Kéry 2003; Bitterlin and Van Buskirk 2014). Indeed, in Europe, long-distance migrants have advanced autumn migration, while short-distance migrants seem to have delayed migration (Jenni and Kéry 2003). Another study showed that species that moult before departure have delayed autumn migration, while those that moult after or during migration have advanced (Bitterlin and Van Buskirk 2014). In addition, large-bodied species and species that feed on seeds, insects, and fruits seem to have delayed autumn migration. Moreover, the date of last departure in autumn has grown later at high latitudes, especially in Europe, and earlier at low latitudes (Bitterlin and Van Buskirk 2014). Finally, multi-brooded species seem to have not begun to migrate later (Jenni and Kéry 2003; Bitterlin and Van Buskirk 2014). This lack of effect is surprising since climate change has caused an extension of the growing season in many parts of the Northern Hemisphere, which should encourage some facultative multi-brooded species to produce an extra brood (Møller et al. 2010). However, this may be due to other ecological and life history constraints mentioned above (e.g., timing of the post-nuptial moult). In addition, it has been argued that an extension of summer weather does not imply that favourable foraging conditions extend into late summer (Bitterlin and Van Buskirk 2014). Thus, extension of the breeding season seems to have been achieved more by earlier arrival to the breeding grounds rather than later departures.

8.3 Mechanisms leading to changes in migration

The arrival date of a bird after its migratory journey is determined by the moment and the place of departure from the original location, as well as by the time needed to accomplish the route, which, in turn, is affected by both the speed of flight and the time devoted to refuel at stopover locations. Proximate mechanisms mediating the climate effect on variation in migration phenology can be direct, when weather directly impacts departure decisions, stopover duration, and flight speed, or indirect, when ecological conditions affect individual body condition, physiology, and survival, both before the departure and during migration.

8.3.1 Changes in climate on wintering and breeding grounds

Arrival dates on the breeding grounds show a generalized advancement (see section 8.2.1) especially in areas characterized by the largest increase in temperature (Usui et al. 2017; Cohen et al. 2018). This suggests that temperature should be considered the main driver of phenological responses. Species that winter and breed in the same biogeographic region (e.g., birds breeding in northern Europe and wintering in continental-southern Europe) are exposed to the same large-scale climatic systems during the entire circannual cycle, and therefore can rely on more abundant and reliable information about the ecological conditions in their breeding sites in order to optimally adjust their migration schedule. For instance, two studies on European species have showed that short-distance migratory species seem able to adjust the length of their migratory journey to winter temperature at the breeding grounds by staying closer to their breeding sites in warm winters (Visser et al. 2009; Ambrosini et al. 2016). Mechanism explaining an advanced timing of migration can also involve a reduction of wintering time and a consequent early departure from the wintering grounds, as documented, for example, in the pink-footed goose (*Anser brachyrhynchus*) population breeding in the Svalbard archipelago (Bauer et al. 2008).

In contrast, the endogenous rhythm of long-distance migrants is thought to be controlled mainly by photoperiod, which follows the same annual pattern independently of climate change. In addition, climate conditions have changed to a lesser extent in the tropics, so long-distance migrants overwintering there should have less information to adjust their departure time. This assumption has been challenged, however, because conditions also vary in the tropics, and birds may be able to use this information to properly time their departure (Gordo et al. 2005; Saino et al. 2007). For instance, temperatures in areas south of the Sahara in February are correlated with those in Europe in March and April (Saino and Ambrosini 2008). In addition, using ringing data, Pancerasa et al. (2018) showed that the temperatures at the European breeding sites of individual barn swallows are more strongly correlated with those at their actual wintering sites and at times of departure than with those at other sites in Africa. Such 'climatic connectivity' can therefore potentially allow long-distance migrators to tune the starting of migration according to the conditions expected at the breeding grounds.

Weather and climate experienced by birds at the wintering grounds can also have indirect effects on departure time, likely mediated by food availability, which proximately affect individual physiological status. For instance, the level of precipitation during wintering predicts migration phenology of both Afro-Palearctic and Neotropical migrants, which advanced migration in years when rainfall, and therefore vegetation, is more abundant (Saino et al. 2004a; Gordo et al. 2005; Both 2010; Tøttrup et al. 2012a; McKellar et al. 2013; but see Tøttrup et al. 2008; Robson and Barriocanal 2011; López-Calderón 2019). This association has been shown in single populations over years (Saino et al. 2004a; Tøttrup et al. 2012a), between different populations of the same species spending the wintering period in geographically distant areas (McKellar et al. 2013), and among different species sharing the same wintering grounds (Gordo et al. 2005). Individual accumulation of fat reserves needed for feather moult and migration is also dependent on the quality of the habitats where overwintering took place (Marra 1998). For example, in the Neotropical migrant American redstart (*Setophaga ruticilla*) individuals wintering in rich environments (i.e., forests) were in better physical condition and departed earlier than individuals spending their wintering period in poor environments (i.e., scrub) (Marra and Holberton 1998).

8.3.2 Changes in climate and food abundance along migration routes

Refuelling opportunities during stopovers are determined by food supply, which, in turn, depends on ecological conditions and can thus vary greatly among and within years. Although the effects of climate change on food abundance at stopover sites have rarely been studied, there is evidence that birds of many taxa are able to track the vegetation greenness throughout their annual cycle, including during the migration periods (McGrath et al. 2009; Robson and Barriocanal 2011; Thorup et al. 2017). For instance, a tracking study showed that three Palearctic-African species, the common cuckoo (*Cuculus canorus*), the red-backed shrike, and the thrush nightingale, adjust their migratory movements with seasonal changes in resource availability over Europe and Africa (Thorup et al. 2017).

According to the so-called 'green-wave hypothesis', during the northward migration, birds can exploit the flush of spring growth of forage plants, and thus insects living therein, at each stopover site along the latitudinal gradient (McGrath et al. 2009; Thorup et al. 2017). Under such circumstances, the feeding rate of herbivores and insectivores should be enhanced in warmer springs (McGrath et al. 2009) because plant and arthropod abundance are linked to ambient temperature. Increasing temperatures can therefore decrease the duration of stopovers because of increased food availability, particularly for first-arriving migrants, which can benefit from their early passage in a rich environment, while late individuals may suffer from a depletion of resources (Moore and Yong 1991; Ottich and Dierschke 2003). However, it is important to note that the positive or negative impact of climate change on food availability during migration is highly variable. For example, the red knot (*Calidris canutus rufa*) population wintering in Tierra del Fuego underwent a conspicuous decrease in size after the depletion of their main food, the eggs of horseshoe crabs (*Limulus polyphemus*), at the final stopover site of their pre-breeding migration in Delaware Bay, USA (Baker et al. 2004).

8.4 Carry-over effects and climate change

Life history stages are intimately linked by processes whereby events at one stage have repercussions on those at subsequent ones. This simple intuitive idea has been crystallized in the concept of ecological carry-over effects (COEs). Authors have diverged to some extent in how inclusive the definition of ecological COEs should be (Harrison et al. 2011; O'Connor et al. 2014). Here, we adopt a broad definition that also encompasses developmental and even trans-generational processes, because these can also mediate cascading effects of climate.

8.4.1 Climate change and carry-over effects at the individual level

Most studies have investigated COEs at the level of single individuals. Merely phenological COEs have been studied in the barn swallow year-round by means of light-level geolocators (Saino et al. 2017). In addition, strong COEs of timing of departure from the wintering grounds on spring migration and arrival date in Europe existed in both sexes, but COEs of spring migration phenology on breeding date and seasonal fecundity were observed only in females. Phenological variation during the non-breeding season, in combination with ecological conditions (NDVI) in the wintering areas, explained as much as 65–70 per cent of variation in subsequent fecundity in females, whereas these figures were lower for males. Similarly, timing of breeding strongly influenced departure time on autumn migration in the Savannah sparrow (*Passerculus sandwichensis*) (Mitchell et al. 2012), but only in females.

In long-distance migrants, variation in habitat quality during the non-breeding season may carry over to major life history traits. For example, the quality of tropical wintering territories, as gauged from feather isotopic fingerprints, greatly affected breeding success of American redstarts breeding at temperate latitudes (Norris et al. 2004). However, sex-, age-, and also population-dependent effects recur and represent a further level of complexity in COEs studies, as shown in the European barn swallows. Indeed, climate change is altering both the timing of breeding and ecological conditions in the sub-Saharan wintering areas of the European populations of this species (Saino et al. 2004a; López-Calderón 2019), and is, therefore, expected to affect their phenology via COEs year-round. Differences among populations seem to exist in the intensity and even in the direction of such COEs. For instance, in a barn swallow population breeding in northern Italy, greater primary productivity (NDVI) in the wintering areas carried over into earlier arrival on the breeding grounds and earlier breeding (Saino et al. 2004b), as well as into higher breeding success (Saino et al. 2004a). In contrast, a similar study on barn swallows breeding in Denmark showed that individuals delayed both arrival to their breeding areas and onset of breeding, but produced larger clutches in the first brood in years when NDVI values in their wintering areas were higher (López-Calderón 2019). In both populations, these responses were age- but not sex-dependent, since response was increasingly stronger as individuals aged. These age-dependent responses are therefore likely due to a plastic response of individuals, rather than to disappearance of poor quality individuals from the populations.

Ecological conditions during the non-breeding season can also carry over onto dispersal decisions, as shown in American redstarts, where the latitude of breeding was affected by habitat occupancy during the preceding non-breeding season. This was likely attained via an effect on timing of departure on spring migration and thus variation in the latitudinal location of the most suitable habitats for breeding at the time of migration (Studds et al. 2008). COEs are not an absolute rule, however. In the Hudsonian godwit (*Limosa haemastica*), phenological COEs were not found throughout the annual cycle and were not consistently observed for the same individuals in different years. In this species, it appeared that differences in arrival on the wintering grounds and, thus, potential COEs were reset by the spring when birds departed for the breeding grounds (Senner et al. 2014).

Experimental studies of COEs are rare and have most often dealt with the effects of food, which can be manipulated with relative ease and made to mimic climate change effects. For example, in a study of snow geese (*Anser caerulescens*) birds were held in captivity with or without access to food during spring staging and their reproductive success was then recorded (Legagneux et al. 2012). This study showed that a longer time spent in captivity markedly reduced reproductive success of individuals, whereas food deprivation had no effect. However, the COEs of captivity disappeared in a year with favourable ecological conditions on the breeding grounds, suggesting that COEs were probably mediated by stress response to captivity rather by an acute episode of food deprivation, because they vanished when conditions were unusually good for breeding. Climate change can thus be expected to have COEs on breeding performance if it elicits a physiological stress response.

8.4.2 How climate change can impact carry-over effects at the population level

Population-level studies of COEs often directly focus on meteorological variables and are therefore directly relevant for the analysis of the consequences of climate change. For instance, Okendon et al. (2013) found that inter-seasonal effects of precipitation on the wintering grounds was weak or non-significant for timing of breeding and clutch size of 19 European migratory species. On the other hand, variation in breeding phenology and success in three long-distance Afro-Palearctic migrants was predicted by temperatures experienced in Mediterranean passage areas more strongly than by conditions at the breeding grounds (Finch et al. 2014). In the greater snow goose (*Anser caerulescens atlanticus*) clutch size and hatching success over 22 years were dependent on temperatures at the breeding sites. However, progression to reproduction was mainly negatively affected by warmer climatic conditions during the non-breeding season (van Oudenhove et al. 2014). In El Niño years, Nearctic migrants overwintering in South America experience drier environments, reach their migration stopover sites in poorer energetic conditions, and perform stopovers in low-quality coastal habitats with a greater density of competitors rather than in more favourable inland forests (Paxton et al. 2014).

Changes in conditions during the non-breeding season can also carry over to affect sexual selection processes during breeding. In the barn swallow, long-term trends in arrival dates from spring migration, possibly driven by climate effects, can affect sexual selection via an effect on the degree of protandry (Møller 2007). In the American redstart, males that spent the winter in favourable tropical habitats gained a sexual selection advantage via lower paternity loss and a higher level of polygyny, partly due to their earlier arrival on the breeding grounds in Canada (Reudink et al. 2009).

8.5 Migratory connectivity and climate change

Migratory connectivity refers to the strength of the linkage of individuals and populations between the areas where they spend different seasons of the annual cycle (Webster et al. 2002; Cohen et al. 2017). It has been defined as strong if, for instance, all or most individuals breeding in one area winter in the same area and vice versa. In contrast, if individuals breeding in one area spread through different areas for wintering, then migratory connectivity is weak (Webster et al. 2002).

Populations showing strong migratory connectivity have been considered more prone to suffer the adverse ecological effects of climate change than those showing weak connectivity. Dolman and Sutherland (1994) used a theoretical model to explore the interactions between habitat loss, population regulation, and evolutionary changes in migration behaviour due to the gradual loss of a wintering site. A key prediction of their model was that breeding populations can decline sharply due to the loss of one winter habitat when migratory connectivity is strong, while when it is weak, the negative effects are much lower. This can occur because the population dynamics of migratory organisms are determined by events that occur at their breeding and wintering grounds, as well as stopover areas used to migrate between the two. A recent review has shown that migratory connectivity is generally low in long-distance migratory birds (Finch et al. 2017), although it can be important for conservation. For example, Martin et al. (2007) showed that knowledge about patterns of migratory connectivity was needed to protect the American redstart across its entire range, while failure to take into account migratory connectivity would doom some regional populations to extinction. A notable exception to the general role that low migratory connectivity can play in somewhat buffering populations against the negative effects of climate change is the spread of pathogens and parasites favoured by new climatic conditions, which can have a larger impact on populations that mix with other ones than in populations that segregate during all phases of the year (Bauer et al. 2015; Chapter 14).

8.6 Case studies

A clear and general picture of the effects of climate change on shifts in phenology is shown in a broad meta-analysis of time series of animal phenology across the planet conducted by Cohen et al. (2018). The authors showed that temperature is closely related to phenological data and the relationship is stronger in areas that have experienced more climate change (Figure 8.1a, b). In addition, this study found that phenological shifts were associated with climatic variables that drive seasonality in different areas of the globe, such as temperatures at mid-latitudes and precipitation in tropical and subtropical zones. Indeed, changes in temperature became more predictive of the magnitude of phenological shifts at increasing latitude in both Northern and Southern Hemispheres, while precipitation became more important as latitude decreased (Figure 8.1c). Finally, the correlation between phenology and temperature in temperate zones was stronger than that between phenology and precipitation in the tropics, probably because temperature change has a greater effect on phenology than precipitation (Cohen et al. 2018). This study encompassed all animal taxa and included different phenological estimates, for instance, arrival and laying dates, which are not

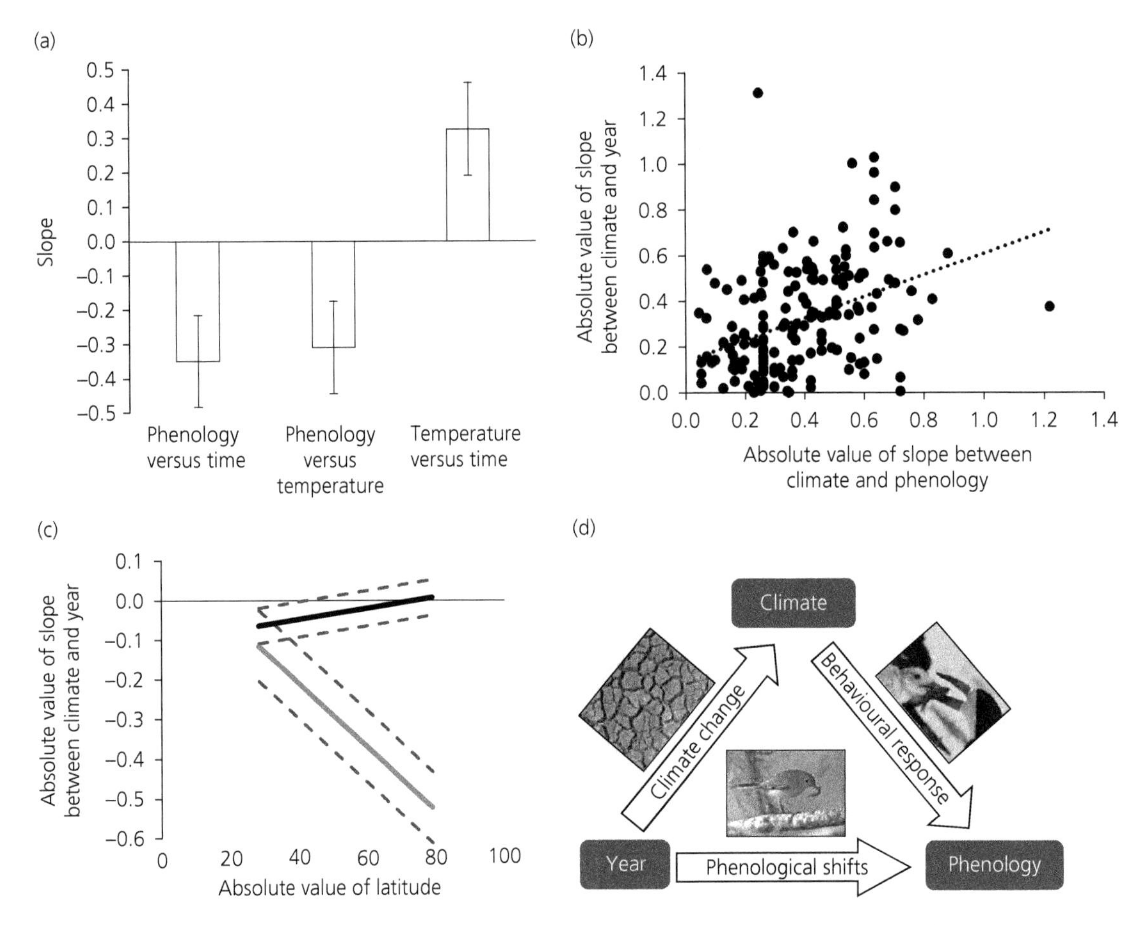

Figure 8.1 Advancements in phenology due to climate change. A: Across 1011 time series, phenology advanced through time as temperature increased and the increases in temperature were negatively correlated with phenology. Error bars represent s.e.m. B: Phenology was more closely linked with mean temperature (*x* axis) in areas with more climate change (*y* axis; $R^2 = 0.152$, d.f. = 175, $P < 0.0001$). C: Precipitation (grey line) becomes more important in driving phenological responses (i.e., values become more negative) towards the Equator, whereas temperature (black line) becomes more important as one moves away from the Equator. Dashed lines indicate 95% confidence band. D: Conceptual figure explaining the meaning of the slope and correlation terms on the other panels, which represent relationships between year, climate, and phenology (redrawn from Cohen et al. 2018).

necessarily influenced by the same factors (Kristensen et al. 2015). However, bird studies provided the majority of the analysed time series, so it is not surprising these results are generally consistent with those of other studies than focused only on birds (Gordo et al. 2005; Saino et al. 2007, 2011; McKellar et al. 2013; Okendon et al. 2013; Usui et al. 2017).

The importance of including estimates of migratory connectivity in studies of population dynamics of migratory birds is well shown by a study of the wood thrush (*Hylocichla mustelina*), a long-distance migratory songbird that breeds throughout eastern North America and winters from southern Mexico to Colombia (Rushing et al. 2016). This study combined information from extensive bird surveys (the North America Breeding Bird Survey—BBS), known patterns of migratory connectivity, and remote sensing data (monthly Enhanced Vegetation Index) to quantify how climate change and habitat loss across the entire range and over the entire annual cycle have contributed to population declines. The authors first used BBS data to delineate 17 geographical populations in breeding areas that differ with regard to both demographic trend and abundance, and five non-breeding regions, linked to different breeding populations by known patterns of migratory connectivity. Then they modelled annual abundance as a function of habitat availability and large-scale climatic conditions on both the breeding and non-breeding grounds. They found that, generally, both habitat loss and climate have contributed to the observed declines in wood thrush breeding abundance, but the relative importance of breeding versus non-breeding factors was population-specific. Importantly, breeding populations where wood thrushes were highly abundant (core populations) appeared to be more limited by habitat loss, whereas low-abundance (peripheral) populations appeared to be limited by climate-driven seasonal interactions.

8.7 Critical needs and future perspectives

Globally there is a huge amount of information available on the effects of climate change on birds. However, most of it is distributed across studies and inference about the generality of phenological shifts results mostly from meta-analytical approaches (see, e.g., Parmesan and Yohe 2003; Usui et al. 2017; Cohen et al. 2018), which are biased towards western Europe and North America, because these areas of the globe have the most long-term datasets (Bitterlin and Van Buskirk 2014; Usui et al. 2017). There is therefore a need for further studies on Southern Hemisphere species because substantial differences exist between the migratory systems of the hemispheres. Most notably, migratory species in the Southern Hemisphere mainly migrate within single continents and few cross the Equator.

The study of climate-driven COEs is also still in its infancy, and there is, therefore, ample scope for studies in this field, particularly for those explicitly designed to test hypotheses on the interplay between climate change and the evolution and expression of COEs, and on how COEs may be involved in adaptive response to climate change. Indeed, most of the studies currently available in the literature simply discuss the implications of COEs for populations as a consequence of climate change.

The last few years have seen a large increase in studies based on novel technologies able to track animals year-round, and even larger improvements can be foreseen in the near future (Bridge et al. 2011). These technological improvements, however, will not reduce the importance of the impressive amount of data already collected by ornithologists, such as ringing data or phenological time series, as these datasets can reveal historical changes in migration patterns. There is also a tremendous need for studies that synthesize information from different sources (e.g., ring recoveries, tracking data, and phenological time series) and couple them with climate information. This will allow the making of generalized statements on the effects of climate change that are urgently needed to inform conservation policies.

References

Ahola, M., Laaksonen, T., Sippola, K., Eeva, T., Rainio, K., and Lehikoinen, E. (2004). Variation in climate warming along the migration route uncouples arrival and breeding dates. *Global Change Biology*, 10, 1610–17.

Åkesson, S., and Hedenström, A. (2000). Wind selectivity of migratory flight departures in birds. *Behavioral Ecology and Sociobiology*, 47, 140–4.

Ambrosini, R., Cuervo, J.J., du Feu, C., Fiedler, W., Musitelli, F., Rubolini, D., Sicurella, B., Spina, F., Saino, N., and Møller, A.P. (2016). Migratory connectivity and effects of winter temperatures on migratory behaviour of the European robin *Erithacus rubecula*: a continent-wide analysis. *Journal of Animal Ecology*, 85, 749–60.

Ambrosini, R., Rubolini, D., Møller, A.P., Bani, L., Clark, J., Karcza, Z., Vangeluwe, D., Du Feu, C., Spina, F., and Saino, N. (2011). Climate change and the long-term northward shift in the African wintering range of the barn swallow *Hirundo rustica*. *Climate Research*, 49, 131–41.

Baker, A.J., González, P.M., Piersma, T., Niles, L.J., de Lima Serranodo Nascimento, I., Atkinson, P.W., Clark, N.A., T Minton, C.D., Peck, M.K., Aarts, G., Oeste, A., and Negro, R. (2004). Rapid population decline in red knots: fitness consequences of decreased refuelling rates and late arrival in Delaware Bay. *Proceedings of the Royal Society of London Series B: Biological Sciences*, 271, 875–82.

Barbraud, C., and Weimerskirch, H. (2006). Antarctic birds breed later in response to climate change. *Proceedings of the National Academy of Sciences of the United States of America*, 103, 6248–51.

Bauer, S., Dinther, M. Van, Høgda, K., Klaassen, M., Madsen, J., and Box, O. (2008). The consequences of climate-driven stop-over sites changes on migration schedules and fitness of Arctic geese. *Journal of Animal Ecology*, 77, 654–60.

Bauer, S., Lisovski, S., and Hahn, S. (2015). Timing is crucial for consequences of migratory connectivity. *Oikos*, 125, 605–12.

Berthold, P. (1991). Patterns of avian migration in light of current global 'greenhouse' effects: a central European perspective. In: S.B. Terrill and P.Z. Antas (eds), *Proceedings of the 22th International Ornithological Congress*, pp. 780–786. New Zealand Ornithological Congress Trust Board, Wellington.

Billerman, S.M., Huber, G.H., Winkler, D.W., Safran, R.J., and Lovette, I.J. (2011). Population genetics of a recent transcontinental colonization of South America by breeding Barn Swallows (*Hirundo rustica*). *The Auk*, 128, 506–13.

Bitterlin, L.R., and Van Buskirk, J. (2014). Ecological and life history correlates of changes in avian migration timing in response to climate change. *Climate Research*, 61, 109–21.

Bolus, R.T., Diehl, R.H., Moore, F.R., Deppe, J.L., Ward, M.P., and Zenzal, T.J.J. (2017). Swainson's thrushes do not show strong wind selectivity prior to crossing the Gulf of Mexico. *Scientific Reports*, 7, 14280.

Both, C. (2010). Flexibility of timing of avian migration to climate change masked by environmental constraints en route. *Current Biology*, 20, 243–8.

Both, C., Van Asch, M., Bijlsma, R.G., Van Den Burg, A.B., and Visser, M.E. (2009). Climate change and unequal phenological changes across four trophic levels: Constraints or adaptations? *Journal of Animal Ecology*, 78, 73–83.

Both, C., and Visser, M.E. (2001). Adjustment to climate change is constrained by arrival date in a long-distance migrant bird. *Nature*, 411, 296–8.

Bridge, E.S., Thorup, K., Bowlin, M.S., Chilson, P.B., Diehl, R.H., Fléron, R.W., Hartl, P., Kays, R., Kelly, J.F., Robinson, W.D., and Wikelski, M. (2011). Technology on the move: Recent and forthcoming innovations for tracking migratory birds. *BioScience*, 61, 689–98.

Brommer, J.E., and Møller, A.P. (2010). Range margins, climate change, and ecology. In: A.P. Møller, W. Fiedler, and P.O. Dunn (eds), *Effects of Climate Change on Birds*, pp. 249–274. Oxford University Press, Oxford, UK.

Bussière, E.M.S., Underhill, L.G., and Altwegg, R. (2015). Patterns of bird migration phenology in South Africa suggest northern hemisphere climate as the most consistent driver of change. *Global Change Biology*, 21, 2179–90.

Chambers, L.E., Beaumont, L.J., and Hudson, I.L. (2014). Continental scale analysis of bird migration timing: Influences of climate and life history traits—a generalized mixture model clustering and discriminant approach. *International Journal of Biometeorology*, 58, 1147–62.

Cohen, E.B., Hostetler, J.A., Hallworth, M.T., Rushing, C.S., Sillett, T.S., and Marra, P.P. (2017). Quantifying the strength of migratory connectivity. *Methods in Ecology and Evolution*, 9, 513–24.

Cohen, J.M., Lajeunesse, M.J., and Rohr, J.R. (2018). A global synthesis of animal phenological responses to climate change. *Nature Climate Change*, 8, 224–8.

Cotton, P.A. (2003). Avian migration phenology and global climate change. *Proceedings of the National Academy of Sciences of the United States of America*, 100, 12219–22.

Dolman, P.M., and Sutherland, W.J. (1994). The response of bird populations to habitat loss. *Ibis*, 137, S38–S46.

Dunn, P.O., and Møller, A.P. (2014). Changes in breeding phenology and population size of birds. *Journal of Animal Ecology*, 83, 729–39.

Finch, T., Butler, S., Franco, A., and Cresswell, W. (2017). Low migratory connectivity is common in long-distance migrant birds. *Journal of Animal Ecology*, 38, 42–9.

Finch, T., Pearce-Higgins, J.W., Leech, D.I., and Evans, K.L. (2014). Carry-over effects from passage regions are more important than breeding climate in determining the breeding phenology and performance of three avian migrants of conservation concern. *Biodiversity and Conservation*, 23, 2427–44.

Franks, S.E., Pearce-Higgins, J.W., Atkinson, S., Bell, J.R., Botham, M.S., Brereton, T.M., Harrington, R., and Leech, D.I. (2018). The sensitivity of breeding songbirds to changes in seasonal timing is linked to population change but cannot be directly attributed to the effects of trophic asynchrony on productivity. *Global Change Biology*, 24, 957–71.

Gilroy, J.J., Gill, J.A., Butchart, S.H.M., Jones, V.R., and Franco, A.M.A. (2016). Migratory diversity predicts population declines in birds. *Ecology Letters*, 19, 308–17.

Gordo, O., Brotons, L., Ferrer, X., and Comas, P. (2005). Do changes in climate patterns in wintering areas affect the timing of the spring arrival of trans-Saharan migrant birds? *Global Change Biology*, 11, 12–21.

Harrison, X.A., Blount, J.D., Inder, R., Norris, D.R., and Bearhop, S. (2011). Carry-over effects as drivers of fitness differences in animals. *Journal of Animal Ecology*, 80, 4–18.

Illan, J.G., Wang, G., Cunningham, F.L., and King, D.T. (2017). Seasonal effects of wind conditions on migration patterns of soaring American white pelican. *PLoS One*, 12, e0186948.

Jenni, L., and Kéry, M. (2003). Timing of autumn bird migration under climate change: advances in long-distance migrants, delays in short-distance migrants. *Proceedings of the Royal Society of London Series B: Biological Sciences*, 270, 1467–71.

Jonzén, N., Lindén, A., Ergon, T., Knudsen, E., Vik, J.O., Rubolini, D., Piacentini, D., Brinch, C., Spina, F., Karlsson, L., Stervander, M., Andersson, A., Waldenström, J., Lehikoinen, A., Edvardsen, E., and Stenseth, N.C. (2006). Rapid advance of spring arrival dates in long-distance migratory birds. *Science*, 312, 1959–61.

Kolářová, E., Matiu, M., Menzel, A., Nekovář, J., Lumpe, P., and Adamík, P. (2017). Changes in spring arrival dates and temperature sensitivity of migratory birds over two centuries. *International Journal of Biometeorology*, 61, 1279–89.

Kristensen, N.P., Johansson, J., Ripa, J., and Jonzén, N. (2015). Phenology of two interdependent traits in migratory birds in response to climate change. *Proceedings of the Royal Society of London Series B: Biological Sciences*, 282, 20150288.

La Sorte, F.A., and Thompson III, F.R. (2007). Poleward shifts in winter ranges of North American birds. *Ecology*, 88, 1803–12.

Legagneux, P., Fast, P.L.F., Gauthier, G., and Bêty, J. (2012). Manipulating individual state during migration provides evidence for carry-over effects modulated by environmental conditions. *Proceedings of the Royal Society of London Series B: Biological Sciences*, 279, 876–83.

López-Calderón, C. (2019). Linking different worlds: migration ecology in two species of Hirundines. PhD thesis, Universidad de Sevilla, Sevilla, Spain.

Marra, P.P. (1998). Linking winter and summer events in a migratory bird by using stable-carbon isotopes. *Science*, 282, 1884–6.

Marra, P.P., and Holberton, R.L. (1998). Corticosterone levels as indicators of habitat quality: effects of habitat segregation in a migratory bird during the non-breeding season. *Oecologia*, 116, 284–92.

Martin, T.G., Chadès, I., Arcese, P., Marra, P.P., Possingham, H.P., and Norris, D.R. (2007). Optimal conservation of migratory species. *PLoS One*, 2, e751.

McGrath, L.J., Van Riper III, C., and Fontaine, J.J. (2009). Flower power: Tree flowering phenology as a settlement cue for migrating birds. *Journal of Animal Ecology*, 78, 22–30.

McKellar, A.E., Marra, P.P., Hannon, S.J., Studds, C.E., and Ratcliffe, L.M. (2013). Winter rainfall predicts phenology in widely separated populations of a migrant songbird. *Oecologia*, 172, 595–605.

Meller, K., Vähätalo, A.V., Hokkanen, T., Rintala, J., Piha, M., and Lehikoinen, A. (2016). Interannual variation and long-term trends in proportions of resident individuals in partially migratory birds. *Journal of Animal Ecology*, 85, 570–80.

Mitchell, G.W., Newman, A.E.M., Wikelski, M., and Norris, D.R. (2012). Timing of breeding carries over to influence migratory departure in a songbird: an automated radiotracking study. *Journal of Animal Ecology*, 81, 1024–33.

Møller, A.P. (2007). Tardy females, impatient males: Protandry and divergent selection on arrival date in the two sexes of the barn swallow. *Behavioral Ecology and Sociobiology*, 61, 1311–19.

Møller, A.P., Flensted-Jensen, E., Klarborg, K., Mardal, W., and Nielsen, J.T. (2010). Climate change affects the duration of the reproductive season in birds. *Journal of Animal Ecology*, 79, 777–84.

Møller, A., Nuttall, R., Piper, S., Szép, T., and Vickers, E. (2011). Migration, moult and climate change in barn swallows *Hirundo rustica* in South Africa. *Climate Research*, 47, 201–5.

Møller, A.P., Rubolini, D., and Lehikoinen, E. (2008). Populations of migratory bird species that did not show a phenological response to climate change are declining. *Proceedings of the National Academy of Sciences of the United States of America*, 105, 16195–200.

Moore, F.R., and Yong, W. (1991). Evidence of food-based competition among passerine migrants during stopover. *Behavioral Ecology and Sociobiology*, 28, 85–90.

Morganti, M. (2014). *The potential of migratory birds to adapt to global change: Lessons from European long-distance migrants and Iberian blackcaps*. PhD Thesis, Universidad Complutense de Madrid, Madrid, Spain.

Nakata, M., Chino, N., Chiba, A., Komatsu, Y., Itoh, Y., Akahara, K., Ichimura, Y., Okino, M., Satoh, H., Tachikawa, K., and Fujisawa, M. (2011). Chronological trends in the timing of spring bird migration and its relationship with temperature in a coastal forest near the city of Niigata. *Japanese Journal of Ornithology*, 60, 63–72.

Newton, I. (2011). *The Ecology of Bird Migration*. Academic Press, London, UK.

Norris, D.R., Marra, P.P., Kyser, T.K., Sherry, T.W., and Ratcliff, L.M. (2004). Tropical winter habitat limits reproductive success on temperate breeding grounds of a migratory bird. *Proceedings of the Royal Society of London Series B: Biological Sciences*, 271, 59–64.

O'Connor, C.M., Norris, D.R., Crossin, G.T., and Cooke, S.J. (2014). Biological carryover effects: linking common concepts and mechanisms in ecology and evolution. *Ecosphere*, 5, 28.

Okendon, N., Leech, D., and Pearce-Higgins, J.W. (2013). Climate effects on breeding grounds are more important drivers of breeding phenology in migrant birds than carry-over effects from wintering grounds. *Biology Letters*, 9, 20130699.

Ottich, I., and Dierschke, V. (2003). Exploitation of resources modulates stopover behaviour of passerine migrants. *Journal of Ornithology*, 144, 307–16.

Pancerasa, M., Ambrosini, R., Saino, N., and Casagrandi, R. (2018). Barn swallows long-distance migration occurs between significantly temperature-correlated areas. *Scientific Reports*, 8, 12359.

Parmesan, C., and Yohe, G. (2003). A globally coherent fingerprint of climate change impacts across natural systems. *Nature*, 421, 37–42.

Paxton, K.L., Cohen, E.B., Paxton, E.H., Németh, Z., and Moore, F.R. (2014). El Niño-Southern Oscillation is linked to decreased energetic condition in long-distance migrants. *PLoS One*, 9, e95383.

Pulido, F. (2011). Evolutionary genetics of partial migration—the threshold model of migration revis(it)ed. *Oikos*, 120, 1776–83.

Reudink, M.W., Marra, P.P., Kyser, T.K., Boag, P.T., Langin, K.M., and Ratcliffe, L.M. (2009). Non-breeding season events influence sexual selection in a long-distance migratory bird. *Proceedings of the Royal Society of London Series B: Biological Sciences*, 276, 1619–26.

Robson, D., and Barriocanal, C. (2011). Ecological conditions in wintering and passage areas as determinants of timing of spring migration in trans-Saharan migratory birds. *Journal of Animal Ecology*, 80, 320–31.

Rushing, C.S., Ryder, T.B., and Marra, P.P. (2016). Quantifying drivers of population dynamics for a migratory bird throughout the annual cycle. *Proceedings of the Royal Society of London Series B: Biological Sciences*, 283, 20152846.

Saino, N., and Ambrosini, R. (2008). Climatic connectivity between Africa and Europe may serve as a basis for phenotypic adjustment of migration schedules of trans-Saharan migratory birds. *Global Change Biology*, 14, 250–63.

Saino, N., Ambrosini, R., Caprioli, M., Romano, A., Romano, M., Rubolini, D., Scandolara, C., and Liechti, F. (2017). Sex-dependent carry-over effects on timing of reproduction and fecundity of a migratory bird. *Journal of Animal Ecology*, 86, 239–49.

Saino, N., Ambrosini, R., Rubolini, D., Von Hardenberg, J., Provenzale, A., Hüppop, K., Hüppop, O., Lehikoinen, A., Lehikoinen, E., Rainio, K., Romano, M., and Sokolov, L. (2011). Climate warming, ecological mismatch at arrival and population decline in migratory birds. *Proceedings of the Royal Society of London Series B: Biological Sciences*, 278, 835–842.

Saino, N., Rubolini, D., von Hardenberg, J., Ambrosini, R., Provenzale, A., Romano, M., and Spina, F. (2010). Spring migration decisions in relation to weather are predicted by wing morphology among trans-Mediterranean migratory birds. *Functional Ecology*, 24, 658–69.

Saino, N., Rubolini, D., Jonzén, N., Ergon, T., Montemaggiori, A., Stenseth, N.C., and Spina, F. (2007). Temperature and rainfall anomalies in Africa predict timing of spring migration in trans-Saharan migratory birds. *Climate Research*, 35, 123–34.

Saino, N., Szép, T., Ambrosini, R., Romano, M., and Møller, A.P. (2004a). Ecological conditions during winter affect sexual selection and breeding in a migratory bird. *Proceedings of the Royal Society of London Series B: Biological Sciences*, 271, 681–6.

Saino, N., Szép, T., Romano, M., Rubolini, D., Spina, F., and Møller, A.P. (2004b). Ecological conditions during winter predict arrival date at the breeding quarters in a trans-Saharan migratory bird. *Ecology Letters*, 7, 21–5.

Samplonius, J.M., Bartošová, L., Burgess, M.D., Bushuev, A.V., Eeva, T., Ivankina, E.V., Kerimov, A.B., Krams, I., Laaksonen, T., Mägi, M., Mänd, R., Potti, J., Török, J., Trnka, M., Visser, M.E., Zang, H., and Both, C. (2018). Phenological sensitivity to climate change is higher in resident than in migrant bird populations among European cavity breeders. *Global Change Biology*, 24, 3780–3790.

Senner, N.R., Hochachka, W.M., Fox, J.W., and Afanasyev, V. (2014). An exception to the rule: carry-over effects do not accumulate in a long-distance migratory bird. *PLoS One*, 9, e86588.

Sinelschikova, A., Kosarev, V., Panov, I., and Baushev, A.N. (2007). The influence of wind conditions in Europe on the advance in timing of the spring migration of the song thrush (*Turdus philomelos*) in the south-east Baltic region. *International Journal of Biometeorology*, 51, 431–40.

Siriwardena, G.M., and Wernham, C. (2002). Synthesis of the migration patterns of British and Irish birds. In: C.V. Wernham, M.P. Toms, J.H. Marchant, J.A. Clark, G.M. Siriwardena, and S.R. Baillie (eds), *The migration atlas: movements of the birds of Britain and Ireland*, pp. 70–102. T. & A.D. Poyser, London, UK.

Smith, A.D., and McWilliams, S.R. (2018). What to do when stopping over: behavioral decisions of a migrating songbird during stopover are dictated by initial change in their body condition and mediated by key environmental conditions. *Behavior Research*, 25, 1423–35.

Studds, C.E., Kyser, T.K., and Marra, P.P. (2008). Natal dispersal driven by environmental conditions interacting across the annual cycle of a migratory songbird. *Proceedings of the National Academy of Sciences of the United States of America*, 105, 2922–33.

Studds, C.E., and Marra, P.P. (2005). Nonbreeding habitat occupancy and population processes: An upgrade experiment with a migratory bird. *Ecology*, 86, 2380–5.

Thorup, K., Tøttrup, A.P., Willemoes, M., Klaassen, R.H.G., Strandberg, R., Vega, M.L., Dasari, H.P., and Araújo, M.B. (2017). Resource tracking within and across continents in long-distance bird migrants. *Science Advances*, 3, e1601360.

Tøttrup, A.P., Klaassen, R.H.G., Kristensen, M.W., Strandberg, R., Vardanis, Y., Lindström, Å., Rahbek, C., Alerstam, T., and Thorup, K. (2012a). Drought in Africa caused delayed arrival of European songbirds. *Science*, 338, 1307.

Tøttrup, A.P., Klaassen, R.H.G., Strandberg, R., Thorup, K., Kristensen, M.W., Jørgensen, P.S., Fox, J.W., Afanasyev, V., Rahbek, C., and Alerstam, T. (2012b). The annual cycle of a trans-equatorial Eurasian-African passerine migrant: different spatio-temporal strategies for autumn and spring migration. *Proceedings of the Royal Society of London Series B: Biological Sciences*, 279, 1008–16.

Tøttrup, A.P., Thorup, K., Rainio, K., Yosef, R., Lehikoinen, E., and Rahbek, C. (2008). Avian migrants adjust migration in response to environmental conditions en route. *Biology Letters*, 4, 685–8.

Usui, T., Butchart, S.H.M., and Phillimore, A.B. (2017). Temporal shifts and temperature sensitivity of avian spring migratory phenology: a phylogenetic meta-analysis. *Journal of Animal Ecology*, 86, 250–61.

van Nus, T., and Neto, J.M. (2017). Urban roost of wintering barn swallows *Hirundo rustica* in Aveiro, Portugal. *Ardea*, 105, 73–8.

van Oudenhove, L., Gauthier, G., and Lebreton, J.D. (2014). Year-round effects of climate on demographic parameters of an Arctic-nesting goose species. *Journal of Animal Ecology*, 84, 1322–33.

Végvári, Z., Bókony, V., Barta, Z., and Kovács, G. (2010). Life history predicts advancement of avian spring migration in response to climate change. *Global Change Biology*, 16, 1–11.

Visser, M.E., Perdeck, A.C., van Balen, J.H., and Both, C. (2009). Climate change leads to decreasing bird migration distances. *Global Change Biology*, 15, 1859–65.

Webster, M.S., Marra, P.P., Haig, S.M., Bensch, S., and Holmes, R.T. (2002). Links between worlds: Unraveling migratory connectivity. *Trends in Ecology & Evolution*, 17, 76–83.

Winkler, D.W., Gandoy, F.A., Areta, J.I., Iliff, M.J., Rakhimberdiev, E., Kardynal, K.J., and Hobson, K.A. (2017). Long-distance range expansion and rapid adjustment of migration in a newly established population of barn swallows breeding in Argentina. *Current Biology*, 27, 1080–4.

CHAPTER 9

Changes in timing of breeding and reproductive success in birds

Peter O. Dunn

9.1 Introduction

Some of the strongest evidence for the effects of climate change on organisms comes from studies of the annual timing of breeding (e.g., Scheffers et al. 2016). In birds, the decision to breed at a particular time is potentially one of the most important determinants of fitness because the date of the first egg (laying date) is often closely associated with clutch size, which establishes an upper limit on reproductive success. Thus, the factors that influence when individuals breed are potentially important in predicting how climate change will affect reproductive success and population growth. Some of these factors are environmental, such as food supply, but other factors may be under genetic control, such as circannual rhythms. Phenology of egg-laying appears to be highly flexible (plastic) in response to some environmental changes, particularly annual differences in temperature, but there is also some evidence for slower, less dramatic changes occurring through selection, which is reviewed in Chapter 11. There is large variation in the rate at which species are responding to climate change (see section 9.3), so it is important to review the proximate mechanisms influencing laying date and their effects on reproductive success.

9.2 What proximate factors influence laying date?

Studies of climate change often focus on temperature, which has an influence on the timing of egg-laying in many species of birds (Dunn 2004), but there are several other proximate factors that can potentially influence the start of breeding, including other weather variables such as precipitation (Skinner et al. 1998; Leitner et al. 2003; Rodriguez and Bustamante 2003) and wind speed (Møller 2013), as well as food abundance, breeding density, photoperiod, and hormones (Dawson 2008; Caro et al. 2013). Many predictive models of the effects of climate change on animals rely solely on temperature, so it is important to determine how the timing of laying is influenced by temperature and how this relationship is modified by other environmental and physiological factors. Figure 9.1 presents an overview of the factors influencing timing of egg-laying.

9.2.1 Photoperiod

For most temperate-breeding birds, the factors influencing laying date are thought to act in a hierarchy starting with increasing daylength in the spring (photoperiod) as the primary cue for gonadal

Dunn, P.O., *Changes in timing of breeding and reproductive success in birds. In: Effects of Climate Change on Birds*. Second Edition. Edited by Peter O. Dunn and Anders Pape Møller: Oxford University Press (2019). © Oxford University Press. DOI: 10.1093/oso/9780198824268.003.0009

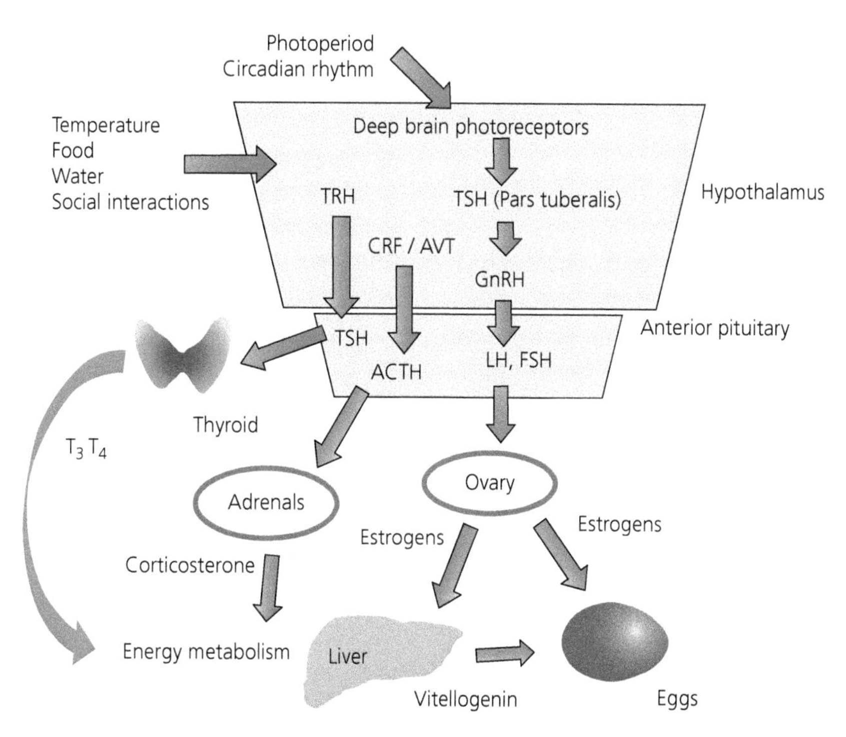

Figure 9.1 Schematic diagram of physiological pathways that are involved in the timing of egg-laying and the production of eggs. Under increasing daylength, photoreceptors in the hypothalamus activate the secretion of thyroid-stimulating hormone (TSH) from the anterior pituitary gland, which triggers secretion of gonadotropin releasing hormone (GnRH). GnRH stimulates the production of gonadotropins (e.g., luteinizing hormone [LH], follicle-stimulating hormone [FSH]) that promote the development of the ovary. TSH also stimulates release of thyroid hormones (T3, T4) that increase metabolic and fattening rates. Corticosterone and vitellogenin have been implicated in some studies of laying date (Hennin et al. 2016). Secretion of corticosterone begins in the hypothalamus with corticotropin releasing factor (CRF) and arginine vasotocin (AVT), which stimulate the pituitary gland to release adrenocorticotropic hormone (ACTH). ACTH travels through the blood to the adrenal glands where corticosteroids are synthesized by the adrenal cortex. Vitellogenin is incorporated into the yolk and is produced by the liver. For simplicity, feedbacks were ignored in the figure, which is based on Caro et al. (2013), Ikegami and Yoshimura (2016), and Lattin et al. (2016).

maturation and release of hormones (Figure 9.1). Although daylength varies less in the tropics, there is evidence that some species, such as antbirds, can respond to differences in daylength as short as 17 min (Hau 2001; reviewed by Dawson 2008) and some photoperiod experiments with African stonechats (*Saxicola torquatus axillaris*) indicate that it is the change in time of day when sunrise and sunset occur (~30 min variation per year) that influences circannual rhythms, rather than total daylength, which is relatively constant (Goymann et al. 2012; see also Shaw 2017). Even some opportunistic species, such as crossbills (*Loxia* spp.), appear to require regular changes in photoperiod to initiate a physiological window during which breeding can occur once food conditions are adequate (Dawson 2008). Increasing photoperiod often initiates reproductive activity, but species and individuals (Watts et al. 2015) differ in terms of the threshold amount of light required to start gonadal growth and they also differ in the rate at which growth proceeds under increasing daylength (Dawson 2008). These interspecific differences in gonadal responses could influence the rate at which various species respond to temperature or other environmental conditions.

9.2.2 Hormones

A variety of hormones are involved in responding to both photoperiod and temperature (Figure 9.1). Studies of quail have found that, under increasing daylength, photoreceptors deep in the brain activate the secretion of thyroid-stimulating hormone (TSH) in the hypothalamus (Pars tuberalis), which initiates local production of triiodothryonine (T3; Ikegami and Yoshimura 2016). T3, in turn, triggers secretion of gonadotropin releasing hormone (GnRH), which leads to the production of gonadotropins (e.g., luteinizing hormone [LH], follicle-stimulating hormone [FSH]; Figure 9.1) that promote the development of reproductive organs. TSH is also produced in the anterior pituitary, and it travels to the thyroid gland to stimulate production of T3, which increases metabolic rate and is associated with breeding relatively early in house sparrows (*Passer domesticus*; Chastel et al. 2003).

Once daylength, or other cues such as temperature and rainfall, lead to the start of breeding, it can still take several days to a few weeks for the development of the follicles and the release of eggs. The period of rapid yolk development can last from 3–4 days per follicle in small passerines up to 25 days in large seabirds (Williams 2012). In controlled aviary experiments with great tits (*Parus major*) the size of follicles in March only explained 1.2 per cent of the variation in laying date (starts in mid-April), and there was also no correlation between laying date and LH (Schaper et al. 2012a). Although there is usually a surge of LH 4–6 hours before the laying of eggs (Johnson 2015), to date, there is little evidence of a strong correlation between laying date and levels of hormones in the weeks to days before laying. For example, corticosterone stimulates foraging behaviour and, in some species, is associated with increases in plasma lipoproteins (triglycerides) that are related to weight gain and production of the egg yolk (Hennin et al. 2016). However, direct evidence for a correlation between laying date and corticosterone is mixed. Some of this variation may be related to how different species acquire energy for laying; i.e., whether they are 'capital' breeders that rely on stored fat reserves or 'income' breeders that acquire the energy needed for laying from food they eat daily. It is also important to consider the relative effects of both corticosterone and rates of fattening on condition (Hennin et al. 2016), which are discussed more below (section 9.2.6).

9.2.3 Effects of temperature—cue or constraint?

The timing of breeding varies between years, even in the same individuals, so photoperiod likely interacts simultaneously with other environmental cues such as temperature, food abundance, and social stimulation to set the physiological window during which egg-laying will occur (Dawson 2008; Schoech and Hahn 2007). Despite many studies showing correlations between laying date and temperature (Dunn 2004), it is still not clear how temperature influences the start of laying. There is evidence that temperature can act as both a cue for improving conditions and an energetic constraint on the timing of laying. Some of the best evidence that temperature is a cue comes from controlled aviary studies. In a six-year experiment with great tits in temperature and photoperiod controlled aviaries, birds in a warm treatment laid an average of 6.3 days earlier than birds in the cold treatment (~4°C difference; Visser et al. 2009). Laying date was also most closely associated with temperature three weeks prior to the date of laying, which suggests that the birds are using temperatures early in the season as a cue for laying, rather than lifting an energetic constraint just prior to laying. A follow-up study indicated that it is the rate of increase in temperature, rather than the mean daily temperature, that is most strongly correlated with laying date (Schaper et al. 2012b). If temperature acts as a longer-term predictive cue, then birds might actually be responding to the increase in vegetation in the spring (e.g., bud burst) or the presence of food, rather than temperature per se (Bourgault et al. 2010). In aviary experiments, great tits exposed to unfolding leaves and caterpillars did not advance their laying date compared with pairs that only saw branches with buds (Schaper et al. 2011). However, the importance of various cues appears to differ between populations. In Corsica, for example, populations of blue tits (*Cyanistes caeruleus*) time their laying based on the phenology of vegetation or insects in deciduous oak (*Quercus humilis*) forest, but in evergreen oak (*Q. ilex*) forests 6–25 km away, females lay their eggs based primarily on temperature cues (Thomas et al. 2010). The scale of analysis may also be important, since tits in England appear to time their laying in relation to bud burst of individual oak trees near the nest (Hinks et al. 2015).

Although temperature might act as a longer-term cue, there is also evidence that temperature is a constraint on the timing of egg-laying. For example, extreme cold (or heat) events are known to delay laying or cause abandonment in some species (Moreno and Møller 2011). Also, in some Arctic (Liebezeit et al. 2014; Saalfeld and Lanctot 2017) and boreal (Lehikoinen et al. 2011) species laying may be prevented by snow which can cover nesting sites or limit the availability of food. Temperature could also have direct effects on thermoregulation of birds or their ability to maintain viable eggs, and, thus, be an energetic constraint on laying date. Experiments that heat and cool aviaries under controlled conditions (Meijer et al. 1999; Salvante et al. 2007; Visser et al. 2009) support the idea that temperature has a direct effect on timing of laying.

Although we often find correlations between laying date and temperature, one of the biggest gaps in our knowledge is how temperature is perceived and then integrated into other physiological processes that lead to laying (Johnson 2015; Caro et al. 2013). The preoptic area (POA) of the hypothalamus is innervated up to 30 per cent by temperature-sensitive neurons and, thus, is an important area for integrating temperature information (Caro et al. 2013). The hypothalamus not only produces releasing hormones for many gonadotropins that lead to ovarian development, but also produces thyrotropin-releasing hormone (TRH), which ultimately leads to production of T3 in the thyroid and increased metabolic and fattening rates that are related to timing of breeding (Chastel et al. 2003; Caro et al. 2013).

9.2.4 Effects of precipitation and wind

Although temperature is the usual focus of studies of climate change, there are two other important environmental factors that can influence the timing of breeding in birds. First, in areas of relatively high constant temperature, the stimulus for laying may be rainfall, which signals the beginning of the growing season, rather than temperature (Dawson 2008). Zebra finches (*Taeniopygia guttata*) are considered one of the classic examples of this opportunistic breeding strategy, but detailed studies suggest that it follows a regular annual schedule of breeding that is influenced by photoperiod as well as rainfall (Hahn et al. 2008). A second, and less well known, factor that appears to be changing the timing of breeding is wind speed. Surface winds over much of the Northern Hemisphere land mass have slowed 5–15 per cent between 1979 and 2008, and the slowing is correlated with increases in vegetation as measured by satellite, which may be due to changes in land use, as well as increased vegetation growth from warmer temperatures (Vautard et al. 2010). In contrast, winds over the oceans have been increasing from 1991 to 2008, leading to higher waves (Young et al. 2011). Wind speed is especially important for aerial foragers, such as seabirds (Weimerskirch et al. 2012; Saraux et al. 2016), swallows and swifts, as higher winds (and waves) may make it more difficult to locate and catch prey. Relatively few studies have examined the effects of wind speed on laying date or breeding success, but three studies of swallows in Denmark (Møller 2013), and Alaska (Irons et al. 2017) and Wisconsin (P. Dunn unpubl. data) USA indicate that wind speeds are decreasing and leading to earlier laying, presumably because it is easier for females to catch food when it is less windy.

9.2.5 Food abundance

Food abundance is thought to be one of the most important of the secondary cues used to fine-tune date of laying. As just noted, in species like swallows that rely on daily food intake to produce eggs (income breeders) the timing of laying may be limited by the availability of food they can acquire prior to laying. On the other hand, capital breeders, especially larger species like some waterfowl, can carry substantial fat reserves on their bodies to their breeding grounds and use them (at least partially) to form eggs (Drent 2006). Although there is a large literature suggesting that food abundance limits the timing of breeding, there are a number of issues that can make interpreting the results difficult (Williams 2012). For example, studies that provide artificial food generally show that it advances laying date (Ruffino et al. 2014), but these studies can be confounded by competition for the supplemental food or differences in quality between the supplemental and natural foods. Some experimental studies have taken the opposite approach and decreased food intake for females, through temporary captivity (Descamps et al. 2011) or feather clipping (Nooker et al. 2005). As expected, these studies find that food restriction leads to later laying. In these studies experimental females also lost body condition (mass adjusted for size), which is often suggested as an important constraint on the timing of laying. However, in income breeders, females do not store large body fat reserves, and thus feather clipping primarily affects laying date by reducing foraging efficiency and the daily food intake, rather than body condition (stored body fat; Nooker et al. 2005). These differences are important because we should not necessarily expect to see a relationship between laying date and body condition in income breeders (Williams 2012). In these species, the energy for food is still important for laying, but it is less likely to come from stored fat. One promising way to monitor energy intake in income breeders is to measure levels of triglycerides, such as vitellogenin (VTG), which are incorporated into eggs and correlated with laying date in some species (Hennin et al. 2016).

9.2.6 Integration of cues for laying

Experiments that examine a variety of factors simultaneously, such as temperature, food supply, physiology, and photoperiod, reveal that laying date is influenced by several factors, and most likely in a hierarchical process, beginning with increasing daylength. Social cues, such as a mate and his singing, are also important, as females will not lay eggs without a male in some aviary studies (Deviche 2015; Watts et al. 2016). We still know little about how or where in the body these cues are integrated over the course of the breeding season, although it is likely that the hypothalamus plays a key role (Davies and Deviche 2014). There is also a large gap in our understanding of the physiological mechanisms that lead to the start of rapid follicular growth and ovulation. Most previous studies have focused on hormones, particularly LH and progesterone, but there are a variety of other hormones and pathways that influence metabolism and condition (e.g., TH, corticosterone, prolactin, leptin, gonadotrophin inhibiting hormone, and neuropeptide Y) and, thus, could also be influential in the timing of laying (Caro et al. 2013; Davies and Deviche 2014; Figure 9.1). One promising avenue of research is combining studies of both physiological fattening (using triglycerides) and energy demand (corticosterone) to produce a more integrated measure of condition. In a recent study of eiders, it was the interaction of triglyceride and corticosterone levels, after controlling for body mass, that best predicted individual variation in laying date (Hennin et al. 2016). In this case, early-laying birds had high levels of triglycerides in the plasma, indicating fattening, and low levels of corticosterone, which suggests lower levels of energy mobilization, allowing for more resources to go into follicular growth. More studies such as this are needed that integrate both energy input and output for condition and egg production.

9.3 Patterns of changing laying date

Many studies have shown long-term trends toward earlier egg-laying, as well as laying earlier in response to warmer temperatures. For example, in a review of 196 studies from 89 species, Dunn and Møller (2014) found advancement in laying of 0.19 days per year and 2.0 days per degree C (based on 71 and 75 species averages, respectively). The same estimate (–0.19 d/yr) for the advancement of laying was found in a survey of UK birds during 1976–2005, but in more recent years (1996–2005) the rate of advancement has increased (–0.27 d/yr; Thackery et al. 2010).

9.3.1 Interspecific variation in laying date

Although birds are generally breeding earlier, there is a great deal of variation between and within species that is not understood well. Species may vary in their response to climate change for a variety of ecological, life history, geographic, and phylogenetic reasons. In particular, migratory species may have a different hierarchy of cues for breeding (e.g., photoperiod and food) than resident species. There could also be differences in geography (temperate vs tropical) that affect the magnitude of response, as there have been much larger increases in temperature in certain regions, particularly in boreal and Arctic regions (Myneni et al. 1997). Indeed, across a variety of taxa, there has been a small but significantly greater response at higher latitudes (Parmesan 2007). Thus, we might predict greater phenological responses by birds at higher latitudes. Similarly, it is often suggested that long-distance migrants will be less responsive to climate change than resident species because they start migration from the wintering grounds without any accurate cues to temperature change on the breeding grounds.

To test these predictions, I updated the analysis of changes in laying date in Dunn and Møller (2014) with new studies published since 2011 (see www.oup.co.uk/companion/dunn&moller). As in the previous study, advances in laying date per year were associated with multiple broods per season, trophic level, and habitat. Species with multiple broods per season were advancing laying date (−0.352 days per year) more than single-brooded species (−0.216). It is thought that multi-brooded species advance the laying date of their first brood more than single-brooded species, because reproductive success for multi-brooded species depends more on the duration of the season than in single-brooded species, and so the timing of the first brood relative to peak food abundance is less important to seasonal reproductive success (Crick et al. 1993). As a consequence, multi-brooded species are more likely to benefit from breeding as soon as possible in the spring. Laying date was also advancing faster for primary (−0.403 days per year) and tertiary (−0.315) consumers than secondary consumers (−0.134; Dunn and Møller 2014). A similar pattern was found among birds in the UK (Thackeray et al. 2016). The type of habitat (agricultural or not), northernmost breeding latitude, migration distance, body mass, and duration of the study had non-significant effects on the change in laying date. Some studies have found that migration distance and other factors were correlated with arrival date (e.g., Salido et al. 2012; Chapter 8). However, laying date was not correlated with arrival date in a comparative analysis (Dunn and Møller 2014), which suggests that some species, such as long-distance migrants, might be compensating for a relatively late arrival by decreasing the time prior to laying (McDermott and DeGroote 2017). On the other hand, Kluen et al. (2017) found that among 26 common species in Finland, resident species were advancing their laying date more strongly than migrants, as predicted above. It is important to note that arrival and egg-laying dates likely evolve independently, because they tend to influence different components of fitness (adult and nestling survival; Kristensen et al. 2015). Thus, if there are high survival costs to adults of laying early, then it may actually be optimal for migrants to lay their eggs later in the season and appear to be 'mismatched' with the food supply (section 9.4.1). As a consequence, observations of mismatches are insufficient evidence to conclude that populations are not responding adaptively to climate change.

It is also important to be cautious about interpreting meta-analyses because they sometimes combine studies from different populations and time periods, and, thus, some biologically important patterns from individual studies may be obscured in a large meta-analysis. This is particularly important if there are opposite trends in different populations or time periods, as seems to occur in some seabirds (Hindell et al. 2012; Chambers et al. 2014). Related to this, a recent study of body condition using 21 years of data from 46 species at 80 sites in the Netherlands found as much variation between populations as between species (McLean et al. 2018). Body condition is often related to laying date, so this implies that we may need to incorporate responses specific to local populations or study sites, rather than assume one average value for a species.

9.3.2 Intraspecific differences in the timing of breeding

Studies of laying date in different populations of the same species control for life history differences between species and can potentially help to reveal some of the ecological causes of variation in response to climate change. As noted above with body condition, these studies have already revealed substantial variation in the response of different local populations to climate change. There are many examples of intraspecific variation in the long-term response to climate change. In these cases, it appears that the most likely explanation for geographic differences in the timing of breeding is local variation in temperature change. In some well-studied species, such as pied flycatchers (Both et al. 2004), great and blue tits (e.g., Visser et al. 2003; see also Sæther et al. 2003) and barn swallows (Møller 2008) there is considerable variation between study sites that is related to local temperature differences. Similar geographic variation in the response of species to temperature has also been reported in several North American species (Torti and Dunn 2005). Some of these differences can occur at very small geographic scales. For example, in Corsica, the laying date of blue tits is about a month later in evergreen than deciduous oak forests that are only 6–25 km apart (Thomas et al. 2010). In both habitats, there are significant correlations between timing of laying and both temperature and insect phenology. However, path analysis has revealed the relative importance of these relationships differs in the two habitats, as well as over the course of the breeding season (Thomas et al. 2010). These differences between habitats are also reflected in the genome, despite high gene flow (Szulkin et al. 2015). Intraspecific differences in phenology appear to be widespread, and if some populations are stable or increasing while others decline, then this variation may help buffer the effects of climate change on overall population size.

9.4 What are the ultimate consequences of changes in laying date?

9.4.1 Is there a mismatch between timing of breeding and food supply?

It is often suggested that climate change is making the current cues for the start of laying (photoperiod, temperature, and food) poorer predictors of food supply later in the season when nestlings require large amounts of food for development. As a consequence, the cues for starting to breed become decoupled from peaks in food abundance, which will potentially lead to mismatches between the start of breeding and the availability of food for nestlings. These mismatches are thought to arise because increasing spring temperatures lead to faster development of plants and insects than the advancement of laying date or development of nestlings. Thus, warmer spring temperatures can lead to earlier peaks in food supply and, consequently, a mismatch between the timing of peak food abundance and the energetic requirements of developing young (Visser et al. 1998; Both 2010).

Even though the mismatch hypothesis has been mentioned hundreds of times since 2000, there have been relatively few studies with measurements of food abundance (or a close index) in relation to timing of breeding. Indeed, a review of the literature reveals only 30 studies with measurements of food abundance (or an index of it), and, of these, 37 per cent (11) conclude that mismatching is associated with reduced reproductive success (Online Supplementary Table 9.1). Most of the studies supporting the mismatch hypothesis examined populations that breed in the Arctic where the breeding season is short or they have a sharp peak in food supply during the nesting season. In general, the mismatch hypothesis has several important assumptions related to: 1) seasonality of the food supply, 2) the strength of selection on synchronized breeding, and 3) the predictability of peaks in food supply later in the season (reviewed by Dunn et al. 2011). Mismatching should be more important in populations that rely on food supplies that show a seasonal peak, as occurs in some tits and flycatchers that primarily feed caterpillars to their young. In contrast, mismatching will be less important for species that rely on food that is relatively abundant throughout the breeding season. For example, in tree swallows (*Tachycineta bicolor*) insect abundance generally continues to increase throughout the nestling period, and there is no sharp peak in abundance (Dunn et al. 2011). Synchronization may also be difficult because at the time of laying there are few environmental cues to predict when the peak of food abundance will occur (Dunn et al. 2011). In other species, such as waterfowl (Drever and Clark 2007) and grouse (Wegge and Rolstad 2017), timing of laying appears to have little influence on reproductive success because it is most strongly influenced by nest predation. In contrast to the potential negative effects of mismatching, some theoretical models suggest that mismatching could be adaptive in certain situations (Visser et al. 2012; Johansson et al. 2015). For example, if there is competition for territories that favours early arrival on the breeding grounds, but early arrival also increases the likelihood of adult mortality (e.g., due to extreme weather), then it may be optimal to breed later than expected based simply on the food supply for nestlings.

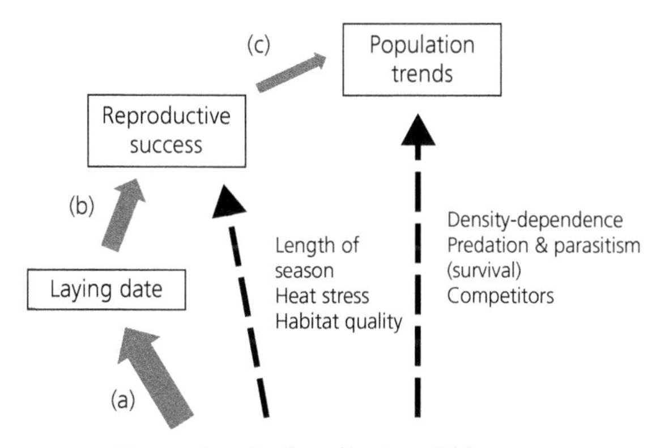

Figure 9.2 Relationships between temperature and laying date, reproductive success and population growth rates. Significant relationships have been found between temperature and laying date (a) as well as laying date and reproductive success (b). However, the relationship between reproductive success and population growth rate (c) is weak in most studies. Temperature effects on reproductive success and population growth tend to be weaker because of indirect effects of temperature on these variables. For example, temperature can have direct effects on reproduction by producing heat stress or by lengthening the season (and increasing opportunities for breeding). Temperature can also have effects on population growth by influencing survival through several factors such as competition from other individuals and predation. Thickness of arrows indicates relative strength of the relationship. Figure is based on McLean et al. (2016).

9.4.2 Are breeding performance and population size changing with laying date?

In many species, individuals that start to lay earlier also have larger clutch sizes, and this often translates into more fledglings. Given that many species show a long-term advancement in laying date, we might also expect an increase in clutch size in these species, and possibly greater reproductive success and larger population sizes (Figure 9.2). However, the results of empirical studies reveal a more complicated situation. For example, recent studies of long-term productivity and population size in common UK species found that breeding is earlier in warmer years, and earlier breeding is related to greater fledging success, as predicted (McLean et al. 2016; Franks et al. 2018). However, the extent of mismatching did not influence fledging success (Franks et al. 2018) and fledging success was not related to population trends (McLean et al. 2016). Although Franks et al. (2018) found that species with the largest advance in laying dates over the last 30 years have more positive population trends, they did not find a link through mismatching, which accounted for just 5 per cent of the annual variation in fledging success. Thus, the simple pathway described above does not seem to connect timing of laying with trends in population size (Figure 9.2). At a larger scale, there was also no relationship between population trends and the advancement of laying date over time in a comparative analysis of 195 study populations (Dunn and Møller 2014) or an update for this chapter with 128 additional study populations (see Supplement Online).

Several explanations have been proposed for the weak links between productivity and population trends. First, several studies have suggested that there might be density-dependent processes that buffer the potential effects of productivity on population size. For example, demographic modelling suggests that survival in great tits is density-dependent, so in years with mismatching and low productivity, the survival of juveniles increases, and vice versa for years when there is little mismatching and productivity is higher (Reed et al. 2013). Second, there could be regional effects of mismatching that disappear when they are averaged out in national and international analyses. As an example, Franks et al. (2018) pointed out that willow warblers (*Phylloscopus trochilus*) have low productivity and are declining in southern Britain, but they have high productivity and are increasing in the north. Lastly, there are many factors that can influence population trends other than through the laying date–reproductive success pathway underlying the mismatch hypothesis (a and b in Figure 9.2). For example, McLean et al. (2016) found a significant direct effect of temperature on reproductive success in their analysis of 27 UK species (path c in Figure 9.2). Extreme weather (both hot and cold) can reduce reproductive success (Chapter 10) as well as affect survival at other times of the year. Indeed, modelling of UK species suggests that population change is influenced more strongly by recruitment and adult survival than reproductive success (Robinson et al. 2014). Climate change can also influence population growth through altered trophic interactions. For example, climate change in Arizona has decreased snowfall, which has allowed elk (*Cervus elaphus*) into higher elevation areas where they over-browsed woody plants. The over-browsing reduced plant density and increased nest predation by small mammals, which led to the decline of six species of songbirds (Martin and Maron 2012).

9.5 Conclusions and future avenues of research

There has been a large increase in the number of studies of avian phenology in the last eight years (Dunn and Møller 2014; online supplement). Despite this increased interest and information, there are still many of the same needs for additional data and experiments. Starting first with the proximate factors that influence timing of laying, there have been many informative studies using controlled aviary experiments, hormonal measurements, and long-term studies using path analysis to tease apart the factors influencing the date of laying. However, as mentioned above, there is still a large gap in our understanding of the physiological mechanisms that influence the timing of egg-laying in the final weeks and days before the first egg is laid (e.g., what initiates the period of rapid follicular growth?). Most previous studies have focused on a relatively small number of hormones, and usually in isolation. Given the potentially complex interactions between hormones and condition, it would be beneficial to study more hormones (e.g., thyroxine) that influence metabolism, and their inter-relationships with triglycerides and the timing of egg-laying. Recent studies of fattening rates and their interaction with corticosterone (Hennin et al. 2016) provide an excellent example of how to take a more integrative approach that focuses on multiple factors that influence body condition, rather than focusing on one or two hormones.

The bigger questions are how does timing of laying affect reproductive success, and does that affect population growth? In the past eight years many researchers have heeded the call for more studies of how food abundance (or an index of it) relates to timing of breeding, reproductive success, and, ultimately, population growth (Online Supplementary Table 9.1). With more studies from a wider variety of species, it has become obvious that there is no clear answer to how reproductive success depends on synchronizing the timing of breeding with the food supply (i.e., the mismatch hypothesis). Although there are some well-studied examples where mismatching appears to be important to reproductive success, there are also many populations where the hypothesis does not seem to apply because: 1) the food supply is abundant during chick rearing, so timing is not important, 2) birds cannot predict the peak of food abundance for their young early in the season when eggs are laid, or 3) reproductive success is primarily limited by predation and extreme weather events that negate any advantages of synchronizing laying with food supplies. There is also increasing evidence that even though events in the breeding cycle (arrival or laying date, fledging success) are related to climate change, they do not subsequently produce the predicted negative (or positive) effects on population growth. In many cases reproductive success is only weakly linked to population growth. Theoretical models indicate that population growth will depend on a variety of factors including the strength of trophic interactions, density dependence, effects on other components of fitness (adult survival), and life history traits that influence survival and fecundity (Johansson et al. 2015). Ultimately, to make long-term predictions, we need to understand how changes in timing of breeding affect population growth. Thus, we need analyses of the full annual cycle to fully understand what drives population changes, regardless of the cause, be it climate change, habitat loss, exploitation, or pollution (Chapter 17).

Acknowledgments

I thank A.P. Møller and L.A. Whittingham for comments on the manuscript.

References

Both, C. (2010). Food availability, mistiming, and climatic change. In: A.P. Møller, W. Fiedler, and P. Berthold (eds), *Effects of climate change on birds*, pp. 129–147. Oxford University Press, Oxford, UK.

Both, C., Artemyev, A.V., Blaauw, B., Cowie, R.J., Dekhuijzen, A.J., Eeva, T., Enemar, A., Gustafsson, L., Ivankina, E.V., Järvinen, A., Metcalfe, N.B., Nyholm, N.E.I., Potti, J., Ravussin, P.-A., Sanz, J.J., Silverin, B., Slater, F.M., Sokolov, L.V., Török, J., Winkel, W., Wright, J., Zang, H., and Visser, M.E. (2004). Large-scale geographical variation confirms that climate change causes birds to lay earlier. *Proceedings of the Royal Society of London Series B: Biological Sciences*, 271, 1657–62.

Bourgault, P., Thomas, D., Perret, P., and Blondel, J. (2010). Spring vegetation phenology is a robust predictor of breeding date across broad landscapes: A multi-site approach using the Corsican blue tit (*Cyanistes caeruleus*). *Oecologia*, 162, 885–92.

Caro, S.P., Schaper, S.V., Hut, R.A., Ball, G.F., and Visser, M.E. (2013). The case of the missing mechanism: How does temperature influence seasonal timing in endotherms? *PLoS Biology*, 11, e1001517.

Chambers, L.E., Dann, P., Cannell, B., and Woehler, E.J. (2014). Climate as a driver of phenological change in southern seabirds. *International Journal of Biometeorology*, 58, 603–12.

Chastel, O., Lacroix, A., and Kersten, M. (2003). Prebreeding energy requirements: Thyroid hormone, metabolism and the timing of reproduction in house sparrows *Passer domesticus*. *Journal of Avian Biology*, 34, 298–306.

Crick, H.Q.P., Gibbons, D.W., and Magrath, R.D. (1993). Seasonal changes in clutch size in British birds. *Journal of Animal Ecology*, 62, 263–73.

Davies, S., and Deviche, P. (2014). At the crossroads of physiology and ecology: Food supply and the timing of avian reproduction. *Hormones and Behavior*, 66, 41–55.

Dawson, A. (2008). Control of the annual cycle in birds: Endocrine constraints and plasticity in response to ecological variability. *Philosophical Transactions of the Royal Society of London Series B: Biological Sciences*, 1497, 1621–33.

Descamps, S., Bêty, J., Love, O.P., and Gilchrist, H.G. (2011). Individual optimization of reproduction in a long-lived migratory bird: A test of the condition-dependent model of laying date and clutch size. *Functional Ecology*, 25, 671–81.

Deviche, P. (2015). Reproductive behavior. In: C.G. Scanes (ed.), *Sturkie's Avian Physiology*, 6th edition. Academic Press, San Diego, CA, USA.

Drent, R.H. (2006). The timing of birds' breeding seasons: The Perrins hypothesis revisited especially for migrants. *Ardea*, 94, 305–22.

Drever, M.C., and Clark, R.G. (2007). Spring temperature, clutch initiation date and duck nest success: A test of the mismatch hypothesis. *Journal of Animal Ecology*, 76, 139–48.

Dunn, P.O. (2004). Breeding dates and reproductive performance. In: A.P. Møller, W. Fiedler, and P. Berthold (eds), *Birds and Climate Change*, vol. 35, pp. 67–85. Elsevier, San Diego, CA, USA.

Dunn, P.O., and Møller, A.P. (2014). Changes in breeding phenology and population size of birds. *Journal of Animal Ecology*, 83, 729–39.

Dunn, P.O., Winkler, D.W., Whittingham, L.A., Hannon, S.J., and Robertson, R.J. (2011). A test of the mismatch hypothesis: How is timing of reproduction related to food abundance in an aerial insectivore? *Ecology*, 92, 450–61.

Franks, S.E., Pearce-Higgins, J.W., Atkinson, S., Bell, J.R., Botham, M.S., Brereton, T.M., Harrington, R., and Leech, D.I. (2018). The sensitivity of breeding songbirds to changes in seasonal timing is linked to population change but cannot be directly attributed to the effects of trophic asynchrony on productivity. *Global Change Biology*, 24, 957–71.

Goymann, W., Helm, B., Jensen, W., Schwabl, I., and Moore, I.T. (2012). A tropical bird can use the equatorial change in sunrise and sunset times to synchronize its circannual clock. *Proceedings of the Royal Society of London Series B: Biological Sciences*, 279, 3527–34.

Hahn, T.P., Cornelius, J.M., Sewall, K.B., Kelsey, T.R., Hau, M., and Perfito, N. (2008). Environmental regulation of annual schedules in opportunistically-breeding songbirds: Adaptive specializations or variations on a theme of white-crowned sparrow? *General and Comparative Endocrinology*, 157, 217–26.

Hau, M. (2001). Timing of breeding in variable environments: Tropical birds as model systems. *Hormones and Behavior*, 40, 281–90.

Hennin, H.L., Bêty, J., Legagneux, P., Gilchrist, H.G., Williams, T.D., and Love, O.P. (2016). Energetic physiology mediates individual optimization of breeding phenology in a migratory Arctic seabird. *The American Naturalist*, 188, 434–45.

Hindell, M.A., Bradshaw, C.J.A., Brook, B.W., Fordham, D.A., Kerry, K., Hull, C., and McMahon, C.R. (2012). Long-term breeding phenology shift in royal penguins. *Ecology and Evolution*, 2, 1563–71.

Hinks, A.E., Cole, E.F., Daniels, K.J., Wilkin, T.A., Nakagawa, S., and Sheldon, B.C. (2015). Scale-dependent phenological synchrony between songbirds and their caterpillar food source. *The American Naturalist*, 186, 84–97.

Ikegami, K., and Yoshimura, T. (2016). Comparative analysis reveals the underlying mechanism of vertebrate seasonal reproduction. *General and Comparative Endocrinology*, 227, 64–8.

Irons, R.D., Harding Scurr, A., Rose, A.P., Hagelin, J.C., Blake, T., and Doak, D.F. (2017). Wind and rain are the primary climate factors driving changing phenology of an aerial insectivore. *Proceedings of the Royal Society of London Series B: Biological Sciences*, 284, 20170412.

Johansson, J., Kristensen, N.P., Nilsson, J.Å., and Jonzén, N. (2015). The eco-evolutionary consequences of interspecific phenological asynchrony—a theoretical perspective. *Oikos*, 124, 102–12.

Johnson, A.L. (2015). Reproduction in the female. In: C.G. Scanes (ed.), *Sturkie's Avian Physiology*, 6th edition. Academic Press, San Diego, CA, USA.

Kluen, E., Nousiainen, R., and Lehikoinen, A. (2017). Breeding phenological response to spring weather conditions in common Finnish birds: Resident species respond stronger than migratory species. *Journal of Avian Biology*, 48, 611–19.

Kristensen, N.P., Johansson, J., Ripa, J., and Jonzén, N. (2015). Phenology of two interdependent traits in migratory birds in response to climate change. *Proceedings of the Royal Society of London Series B: Biological Sciences*, 282, 20150288.

Lattin, C.R., Breuner, C.W., and Romero, L.M. (2016). Does corticosterone regulate the onset of breeding in free-living birds? The cort-flexibility hypothesis and six potential mechanisms for priming corticosteroid function. *Hormones and Behavior*, 78, 107–20.

Lehikoinen, A., Ranta, E., Pietiainen, H., Byholm, P., Saurola, P., Valkama, J., Huitu, O., Henttonen, H., and Korpimäki, E. (2011). The impact of climate and cyclic food abundance on the timing of breeding and brood size in four boreal owl species. *Oecologia*, 165, 349–355.

Leitner, S., Van't Hof, T.J., and Gahr, M. (2003). Flexible reproduction in wild canaries is independent of photoperiod. *General and Comparative Endocrinology*, 130, 102–8.

Liebezeit, J.R., Gurney, K.E.B., Budde, M., Zack, S., and Ward, D. (2014). Phenological advancement in Arctic bird species: Relative importance of snow melt and ecological factors. *Polar Biology*, 37, 1309–20.

Martin, T.E., and Maron, J.L. (2012). Climate impacts on bird and plant communities from altered animal-plant interactions. *Nature Climate Change*, 2, 195–200.

McDermott, M.E., and Degroote, L.W. (2017). Linking phenological events in migratory passerines with a changing climate: 50 years in the laurel highlands of Pennsylvania. *PLoS One*, 12, e0174247.

McLean, N., Lawson, C.R., Leech, D.I., and van de Pol, M. (2016). Predicting when climate-driven phenotypic change affects population dynamics. *Ecology Letters*, 19, 595–608.

McLean, N., van der Jeugd, H.P., and van de Pol, M. (2018). High intra-specific variation in avian body condition responses to climate limits generalisation across species. *PLoS One*, 13, e0192401.

Meijer, T., Nienaber, U., Langer, U., and Trillmich, F. (1999). Temperature and timing of egg-laying of European starlings. *Condor*, 101, 124–32.

Møller, A.P. (2008). Climate change and micro-geographic variation in laying date. *Oecologia*, 155, 845–57.

Møller, A.P. (2013). Long-term trends in wind speed, insect abundance and ecology of an insectivorous bird. *Ecosphere*, 4, 6.

Moreno, J., and Møller, A. P. (2011). Extreme climatic events in relation to global change and their impact on life histories. *Current Zoology*, 57, 375–89.

Myneni, R.B., Keeling, C.D., Tucker, C.J., Asrar, G., and Nemani, R.R. (1997). Increased plant growth in the northern high latitudes from 1981 to 1991. *Nature*, 386, 698–702.

Nooker, J.K., Dunn, P.O., and Whittingham, L.A. (2005). Effects of food abundance, weather and female condition on reproductive success in tree swallows (*Tachycineta bicolor*). *Auk*, 122, 1225–38.

Parmesan, C. (2007). Influences of species, latitudes and methodologies on estimates of phenological response to global warming. *Global Change Biology*, 13, 1860–72.

Reed, T.E., Grotan, V., Jenouvrier, S., Sæther, B.-E., and Visser, M.E. (2013). Population growth in a wild bird is buffered against phenological mismatch. *Science*, 340, 488–91.

Robinson, R.A., Morrison, C.A., Baillie, S.R., and Francis, C. (2014). Integrating demographic data: Towards a framework for monitoring wildlife populations at large spatial scales. *Methods in Ecology and Evolution*, 5, 1361–72.

Rodriguez, C., and Bustamante, J. (2003). The effect of weather on lesser kestrel breeding success: Can climate change explain historical population declines? *Journal of Animal Ecology*, 72, 793–810.

Ruffino, L., Salo, P., Koivisto, E., Banks, P.B., and Korpimäki, E. (2014). Reproductive responses of birds to experimental food supplementation: A meta-analysis. *Frontiers in Zoology*, 11, 80.

Saalfeld, S.T., and Lanctot, R.B. (2017). Multispecies comparisons of adaptability to climate change: A role for life-history characteristics? *Ecology and Evolution*, 7, 10492–502.

Sæther, B.E., Engen, S., Møller, A.P., Matthysen, E., Adriaensen, F., Fielder, W., Leivits, A., Lambrechts, M.M., Visser, M.E., Anker-Nilssen, T., Both, C., Dhondt, A.A., McCleery, R.H., McMeeking, J., Potti, J., Røstad, O.W., and Thomson, D. (2003). Climate variation and regional gradients in population dynamics of two hole-nesting passerines. *Proceedings of the Royal Society of London Series B: Biological Sciences*, 270, 2397–2404.

Salido, L., Purse, B.V., Marrs, R., Chamberlain, D.E., and Shultz, S. (2012). Flexibility in phenology and habitat use act as buffers to long-term population declines in UK passerines. *Ecography*, 35, 604–13.

Salvante, G.K., Walzem, L.R., and Williams, D.T. (2007). What comes first, the zebra finch or the egg: Temperature-dependent reproductive, physiological, and behavioural plasticity in egg-laying zebra finches. *Journal of Experimental Biology*, 210, 1325–34.

Saraux, C., Chiaradia, A., Salton, M., Dann, P., and Viblanc, V.A. (2016). Negative effects of wind speed on individual foraging performance and breeding success in little penguins. *Ecological Monographs*, 86, 61–77.

Schaper, S.V., Dawson, A., Sharp, P.J., Caro, S.P., and Visser, M.E. (2012a). Individual variation in avian reproductive physiology does not reliably predict variation in laying date. *General and Comparative Endocrinology*, 179, 53–62.

Schaper, S.V., Dawson, A., Sharp, P.J., Gienapp, P., Caro, S.P., and Visser, M.E. (2012b). Increasing temperature, not mean temperature, is a cue for avian timing of reproduction. *American Naturalist*, 179, E55–E69.

Schaper, S.V., Rueda, C., Sharp, P.J., Dawson, A., and Visser, M.E. (2011). Spring phenology does not affect timing of reproduction in the great tit (*Parus major*). *Journal of Experimental Biology*, 214, 3664–71.

Scheffers, B.R., De Meester, L., Bridge, T.C.L., Hoffmann, A.A., Pandolfi, J.M., Corlett, R.T., Butchart, S.H.M., Pearce-Kelly, P., Kovacs, K.M., Dudgeon, D., Pacifici, M., Rondinini, C., Foden, W.B., Martin, T.G., Mora, C., Bickford, D., and Watson, J.E.M. (2016). The broad footprint of climate change from genes to biomes to people. *Science*, 354, 719.

Schoech, S.J., and Hahn, T.P. (2007). Food supplementation and timing of reproduction: Does the responsiveness to supplementary information vary with latitude? *Journal of Ornithology*, 148, S625–S632.

Shaw, P. (2017). Rainfall, leafing phenology and sunrise time as potential zeitgeber for the bimodal, dry season laying pattern of an African rain forest tit (*Parus fasciiventer*). *Journal of Ornithology*, 158, 263–75.

Skinner, W.R., Jefferies, R.L., Carleton, T.J., Rockwell, R.F., and Abraham, K.F. (1998). Prediction of reproductive success and failure in lesser snow geese based on early season climatic variables. *Global Change Biology*, 4, 3–16.

Szulkin, M., Gagnaire, P.A., Bierne, N., and Charmantier, A. (2015). Population genomic footprints of fine-scale differentiation between habitats in Mediterranean blue tits. *Molecular Ecology*, 25, 542–58.

Thackeray, S.J., Sparks, T.H., Frederiksen, M., Burthe, S., Bacon, P.J., Bell, J.R., Botham, M.S., Brereton, T.M., Bright, P.W., Carvalho, L., Clutton-Brock, T.H., Dawson, A., Edwards, M., Elliott, J.M., Harrington, R., Johns, D., Jones, I.D., Jones, J.T., Leech, D.I., Roy, D.B., Scott, W.A., Smith, M., Smithers, R.J., Winfield, I.J., and Wanless, S. (2010). Trophic level asynchrony in rates of phenological change for marine, freshwater and terrestrial environments. *Global Change Biology*, 16, 3304–13.

Thomas, D.W., Bourgault, P., Shipley, B., Perret, P., and Blondel, J. (2010). Context-dependent changes in the weighting of environmental cues that initiate breeding in a temperate passerine, the Corsican blue tit (*Cyanistes caeruleus*). *The Auk*, 127, 129–39.

Torti, V.M., and Dunn, P.O. (2005). Variable effects of climate change on six species of North American birds. *Oecologia*, 145, 486–95.

Vautard, R., Cattiaux, J., Yiou, P., Thepaut, J.-N., and Ciais, P. (2010). Northern hemisphere atmospheric stilling partly attributed to an increase in surface roughness. *Nature Geoscience*, 3, 756–61.

Visser, M.E., Adriaensen, F., van Balen, J.H., Blondel, J., Dhondt, A.A., van Dongen, S., du Feu, C., Ivankina, E.V., Kerimov, A.B., de Laet, J., Matthysen, E., McCleery, R., Orell, M., and Thomson, D.L. (2003). Variable responses to large-scale climate change in European *Parus* populations. *Proceedings of the Royal Society of London Series B: Biological Sciences*, 270, 367–72.

Visser, M.E., Holleman, L.J.M., and Caro, S.P. (2009). Temperature has a causal effect on avian timing of reproduction. *Proceedings of the Royal Society of London Series B: Biological Sciences*, 276, 2323–31.

Visser, M.E., te Marvelde, L., and Lof, M.E. (2012). Adaptive phenological mismatches of birds and their food in a warming world. *Journal of Ornithology*, 153, S75–S84.

Visser, M.E., van Noordwijk, A.J., Tinbergen, J.M., and Lessells, C.M. (1998). Warmer springs lead to mistimed reproduction in great tits (*Parus major*). *Proceedings of the Royal Society of London Series B: Biological Sciences*, 265, 1867–70.

Watts, H.E., Edley, B., and Hahn, T.P. (2016). A potential mate influences reproductive development in female, but not male, pine siskins. *Hormones and Behavior*, 80, 39–46.

Watts, H.E., Macdougall-Shackleton, S.A., and Hahn, T.P. (2015). Variation among individuals in photoperiod responses: Effects of breeding schedule, photoperiod, and age-related photoperiodic experience in birds. *Journal of Experimental Zoology Part A: Ecological Genetics and Physiology*, 323, 368–74.

Wegge, P., and Rolstad, J. (2017). Climate change and bird reproduction: Warmer springs benefit breeding success in boreal forest grouse. *Proceedings of the Royal Society of London Series B: Biological Sciences*, 284, 20171528.

Weimerskirch, H., Louzao, M., De Grissac, S., and Delord, K. (2012). Changes in wind pattern alter albatross distribution and life-history traits. *Science*, 335, 211–14.

Williams, T.D. (2012). *Physiological adaptations for breeding in birds*. Princeton University Press, Princeton, NJ.

Young, I., Zieger, S., and Babanin, A. (2011). Global trends in wind speed and wave height. *Science*, 332, 451–5.

CHAPTER 10

Physiological and morphological effects of climate change

Andrew E. McKechnie

10.1 Introduction and overview of avian thermoregulation

Birds occur everywhere from the coldest to the hottest places on Earth, and understanding the fluxes of energy, nutrients, and water between birds and their surroundings is vital for modelling their responses to climate change. Like mammals, birds are endothermic homeotherms that evolved mechanisms of endogenous heat production to defend body temperature (T_b) at approximately constant levels. Avian T_b values are higher than those of mammals, with the active-phase resting T_b of birds averaging ~41 °C, and there is far less variation in normothermic T_b across avian taxa than is the case among mammals (Prinzinger et al. 1991; Lovegrove 2012).

Thermoregulation in birds often, but not always, conforms to the classic Scholander–Irving model of endothermic homeothermy (Figure 10.1), which involves a thermoneutral zone (TNZ) below which increases in internal heat production compensate for heat loss, and above which the metabolic rate typically increases approximately linearly with increasing environmental temperature as a result of the energetic cost of heat dissipation processes such as panting (Scholander et al. 1950; Dawson and Whittow 2000). Evaporative water loss (EWL) is low and constant below the TNZ and begins to increase approximately linearly above an inflection point that may be below or above the upper limit of the TNZ (Dawson and Whittow 2000). A concept rarely encountered in the ornithological literature but regularly used in poultry science is the zone of least thermoregulatory effort: the range of environmental temperatures between the lower boundary of the TNZ (lower critical limit of thermoneutrality, T_{lc}) and the inflection in EWL (Etches et al. 2008; Figure 10.1).

Heat fluxes between birds and their environments occur via four avenues: radiation, convection, conduction, and evaporation. Major radiative fluxes involve the visible and short-wave infrared components of solar radiation, which for small species in exposed microsites may result in operative temperatures (see section 10.2.1) 15 °C or more above air temperature (T_a), with profound consequences for thermoregulation and behaviour in hot environments (Bakken 1976; Robinson et al. 1976). Heat fluxes via convection, conduction, and long-wave infrared radiation may involve either heat gains or losses, depending on the characteristics of a bird's thermal environment. Evaporation of water from a bird's respiratory system and skin is a critical avenue for heat loss, and indeed is the only avenue of heat dissipation available when environmental temperature exceeds T_b (Dawson and Whittow 2000).

Like other animals, birds use multiple physiological and behavioural processes to modify rates of heat gain or loss. Physiological processes include

McKechnie, A.E., *Physiological and morphological effects of climate change*. In: *Effects of Climate Change on Birds*. Second Edition. Edited by Peter O. Dunn and Anders Pape Møller: Oxford University Press (2019). © Oxford University Press.
DOI:10.1093/oso/9780198824268.003.0010

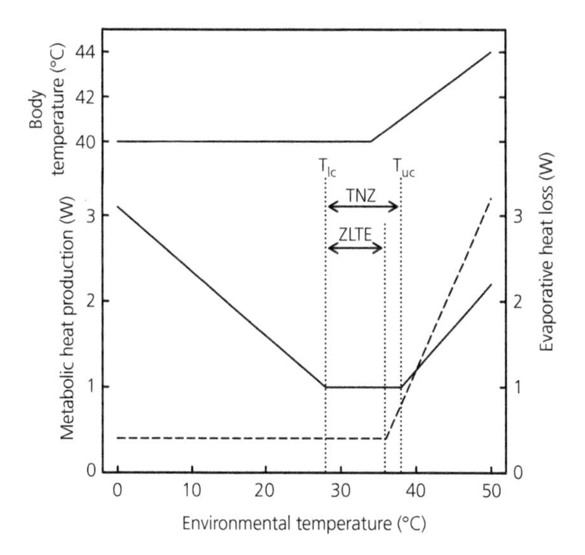

Figure 10.1 Approximate relationships between avian body temperature (upper panel), metabolic heat production (MHP, solid line, lower panel), and evaporative heat loss (EHL, dashed line, lower panel) at moderate to hot environmental temperatures. At high environmental temperatures, EHL/MHP > 1, providing the basis for defence of body temperature below environmental temperature. The thermoneutral zone (TNZ) is the range of environmental temperatures between the lower and upper critical limits of thermoneutrality (T_{lc} and T_{uc}, respectively). Also shown is the zone of least thermoregulatory effort (ZLTE), a concept frequently used in poultry science but rarely in ornithological literature. The patterns of increasing body temperature and metabolic heat production shown are those expected for taxa such as passerines in which panting is the major avenue of evaporative heat dissipation; taxa in which gular flutter or cutaneous water loss predominate generally show much more modest increases in resting metabolic rate above the thermoneutral zone, and in some cases none at all.

metabolic thermogenesis in response to elevated rates of heat loss, achieved primarily using both shivering and non-shivering thermogenesis (Dawson and Whittow 2000). Rates of heat loss can be greatly reduced through behaviours such as communal roosting or the construction of insulated roost sites. Under hot conditions, rates of heat gain can be minimized by selecting cool, shaded microsites and reducing activity and the associated metabolic heat production, and rates of heat loss can be increased by behaviours such as panting and wing-drooping. Additional processes to increase heat loss include thermal windows, with recent studies revealing that the avian bill acts as an important heat radiator, particularly in large-beaked taxa like toucans (Tattersall et al. 2009) and hornbills (van de Ven et al. 2016).

In the context of climate change, the direct effects of warming are easiest to discern in environments where high environmental temperatures constrain fitness components such as survival or reproduction through processes such as trade-offs between thermoregulation and foraging. For this reason, this chapter has a strong focus on arid-zone bird communities, in whose habitats a combination of high environmental temperatures and scarce, unpredictable food and water resources create conditions under which there is a high probability of mismatches between supply and demand. The direct, proximate effects of warming can be thought of as representing a continuum from acute effects manifested over timescales of minutes to hours, chronic effects manifested over timescales of days to weeks, through to longer-term morphological or physiological changes over periods of years to decades. In the following sections, I provide a brief overview of these various categories of impacts.

10.2 Acute effects of increasing temperatures (minutes–hours)

In habitats where birds routinely experience very high T_a or intense solar radiation, each day brings the risks of acute hyperthermia and dehydration. These risks are dramatically illustrated by historical and recent accounts of catastrophic mortality events during extreme heat waves (reviewed by McKechnie et al. 2012). Such events can have severe impacts on threatened species with small populations; Saunders et al. (2011) reported the deaths in January 2010 of 208 endangered Carnaby's black-cockatoos (*Calyptorhynchus latirostris*), a species whose slow breeding rate and long generation time make it highly sensitive to the loss of so many individuals. Mortality events such as these, as well as those reported for birds in other regions and for pteropodid bats (Welbergen et al. 2008), underscore the point that extreme heat can push birds and other animals beyond their physiological tolerances, and heat waves thus have the potential to result in major population declines over very short timescales.

10.2.1 Hyperthermia

In many hot subtropical deserts, T_a values in summer may exceed avian normothermic T_b for 6–8 hr per day, creating conditions under which birds must increase rates of evaporative heat loss to defend T_b below lethal limits. Under such conditions, however, T_a is a minimum index of heat load. Solar radiation in subtropical deserts can exceed 1000 W m^{-2} and result in birds experiencing operative temperatures (Bakken 1976; Robinson et al. 1976) exceeding 50 °C for several hours each day. For example, a female rufous-cheeked nightjar (*Caprimulgus rufigena*) incubating eggs in a completely sunlit site on red sand in the southern African arid zone experienced an average operative temperature of 51 °C around midday (O'Connor et al. 2018), and black bulb temperatures at the nest of an incubating lesser nighthawk (*Chordeiles acutipennis*) in California's Salton Sea sometimes reached ~58 °C (Grant 1982).

Whereas the maximum T_b that birds can tolerate appears to be largely invariant across taxa (typically ~45–46 °C), the highest environmental temperatures at which they can defend T_b at sub-lethal levels (heat tolerance limit, HTL) varies widely among taxa (e.g., Smith et al. 2015; Whitfield et al. 2015; O'Connor et al. 2017a; Smit et al. 2018). In some regions, maximum T_a values already approach and occasionally exceed avian HTLs. For instance, Smith et al. (2017) reported HTL = 50 °C for six of seven Sonoran Desert passerines; on 20 June 2017 T_a reached 47–49 °C in the Phoenix area (https://cals.arizona.edu/azmet/). On 8 January 2013, T_a = 52 °C was forecast in the interior of South Australia (http://www.bom.gov.au/), well above the HTLs of several passerines occurring in this region (McKechnie et al. 2017b). These examples reiterate the risk of direct mortality from hyperthermia during extremely hot weather, a scenario that will become more likely as absolute maximum T_a values increase (IPCC 2011).

Another group in which hyperthermia risk has emerged as important is high-latitude seabirds (Gaston et al. 2002; Oswald and Arnold 2012; Oswald et al. 2008, 2011). Oswald and colleagues have noted that the spatial separation between foraging areas at sea and breeding areas on land make these seabirds an ideal model taxon for partitioning direct effects of heat stress versus indirect impacts of warming manifested through subtle and complex shifts in the marine ecosystems in which these birds are consumers. In great skuas (*Catharacta skua*), heat stress becomes apparent whenever T_a exceeds 14 °C and is manifested as bathing and consequent reduced nest attendance by adults (Oswald et al. 2008). Bathing in this species appears to have a thermoregulatory function and the probability of both parents being away from breeding territories increased rapidly as operative temperature approached 20 °C, with a concomitant increase in the risk of chicks being killed and eaten by conspecifics (Oswald et al. 2008). A subsequent analysis of breeding ranges and climate variables for 13 seabird species suggested an inverse correlation between foraging distance and the importance of heat stress as a constraint at breeding sites (Oswald et al. 2011).

10.2.2 Dehydration

A second way in which extremely hot weather can negatively affect birds over short timescales is through dehydration. Even if they can dissipate heat sufficiently rapidly to maintain T_b at sub-lethal levels, rapid rates of evaporative water loss may lead to severe dehydration in a matter of hours, particularly in small species. A 7-g verdin (*Auriparus flaviceps*), for instance, loses the equivalent of ~7 per cent of its body mass per hour at T_a approaching 50 °C (Wolf and Walsberg 1996). Dehydration risk is exacerbated by reductions in activity during very hot weather; small birds respond to high T_a by curtailing activity and resting in the coolest microsites they can find, which are often shaded sites that reduce radiative heat loads but provide little or no buffering from T_a. Under these conditions, water loss via evaporation is not offset by water gains via drinking or foraging, with the result that dehydration tolerance limits can be approached rapidly during extreme heat events (Figure 10.2).

A recent model of the risk of lethal dehydration faced by five North American passerines revealed that, by the end of this century, arid-zone specialists such as cactus wren (*Campylorhynchus brunneicapillus*) and curve-billed thrasher (*Toxostoma curvirostre*) will regularly experience a significant risk of lethal dehydration in some parts of their ranges, particularly southwestern Arizona (Albright et al. 2017). Increases in the risk of lethal dehydration are strongly mass-dependent, with small species experiencing the greatest fractional increases in evaporative cooling

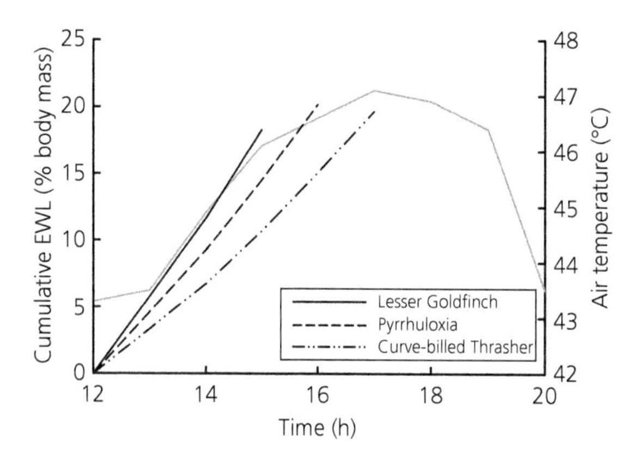

Figure 10.2 Small birds can rapidly approach lethal dehydration tolerance limits during extreme heat events, as shown by estimated cumulative water losses for three Sonoran Desert passerines on 20 June 2017, when air temperatures reached 47–49 °C in the Phoenix, Arizona area. The rates of water loss for lesser goldfinches (*Spinus psaltria*, 10 g), pyrrhuloxias (*Cardinalis sinuatus*, 34 g) and curve-billed thrashers (*Toxostoma curvirostre*, 71 g) are calculated from measurements reported by Smith et al. (2017), and the hourly air temperature data (12:00–20:00) for 20 June 2017 were obtained from https://cals.arizona.edu/azmet/ (data for Phoenix Greenway weather station). The same day saw major disruptions of flights out of Phoenix Sky Harbor International Airport on account of extreme air temperatures (http://www.bbc.com/news/world-us-canada-40339730).

requirements, and consequently the largest reductions in survival time during extremely hot weather (Albright et al. 2017; McKechnie and Wolf 2010). These models suggest that hot subtropical deserts may lose substantial numbers of species in coming decades, with small species most severely affected on account of their higher rates of mass-specific EWL and more rapid depletion of body water (Wolf 2000; McKechnie and Wolf 2010; Albright et al. 2017).

In hot environments, dehydration risk is particularly pronounced for nocturnal species that do not drink between sunrise and sunset. These risks are illustrated by a recent study of rufous-cheeked nightjars, which often roost in only partially shaded sites, despite summer T_a values exceeding 30 °C (O'Connor et al. 2018). Consequently, individual nightjars may experience operative temperatures approaching 50 °C for several hours per day, and evaporative water requirements may be three- to four-fold higher than if the birds roosted in completely shaded microsites (O'Connor et al. 2018). Daily evaporative water losses estimated by these authors for the nightjars were equivalent to up to 19–20 per cent of their body mass. Under a 4 °C warming scenario, the equivalent daily water requirements may become as high as 26 per cent of body mass each day, a value very likely above the dehydration tolerance limits of these nightjars (O'Connor et al. 2018).

10.3 Medium-term effects of increasing temperature (days–weeks)

The acute effects of extremely hot weather discussed in the preceding section involve birds exceeding physiological tolerance limits and direct mortality of individuals during extreme heat events. However, many of the effects of warming are manifested through more subtle processes over longer timescales, but which are no less consequential for survival or breeding success. Detecting these impacts generally requires intensive, species-specific studies in which the behaviour and condition of individuals are carefully monitored. In this section, I briefly review recent studies that shed light on how hot weather can drive losses of condition over timescales of days to weeks.

10.3.1 Chronic body mass loss

Temperature is a major determinant of avian behaviour, particularly for species in hot, arid environments (e.g., Ricklefs and Hainsworth 1968; Austin 1976). A study of southern pied babblers (*Turdoides bicolor*) in the Kalahari Desert of southern Africa revealed how trade-offs between foraging and heat dissipation behaviours can result in progressive loss of body condition during periods of hot weather. Du Plessis et al. (2012) found that the proportion of time spent foraging by non-breeding babblers habituated to human observers was largely unaffected by T_a. However, foraging efficiency (i.e., the mass of food acquired per unit time) decreased with increasing T_a, because foraging efficiency was substantially lower during heat dissipation behaviours (HDB) such as panting and wing-spreading (du Plessis et al. 2012). For example, on days when maximum T_a was 34 °C, babblers typically spent ~10 per cent of their time engaged in HDB. On days with maximum T_a > 40 °C, however, the proportion of time spent engaged in HDB increased to ~40 per cent, with a concomitant decrease in foraging efficiency. This interplay between T_a, behaviour, and foraging efficiency was manifested as a negative relationship between diurnal body mass gain and daily maximum T_a (Figure 10.3); on days with maximum T_a = 34 °C, babblers gained ~5 per cent M_b between sunrise and sunset, but on days with maximum T_a > 40 °C diurnal M_b gain was zero. Combined with an average overnight M_b loss of ~4 per cent, this creates a situation where babblers experience a nett loss of body condition over each 24-hr cycle on days with maximum T_a above 35.5 °C (Figure 10.3), and hence the consequences of protracted periods of hot weather can readily be modelled.

Quantitatively similar effects of high T_a have recently been shown for another, larger species resident in the Kalahari, the southern yellow-billed hornbill (*Tockus leucomelas*). In this species, as is typical for Bucerotidae, females seal themselves into nest cavities while breeding, and are completely dependent on males for food provisioning during this period (Kemp 2005). For male hornbills, diurnal mass gain showed a negative relationship with daily maximum T_a, from ~5 per cent M_b when T_a = 25 °C to ~0 per cent when T_a = 40 °C (van de Ven 2017). The temperature-dependence of diurnal mass gain in the male hornbills combines with an average overnight mass loss of ~4.5 per cent to result in a nett 24-hr mass loss on most days with maximum T_a > 25 °C. However, the rate of mass loss increases rapidly with T_a such that on very hot days the male hornbills lose ~5 per cent body mass per day (Figure 10.3). A similar pattern was evident for female hornbills in nests, with each 24-hr day with maximum T_a around 40 °C also associated with body mass loss of approximately 5 per cent (van de Ven 2017).

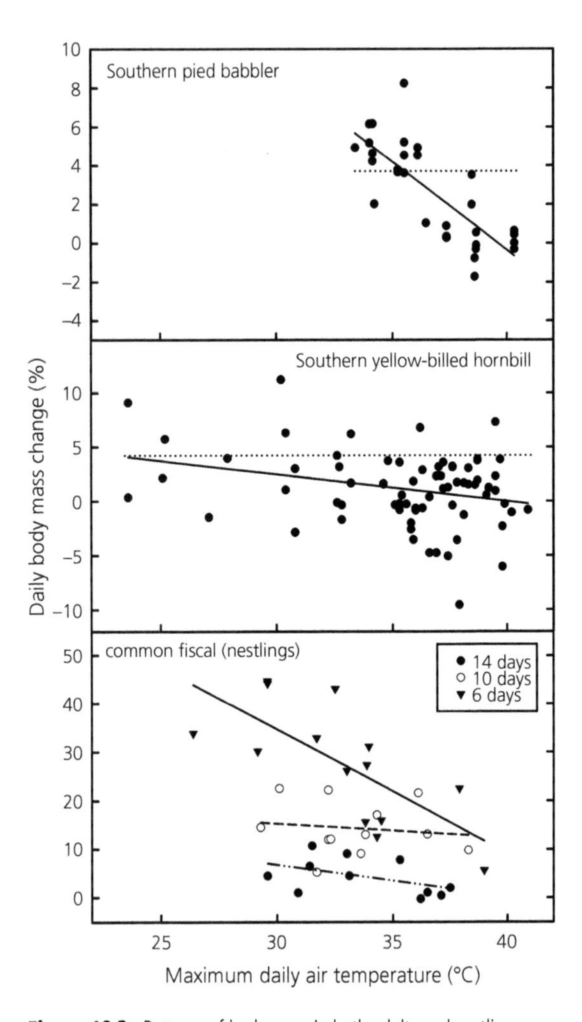

Figure 10.3 Patterns of body mass in both adults and nestlings are dependent on daily maximum air temperature in both adults and nestlings of arid-zone birds. In non-breeding adult southern pied babblers (*Turdoides bicolor*; top panel) and male southern yellow-billed hornbills (*Tockus leucomelas*; middle panel) provisioning females and chicks during summer in the Kalahari Desert, daytime body mass gain in adults decreased with increasing air temperature during summer (du Plessis et al. 2012; van de Ven, 2017). In the case of the babblers, air temperatures above 35.5 °C were associated with night-time mass loss (dotted horizontal line) exceeding daytime mass gain, whereas in male hornbills provisioning nests nearly all air temperatures above 25 °C were associated with a net loss of mass. In the case of common fiscal (*Lanius collaris*) nestlings (bottom panel), daily maximum air temperature did not significantly influence the growth rates of 10- or 14-day old nestlings, but had a significantly negative influence on growth rates of 6-day old nestlings (Cunningham et al. 2013b). The data for southern pied babblers, southern yellow-billed hornbills, and nestling common fiscals were replotted from du Plessis et al. (2012), van de Ven (2017), and Cunningham et al. (2013), respectively.

10.3.2 Reductions in nestling growth rates and fledging mass

The rates at which breeding birds provision nestlings or incubating mates decrease during very hot weather (e.g., Luck 2001). Cunningham et al. (2013b) found that in common fiscals (*Lanius collaris*) there was a strong negative effect of maximum T_a on daily mass gain in six-day old nestlings, but not in older nestlings (Figure 10.3). Moreover, the numbers of days on which maximum T_a exceeded threshold values emerged as strong predictors of fledging body mass (threshold T_a = 33 °C) and tarsus length (37 °C), and age at fledge (35 °C) (Cunningham et al. 2013b). In other words, the more frequently T_a exceeded these thresholds during the nestling period, the smaller fledglings were and the longer they remained in the nest before fledging. Thus, high T_a affected both the likelihood of nestlings surviving to fledge via increasing the probability of nest predation, and potentially affected lifetime fitness of individuals that successfully fledged (Cunningham et al. 2013b). The authors noted that the mechanism driving these temperature effects could have involved trade-offs between foraging and thermoregulation by adults, or reductions in prey availability on hot days. The existence of trade-offs between foraging and thermoregulation in this species was subsequently documented by the same authors (Cunningham et al. 2015).

Similar patterns were recently reported for southern yellow-billed hornbills (van de Ven, 2017). The biomass provisioned by male hornbills to females and nestlings decreased with increasing daily maximum T_a, and, as a consequence, nestlings gained less body mass on hot days, and experienced 24-hr mass gain of zero or less on days hotter than 40.6 °C. The body mass and tarsus length of chicks at fledging were lower and the age at fledging greater when daily maximum T_a was higher during the nestling period. The effect sizes were large; most chicks that experienced a mean daily maximum T_a < 34 °C weighed 200–250 g, but chicks that experienced means of 37–38 °C typically weighed less than 150 g, and in one instance less than 100 g (van de Ven 2017).

The studies discussed above reveal how more frequent hot days and the associated higher mean daily T_a maxima directly result in smaller fledglings, likely with knock-on effects for lifetime fitness (Lindström 1999; Schwagmeyer and Mock 2008). For ground-nesting species that nest in exposed microsites, however, these effects may be compounded by changes in the dynamics of trade-offs between nest microclimate and predation risk. Tieleman et al. (2008) tested the hypothesis that nest site selection by birds breeding in hot deserts is driven by a trade-off between microclimate and predation risk, whereby predation risk is lower for nests in exposed microsites away from vegetation but thermoregulation is more demanding than in shaded microsites under vegetation. Hoopoe larks (*Alaemon alaudipes*) breeding in the Arabian Desert showed a shift from building nests mainly in exposed sites early in the breeding season to nesting under cover later in the season when T_a values were higher (Tieleman et al. 2008). These authors argued that the adult larks favoured sites with lower predation risk to themselves early in the season, but as temperatures increased were forced to select more risky sites in vegetation. Climate change will directly affect such trade-offs, and more frequent hot weather will likely force ground-nesting species to use shaded sites where they face a greater predation risk. Thus, in ground-nesting species, climate change will result in increased risk of predation during breeding by a) increasing age at fledging and hence cumulative predation risk during the nestling stage, and b) forcing birds to breed in sites with a higher likelihood of predation per unit time.

10.4 Long-term effects of increasing temperature (years–decades)

Over long-term timescales, the impacts of climate change may involve shifts in body size or appendage size. Declines in body size have been suggested to represent a third universal response to global warming, along with changes in phenology and shifts in distribution (Daufresne et al. 2009; Gardner et al. 2011). Shifts in body size and appendage size with advancing climate change can be predicted a priori from Bergmann's Rule and Allen's Rule, respectively, supporting the notion that these are general responses across a multitude of taxa.

10.4.1 Decreasing body size

Among birds, temporal changes in body mass have been variable in direction and magnitude, with studies based on museum specimens supporting a general, but not universal, trend towards smaller body sizes (Yom-Tov 2001; Yom-Tov et al. 2002; Gardner et al. 2009; Gardner et al. 2014a). Long-term studies of single species or ringing data reveal similar variability in body mass responses, with many species showing decreases over recent decades (e.g.,Van Buskirk et al. 2010), but others showing no change or increases (e.g., Kaňuščák et al. 2004; Gardner et al. 2014b). In some cases, declines in body mass have been very rapid. In a population of burrowing owls (*Athene cunicularia*) in New Mexico, between 1998 and 2013 adult body mass declined by 8 per cent in males and 11 per cent in females, despite no apparent changes in parameters such as wing and tarsus length, while nestling mass decreased by 20 per cent over just eight years (Cruz-McDonnell and Wolf 2016). These declines in mass accompanied a catastrophic 98 per cent population decline, with reduced reproductive output strongly associated with higher T_a and decreased precipitation (Cruz-McDonnell and Wolf 2016).

The processes driving the general pattern of declining body size remain hotly debated and largely unclear, with most arguments focusing on the mechanisms thought to underpin Bergmann's Rule, the general positive relationship between body size and latitude (Bergmann 1847). For birds inhabiting hot environments, one possible mechanism for a decline in body mass with warming involves the proximate effects of high temperatures on fledging body size identified in common fiscals and southern yellow-billed hornbills, where reduced provisioning rates and slower growth resulted in smaller fledglings (Cunningham et al. 2013b; van de Ven 2017). In these species, a decrease in average fledgling size as climate change advances would presumably lead to gradual declines in adult body mass, as smaller birds recruit into the adult population (Cunningham et al. 2013b).

Other studies of arid-zone species, however, reveal increasing body size over time. A particularly instructive example is a recent investigation of individual and demographic consequences of high temperatures in a population of white-plumed honeyeaters (*Ptilotula penicillatus*) in a semiarid habitat (Gardner et al. 2016). These authors found that, between 1986 and 2012, multi-day exposure to T_a exceeding 35 °C resulted in a loss of body mass of 3 per cent per day, but only during dry conditions with no rainfall in the preceding 30 days (Gardner et al. 2016). Heat exposure resulted in increased mortality for females and for smaller males, which the authors suggested may have driven the overall increase in mean body size in this population over a 23-year period. A particularly interesting aspect of Gardner et al.'s study concerned the way in which rainfall ameliorated the effects of hot weather on body condition, an observation that reiterates the importance of considering both temperature and precipitation in models of avian responses to warming.

The importance of the number of days on which T_a exceeded a threshold value (rather than mean maximum temperature) that emerged from Gardner et al.'s (2016) study is consistent with the pattern previously identified in common fiscals (Cunningham et al. 2013a; Cunningham et al. 2013b). These findings support the view that analyses of temperature effects on arid-zone birds should focus on threshold T_a values rather than average values (Cunningham et al. 2013a; Cunningham et al. 2013b). Similar patterns have also been documented in species in cooler, more temperate environments; for instance, consecutive days over 30 °C resulted in decreases in body mass in red-winged fairywrens (*Malurus elegans*), but not in sympatric white-browed scrubwrens (*Sericornis frontalis*) (Gardner et al. 2018).

10.4.2 Appendage size

In addition to changes in overall body size, the size of appendages such as the bill may be responsive to warming in accordance with the predictions of Allen's rule (Allen 1877; Campbell-Tennant et al. 2015). Campbell-Tennant et al. (2015) tested the hypothesis that recent warming has driven increases in bill size using museum specimens, collected between 1871 and 2008, of five Australian parrots. After controlling for body mass and other factors, these authors found increases in bill surface area of 4–10 per cent for mulga parrots (*Psephotus varius*), gang-gang cockatoos (*Callocephalon fimbriatum*), red-rumped parrots (*Psephotus haematonotus*), and male crimson rosellas (*Platycercus elegans*). Potentially confounding factors, such as changes in food supply or foraging habitat, make it difficult to conclude that bill size increased as a direct response to temperature, but it is consistent with recent insights into the thermoregulatory significance of avian bills.

There is increasing evidence that active regulation of heat loss via circulatory adjustments has played a major role in the evolution of avian bill morphology (Tattersall et al. 2009, 2017, 2018; van de Ven et al. 2016), and bill size is correlated with environmental temperature both among and within species (Tattersall et al. 2017). One potential functional explanation for the increases in bill surface area documented by Campbell-Tennant et al. (2015) is selection for greater heat dissipation across the bill surface, although testing this hypothesis will be challenging.

10.5 Physiological modelling approaches

Physiological models provide a mechanistic, process-based approach to predicting avian responses to climate change that contrasts with the correlative, pattern-based climate envelope approaches employed when interest in this field first developed (e.g., Peterson 2001; Erasmus et al. 2002; Simmons et al. 2005). However, physiological models have the disadvantage of requiring intensive, time-consuming, and often expensive species-specific studies, whereas climate envelope models allow for the rapid generation of predictions for large numbers of species. In this section, I outline several physiological approaches that have been proposed in recent years.

10.5.1 Models of lethal dehydration risk during acute heat exposure

The survival time of inactive birds in shaded microsites during extremely hot weather can be readily modelled using EWL data collected under laboratory conditions, by estimating the time taken for cumulative evaporative losses to reach dehydration tolerance limits (Wolf 2000; McKechnie and Wolf 2010; Albright et al. 2017). These models assume that birds are in completely shaded and windless microsites and experience operative temperatures equal to T_a; their predictions of lethal dehydration risk are thus conservative and represent the best-case scenario. When historical climate data and future climate projections are incorporated in such models, they can provide detailed insights into future range losses by arid-zone birds (Albright et al. 2017).

10.5.2 Linking thermal tolerance limits to species declines

A recent study in southern Africa's Fynbos biome tested for mechanistic links between interspecific variation in thermal physiology and observed range contractions and population declines (Milne et al. 2015). These authors showed that population declines and reductions in range were most pronounced in species occupying the coolest areas and which experienced the greatest increases in T_a over the last 25 years. However, there was no strong evidence for a mechanistic link between these changes and interspecific variation in thermal physiology; the effects of warming were not necessarily greater in species in which evaporative water loss increased more rapidly at high T_a values, or those with lower inflection T_a values for EWL (Milne et al. 2015). However, one species in which there did appear to be a strong link between thermal physiology and recent range contractions was the Cape rockjumper (*Chaetops frenatus*). In this species, increases in both EWL and T_b with increasing T_a were more rapid than expected, leading the authors to argue that, in this species, modest evaporative cooling capacity is a major contributing factor to recent declines and moreover makes this species particularly vulnerable to warming (Milne et al. 2015).

10.5.3 Behavioural indices of vulnerability to high temperatures

In an attempt to find middle ground between time-consuming, expensive species-specific studies and correlative climate envelope models, Smit et al. (2016) suggested that readily quantifiable behavioural indices of heat dissipation behaviour can be used for broad assessments of how species' vulnerabilities to high temperatures vary within arid-zone bird communities. The feasibility of this approach is supported by the trade-offs between foraging and behaviours such as wing-drooping observed in some Kalahari Desert species (du Plessis et al. 2012; van de Ven 2017), although understanding the functional links between heat dissipation behaviours and the underlying patterns of T_b will be a prerequisite to using this approach.

10.5.4 Biophysical modelling: Niche Mapper

One of the most exciting approaches to building mechanistic models of the impacts of higher temperatures on birds to emerge in recent years is the application of the *Niche Mapper* biophysical modelling package developed by Warren Porter and Michael Kearney (Kearney and Porter 2004,

2009, 2017; Mathewson et al. 2017) to endotherms. Consisting of coupled mass and energy flux equations parameterized with empirical physiological and behavioural data, *Niche Mapper* facilitates the modelling of energy and water fluxes between organisms and their environments with high temporal resolution, providing detailed insights into interactions between organisms and the thermal landscapes in which they operate.

An example of how *Niche Mapper* can be applied to birds was provided by Kearney et al. (2016), who modelled the water balance of Australia's recently rediscovered night parrot (*Pezoporus occidentalis*). The authors developed a biophysical model for the parrot, which they parameterized by scaling up physiological data for a closely related species. Combining this model with microclimate data for the site where the parrot was rediscovered, they predicted how the species' reliance on succulent plants and surface water will increase in coming decades. Under current conditions, night parrots can maintain water balance during winter on a diet of seeds, but require access to succulent plants containing at least 55 per cent water to completely avoid lethal dehydration on extremely hot days (Kearney et al. 2016). By 2070, however, access to succulent plant matter will no longer be adequate to maintain water balance during hot periods, and the parrots will become reliant on surface water for survival in summer (Kearney et al. 2016).

The analysis by Kearney and colleagues provides a good example of how biophysical models combined with microclimate and physiological data can reveal how the impacts of climate change will affect the capacity of threatened species to persist under warmer conditions, and the management interventions that may be required for them to do so. One point reiterated by this study is that the provision of artificial water sources may be critical for avoiding species losses in many arid systems. This management intervention, however, is reliant on a better understanding of the behavioural and physiological trade-offs associated with water holes; for instance, birds drinking at isolated water holes in deserts may well face strong trade-offs between dehydration risk and predation risk, and in the case of diurnal species, between dehydration risk and hyperthermia risk (Abdu et al. 2018).

10.5.5 Thermoneutrality-based models

Recent years have seen a number of studies using the upper and lower critical limits of thermoneutrality, also referred to as upper critical temperature (UCT) and lower critical temperature (LCT), respectively, as measures of endotherms' upper and lower thermal tolerances (Araújo et al. 2013; Khaliq et al. 2014, 2015; Buckley and Huey 2016). In addition to serious problems with the quality of data included in some of these analyses of UCT values (McKechnie et al. 2017a; Wolf et al. 2017), a more fundamental shortcoming with this approach is that the interpretation of upper and lower boundaries of the TNZ as indices of thermal tolerance limits is not defensible from a thermal physiology perspective.

One reason is that many birds, like mammals, spend large fractions of their lives at T_a values outside of their TNZ (Swanson et al. 2017; Mitchell et al. 2018). A second is that the inflection point in resting metabolic rate at high T_a is strongly dependent on the avenue of evaporative cooling used by a particular taxon. Some caprimulgids, for instance, show no upper inflection in resting metabolic rate (i.e., UCT; O'Connor et al. 2017b). Third, even within taxa that do rely heavily on panting, interspecific variation in the UCT is not always directly related to the T_a associated with the onset of panting (e.g., McKechnie et al. 2017b). For these reasons, the upper boundary of the avian TNZ should not be interpreted as a measure of upper thermal tolerance limits.

10.6 Critical needs and future directions

In many instances the impacts of higher temperatures on birds, particularly those occupying hot environments, occur via relatively well-understood physiological processes involved in thermoregulation. However, a host of questions remain, and there are several emerging fields of enquiry that will prove critical for our ability to adequately model the responses of birds to climate change. I discuss three such areas below.

10.6.1 Phenotypic plasticity and epigenetic effects

The upper limits to avian T_b are constrained by fundamental biochemical constraints related to protein function. However, our knowledge of phenotypic flexibility in traits related to heat tolerance and the limits to that flexibility remains extremely limited. In a southern African passerine, for instance, a population in a hot desert habitat significantly increased heat tolerance during summer, whereas no seasonal variation was evident in two conspecific populations from cooler areas (Noakes et al. 2016).

The direct effects of warmer temperatures on nestlings have also received little attention. Experimental heating of blue tit (*Cyanistes caeruleus*) nests in southern Sweden to temperatures as high as 50 °C led to significantly higher T_b (up to 43.8 °C) without increases in mortality, although rates of body mass gain were lower compared to unheated control nests (Andreasson et al. 2018). A study on great tits (*Parus major*) in eastern Spain in which the average temperature of experimentally heated nests was ~40 °C revealed that higher nest temperatures resulted in smaller fledglings, and there was some evidence for lower post-fledging survival (Rodríguez et al. 2016).

A remarkable insight into the potential importance of developmental plasticity for avian responses to climate change concerns recent observations that zebra finches (*Taeniopygia guttata*) use prenatal acoustic communication to program offspring for higher temperatures (Mariette and Buchanan 2016). The authors of this study showed that vocalizations given by incubating females on hot days resulted in offspring with more heat-tolerant phenotypes, which influenced the offspring's subsequent thermal preferences and reproductive success.

10.6.2 High temperatures as stressors: mechanisms and consequences

The functional significance of neuroendocrine stress pathways for avian responses to heat waves remains little-explored. Whereas there is a rich literature on the ways in which stress pathways mediate responses to severe cold weather events in north-temperate climates, far less work has examined the role of stress physiology in the context of heat waves (Wingfield et al. 1992; Xie et al. 2017). Studies of the roles of acute and chronic stress responses in mediating the effects of hot weather, and how interspecific variation in stress physiology interacts with thermoregulatory performance at high T_a values, will likely prove very important for predicting species' relative vulnerabilities to climate change.

10.6.3 Physiological aspects of migration

This chapter has focused on birds resident in hot, arid environments. However, many migratory species that occupy cooler environments for much of the year traverse hot deserts during their annual migrations, for instance species that cross the Sahara Desert while migrating between the Palearctic and Africa. During such desert crossings, periods of extreme heat can result in severe loss of body condition, sometimes leading to death (Miller 1963). For these long-distance migrants, management of water balance is just as critical as that of fuel stores (Carmi et al. 1992; Klaassen 1995; Gerson and Guglielmo 2011), and hotter future climates may thus have severe consequences for the ability of these migratory species to successfully complete long desert crossings.

10.7 Conclusions

The direct impacts of global warming on birds occur via several physiological and behavioural processes. Whereas these impacts are currently best-understood in species inhabiting hot, arid environments, they are no less consequential in other habitats, as evidenced by the work of Oswald and colleagues on high-latitude seabirds (Oswald and Arnold 2012; Oswald et al. 2008; Oswald et al. 2011). Constraints on evaporative cooling and trade-offs between thermoregulation and foraging similar to those that occur in arid-zone birds may, a priori, be predicted for birds in many other regions too. For instance, in tropical habitats, where most of the Earth's avian diversity is found, high atmospheric humidity likely constrains evaporative cooling even at moderate T_a (Gerson et al. 2014), and similar processes to those operating in desert birds likely affect survival and reproduction (e.g., Woodworth et al. 2018). There is an urgent need for research into the direct impacts of warming in tropical regions, and for conservation planning for threatened species to explicitly incorporate climate change.

Acknowledgements

I am grateful to Susie Cunningham for her insightful comments on an earlier version of the manuscript.

References

Abdu, S., McKechnie, A.E., Lee, A.T., and Cunningham, S.J. (2018). Can providing shade at water points help Kalahari birds beat the heat? *Journal of Arid Environments*, 152, 21–7.

Albright, T.P., Mutiibwa, D., Gerson, A.R., Smith, E.K., Talbot, W.A., McKechnie, A.E., and Wolf, B.O. (2017). Mapping evaporative water loss in desert passerines reveals an expanding threat of lethal dehydration. *Proceedings of the National Academy of Sciences of the United States of America*, 114, 2283–8.

Allen, J.A. (1877). The influence of physical conditions in the genesis of species. *Radical Review*, 1, 108–40.

Andreasson, F., Nord, A., and Nilsson, J.-Å. (2018). Experimentally increased nest temperature affects body temperature, growth and apparent survival in blue tit nestlings. *Journal of Avian Biology*, 49, jav-01620.

Araújo, M.B., Ferri-Yáñez, F., Bozinovic, F., Marquet, P.A., Valladares, F., and Chown, S.L. (2013). Heat freezes niche evolution. *Ecology Letters*, 16, 1206–19.

Austin, G.T. (1976). Behavioral adaptations of the verdin to the desert. *Auk*, 93, 245–62.

Bakken, G.S. (1976). A heat transfer analysis of animals: unifying concepts and the application of metabolism chamber data to field ecology. *Journal of Theoretical Biology*, 60, 337–84.

Bergmann, C. (1847). Über die Verhältnisse der Warmeökonomie die Thiere zu ihrer Grösse. *Göttinger Studien*, 3.

Buckley, L.B., and Huey, R.B. (2016). Temperature extremes: geographic patterns, recent changes, and implications for organismal vulnerabilities. *Global Change Biology*, 22, 3829–42.

Campbell-Tennant, D.J.E., Gardner, J.L., Kearney, M.R., and Symonds, M.R.E. (2015). Climate-related spatial and temporal variation in bill morphology over the past century in Australian parrots. *Journal of Biogeography*, 42, 1163–75.

Carmi, N., Pinshow, B., Porter, W.P., and Jaeger, J. (1992). Water and energy limitations on flight duration in small migrating birds. *The Auk*, 109, 268–76.

Cruz-McDonnell, K.K., and Wolf, B.O. (2016). Rapid warming and drought negatively impact population size and reproductive dynamics of an avian predator in the arid southwest. *Global Change Biology*, 22, 237–53.

Cunningham, S.J., Kruger, A.C., Nxumalo, M.P., and Hockey, P.A.R. (2013a). Identifying biologically meaningful hot-weather events using threshold temperatures that affect life-history. *PLoS One*, 8, e82492.

Cunningham, S.J., Martin, R.O., and Hockey, P.A. (2015). Can behaviour buffer the impacts of climate change on an arid-zone bird? *Ostrich*, 86, 119–26.

Cunningham, S.J., Martin, R.O., Hojem, C.L., and Hockey, P.A.R. (2013b). Temperatures in excess of critical thresholds threaten nestling growth and survival in a rapidly-warming arid savanna: a study of common fiscals. *PLoS One*, 8, e74613.

Daufresne, M., Lengfellner, K., and Sommer, U. (2009). Global warming benefits the small in aquatic ecosystems. *Proceedings of the National Academy of Sciences of the United States of America*, 106, 12788–93.

Dawson, W.R., and Whittow, G.C. (2000). Regulation of body temperature. In: G.C. Whittow (ed.), *Sturkie's Avian Physiology*, pp. 343–390. Academic Press, New York.

du Plessis, K.L., Martin, R.O., Hockey, P.A.R., Cunningham, S.J., and Ridley, A.R. (2012). The costs of keeping cool in a warming world: implications of high temperatures for foraging, thermoregulation, and body condition of an arid-zone bird. *Global Change Biology*, 18, 3063–70.

Erasmus, B.F.N., van Jaarsveld, A.S., Chown, S.L., Kshatriya, M., and Wessels, K.J. (2002). Vulnerability of South African taxa to climate change. *Global Change Biology*, 8, 679–93.

Etches, R.J., John, T.M., and Verrinder Gibbins, A.M. (2008). Behavioural, physiological, neuroendocrine and molecular responses to heat stress. In: N.J. Daghir (ed.), *Poultry production in hot climates*, 2nd edition, pp. 48–79. CAB International, Wallingford, Oxon, UK.

Gardner, J.L., Amano, T., Backwell, P.R., Ikin, K., Sutherland, W.J., and Peters, A. (2014a). Temporal patterns of avian body size reflect linear size responses to broadscale environmental change over the last 50 years. *Journal of Avian Biology*, 45, 529–35.

Gardner, J.L., Amano, T., Mackey, B.G., Sutherland, W.J., Clayton, M., and Peters, A. (2014b). Dynamic size responses to climate change: prevailing effects of rising temperature drive long-term body size increases in a semi-arid passerine. *Global Change Biology*, 20, 2062–75.

Gardner, J.L., Amano, T., Sutherland, W.J., Clayton, M., and Peters, A. (2016). Individual and demographic consequences of reduced body condition following repeated exposure to high temperatures. *Ecology*, 97, 786–95.

Gardner, J.L., Heinsohn, R., and Joseph, L. (2009). Shifting latitudinal clines in avian body size correlate with global warming in Australian passerines. *Proceedings of the Royal Society of London Series B: Biological Sciences*, 276, rspb20091011.

Gardner, J.L., Peters, A., Kearney, M.R., Joseph, L., and Heinsohn, R. (2011). Declining body size: a third universal response to warming? *Trends in Ecology and Evolution*, 26, 285–91.

Gardner, J.L., Rowley, E., de Rebeira, P., de Rebeira, A., and Brouwer, L. (2018). Associations between changing climate and body condition over decades in two southern hemisphere passerine birds. *Climate Change Responses*, 5, 2.

Gaston, A.J., Hipfner, J.M., and Campbell, D. (2002). Heat and mosquitoes cause breeding failures and adult mortality in an Arctic-nesting seabird. *Ibis*, 144, 185–91.

Gerson, A.R., and Guglielmo, C.G. (2011). Flight at low ambient humidity increases protein catabolism in migratory birds. *Science*, 333, 1434–36.

Gerson, A.R., Smith, E.K., Smit, B., McKechnie, A.E., and Wolf, B.O. (2014). The impact of humidity on evaporative cooling in small desert birds exposed to high air temperatures. *Physiological and Biochemical Zoology*, 87, 782–95.

Grant, G.S. (1982). Avian incubation: egg temperature, nest humidity, and behavioral thermoregulation in a hot environment. *Ornithological Monographs*, 30, 1–100.

IPCC (2011). *Intergovernmental Panel on Climate Change special report on managing the risks of extreme events and disasters to advance climate change adaptation.* Cambridge University Press, Cambridge, UK.

Kaňuščák, P., Hromada, M., Tryjanowski, P., and Sparks, T. (2004). Does climate at different scales influence the phenology and phenotype of the River Warbler *Locustella fluviatilis*? *Oecologia*, 141, 158–63.

Kearney, M.R., and Porter, W. (2004). Mapping the fundamental niche: physiology, climate, and the distribution of a nocturnal lizard. *Ecology*, 85, 3119–31.

Kearney, M.R., and Porter, W. (2009). Mechanistic niche modelling: combining physiological and spatial data to predict species' ranges. *Ecology Letters*, 12, 334–50.

Kearney, M.R., and Porter, W.P. (2017). NicheMapR—an R package for biophysical modelling: the microclimate model. *Ecography*, 40, 664–74.

Kearney, M.R., Porter, W.P., and Murphy, S.A. (2016). An estimate of the water budget for the endangered night parrot of Australia under recent and future climates. *Climate Change Responses*, 3, 14.

Kemp, A.C. (2005). Southern Yellow-billed Hornbill. In: P.A.R. Hockey, W.R.J. Dean, and P.G. Ryan (eds), *Roberts birds of southern Africa*, pp. 152–153. The Trustees of the John Voelcker Bird Book Fund, Cape Town.

Khaliq, I., Fritz, S.A., Prinzinger, R., Pfenninger, M., Böhning-Gaese, K., and Hof, C. (2015). Global variation in thermal physiology of birds and mammals: evidence for phylogenetic niche conservatism only in the tropics. *Journal of Biogeography*, 42, 2187–96.

Khaliq, I., Hof, C., Prinzinger, R., Böhning-Gaese, K., and Pfenninger, M. (2014). Global variation in thermal tolerances and vulnerability of endotherms to climate change. *Proceedings of the Royal Society of London Series B: Biological Sciences*, 281, 20141097.

Klaassen, M. (1995). Water and energy limitations on flight range. *The Auk*, 112, 260–2.

Lindström, J. (1999). Early development and fitness in birds and mammals. *Trends in Ecology & Evolution*, 14, 343–8.

Lovegrove, B.G. (2012). The evolution of endothermy in Cenozoic mammals: a Pleisiomorphic–Apomorphic continuum. *Biological Reviews*, 87, 128–62.

Luck, G.W. (2001). Variability in provisioning rates to nestlings in the cooperatively breeding rufous treecreeper, *Climacteris rufa*. *Emu*, 101, 221–4.

Mariette, M.M., and Buchanan, K.L. (2016). Prenatal acoustic communication programs offspring for high posthatching temperatures in a songbird. *Science*, 353, 812–14.

Mathewson, P.D., Moyer-Horner, L., Beever, E.A., Briscoe, N.J., Kearney, M., Yahn, J.M., and Porter, W.P. (2017). Mechanistic variables can enhance predictive models of endotherm distributions: the American pika under current, past, and future climates. *Global Change Biology*, 23, 1048–64.

McKechnie, A.E., Coe, B.H., Gerson, A.R., and Wolf, B.O. (2017a). Data quality problems undermine analyses of endotherm upper critical temperatures. *Journal of Biogeography*, 44, 2424–6.

McKechnie, A.E., Gerson, A.R., McWhorter, T.J., Smith, E.K., Talbot, W.A., and Wolf, B.O. (2017b). Avian thermoregulation in the heat: evaporative cooling in five Australian passerines reveals within-order biogeographic variation in heat tolerance. *Journal of Experimental Biology*, 220, 2436–44.

McKechnie, A.E., and Wolf, B.O. (2010). Climate change increases the likelihood of catastrophic avian mortality events during extreme heat waves. *Biology Letters*, 6, 253–6.

McKechnie, A.E., Hockey, P.A.R., and Wolf, B.O. (2012). Feeling the heat: Australian landbirds and climate change. *Emu*, 112, i–vii.

Miller, A.H. (1963). Desert adaptations in birds. In: *Proceedings of the XIII International Ornithological Congress*, pp. 666–74.

Milne, R., Cunningham, S.J., Lee, A.T.K., and Smit, B. (2015). The role of thermal physiology in recent declines of birds in a biodiversity hotspot. *Conservation Physiology*, 3. doi:10.1093/conphys/cov048

Mitchell, D., Snelling, E.P., Hetem, R.S., Maloney, S.K., Strauss, W.M., and Fuller, A. (2018). Revisiting concepts of thermal physiology: predicting responses of mammals to climate change. *Journal of Animal Ecology*, 87, 956–73. doi: 10.1111/1365-2656.12818

Noakes, M.J., Wolf, B.O., and McKechnie, A.E. (2016). Seasonal and geographical variation in heat tolerance and evaporative cooling capacity in a passerine bird. *Journal of Experimental Biology*, 219, 859–69.

O'Connor, R.S., Brigham, R.M., and McKechnie, A.E. (2017a). Diurnal body temperature patterns in free-ranging populations of two southern African arid-zone nightjars. *Journal of Avian Biology*, 48, 1195–204.

O'Connor, R.S., Brigham, R.M., and McKechnie, A.E. (2018). Roosting in exposed microsites by a nocturnal bird, the rufous-cheeked nightjar: implications for water balance under current and future climate conditions. *Canadian Journal of Zoology*, 96, 1122–9.

O'Connor, R.S., Wolf, B.O., Brigham, R.M., and McKechnie, A.E. (2017b). Avian thermoregulation in the heat: efficient evaporative cooling in two southern African nightjars. *Journal of Comparative Physiology B*, 187, 477–91.

Oswald, S.A., and Arnold, J.M. (2012). Direct impacts of climatic warming on heat stress in endothermic species: seabirds as bioindicators of changing thermoregulatory constraints. *Integrative Zoology*, 7, 121–36.

Oswald, S.A., Bearhop, S., Furness, R.W., Huntley, B., and Hamer, K.C. (2008). Heat stress in a high-latitude seabird: effects of temperature and food supply on bathing and nest attendance of great skuas *Catharacta skua*. *Journal of Avian Biology*, 39, 163–9.

Oswald, S.A., Huntley, B., Collingham, Y.C., Russell, D.J., Anderson, B.J., Arnold, J.M., Furness, R.W., and Hamer, K.C. (2011). Physiological effects of climate on distributions of endothermic species. *Journal of Biogeography*, 38, 430–8.

Peterson, A.T. (2001). Predicting species' geographical distributions based on ecological niche modeling. *Condor*, 103, 599–605.

Prinzinger, R., Preßmar, A., and Schleucher, E. (1991). Body temperature in birds. *Comparative Biochemistry and Physiology*, 99A, 499–506.

Ricklefs, R.E., and Hainsworth, F.R. (1968). Temperature dependent behavior of the cactus wren. *Ecology*, 49, 227–33.

Robinson, D.E., Campbell, G.S., and King, J.R. (1976). An evaluation of heat exchange in small birds. *Journal of Comparative Physiology B*, 105, 153–66.

Rodríguez, S., Diez-Méndez, D., and Barba, E. (2016). Negative effects of high temperatures during development on immediate post-fledging survival in great tits *Parus major*. *Acta Ornithologica*, 51, 235–44.

Saunders, D.A., Mawson, P., and Dawson, R. (2011). The impact of two extreme weather events and other causes of death on Carnaby's black cockatoo: a promise of things to come for a threatened species? *Pacific Conservation Biology*, 17, 141–8.

Scholander, P.F., Hock, R., Walters, V., Johnson, F., and Irving, L. (1950). Heat regulation in some Arctic and tropical mammals and birds. *Biological Bulletin*, 99, 237–58.

Schwagmeyer, P., and Mock, D.W. (2008). Parental provisioning and offspring fitness: size matters. *Animal Behaviour*, 75, 291–8.

Simmons, R.E., Barnard, P., Dean, W.R.J., Midgley, G.F., Thuiller, W., and Hughes, G. (2005). Climate change and birds: perspectives and prospects from southern Africa. *Ostrich*, 75, 295–308.

Smit, B., Whitfield, M.C., Talbot, W.A., Gerson, A.R., McKechnie, A.E., and Wolf, B.O. (2018). Avian thermoregulation in the heat: phylogenetic variation among avian orders in evaporative cooling capacity and heat tolerance. *Journal of Experimental Biology*, 221, jeb174870.

Smit, B., Zietsman, G., Martin, R.O., Cunningham, S.J., McKechnie, A.E., and Hockey, P.A.R. (2016). Can behaviour provide a basis for rapid assessment of relative vulnerability of desert birds to climate change? *Climate Change Responses*, 3, 9.

Smith, E.K., O'Neill, J., Gerson, A.R., and Wolf, B.O. (2015). Avian thermoregulation in the heat: resting metabolism, evaporative cooling and heat tolerance in Sonoran Desert doves and quail. *Journal of Experimental Biology*, 218, 3636–46.

Smith, E.K., O'Neill, J.J., Gerson, A.R., McKechnie, A.E., and Wolf, B.O. (2017). Avian thermoregulation in the heat: resting metabolism, evaporative cooling and heat tolerance in Sonoran Desert songbirds. *Journal of Experimental Biology*, 220, 3290–300.

Swanson, D.L., McKechnie, A.E., and Vézina, F. (2017). How low can you go? An adaptive energetic framework for interpreting basal metabolic rate variation in endotherms. *Journal of Comparative Physiology B*, 187, 1039–56.

Tattersall, G.J., Andrade, D.V., and Abe, A.S. (2009). Heat exchange from the toucan bill reveals a controllable vascular thermal radiator. *Science*, 325, 468–70.

Tattersall, G.J., Arnaout, B., and Symonds, M.R. (2017). The evolution of the avian bill as a thermoregulatory organ. *Biological Reviews*, 92, 1630–56.

Tattersall, G.J., Chaves, J.A., and Danner, R.M. (2018). Thermoregulatory windows in Darwin's finches. *Functional Ecology*, 32, 358–68.

Tieleman, B.I., van Noordwijk, H.J., and Williams, J.B. (2008). Nest site selection in a hot desert: trade-off between microclimate and predation risk? *Condor*, 110, 116–24.

Van Buskirk, J., Mulvihill, R.S., and Leberman, R.C. (2010). Declining body sizes in North American birds associated with climate change. *Oikos*, 119, 1047–55.

van de Ven, T.M.F.N., Martin, R.O., Vink, T., McKechnie, A.E., and Cunningham, S.J. (2016). Regulation of heat exchange across the hornbill beak: Functional similarities with toucans? *PLoS One*, 11, e0154768.

van de Ven, T.M.F.N. (2017). Implications of climate change on the reproductive success of the southern yellow-billed hornbill *Tockus leucomelas*. PhD Thesis, University of Cape Town.

Welbergen, J.A., Klose, S.M., Markus, N., and Eby, P. (2008). Climate change and the effects of temperature extremes on Australian flying-foxes. *Proceedings of the Royal Society of London Series B: Biological Sciences*, 275, 419–25.

Whitfield, M.C., Smit, B., McKechnie, A.E., and Wolf, B.O. (2015). Avian thermoregulation in the heat: scaling of heat tolerance and evaporative cooling capacity in three southern African arid-zone passerines. *Journal of Experimental Biology*, 218, 1705–14.

Wingfield, J.C., Vleck, C.M., and Moore, M.C. (1992). Seasonal changes of the adrenocortical response to stress in birds of the Sonoran Desert. *Journal of Experimental Zoology Part A: Ecological Genetics and Physiology*, 264, 419–28.

Wolf, B.O. (2000). Global warming and avian occupancy of hot deserts: a physiological and behavioral perspective. *Revista Chilena de Historia Natural*, 73, 395–400.

Wolf, B.O., Coe, B.H., Gerson, A.R., and McKechnie, A.E. (2017). Comment on an analysis of endotherm thermal tolerances: systematic errors in data compilation undermine its credibility. *Proceedings of the Royal Society of London Series B: Biological Sciences*, 284, 20162523.

Wolf, B.O., and Walsberg, G.E. (1996). Respiratory and cutaneous evaporative water loss at high environmental temperatures in a small bird. *Journal of Experimental Biology*, 199, 451–7.

Woodworth, B.K., Norris, D.R., Graham, B.A., Kahn, Z.A., and Mennill, D.J. (2018). Hot temperatures during the dry season reduce survival of a resident tropical bird. *Proceedings of the Royal Society of London Series B: Biological Sciences*, 285, 20180176.

Xie, S., Romero, L.M., Htut, Z.W., and McWhorter, T.J. (2017). Stress responses to heat exposure in three species of Australian desert birds. *Physiological and Biochemical Zoology*, 90, 348–58.

Yom-Tov, Y. (2001). Global warming and body mass decline in Israeli passerine birds. *Proceedings of the Royal Society of London Series B: Biological Sciences*, 268, 947–52.

Yom-Tov, Y., Benjamini, Y., and Kark, S. (2002). Global warming, Bergmann's rule and body mass–are they related? The chukar partridge (*Alectoris chukar*) case. *Journal of Zoology*, 257, 449–55.

CHAPTER 11

Evolutionary consequences of climate change in birds

Céline Teplitsky and Anne Charmantier

11.1 Introduction

There is now overwhelming evidence that the recent rapid climate change has multiple consequences for birds, at all levels of observation, from individual birds to populations, species, and communities. For instance, individual birds have altered their morphology, physiology, behaviour, and reproductive traits (e.g., Chapters 9, 10, and 15). Populations have shown changes in their demography and their connectivity (Chapters 8 and 12). Finally, anthropogenic climate change results in poleward range expansions for many bird species, sometimes concomitant with adjustments in species' ecological niches (Virkkala and Lehikoinen 2014; Chapter 13). While these changes have been described in some cases in great detail, there is still much to do to understand the processes behind these changes, as well as their ecological, evolutionary, demographical, and conservation consequences. The present chapter will focus on the evolutionary processes involved in the changes that have been observed in avian traits linked with climate change; that is, genetic changes in traits that are heritable and for which selection is climate-dependent.

One particularly well-known case study is the observed advancement in the timing of seasonal cycles in birds, and in particular the earlier timing in reproduction (as well as in migration) for temperate insectivorous passerines. Because such change in avian breeding phenology has been the subject of many published studies in the last two decades, it will be a *leitmotiv* case study in our chapter. We will however attempt to cover other types of phenotypic responses where evolutionary processes have been investigated.

Following the multiple reports on phenotypic changes in birds matching changes in the local or global climate, two fundamental questions arise for evolutionary ecologists. First, what is the origin of the change? Second, is the observed change adaptive? Responses to these two questions are often not trivial and require sophisticated statistical analyses on large datasets. To answer the first question, it is crucial to obtain individual-based repeated data to understand how much of the changes observed, e.g. in the birds' timing of egg laying, can be attributed to individual adjustments, i.e. plasticity, and how much can be attributed to a change in the population genetic background because of genetic drift or following an evolutionary response to selection. The second question calls for insight into the direction and force of natural (and sexual) selection acting on the focal changing trait. As the reader will discover in the next section, while selection acting on key adaptive traits such as timing of breeding has been repeatedly estimated in birds, attempts to link fluctuations in such selection with climate fluctuations are in fact rare.

Teplitsky, C., and Charmantier, A., *Evolutionary consequences of climate change in birds.* In: *Effects of Climate Change on Birds.* Second Edition. Edited by Peter O. Dunn and Anders Pape Møller: Oxford University Press (2019). © Oxford University Press. DOI: 10.1093/oso/9780198824268.003.0011

11.2 Is climate change a new force of selection?

Because climate change leads to substantial alterations in the mean and the variance of climate variables (e.g., temperature and rainfall) of all ecosystems around the world, it may affect the physiology of species at all trophic levels, thereby leading to extensive modifications of biotic and abiotic interactions. These changes can alter selection pressures, if they affect individuals differentially according to their phenotypes. Indeed, selection acting on a focal trait is classically measured as the statistical effect of this trait on individual fitness, i.e. a linear regression for directional selection and an additional quadratic effect for nonlinear selection (Lande and Arnold 1983). Such selection models make the underlying assumption that the trait has a causative effect on fitness (Morrissey and Sakrejda 2013). To demonstrate that climate change induces new selection forces, it is necessary to show that the relationship between fitness and the trait is modified by climatic conditions (e.g., selection for lay date is stronger in warmer years (Marrot et al. 2018)) and that these conditions are changing at the local scale.

A recent meta-analysis of selection gradients across animals and plants, suggests that mean annual precipitation is a strong driver of selection, even stronger than mean annual temperature (Siepielski et al. 2017). However, in the shorter term, selection is also affected by potential evapotranspiration, an index integrating temperature, humidity, and radiation (Siepielski et al. 2017). In birds, most studies have focused on selection on phenology (laying date, duration of breeding season, migration timing) and morphology, and climate change is most often measured as temperature warming.

11.2.1 Selection on phenology

Studies across a range of avian taxa have repeatedly demonstrated strong negative directional selection on phenology of avian breeding and migration: in many avian species reproductive success is maximized by breeding early (Charmantier and Gienapp 2014; Verhulst and Nilsson 2008). Because timing of breeding and migration have previously been related to the phenology of the local ecosystem, which is usually directly influenced by ambient temperature (van Noordwijk et al. 1995), selection for earlier phenology is expected to intensify under global warming (Gienapp et al. 2014). This section reviews evidence for such intensification of selection linked to warming in birds.

The most intensively studied trait is timing of breeding, measured as the calendar date for an individual's seasonal first egg laying. The now classic example of climate-induced selection pressure is the intensification of selection for earlier laying with increasing temperatures, shown in pied flycatchers *Ficedula hypoleuca* (Visser et al. 2015), great tits *Parus major* (Chevin et al. 2015; Husby et al. 2011; Reed et al. 2013; Vedder et al. 2013) and blue tits *Cyanistes caeruleus* (Marrot et al. 2018; Phillimore et al. 2016). A likely hypothesis to explain this pattern is that as warmer spring temperatures are speeding up caterpillar development, these avian species need to breed sufficiently early to match the peak of nestling needs with the spring food peak (Thomas et al. 2001; Visser et al. 1998; Figure 11.1; but see Dunn et al. 2011). However, note that climate may be affecting optimal laying date even in the absence of a correlation between temperature and selection if individuals are successfully tracking the optimal breeding time. A powerful way to assess the environmental sensitivity of selection is thus to estimate the changes in the optimal phenotype itself induced by climatic fluctuations (Chevin et al. 2015). Using a new method based on random regressions to estimate selection, Chevin and colleagues estimate the annual optimum laying date in a great tit population and show that optimum breeding time is predicted by temperature as expected. However, the optimum fluctuates beyond this expectation, following temporally autocorrelated random fluctuations, suggesting other environmental factors are also important in determining optimal laying date.

While the in-depth study of climate-driven selection fluctuations for phenology of the three cited species are remarkable case studies, it can in no way be extrapolated to other birds, or even other passerines, without similar scrutiny. Indeed, the apparent ubiquity of selection for earlier laying date reported in the literature most certainly hides large variability in selection pressures and in the drivers of selection. Depending on their ecology, some

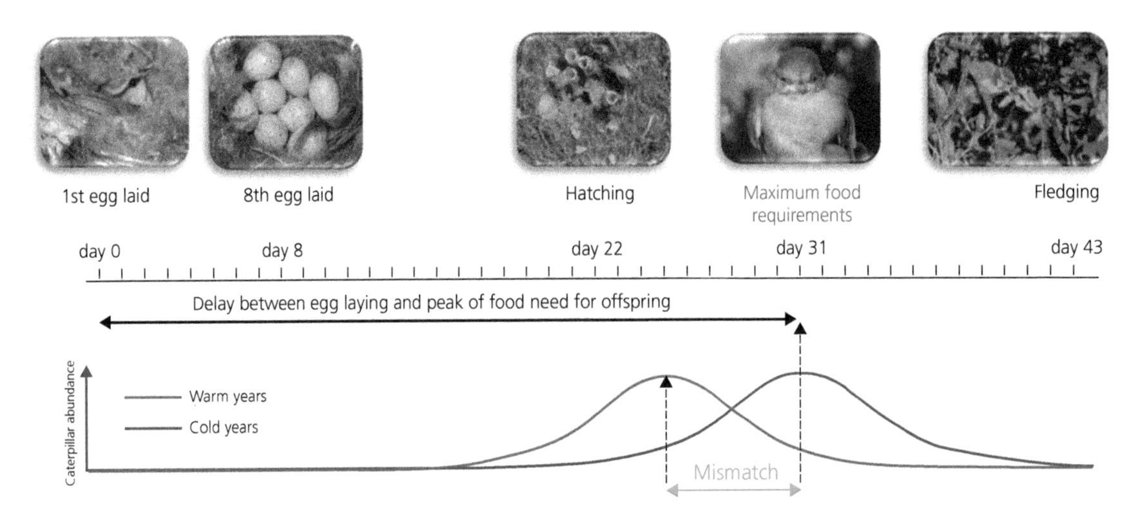

Figure 11.1 Illustration of a mismatch in blue tits. Day 0 is laying date (varying among individuals and depending on temperature). In cool years (blue), the peak of caterpillar abundance matches the peak of offspring need (caterpillar abundance highest when food requirement is maximal), while during warmer years (red), the peak of caterpillar abundance precedes that of offspring needs. Photo credit: C. Doutrelant, A. Charmantier, S. Tillo.

species may be more or less sensitive to climate. For example, on the Swedish island of Öland, selection for earlier breeding is stronger on collared (*Ficedula albicollis*) than on pied flycatcher (*F. hypoleuca*), probably because of the broader diet of the pied flycatcher (Sirkiä et al. 2018). Also, while in temperate regions, temperature may be the main climatic cue triggering breeding for many insectivorous passerines, wind is also a possible driver (Irons et al. 2017; Møller 2013) while rain is a more likely key determinant of breeding in the tropics (Oppel et al. 2013; Senapathi et al. 2011).

11.2.2 Selection on morphological attributes

Another response to climate change that received much attention in the avian literature is decreased body size. It has been described as the third main response to climate change, aside from changes in species distribution and in phenology (Gardner et al. 2011; Sheridan and Bickford 2011; Chapter 10). One line of reasoning is that Bergmann's rule predicts smaller body size for populations or species living under warmer conditions. Substituting time for space can thus lead to the prediction that body size should decline when temperature is increasing under climate change. In birds, a vast majority of studies report a negative correlation between temperature and body size (Teplitsky and Millien 2014), but these correlations tend to disappear when controlling for time (Radchuk et al., in press). Moreover, temporal trends in temperature are much smaller than spatial gradients in temperature along which Bergmann's rule is usually observed, so that the classical heat dissipation hypothesis is unlikely (Teplitsky and Millien 2014). Very few studies have tested for selection on body size, and none found climate-induced selection for smaller size (Gardner et al. 2014; Teplitsky and Millien 2014; van Gils et al. 2016). On the contrary, in red knots *Calidris canutus*, body size and bill length are decreasing because of lower food availability on the breeding grounds, leading to selection for longer bills on the wintering grounds where short-billed individuals cannot reach deeper buried prey (van Gils et al. 2016). More studies are needed to understand if and why climate change should be expected to affect patterns of selection on size in general.

11.2.3 Selection in other trait types

The focus on shrinkage as a major ecological response to climate change, although worth exploring, may have shifted attention from less obvious but maybe more likely climate-induced changes in selection on other avian characteristics. The first demonstration in a wild bird population that recent climate change alters the force of natural selection is the case of colour morphs in the tawny owl (*Strix aluco*; see section 11.4.3 for more details on this example). Climate change is also likely to affect sexual selection because of the condition-dependent nature of sexually selected traits (Qvarnström et al. 2016). For example, sexual selection is stronger in wet favourable years in the superb fairywren (*Malurus cyaneus*; Cockburn et al. 2008). A study of collared flycatchers even suggests that climate change may reverse selection on ornamentation: the size of the white forehead patch that used to be under positive selection in males is under negative selection when temperature increases (Evans and Gustafsson 2017). Many other avian traits that are environmentally dependent, such as personalities (Dubuc-Messier et al. 2017), may also be under differential selection due to climate change. Finally, since climate change often drastically modifies prey abundance and timing, nestling growth is a likely candidate trait for climate-sensitive selection (e.g., Divoky et al. 2015).

Overall, for phenological traits, intuitive reasons why climate change should affect selection pressures have been supported by many empirical studies. In turn, understanding why and when morphology should be selected differently under climate change is more difficult because morphology is not selected at one time point in the life cycle, but throughout the whole life of individuals, and selection can vary across life stages. The existence of non-intuitive selection patterns arising from climate change also calls for investigating how it can affect selection on other trait types such as ornaments, or physiological or behavioural traits.

11.3 Evolutionary potential of avian adaptive traits

11.3.1 The animal model

The estimation of evolutionary potential in wild populations has been flourishing since the introduction of the *animal model* (Charmantier et al. 2014; Henderson 1973). This powerful and flexible framework has made it possible to estimate the genetic variance of quantitative (continuous) traits in nature (Kruuk 2004). The *animal model* is described in very accessible ways to ecologists and ornithologists elsewhere (Kruuk 2004; Postma and Charmantier 2007; Wilson et al. 2010) and will not be detailed here. Roughly, it can be seen as an extension of the parent–offspring regression where information about all the known relatedness relationships among individuals in the population is used. In wild bird populations, this information is usually derived from a social pedigree. In more statistical terms, the *animal model* is an individual-based mixed model where the information from the pedigree is converted into a matrix of co-ancestry to model the expected genetic covariance between two individuals (Henderson 1973). The flexibility of the *animal model* allows one to investigate complex modes of inheritance such as the genetic covariances among traits, maternal effects, or interactions between genes and environments (Kruuk 2004).

11.3.2 Heritability of phenology and morphology

One of the most famous equations in quantitative genetics is the breeders' equation (Falconer and Mackay 1996):

$$R = h^2 S$$

where R is the predicted response to selection over one generation for a focal trait, h^2 is the trait heritability (the ratio between additive genetic variance V_A and phenotypic variance V_P), and S is the selection differential measuring selection acting on this trait. This equation suggests that heritability is a crucial estimate of evolutionary potential, and many studies have focused on estimating heritabilities for various morphological and life history traits in wild populations (Charmantier et al. 2014). A review of estimates in the wild (Postma 2014) showed that in birds, most measured traits are heritable and heritability is larger for morphological traits (e.g., tarsus length mean ± se: 0.54 ± 0.033) than for life history traits (e.g., clutch size: 0.25 ± 0.049, timing of breeding: 0.21 ± 0.016). Hence in general, there is no absolute genetic constraint for these traits to evolve under natural selection, as would be the case in the absence of additive genetic variance ($h^2 = 0$). In a first approach, heritability seems a convenient descriptor of evolutionary potential, notably because it is bounded between 0 and 1, allowing a quick assessment of whether heritability is comparatively high or low among traits and species. However, many criticisms of heritability have emerged (Hansen et al. 2011; Houle 1992; Postma 2014; Walsh and Blows 2009): listing all the issues and potential biases is beyond the scope of this chapter, and we will only focus on emphasizing aspects of why evolutionary potential is a complex measure.

11.3.3 More complex genetic architecture

We will illustrate three intriguing routes through which genetic architecture may shape evolutionary responses in ways not predictable based on heritability alone. Because of lack of space, we only focus on avian laying date, a key trait responding to climate change where several studies suggest a lack of response of laying date to selection (no microevolution) in spite of significant directional selection and heritability (see section 11.4.2).

First, traits within organisms are not independent but integrated (Olson and Miller 1958), so that genetic covariances among traits may have major impacts on their responses to climate-induced selection. For example, a negative phenotypic correlation between laying date and clutch size is common, i.e. across many species birds laying earlier lay larger clutches (Klomp 1970). If this phenotypic correlation reflects an underlying negative genetic correlation, selection for earlier laying during warm springs will lead to a correlated response for increased clutch size. However, if clutch size is under stabilizing selection (Lack 1947), it could constrain the phenological response to selection. Evidence for a genetic correlation between laying date and clutch size is mixed, as some studies find evidence of a negative genetic correlation, while some others do not (Online Supplementary Table 11.1). Understanding how laying date is genetically correlated with other trait types (e.g., morphology, behaviour) may help understand whether genetic correlations could prevent microevolution of laying date (see 11.4.2).

Second, organisms are interacting with conspecifics (e.g., parents, competitors, mating partners) so that their phenotypes may be partly determined by genes of other individuals (indirect genetic effects; Wolf et al. 1998). For example, if the condition of females before breeding depends on their mate providing food or on the quality of the territory, female laying date may be affected by male genes (Germain et al. 2016; Teplitsky et al. 2010). Brommer and Rattiste (2008) showed in the common gull (*Larus canus*) the existence of a negative genetic covariance between direct (female) and indirect (male) genetic effects for laying date, i.e. genes associated with earlier laying when expressed in females are associated with later laying when expressed in males. In the climate change context, such conflicts can impose a constraint on the response to selection for earlier laying.

Third, available genetic variance may depend upon the environment, for example, additive genetic variance may decrease or increase under stressful conditions (Charmantier and Garant 2005). Available estimates of gene by environment interaction (G*E) for clutch size and laying date in response to temperature (significant or not) suggest that variance explained by G*E could represent up to 60 per cent of the among individual variance in responding to the environment (Gienapp and Brommer 2014). Because G*E analyses are highly demanding in terms of data (van de Pol 2012), only one study so far (Husby et al. 2010) has found significant evidence for G*E in clutch size (not in laying date) with increased additive genetic variance under warmer temperature. More studies are needed before a general conclusion can be reached on how the evolutionary potential of key breeding traits is temperature-dependent. However, it seems that, in general, there is no environmental coupling of heritability and selection (Ramakers et al. 2018), so that patterns such as low heritability when selection is strong should not represent a frequent constraint.

11.3.4 Evolutionary potential of plasticity

Because temperature ranges are changing rapidly in many ecosystems, a challenging question is the evolution of the ability of individuals to respond to temperature fluctuations, and in particular warming. While we know almost nothing about the heritability of laying date plasticity (but see Nussey et al. 2005; Ramakers et al. 2019), some studies have investigated the prerequisite for evolution of plasticity: its variability across females and populations (e.g., Porlier et al. 2012). A comparison of two of the longest individual-based studies of great tits (Charmantier et al. 2008; Nussey et al. 2005) reveals that, at the population level, laying date is more strongly related to spring temperature in Wytham Woods (UK) compared to Hoge Weluwe (Netherlands) (Husby et al. 2010). This leads to a greater advancement in average laying date with climate warming in UK birds, which have advanced their laying dates by more than 3.5 days on average per decade, versus less than 2 days for Dutch birds. Interestingly, this difference is partly explained by higher individual heterogeneity in plasticity for Dutch great tits: while all UK females similarly adjust towards earlier laying during warm springs, Dutch females show variation in their degree of plasticity. The explanation behind these population differences in the level of plasticity and in the between-individual variation in plasticity has still not been elucidated. An interesting hypothesis worth pursuing is that birds from different populations use different cues to trigger egg laying (i.e. relative importance of photoperiod, climatic variables, tree phenology; Bonamour et al. 2019; Chapter 9), or that the reliability of cues is altered differentially with climate change between these two populations (Charmantier et al. 2008). Identifying which environmental cues are drivers of plastic responses to climate change in birds, and how these cues and their use vary among populations and species, is one of the most exciting yet challenging goals in this context of avian phenological adjustment to climate change.

11.4 Responses to selection

11.4.1 Detecting plasticity vs microevolution

There are several ways to assess whether an observed time trend is of plastic or genetic origin. These methods and their caveats have already been discussed in detail elsewhere (Hadfield et al. 2010; Merilä and Hendry 2014; Postma 2006), so we present very briefly those most amenable for wild bird populations.

In theory, the most powerful approach to evaluate whether a response is evolutionary or plastic is to directly assess whether genetic values are changing over time or cohorts, using data from long-term monitoring. This could be accomplished with an animal model, since this model outputs predicted values for each individual (Best Linear Unbiased Predictors, BLUPS), interpreted as predicted breeding values (PBV, akin to individual genetic value). However, several major issues are associated with the analyses of PBV: (i) the confidence intervals around PBV are large and need to be taken into account, (ii) PBV are biased towards phenotypic values when little information is available from the pedigree, and (iii) prediction errors are correlated across relatives, leading to temporal autocorrelation (Hadfield et al. 2010; Postma 2006). To circumvent these issues one may (a) include the time trend as a fixed effect (Postma 2006); (b) use Bayesian methods that carry the information on estimation errors in all steps of the analysis; and (c) compare observed trend expectations under drift to evaluate whether the change is larger than expected (Hadfield et al. 2010).

Other less direct methods involve:

1) Examining whether annual means are better predicted by an environmental variable rather than an annual trend. The premise of this approach is that plastic responses involve a response to the environment, while microevolution is a change in the genetic composition of the population over generations. One key issue for the success of this approach is to identify the environmental variable predicting the trait of interest (Chapter 5).
2) Assessing individual plasticity from repeated measurements on the same individuals. A comparison of individual slopes to population level plasticity can help understand the origin of the population level response (Charmantier et al. 2008). If individual plastic responses are of the same magnitude as the population response, then plasticity may be the main response mechanism.
3) Comparing the observed phenotypic trends to expectations based on estimates of selection and evolutionary potential. However, this only allows one to rule out evolution when the phenotypic trends are in the opposite direction to or much steeper than predicted evolutionary responses (e.g., Dobson et al. 2017).

More recently the use of genomic methods has opened up the possibility of identifying microevolutionary selective events, e.g. by detecting regions of reduced genetic diversity, including within genes that are involved in specific traits of interest for climatic adaptation (Franks and Hoffmann

2012). However, while such methodologies seem promising in plants, where candidate genes for stress response to thermal variation have been identified, detecting genes involved in phenotypically plastic traits involved in avian response to climate change is still very challenging (Gienapp et al. 2017). A recent study of yellow warblers (*Setophaga petechia*) in North America investigated the adaptive potential of this species across its breeding range by identifying genomic variation associated with local climate variation (Bay et al. 2018). This mapping of genotype–environment relationships allowed the authors to predict the 'genomic vulnerability' or the mismatch between current and predicted future genomic variation (under climate change scenarios), and link this metric to population trends. This novel approach (see also Ruegg et al. 2018) suggests that genomic tools could provide important predictive evolutionary insights that could influence conservation efforts.

11.4.2 What does the literature tell us about the prevalence of plastic versus genetic responses of birds?

Climate anomalies presently recorded can be considered as natural, large-scale experiments revealing how populations can adapt to rapid environmental changes. In particular, we have outlined above how long-term individual-based studies can contribute to evaluating the respective roles of drift, evolution, and plasticity in how bird populations adapt to global climate change. A literature survey of all studies published until April 2013, and using individual data to test for a plastic or evolutionary response in avian phenology in response to climate fluctuations, revealed the striking deficiency of studies on this question (Charmantier and Gienapp 2014). This lack of investigation is still topical in 2018, and is mainly due to the challenge, as outlined above, of correctly testing for an underlying genetic change in life history traits even when a temporal or climatic-related phenotypic change is observed at the population or individual level. So far, populations have been submitted to an adequate test of microevolution in laying date in populations of three species: the Gotland collared flycatcher (Brommer et al. 2005a; Przybylo et al. 2000; Sheldon et al. 2003), the Hoge Weluwe great tit (Gienapp et al. 2006; Nussey et al. 2005) and the Kaikoura red-billed gull (Teplitsky et al. 2010). All these studies have failed to provide evidence for an evolutionary response of laying date to climate, and the same lack of evidence (but also lack of studies) is reported when exploring all types of traits and all types of taxa (Merilä 2012; Merilä and Hendry 2014). This is in contrast with the very large number of studies that have provided evidence for strong plasticity in the timing of breeding (and migration; Charmantier and Gienapp 2014; Gienapp et al. 2008), a conclusion concordant with the importance of plasticity in biodiversity responses to human-disturbed contexts (Hendry et al. 2008). We emphasize, however, that the increasing evidence for a role of plasticity does not preclude a concomitant evolutionary response, since the tests for the latter are still too scarce to conclude about the respective roles of each (Merilä 2012; Merilä and Hendry 2014).

11.4.3 Detailed case study on plumage coloration

An interesting example of microevolution in response to climate change outside phenology is that of the plumage coloration in tawny owls (*Strix aluco*) in Finland (Figure 11.2). A long-term monitoring study of birds breeding in nest-boxes since 1977, combined with measurements of museum specimens collected across the twentieth century, revealed that the pheomelanin pigmentation of tawny owls, which can be either grey or brown (Figure 11.2a and b), has changed dramatically over the years, with many more brown morphs in later years (Karell et al. 2011). The colour morphs are highly heritable (Brommer et al. 2005b), which allows for a rapid evolutionary response of the trait. While viability selection historically favoured the grey plumage during snowy winters (perhaps because of lower predation risk, or a physiological advantage; Figure 11.2c), climate warming (Figure 11.2d) has altered this selection, favouring the brown morph. Considering the simple Mendelian inheritance of the colour morphs and the absence of plasticity in colour, the temporal shift towards a higher proportion of brown owls in the population is most likely a rare example of microevolutionary change in response to climate change.

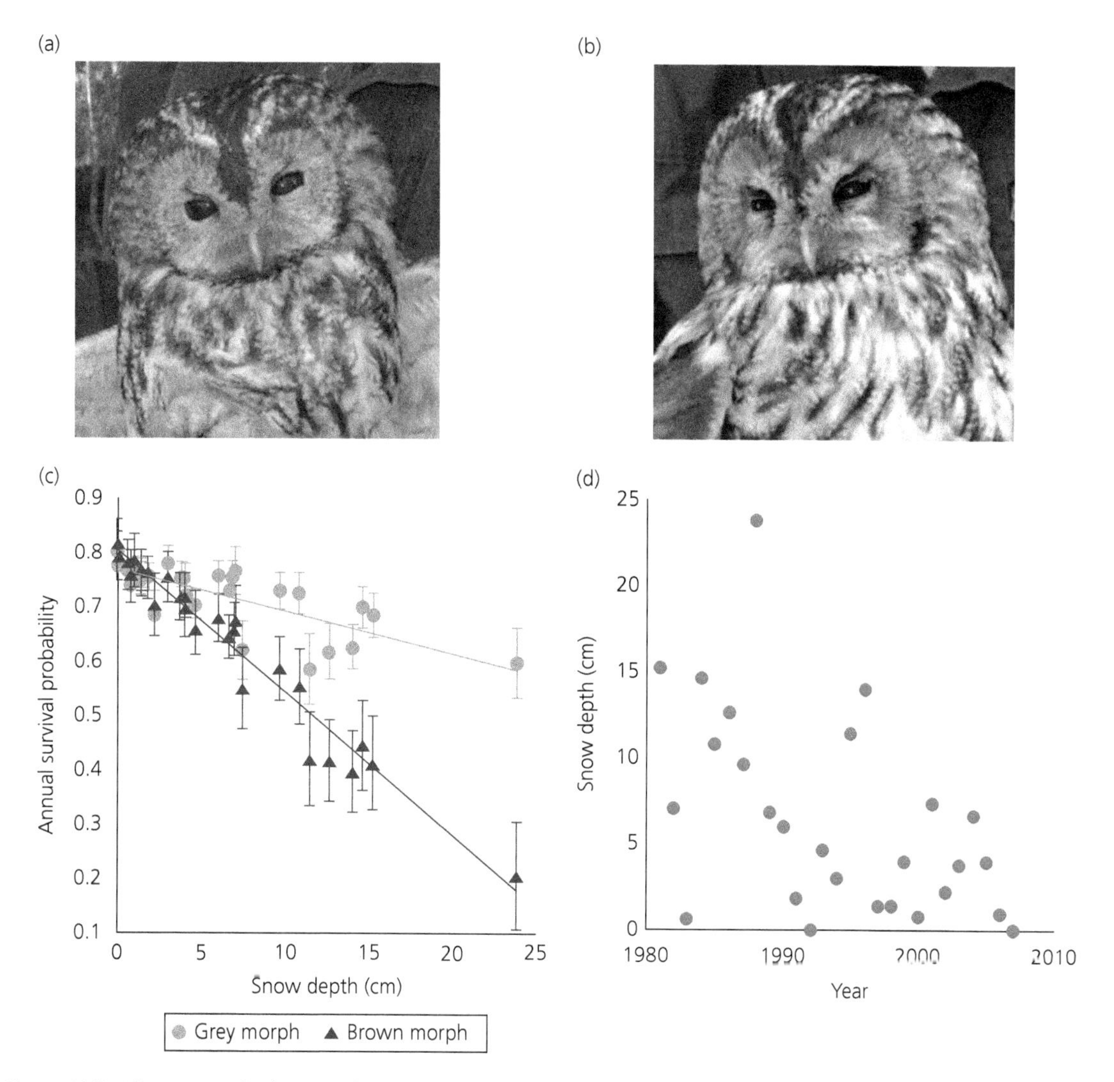

Figure 11.2 A long term study of tawny owl *Strix aluco* in Southern Finland revealed that the frequency of grey morphs (a) compared to brown morphs (b) has declined across time. This temporal change is explained by snow depth decreasing substantially over time (c) thereby decreasing the fitness advantage of grey morphs which survive better than brown ones during snowy winters (d) because of crypsis (adapted from Karell et al. 2011). © Pictures from Finnish Museum of Natural History website.

11.4.4 Contrasting results on general trends with extreme climatic events

Global climatic changes are characterized not only by pervasive global warming, but also by increased climatic variability, and increased frequency in extreme climatic events (ECE) such as heat waves, floods, or hurricanes (Easterling et al. 2000). In contrast to the large focus over the last decades on how natural populations adapt (or do not adapt) to climate warming, very little attention has been paid to the effects of ECEs until recently (van de Pol et al. 2017). Since ECEs are classically defined based on their rarity, it is obvious that studying them will be a challenge in terms of obtaining sufficient data and power to statistically test their effects. This difficulty also explains why most studies that have explored the evolutionary consequences of ECEs have focused on a single catastrophic event (Altwegg et al. 2017). For example, the emblematic study of Darwin's finches *Geospiza* spp. revealed a predictable evolutionary response in beak size following a severe El

Niño event (Grant and Grant 1993), but even in this case a 30-year perspective later revealed unpredictable long-term trends (Grant and Grant 2002).

ECEs are major ecological events with often drastic consequences on organisms' life histories, behaviour, survival, and demography (Fey et al. 2015; Møller 2011; Moreno and Møller 2011). For this reason, it is classically assumed that ECEs have major evolutionary impacts. While this is most certainly true at the macroevolutionary scale (Jablonski 1986), it is not so obvious at the microevolutionary one. In fact, some ECEs might have very little impact on a species' evolution. First, because ECEs are rare, and often restricted to remote areas within a species' distribution range. Second, because an ECE affecting all individuals of a population in the same way (e.g., reducing reproductive success or increasing mortality of all individuals) irrespective of individual phenotypes, will not result in selection (i.e. relationship between fitness and phenotype). It is hence crucial to understand whether natural selection is virtually absent, or on the contrary very strong, following ECEs, to assess the evolutionary consequences of these extraordinary events.

Avian taxa have recently provided insightful examples of the evolutionary consequences of ECEs. In the long-lived Eurasian oystercatcher *Haematopus ostralegus*, subjected to increased frequency of tidal floods during the breeding season, no evidence for evolution or plasticity was found over two decades in the placement of nests above water (Bailey et al. 2017), despite dramatic consequences of the floods on chick survival. The authors of the study postulate that the absence of adjustment of nest elevation by the birds was most probably explained by the low predictability of the floods. Another recent study on the short-lived blue tit *Cyanistes caeruleus* specifically tested whether ECEs translate into stronger selection (Marrot et al. 2017). The analysis of multiple climatic variables over a quarter century reveals that heat waves during the nestling stage substantially strengthened the directional selection favouring earlier breeding. Specifically, the force of natural selection on timing of reproduction was increased by 20 per cent in years with spring heat waves, independently of the effect of mean spring warming. These studies are first steps towards unravelling the evolutionary consequences of ECEs, yet the question of whether catastrophic climatic events can act as selective agents on adaptive traits and on their plasticity (Chevin and Hoffmann 2017) remains largely open.

11.5 Critical needs and future directions

There is still very little information on the nature (adaptive or not) and origin (plasticity vs evolution) of the observed responses to climate change in wild birds. The vast majority of detailed evolutionary studies are based on a restricted number of long-term studies in temperate climates, mostly on passerine laying dates. More diverse studies are urgently needed to improve our understanding of the various effects of climate change, not only in terms of ecological conditions (e.g., in the tropics) but also including a larger range of species, both in terms of taxa and of life history (e.g., long versus short-lived species, specialist vs generalist) and a larger set of traits. Documenting this diversity will open the way for comparative approaches and upscaling from trait changes to populations and communities dynamics (McLean et al. 2016; Phillimore et al. 2016). Further avian studies on responses to climate change will also need to integrate how other components of global change may interact with responses to climate change (e.g., is the ability to express plasticity similar in urban and forest environments?).

One of the most critical needs is to combine evolutionary and demographic models to understand how it matters for species/population persistence. For example, Vedder et al. (2013) used a mechanistic population model developed by Chevin et al. (2010) to predict how great tits from the Wytham Wood population could adapt and persist in an environment with an average yearly warming of 0.5°C, which is above all climatic predictions for the next 50 years. The predictions of long-term population persistence depended dramatically on the inclusion of a plastic individual adjustment of laying date: without plasticity, the population had a 60 per cent risk of extinction. These models predicted little impact of evolutionary responses as opposed to the important role of plasticity. However, because our predictions of response to selection are often wrong, improving our ability to predict response to selection will be an important breakthrough.

References

Altwegg, R., Visser, V., Bailey, L.D., and Erni, B. (2017). Learning from single extreme events. *Philosophical Transactions of the Royal Society of London Series B: Biological Sciences*, 372, 20160141.

Bailey, L.D., Ens, B.J., Both, C., Heg, D., Oosterbeek, K., and van de Pol, M. (2017). No phenotypic plasticity in nest-site selection in response to extreme flooding events. *Philosophical Transactions of the Royal Society of London Series B: Biological Sciences*, 372, 20160139.

Bay, R.A., Harrigan, R.J., Underwood, V. Le, Gibbs, H.L., Smith, T.B., and Ruegg, K. (2018). Genomic signals of selection predict climate-driven population declines in a migratory bird. *Science*, 359, 83–6.

Bonamour, S., Chevin, L.-M., Charmantier, A., and Teplitsky, C. (2019). Phenotypic plasticity in response to climate change: the importance of cue variation. *Philosophical Transactions of the Royal Society of London Series B: Biological Sciences*, 374, 20180178.

Brommer, J.E., Ahola, K., and Karstinen, T. (2005b). The colour of fitness: plumage coloration and lifetime reproductive success in the tawny owl. *Proceedings of the Royal Society of London Series B: Biological Sciences*, 272, 935–40.

Brommer, J.E., Merilä, J., Sheldon, B.C., and Gustafsson, L. (2005a). Natural selection and genetic variation for reproductive reaction norms in a wild bird population. *Evolution*, 59, 1362–71.

Brommer, J.E., and Rattiste, K. (2008). 'Hidden' reproductive conflict between mates in a wild bird population. *Evolution*, 62, 2326–33.

Charmantier, A., and Garant, D. (2005). Environmental quality and evolutionary potential: lessons from wild populations. *Proceedings of the Royal Society of London Series B: Biological Sciences*, 272(1571), 1415–25.

Charmantier, A., Garant, D., and Kruuk, L.E.B. (2014). *Quantitative genetics in the wild*. Oxford University Press, Oxford, UK.

Charmantier, A., and Gienapp, P. (2014). Climate change and timing of avian breeding and migration: evolutionary versus plastic changes. *Evolutionary Applications*, 7, 15–28.

Charmantier, A., McCleery, R.H., Cole, L.R., Perrins, C., Kruuk, L.E.B., and Sheldon, B. C. (2008). Adaptive phenotypic plasticity in response to climate change in a wild bird population. *Science*, 320, 800–3.

Chevin, L.-M., and Hoffmann, A.A. (2017). Evolution of phenotypic plasticity in extreme environments. *Philosophical Transactions of the Royal Society of London Series B: Biological Sciences*, 372, 20160138.

Chevin, L.-M., Lande, R., and Mace, G.M. (2010). Adaptation, plasticity, and extinction in a changing environment: towards a predictive theory. *PLoS Biology*, 8, e1000357.

Chevin, L.-M., Visser, M.E., and Tufto, J. (2015). Estimating the variation, autocorrelation, and environmental sensitivity of phenotypic selection. *Evolution*, 69, 2319–32.

Cockburn, A., Osmond, H.L., and Double, M.C. (2008). Swingin' in the rain: condition dependence and sexual selection in a capricious world. *Proceedings of the Royal Society of London Series B: Biological Sciences*, 275, 605–12.

Divoky, G.J., Lukacs, P.M., and Druckenmiller, M.L. (2015). Effects of recent decreases in arctic sea ice on an ice-associated marine bird. *Progress in Oceanography*, 136, 151–61.

Dobson, F.S., Becker, P.H., Arnaud, C.M., Bouwhuis, S., and Charmantier, A. (2017). Plasticity results in delayed breeding in a long-distant migrant seabird. *Ecology and Evolution*, 7, 3100–9.

Dubuc-Messier, G., Réale, D., Perret, P., and Charmantier, A. (2017). Environmental heterogeneity and population differences in blue tits personality traits. *Behavioral Ecology*, 28, 448–59.

Dunn, P.O., Winkler, D.W., Whittingham, L.A., Hannon, S.J., and Robertson, R.J. (2011). A test of the mismatch hypothesis: How is timing of reproduction related to food abundance in an aerial insectivore? *Ecology*, 92, 450–61.

Easterling, D.R., Meehl, G.A., Parmesan, C., Changnon, S.A., Karl, T.R., and Mearns, L.O. (2000). Climate extremes: observations, modeling, and impacts. *Science*, 289, 2068–74.

Evans, S.R., and Gustafsson, L. (2017). Climate change upends selection on ornamentation in a wild bird. *Nature Ecology and Evolution*, 1, 1–5.

Falconer, D.S., and Mackay, T.F.C. (1996). *Introduction to quantitative genetics*, 4th edition. Longmans Green, London.

Fey, S.B., Siepielski, A.M., Nusslé, S., Cervantes-Yoshida, K., Hwan, J.L., Huber, E.R., and Carlson, S.M. (2015). Recent shifts in the occurrence, cause, and magnitude of animal mass mortality events. *Proceedings of the National Academy of Sciences of the United States of America*, 112, 1083–8.

Franks, S.J., and Hoffmann, A.A. (2012). Genetics of climate change adaptation. *Annual Review of Genetics*, 46, 185–208.

Gardner, J.L., Amano, T., Mackey, B.G., Sutherland, W.J., Clayton, M., and Peters, A. (2014). Dynamic size responses to climate change: Prevailing effects of rising temperature drive long-term body size increases in a semi-arid passerine. *Global Change Biology*, 20, 2062–75.

Gardner, J.L., Peters, A., Kearney, M.R., Joseph, L., and Heinsohn, R. (2011). Declining body size: A third universal response to warming? *Trends in Ecology & Evolution*, 26, 285–91.

Germain, R.R., Wolak, M.E., Arcese, P., Losdat, S., and Reid, J.M. (2016). Direct and indirect genetic and fine-scale location effects on breeding date in song sparrows. *Journal of Animal Ecology*, 85, 1613–24.

Gienapp, P., and Brommer, J.E. (2014). Evolutionary dynamics in response to climate change. In: A. Charmantier, D. Garant, and L.E.B. Kruuk (eds), *Quantitative genetics in the wild*, pp. 254–73. Oxford University Press, Oxford, UK.

Gienapp, P., Laine, V.N., Mateman, A.C., van Oers, K., and Visser, M.E. (2017). Environment-dependent genotype-phenotype associations in avian breeding time. *Frontiers in Genetics*, 8, 102.

Gienapp, P., Postma, E., and Visser, M.E. (2006). Why breeding time has not responded to selection for earlier breeding in a songbird population. *Evolution*, 60, 2381–8.

Gienapp, P., Reed, T.E., and Visser, M.E. (2014). Why climate change will invariably alter selection pressures on phenology. *Proceedings of the Royal Society of London Series B: Biological Sciences*, 281, 20141611.

Gienapp, P., Teplitsky, C., Alho, J.S., Mills, J.A., and Merilä, J. (2008). Climate change and evolution: disentangling environmental and genetic responses. *Molecular Ecology*, 17, 167–78.

Grant, B.R., and Grant, P.R. (1993). Evolution of Darwin's finches caused by a rare climatic event. *Proceedings of the Royal Society of London Series B: Biological Sciences*, 251, 111–17.

Grant, P.R., and Grant, B.R. (2002). Unpredictable evolution in a 30-year study of Darwin's finches. *Science*, 296, 707–17.

Hadfield, J.D., Wilson, A.J., Garant, D., Sheldon, B.C., and Kruuk, L.E.B. (2010). The misuse of BLUP in ecology and evolution. *American Naturalist*, 175, 116–25.

Hansen, T.F., Pélabon, C., and Houle, D. (2011). Heritability is not evolvability. *Evolutionary Biology*, 38, 258–77.

Henderson, C.R. (1973). Sire evaluation and genetic trends. *Journal of Animal Science*, 1, 10–41.

Hendry, A.P., Farrugia, T.J., and Kinnison, M.T. (2008). Human influences on rates of phenotypic change in wild animal populations. *Molecular Ecology*, 17, 20–9.

Houle, D. (1992). Comparing evolvability and variability of quantitative traits. *Genetics*, 130, 195–204. doi:citeulike-article-id:10041224

Husby, A., Nussey, D.H., Visser, M.E., Wilson, A.J., Sheldon, B.C., and Kruuk, L.E.B. (2010). Contrasting patterns of phenotypic plasticity in reproductive traits in two great tit (*Parus major*) populations. *Evolution*, 64, 2221–37.

Husby, A., Visser, M.E., and Kruuk, L.E.B. (2011). Speeding up microevolution: the effects of increasing temperature on selection and genetic variance in a wild bird population. *PLoS Biology*, 9, e1000585.

Irons, R.D., Scurr, A.H., Rose, A.P., Hagelin, J.C., Blake, T., and Doak, D.F. (2017). Wind and rain are the primary climate factors driving changing phenology of an aerial insectivore. *Proceedings of the Royal Society of London Series B: Biological Sciences*, 284, 20170412.

Jablonski, D. (1986). Background and mass extinctions: the alternation of macroevolutionary regimes. *Science*, 231, 129–33.

Karell, P., Ahola, K., Karstinen, T., Valkama, J., and Brommer, J.E. (2011). Climate change drives microevolution in a wild bird. *Nature Communications*, 2, 207–8.

Klomp, H. (1970). The determination of clutch size in birds: a review. *Ardea*, 58, 1–124.

Kruuk, L.E.B. (2004). Estimating genetic parameters in natural populations using the 'animal model.' *Philosophical Transactions of the Royal Society of London Series B: Biological Sciences*, 359, 873–90.

Lack, D. (1947). The significance of clutch size. *Ibis*, 89, 302–52.

Lande, R., and Arnold, S.J. (1983). The measurement of selection on correlated characters. *Evolution*, 37(6), 1210–26.

Marrot, P., Charmantier, A., Blondel, J., and Garant, D. (2018). Current spring warming as a driver of selection on reproductive timing in a wild passerine. *Journal of Animal Ecology*, 87, 754–64.

Marrot, P., Garant, D., and Charmantier, A. (2017). Multiple extreme climatic events strengthen selection for earlier breeding in a wild passerine. *Philosophical Transactions of the Royal Society of London Series B: Biological Sciences*, 372, 20160372.

McLean, N., Lawson, C.R., Leech, D.I., and van de Pol, M. (2016). Predicting when climate-driven phenotypic change affects population dynamics. *Ecology Letters*, 19, 595–608.

Merilä, J. (2012). Evolution in response to climate change: In pursuit of the missing evidence. *BioEssays*, 34, 811–18.

Merilä, J., and Hendry, A.P. (2014). Climate change, adaptation, and phenotypic plasticity: The problem and the evidence. *Evolutionary Applications*, 7, 1–14.

Møller, A.P. (2011). Behavioral and life history responses to extreme climatic conditions: Studies on a migratory songbird. *Current Zoology*, 57, 351–62.

Møller, A.P. (2013). Long-term trends in wind speed, insect abundance and ecology of an insectivorous bird. *Ecosphere*, 4, 6.

Moreno, J., and Møller, A.P. (2011). Extreme climatic events in relation to global change and their impact on life histories. *Current Zoology*, 57, 375–89.

Morrissey, M.B., and Sakrejda, K. (2013). Unification of regression-based methods for the analysis of natural selection. *Evolution*, 67, 2094–100.

Nussey, D.H., Postma, E., Gienapp, P., and Visser, M.E. (2005). Selection on heritable phenotypic plasticity in a wild bird population. *Science*, 310, 304–6.

Olson, E.C., and Miller, R.J. (1958). *Morphological integration*. University of Chicago Press, Chicago.

Oppel, S., Hilton, G.M., Allcorn, R., Fenton, C., Matthews, A.J., and Gibbons, D.W. (2013). The effects of rainfall on different components of seasonal fecundity in a tropical forest passerine. *Ibis*, 155, 464–75.

Phillimore, A.B., Leech, D.I., Pearce-Higgins, J.W., and Hadfield, J.D. (2016). Passerines may be sufficiently plastic to track temperature-mediated shifts in optimum lay date. *Global Change Biology*, 22, 3259–72.

Porlier, M., Charmantier, A., Bourgault, P., Perret, P., Blondel, J., and Garant, D. (2012). Variation in phenotypic plasticity and selection patterns in blue tit breeding time: between- and within-population comparisons. *Journal of Animal Ecology*, 81, 1041–51.

Postma, E. (2006). Implications of the difference between true and predicted breeding values for the study of natural selection and micro-evolution. *Journal of Evolutionary Biology*, 19, 309–20.

Postma, E. (2014). Four decades worth of heritability estimates: What we have learned about the relative role of genes and the environment. In: A. Charmantier, D. Garant, and L.E.B. Kruuk (eds), *Quantitative genetics in the wild*, pp. 16–33. Oxford University Press, Oxford, UK.

Postma, E., and Charmantier, A. (2007). What 'animal models' can and cannot tell ornithologists about the genetics of wild populations. *Journal of Ornithology*, 148, 633–42.

Przybylo, R., Sheldon, B.C., and Merilä, J. (2000). Climatic effects on breeding and morphology: evidence for phenotypic plasticity. *Journal of Animal Ecology*, 69, 395–403.

Qvarnström, A., Ålund, M., Mcfarlane, S.E., and Sirkiä, P.M. (2016). Climate adaptation and speciation. Particular focus on reproductive barriers in *Ficedula* flycatchers. *Evolutionary Applications*, 9, 119–34.

Radchuk, V., Reed, T., Teplitsky, C. et al. (2019). Adaptive responses of animals to climate change: not universal, likely insufficient. *Nature Communications*, In press.

Ramakers, J.J.C., Culina, A., Visser, M.E., and Gienapp, P. (2018). Environmental coupling of heritability and selection is rare and of minor evolutionary significance in wild populations. *Nature Ecology and Evolution*, 2, 1093–103.

Ramakers, J. J. C., Gienapp, P., and Visser, M.E. (2019). Phenological mismatch drives selection on elevation, but not on slope, of breeding time plasticity in a wild songbird. *Evolution*, 73, 175–187.

Reed, T. E., Grotan, V., Jenouvrier, S., Sæther, B., and Visser, M. E. (2013). Population growth in a wild bird is buffered against phenological mismatch. *Science*, 340, 488–91.

Ruegg, K., Bay, R.A., Anderson, E.C., Saracco, J.F., Harrigan, R.J., Whitfield, M., Paxton, E.H., and Smith, T.B. (2018). Ecological genomics predicts climate vulnerability in an endangered southwestern songbird. *Ecology Letters*, 21, 1085–96.

Senapathi, D., Nicoll, M.A.C., Teplitsky, C., Jones, C.G., and Norris, K. (2011). Climate change and the risks associated with delayed breeding in a tropical wild bird population. *Proceedings of the Royal Society of London Series B: Biological Sciences*, 278, 3184–90.

Sheldon, B.C., Kruuk, L.E.B., and Merilä, J. (2003). Natural selection and inheritance of breeding time and clutch size in the collared flycatcher. *Evolution*, 57, 406–20.

Sheridan, J.A., and Bickford, D. (2011). Shrinking body size as an ecological response to climate change. *Nature Climate Change*, 1, 401–6.

Siepielski, A.M., Morrissey, M.B., Buoro, M., Carlson, S.M., Caruso, C.M., Clegg, S.M., Coulson, T., Dibattista, J., Gotanda, K.M., Francis, C.D., Hereford, J., Kingsolver, J.G., Sletvold, N., Svensson, E.I., Wade, M.J., and Maccoll, A.D.C. (2017). Precipitation drives global variation in natural selection. *Science*, 355, 959–62.

Sirkiä, P.M., McFarlane, S.E., Jones, W., Wheatcroft, D., Ålund, M., Rybinski, J., and Qvarnström, A. (2018). Climate-driven build-up of temporal isolation within a recently formed avian hybrid zone. *Evolution*, 72, 363–74.

Teplitsky, C., and Millien, V. (2014). Climate warming and Bergmann's rule through time: Is there any evidence? *Evolutionary Applications*, 7, 156–68.

Teplitsky, C., Mills, J.A., Yarrall, J.W., and Merilä, J. (2010). Indirect genetic effects in a sex-limited trait: The case of breeding time in red-billed gulls. *Journal of Evolutionary Biology*, 23, 935–44.

Thomas, D., Blondel, J., Perret, P., Lambrechts, M.M., and Speakman, J.R. (2001). Energetic and fitness costs of mismatching resource supply and demand in seasonally breeding birds. *Science*, 291, 2598–600.

van de Pol, M. (2012). Quantifying individual variation in reaction norms: how study design affects the accuracy, precision and power of random regression models. *Methods in Ecology and Evolution*, 3, 268–80.

van de Pol, M., Jenouvrier, S., Cornelissen, J.H.C., and Visser, M.E. (2017). Behavioural, ecological and evolutionary responses to extreme climatic events: challenges and directions. *Philosophical Transactions of the Royal Society of London Series B: Biological Sciences*, 372, 20160134.

van Gils, J.A., Lisovski, S., Lok, T., Meissner, W., Ozarowska, A., de Fouw, J., Rakhimberdiev, E., Soloviev, M.Y., Piersma, T., and Klaassen, M. (2016). Body shrinkage due to Arctic warming reduces red knot fitness in tropical wintering range. *Science*, 352, 819–21.

van Noordwijk, A.J., McCleery, R.H., and Perrins, C. (1995). Selection for the timing of great tit breeding in relation to caterpillar growth and temperature. *Journal of Animal Ecology*, 64, 451.

Vedder, O., Bouwhuis, S., and Sheldon, B.C. (2013). Quantitative assessment of the importance of phenotypic plasticity in adaptation to climate change in wild bird populations. *PLoS Biology*, 11, e1001605.

Verhulst, S., and Nilsson, J.A. (2008). The timing of birds' breeding seasons: a review of experiments that manipulated timing of breeding. *Philosophical Transactions of the Royal Society of London Series B: Biological Sciences*, 363(1490), 399–410.

Virkkala, R. and Lehikoinen, A. (2014). Patterns of climate-induced density shifts of species: Poleward shifts faster in northern boreal birds than in southern birds. *Global Change Biology*, 20, 2995–3003.

Visser, M.E., Gienapp, P., Husby, A., Morrissey, M.B., Hera, I., De, Pulido, F., and Both, C. (2015). Effects of spring temperatures on the strength of selection on timing of reproduction in a long-distance migratory bird. *PLoS Biology*, 13, e1002120.

Visser, M.E., van Noordwijk, A.J., Tinbergen, J.M., and Lessells, C.M. (1998). Warmer springs lead to mistimed reproduction in great tits (*Parus major*). *Proceedings of the Royal Society of London Series B: Biological Sciences*, 265, 1867–70.

Walsh, B., and Blows, M.W. (2009). Abundant genetic variation + strong selection = multivariate genetic constraints: A geometric view of adaptation. *Annual Review of Ecology, Evolution, and Systematics*, 40, 41–59.

Wilson, A.J., Réale, D., Clements, M.N., Morrissey, M.B., Postma, E., Walling, C.A., Kruuk, L.E.B., and Nussey, D.H. (2010). An ecologist's guide to the animal model. *Journal of Animal Ecology*, 79, 13–26.

Wolf, J.B., Brodie, E.D.I., Cheverud, J.M., Moore, A.J., and Wade, M.J. (1998). Evolutionary consequences of indirect genetic effects. *Trends in Ecology & Evolution*, 13, 64–9.

CHAPTER 12

Projected population consequences of climate change

David Iles and Stéphanie Jenouvrier

12.1 Introduction

Many atmospheric and oceanic climate processes will change in the coming century (IPCC 2013). These processes affect vital rates of individuals (such as survival, growth, reproduction, and dispersal) that govern the dynamics of natural populations (Chapter 7). Appropriate decision-making and wildlife policy requires anticipating the future state of populations in a changing climate and a rigorous accounting of uncertainty in those future states (Clark et al. 2001; Dietze 2017a).

Population forecasts estimate the future state of populations (Dietze 2017a) and can range from local to global in spatial extent, and from days to centuries in time. Population predictions are forecasts that are based solely on current knowledge of a system and are typically focused on the near term (Dietze et al. 2018). At longer timescales, processes such as technological innovation, socioeconomic change, and policy development cannot be fully anticipated but may have strong effects on the overall behaviour of the system. Yet, it is at these longer timescales that the effects of climate change will be most pronounced (Hawkins and Sutton 2009). Longer-term population forecasts that are contingent on particular scenarios of future change are called projections.

Projecting avian population dynamics in response to climate change requires several integrated steps (reviewed in Jenouvrier 2013). The first step is measuring the effect of climate on the complete life cycle of the studied species, thereby accounting for multiple seasonal and carry-over effects of climate. At this step, variation in vital rates is partitioned into components owing to measured climate variables along with vital rate variation owing to unexplained (i.e., unmeasured) factors, while removing spurious variation owing to imperfect detection or other observation error (Kéry and Schaub 2012). This first step has been the focus of hundreds of studies (see Chapters 5 and 7).

The second step is examining the demographic pathways through which climate influences overall population dynamics. This step requires integrating the statistical relationships between climate and vital rates (found in step one) into population models. At this step, population models become climate-dependent. Climate can be modelled as deterministic or stochastic, while vital rates can be entirely driven by climate or include additional 'unexplained' variation. Resulting models can then be used to evaluate the short- or long-term dynamics resulting from different climate scenarios. For example, they can be used to project population dynamics resulting from climate that is 2 °C warmer on average (even if climate is modelled deterministically; Dybala et al. 2013), or to project the respective role of a change in climate average versus variability, including extreme events (Jenouvrier et al. 2012; Pardo et al. 2017). Simultaneously, perturbation analyses can provide insights to the relative influence of climate properties, channelled through

Iles, D., and Jenouvrier, S., *Projected population consequences of climate change.* In: *Effects of Climate Change on Birds.* Second Edition. Edited by Peter O. Dunn and Anders Pape Møller: Oxford University Press (2019). © Oxford University Press.
DOI: 10.1093/oso/9780198824268.003.0012

different demographic pathways, and the potential for trade-offs among life cycle components (Caswell 2001; McLean et al. 2016; Jenouvrier et al. 2018b).

The final step is fusing climate-dependent population models with projections of future climate from IPCC-class atmospheric–oceanic global circulation models (AOGCMs). AOGCMs project (often nonlinear) changes in climate over time, and critically, provide quantitative estimates of uncertainty in future climate change for multiple climate variables (Hawkins and Sutton 2009). AOGCMs thereby provide a means for uncertainty in future climate to be fully propagated to population forecasts.

Here, we focus on this third step of projecting avian population responses to climate change by linking climate-dependent population models with projections of future climate from IPCC-class AOGCMs; Chapter 7 discusses the related issue of estimating climate effects on vital rates and incorporating them into population models (step 2 in our approach), along with several aspects of forecasting. We first discuss biological considerations of this approach: characterizing the full life cycle (section 12.2) and considering spatial heterogeneity and dispersal processes (section 12.3). We then highlight important methodological challenges: matching the scale of ecological processes with the scale of climate projections (section 12.4) and fully propagating climate and demographic uncertainty to population forecasts (section 12.5). Throughout, we use examples from a long-term study of emperor penguins (*Aptenodytes forsteri*) at Terre Adélie to illustrate key points. Finally, we conduct a literature search to compile a list of studies that have linked IPCC-class climate projections with avian population models (section 12.6). We conclude by discussing commonalities and differences among these studies, along with future prospects and challenges associated with forecasting avian population dynamics under climate change.

12.2 Biological considerations

12.2.1 Consideration of the full life cycle and relevant dimensions of population structure

Climate change will affect birds across their entire life cycle (Carey 2009; Jenouvrier 2013). Failure to account for effects on multiple life cycle stages can severely misrepresent the effect of climate on population dynamics (Ådahl et al. 2006). Climate effects on vital rates can be contrasted and/or delayed between various states of the life cycle, while evolutionary pressures, trade-offs, and physiological constraints can cause vital rates to respond differently to the same climate variables. For example, sea ice conditions impact survival and fecundity of emperor penguins in opposite ways (Barbraud and Weimerskirch 2001). During years with extensive winter sea ice, food is likely more abundant the following summer, increasing adult survival. However, foraging trips are longer in these extensive sea ice years resulting in fewer hatched eggs.

Population models must therefore account for relevant dimensions of population structure, such as differences in vital rates across age, stage, or size classes (Caswell 2001), and sex differences (Jenouvrier et al. 2010). Avian life histories vary considerably (Sæther and Bakke 2000; Sibly et al. 2012), ranging from fast-paced life histories such as house sparrows (*Passer domesticus*) that mature in a single year and produce up to four broods per season, to those with a slow pace of life such as wandering albatross (*Diomedea exulans*) that require up to 11 years to reach reproductive maturity and only produce one egg every two years. All bird populations consist of overlapping generations, but a slower pace of life generates a higher degree of population (st)age structure, which in turn, can strongly mediate the population consequences of climate change. Below, we discuss several key dimensions of population structure that affect avian responses to climate.

Climate often affects younger birds differently than older birds. For example, Oro et al. (2010) found that survival of young blue-footed boobies (*Sula nebouxii*) responded negatively to winter sea surface temperature (SST), while survival of older individuals showed no response. In a black-browed albatross (*Thalassarche melanophris*) population, Pardo et al. (2017) detected strongly nonlinear effects of SST on young and old individuals, but only a weak linear effect on prime-aged individuals. Simultaneously, reproductive success responded differently to SST across age classes. As a result, extreme climate events altered the stage structure of the population with a predicted increase in the proportion of juveniles in the population in a warmer climate with more frequent extreme events.

Body size is a key correlate of vital rates in birds and other animals (Stearns 1992). Body size influences energetic requirements, thermal tolerances, and predation pressure, and is often a reliable indicator of individual quality (Blanckenhorn 2000). For instance, wing length influences the survival and reproduction of black-browed albatross by likely reducing energetic costs incurred during flight. Accordingly, larger-winged individuals have higher fitness and are better able to cope with sub-optimal sea surface temperatures (Jenouvrier et al. 2018b). However, because body size both influences environmental sensitivity (e.g., Lindström, 1999) and simultaneously responds plasticly to environmental conditions (Cooch et al. 1991a, b), it is difficult to predict general relationships between avian body size and environmental sensitivity. Accordingly, studies have reported increasing, decreasing, or equivocal responses of avian body size to climate warming (Gardner et al. 2011; Sheridan and Bickford 2011). Reported effects are hypothesized to occur through a variety of pathways, including direct effects on physiology, changes in food availability, and shifts in predation pressure, among others. In cases where body size influences climate sensitivity, a consideration of population size structure may be required to reliably forecast population dynamics.

Breeding structure of the population can also strongly influence population dynamics. This is especially the case for long-lived species that often have extensive pre-breeding and non-breeding components of the population that occupy different habitats, are subjected to different climate, and experience different energetic constraints than breeders. In southern fulmar (*Fulmarus glacialoides*), approximately 40 per cent may skip breeding, and this proportion varies in response to climate fluctuations (Jenouvrier et al. 2005a).

Finally, sex structure (i.e., the ratio of males to females in a population) is relevant to population forecasting when (1) skewed sex ratios cause population growth to be limited by an inability of females to find mates, and (2) the vital rates of each sex differ or respond to climate differently. Both have been observed in birds. Bird populations often exhibit skewed sex ratios, caused by sex differences in multiple life cycle processes. For example, in wandering albatross, older mothers produce more female hatchlings, while higher quality mothers (those that have higher breeding success) tend to produce more male hatchings (Weimerskirch et al. 2005). Subsequently, mortality risk in many bird species is greater for juvenile males than females, owing to larger body sizes and higher energetic constraints (Clutton-Brock 1986). Yet, mortality risk for adult breeders is highest for the sex that incubates eggs (i.e., females in most species). This leads to male-skewed sex ratios in most bird species (Donald 2007), and females are thus rarely limited by lack of suitable mates. Emperor penguins offer a striking counter-example in which males incubate eggs during the Antarctic winter. As a result, adult males experience higher energetic demands, have lower survival, and are more sensitive to climate (particularly winter sea ice extent) than females (Jenouvrier et al. 2005b). The increased sensitivity of males to winter sea ice was responsible for a large decline in the Terre Adélie population of emperor penguins during a climate regime shift in the 1970s. In this case, a two-sex population model provides better descriptions of overall population dynamics by accounting for breeding limitation when operational sex ratios are uneven (Jenouvrier et al. 2010).

Population structure (owing to age, stage, sex, or other dimensions) generates transient dynamics, such that short-term population growth rates and long-term population abundances depend on the initial structure of the population (Stott et al. 2011). Transient dynamics can also interact with other population processes (e.g., demographic stochasticity) to influence extinction risk (Iles et al. 2016). Thus, even with perfect knowledge of the initial abundance of the population, uncertainty in initial stage structure can yield uncertain forecasts (also see section 12.4.2).

12.2.2 Consideration of spatial heterogeneity and dispersal

Projected changes in climate vary across the globe (see Chapter 2). Simultaneously, uncertainty in the amount of future climate change varies strongly across regions (see section 12.3.2). Populations separated in space will therefore experience different levels of climate change in the future, relative to their respective baselines, affecting the level of spatial heterogeneity in the environment. For example, in Antarctica, AOGCMs project marked regional and seasonal patterns of sea ice change by the end of the century (see Figure 5 in supplementary information of Jenouvrier et al. 2014). This has important consequences for emperor penguins, which breed in colonies scattered around the Antarctic continent. A species-level threat assessment study (Jenouvrier et al. 2014) projected that the most threatened colonies are located in Dronning Maud and Enderby Lands, where projected sea ice concentration (SIC) declines are largest and conditions are most variable. Conversely, colonies in the Ross Sea will experience the least loss of sea ice and are projected to increase relative to their present size by 2100. Overall, at least two-thirds of the colonies are projected to become endangered by sea ice decline by 2100, and the global population is projected to decline by at least 19 per cent (Figure 12.1).

Climate change may also alter the spatial distribution of other factors such as disease prevalence, with concomitant effects on population growth.

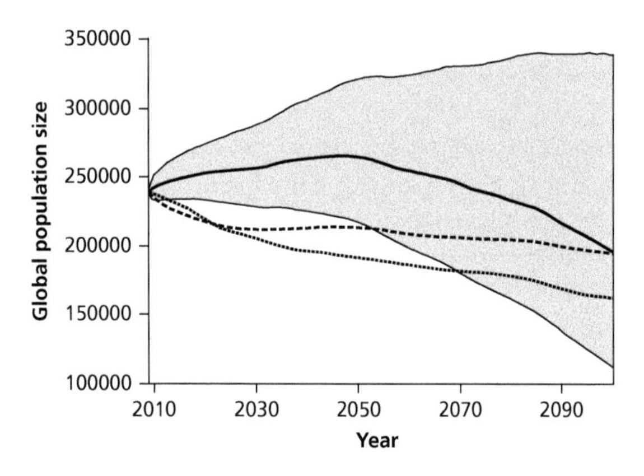

Figure 12.1 Forecast for global population size of emperor penguins using three different climate model selection criteria. Solid black line is median forecast (and associated 90% confidence intervals) resulting from selection of climate models that best reproduce local conditions at each emperor penguin colony. Thus, each colony is forecast using a different ensemble of climate models. Thin hashed grey line is median forecast based on four climate models that are known to accurately reproduce Pan-Antarctic sea ice conditions, but may not reproduce historical climate well for some individual emperor penguin colonies. Thick hashed grey line is median forecast using all AOGCMs.

Liao et al. (2015) projected the dynamics of three Hawaiian honeycreepers under future climate change, while simultaneously modelling the future prevalence of mosquito-borne malaria. Their forecasts indicated that malaria transmission to native birds would increase strongly under mid and high emissions scenarios, with accompanying declines in bird populations. Furthermore, while high elevation habitats historically provided refuge from malaria, their models indicated that climate change would drastically increase malaria transmission in these areas.

Accounting for spatial heterogeneity in avian population forecasts is particularly challenging for at two reasons: 1) demographic responses to climate can differ across space (e.g., owing to local adaptation or species interactions), and 2) dispersal rates can be difficult to estimate but can strongly influence population responses to climate change. We discuss these challenges below.

First, vital rates and their responses to climate can differ across space (Drever et al. 2012; Peery et al. 2012; Zhao et al. 2016), introducing spatial variation in population parameters and their uncertainty (we discuss parameter uncertainty more fully in section 12.3.2). For example, Peery et al. (2012) found that mean reproductive output of spotted owls (*Strix occidentalis*) was negatively correlated with nesting season temperature for a population in Arizona, but positively correlated for a population in southern California. Interestingly, these divergent responses occurred across the same range of climate variation. Differences in population sensitivity to the same climate variation can be driven by multiple processes. First, different populations can be adapted to local conditions, generating different reaction norms to the same environmental drivers. For example, a common garden experiment revealed that blue tits (*Parus caeruleus*) from two populations experienced large differences in the onset of laying date, owing to different adaptive responses to photoperiod (Lambrechts et al. 1997). Second, the strongest effects of climate are often indirect, channelled through effects on species interactions (Ockendon et al. 2014). Demographic responses to climate will therefore often depend on the larger ecological community, which varies across space. For example, in lesser snow geese (*Anser caerulescens caerulescens*), reproductive success was much more responsive to seasonal temperature in breeding habitats that contained higher plant diversity (Iles et al. 2018). Simultaneously, unexplained process variation in reproductive success was three to five times higher in those habitats, suggesting that plant diversity also affected snow goose sensitivity to unmeasured climate variables.

Second, projecting population responses to climate change requires full consideration of the ability of species to disperse (Travis et al. 2012; Ehrlén and Morris 2015), especially when the population decline is driven by climate changes that exceed the tolerance of a species or when acclimation and adaptation are insufficient to allow species persistence (Visser 2008). For example, the Southern Ocean Oscillation Index was a strong predictor of local demography for a population of Scopoli's shearwater (*Calonectris diomedea*), but immigration from other populations almost completely counterbalanced the net effects on population growth (Tavecchia et al. 2016). For emperor penguins, dispersal behaviours can either offset or accelerate climate-driven global population declines relative to a scenario without dispersal (Jenouvrier et al. 2017). Specifically, dispersal may increase the global population by up to 31 per cent or decrease it by 65 per cent, depending on the rate of emigration and distance individuals disperse. Thus, in some cases dispersal can act as an 'ecological rescue' mechanism to offset the global population decline of species endangered by climate change.

12.3 Methodological challenges

12.3.1 Match between scale of ecological process and climate projection

Climate models simulate historical and future climate on a three-dimensional lattice around the globe, typically at resolutions between 1 and 5 degrees latitude and longitude. In contrast, ecological studies commonly focus on the effects of local weather on populations, given that the local environment is the proximate cause of demographic variation (van de Pol et al. 2013). Thus, climate projections are made at coarse spatial scales relative to those at which the factors affecting population processes are typically measured. This disparity introduces several complications for using AOGCM outputs in population forecasts.

First, the effects of local-scale climate on populations can differ from the coarse-scale climate outputs that are available from AOGCM simulations. To deal with this problem, studies can estimate the effect of coarse-scale (rather than local-scale) climate on populations (Hallett et al. 2004). However, this approach may obscure the proximate causes of population fluctuations and exaggerate spatial differences in population sensitivity to climate (van de Pol et al. 2013). Alternatively, climate model outputs can be downscaled using statistical relationships between local-scale weather and coarse-scale climate, or dynamically by linking global AOGCMs with finer-resolution regional climate models (Snover et al. 2013). For example, Wolf et al. (2010) generated downscaled projections of coastal upwelling based on regional AOGCM projections of wind along the California coast to forecast the population dynamics of Cassin's auklets (*Ptychoramphus aleuticus*).

The second complication for using AOGCM outputs in population forecasts is that grid resolutions vary among AOGCMs (Figure 12.2a), and in many cases, the entire study area may be subsumed within a single AOGCM pixel. Individual AOGCM pixels are subject to high forecast uncertainty and bias (Hawkins and Sutton 2009; also see section 12.3.2). To ameliorate this issue, AOGCM outputs should be averaged over multiple pixels. Correlation maps can be used to evaluate the spatial resolution over which climate can be safely aggregated without losing fidelity to demographic responses to local climate. For emperor penguins, Jenouvrier et al. (2012, 2014) used sea ice concentration averaged across several pixels of an AOGCMs grid (Figure 12.2b) to improve the accuracy of sea ice forecasts. This large sector includes the foraging area of emperor penguins, defined by the maximum foraging distances from the colony, of about 100 km during the breeding season and at least 650 km during the non-breeding season. Figure 12.2b illustrates that sea ice concentrations for many pixels in this large spatial sector are strongly correlated to the local sea ice concentration within the foraging area, and can be aggregated to reduce uncertainty in sea ice forecasts while maintaining strong correlations with local dynamics.

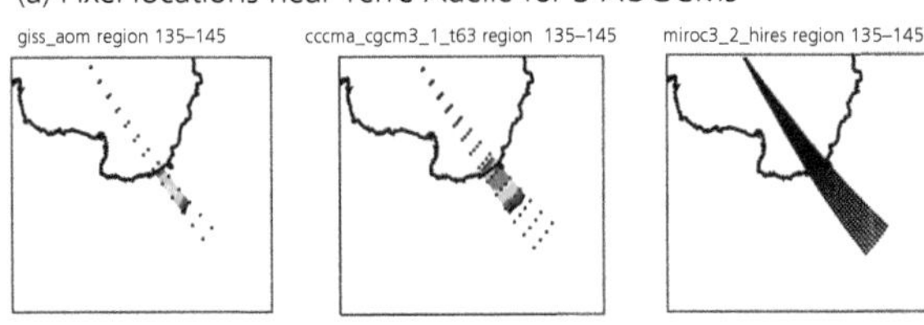

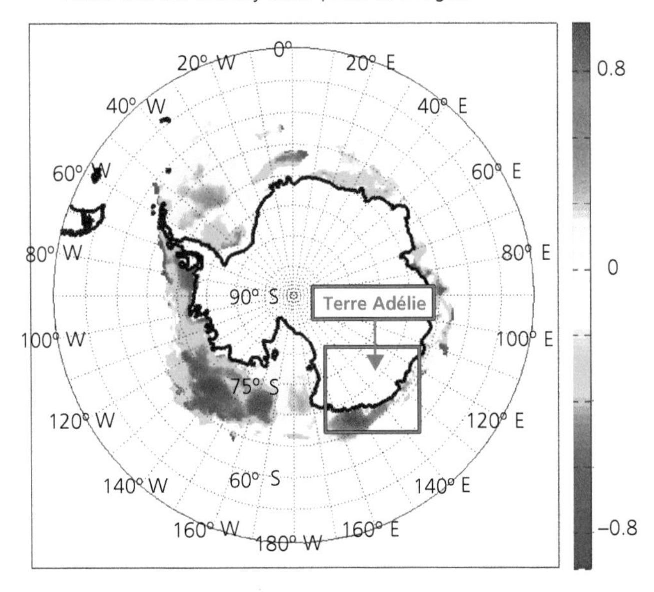

Figure 12.2 AOGCM output for the area surrounding Terre Adélie, Antarctica. (a) AOGCMs differ in their spatial resolution, and thus in the number of pixels encompassing the study area (grid for CMIP3). (b) Jenouvrier et al. (2012, 2014) used a correlation map to estimate the correlation between a time series for sea ice at the location of the local population and at other areas around Antarctica. From this, they determined that local sea ice was correlated with conditions across a much larger area. This provided justification for averaging AOGCM predictions of sea ice across larger spatial extents, reducing forecast uncertainty.

12.3.2 Uncertainties must be fully propagated to population forecasts

Uncertainty is central to population forecasting and enters in every step, from climate modelling (e.g., uncertainty in current climate conditions and future climate change) to demographic modelling (e.g., uncertainty in the responses of vital rates to climate parameters, residual vital rate covariation, etc.) to population modelling (e.g., uncertainty in population abundance and stage structure). Depending on their origin, uncertainties propagate to forecasts differently (Dietze 2017b). Accordingly, forecast uncertainty is dominated by different processes across various spatial and temporal horizons. For example, in climate models, uncertainty in local-scale, short-term forecasts is dominated by uncertainty in initial conditions. Yet, uncertainty in large-scale, 100-year forecasts is dominated by structural uncertainties among climate models and emissions scenarios (Hawkins and Sutton 2009). Similarly, the magnitude of various sources of uncertainty in avian population forecasts depends on a variety of factors, including the spatial and temporal scale of forecasts, sample size, model complexity, and life history of the species. Forecasts that fail to include these key sources of uncertainty will be falsely overconfident, eroding trust in ecological science and hindering ecological understanding (Dietze 2017a). Forecasting therefore requires a quantification of uncertainty in each model component, and importantly, full propagation of these uncertainties to forecasts. Below, we highlight the key sources of uncertainty in avian population forecasts.

Uncertainty in simulations of future climate arises from three sources (Figure 12.3; Hawkins and Sutton 2009): (1) imperfect knowledge of initial climate conditions, (2) structural differences among climate models, and (3) uncertainty in future climate emissions. Climate models and their associated uncertainties are also discussed in Chapter 2. Here, we briefly review these uncertainties with respect to their relevance for avian population forecasts.

Weather dynamics are chaotic. Thus, even infinitesimal uncertainty in the initial state of climate simulations will ultimate grow over time until it approximates the long-term statistical distribution of each climate variable. This component of climate forecasting is called 'internal variability' and typically dominates uncertainty in short-term to medium-term forecasts, especially at small spatial scales (Hawkins and Sutton 2009). However, uncertainty due to internal climate variability reaches an eventual asymptote, usually by 10 years into the future (Figure 12.3a and d; identical uncertainty at 15 and 80 years). Thus, other sources of uncertainty eventually dominate longer forecasts. To measure uncertainty due to internal variability, the same climate model must be simulated many times with initial conditions on each run drawn from probability distributions describing the observation error for climate variables. Alternatively, uncertainty due to internal climate variability can be approximated by using a post hoc approach to estimate the statistical properties of climate variables from a single run. For example, Jenouvrier et al. (2012, 2014) estimated a smoothed mean and variance from a single run of each climate model, and used these estimated

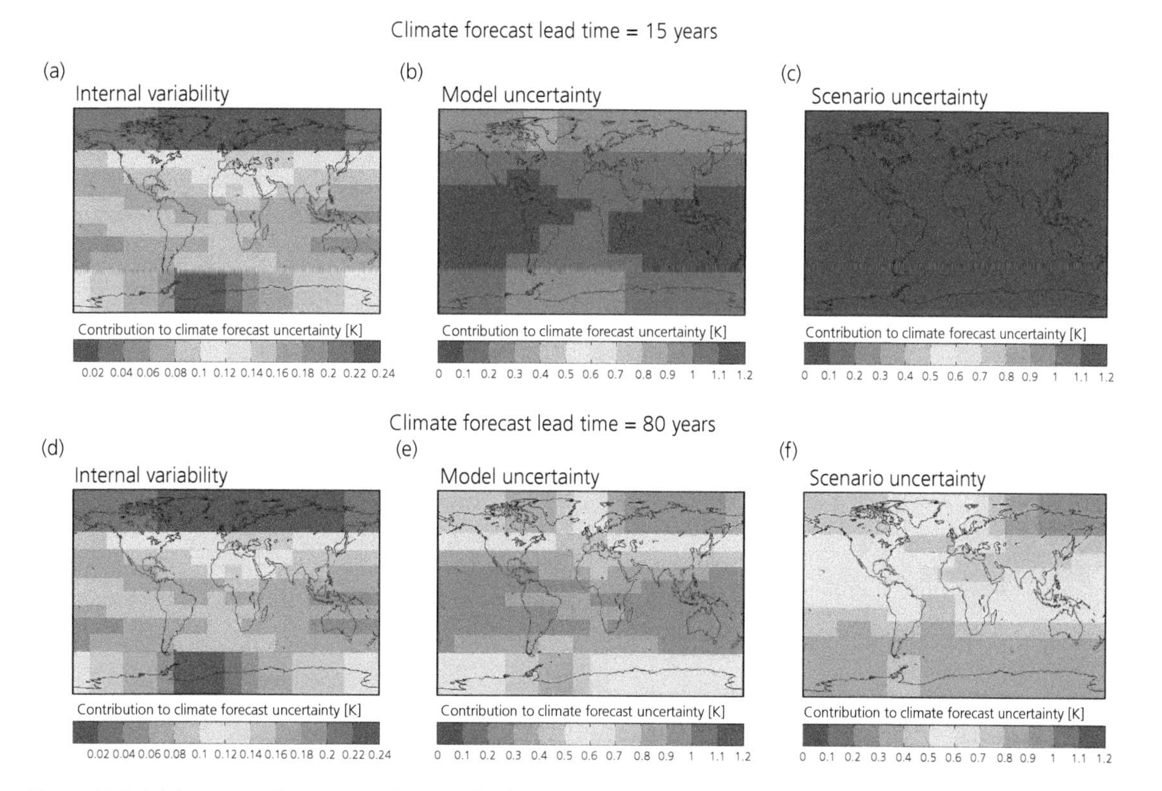

Figure 12.3 Relative sources of uncertainty in forecasts of surface temperature for 15 years into the future (a–c) and 80 years into the future (d–f). Figure is reproduced with permission from the interactive climate uncertainty visualization tool (available at http://ncas-climate.nerc.ac.uk/research/uncertainty), as described in Hawkins and Sutton (2009).

parameters to generate repeated stochastic simulations of climate for each model.

Climate models contributing to IPCC projections are produced by over 20 different research groups from institutions across the world. There is no 'best' climate model; rather, models produced by each group differ slightly in their biases, inclusion of particular physical processes, and numerical approximations (Knutti et al. 2013). Differences between these models give rise to 'model uncertainty' in climate projections. This source of uncertainty can be substantial for both short- and long-term climate projections (Hawkins and Sutton 2009). Quantification of model uncertainty requires simulating future climate using multiple models. Furthermore, models differ in their ability to recreate particular aspects of historical climate. Thus, it is often prudent to select a subset of models that are known to reproduce the climate variables of interest (Jenouvrier 2013; Snover et al. 2013) or perform various model selection methods (Jenouvrier et al. 2017). Notably, selection of climate models can strongly influence population forecasts (Figure 12.1), and formal methods for climate model selection for ecological forecasting remains an active area of research.

Uncertainty in future greenhouse gas concentrations are encapsulated by four representative concentration pathway (RCP) scenarios in the fifth IPCC report (van Vuuren et al. 2011). Each scenario represents a different 'plausible future', depending on socioeconomic change, technological innovation, land use modification, and carbon emissions. This source of uncertainty is negligible in short-term climate forecasts because near-term emissions are similar under any plausible future. Yet, scenario uncertainty (along with model uncertainty) dominates long-term climate forecasts (Hawkins and Sutton 2009). Scenario uncertainty can be quantified by comparing forecasts from multiple RCPs or earlier equivalents, such as the A1, A2, B1, and B2 scenarios from the Special Report on Emissions Scenarios used in the fourth IPCC report.

Population models rely on (often noisy and biased) data to derive statistical relationships between vital rates and climate variables. Sampling variation and imperfect detection of individuals can introduce substantial bias and noise into raw data, which must be corrected statistically. The goal of demographic analysis is to remove these sources of spurious error and generate estimates of true population parameters, such as numbers of individuals, current structure of the population, parameters describing vital rate distributions, relationships between vital rates and environmental variables (e.g., climate effects and density-dependence), and unexplained 'residual' variation in (and covariation among) vital rates. As with uncertainty in climate forecasts, each source of demographic uncertainty contributes differently to population forecasts (Figure 12.4).

In general, the effect of initial condition uncertainty, owing to imperfect knowledge of population sizes or structure, will depend on the details of the population model. For example, the effect of uncertainty in initial population abundance (Figure 12.4b) will depend on population growth rates (uncertainty in abundance will compound faster when population growth rates are high), whether population dynamics are density dependent (and how fast carrying capacity is reached), and whether population dynamics are chaotic or internally stable. Similarly, uncertainty in initial population structure will generate higher forecast uncertainty if populations are capable of more extreme transient dynamics. However, the effect of uncertainty in initial population structure may only manifest after a certain amount of time, depending on the nature of transient dynamics (Figure 12.4c).

Uncertainty in estimates of vital rate parameters, their relationships with environmental variables (e.g., climate, habitat, other species, etc.), or the effect of density dependence upon them is grouped under parameter uncertainty. Parameter uncertainty compounds over time (Figure 12.4d) and often dominates uncertainty in long-term demographic forecasts. For example, Gauthier et al. (2016) forecasted the abundance of greater snow geese under future climate change. At 40 years in the future, almost 90 per cent of prediction variance was due to parameter uncertainty. Fortunately, this source of uncertainty can be reduced by collecting more data, which in turn, can generate more precise parameter estimates.

Process variance (sometimes called process error) is the residual (co)variance in vital rates that is not explained by explicit variables. Vital rate (co) variation generates stochastic population dynamics (Figure 12.4e), which have been the subject of much

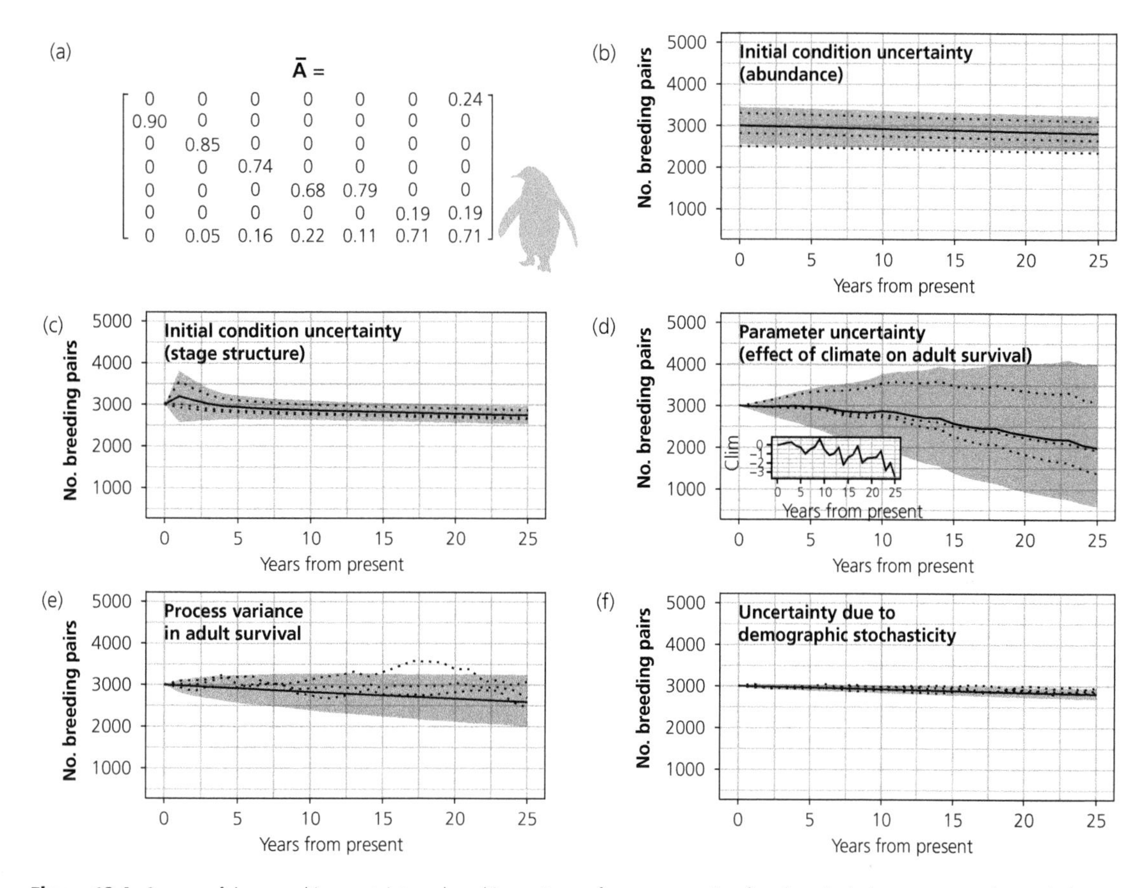

Figure 12.4 Sources of demographic uncertainty and resulting patterns of error propagation for a hypothetical emperor penguin population forecast. (a) Matrix model and mean vital rates were adapted from Jenouvrier et al. (2005a). (b) to (f) illustrate the uncertainty in population forecasts resulting from different sources of uncertainty in the model. (b) Uncertainty in initial population abundance, but perfect knowledge of all other parameters, including initial population structure. (c) Uncertainty in initial population structure, but perfect knowledge of initial total abundance. (d) Uncertainty in the linear relationship between adult survival and a climate variable, but assuming perfect knowledge of future climate. (e) Process variance in adult survival, modelled as a logit-normal process with no temporal autocorrelation. (f) Demographic stochasticity; treating survival and breeding probability as binomial processes, and fertility as a Poisson process. Black line depicts median estimate, grey ribbons depict 95% credible envelope. Dotted lines depict three separate realizations of the population model.

theoretical study (Lande 1993; Doak et al. 2005). It is now well appreciated that process (co)variance in vital rates can strongly influence population growth. Like parameter uncertainty, process variance compounds over time. However, unlike parameter uncertainty, process variance does not decline with larger sample sizes; rather, process variance can only be reduced by including better explanatory covariates for vital rates. Che-Castaldo et al. (2017) detected extremely high process variance in local population growth rates of Adélie penguins, precluding accurate short-term forecasts. Yet, this source of uncertainty could be ameliorated by aggregating abundance over larger spatial extents (i.e., across multiple sub-populations).

Demographic stochasticity introduces uncertainty into forecasts by applying probabilistic vital rates to discrete numbers of individuals; ultimately, this represents a biological consequence of sampling variation in vital rates. When populations are large, uncertainty due to demographic stochasticity is negligible (Figure 12.4f). However, demographic stochasticity can exert an overwhelming influence on small populations, leading to 'chance extinctions' in otherwise viable populations (Lande 1993; Lande et al. 2003; Iles et al. 2016).

12.4 Current state of science and roadmap for the future

12.4.1 Literature search for IPCC-dependent avian population forecasts

We conducted a literature search to compile a list of studies that have made avian population projections by coupling demographic models with IPCC projections. Using the Web of Science search engine, we searched for all papers with topics: *(bird* OR avian) AND (population* OR abundan* OR demograph* OR vital rate*) AND (ensemble* OR climat* OR *GCM* OR IPCC OR climate chang* OR AR5 OR AR4 OR RCP* OR SRES) AND (project* OR forecast*)*. We included several other relevant studies we knew of that were not returned by this search. We also searched all papers that had cited any of the suitable studies we located. We only included studies that estimated explicit relationships between vital rates and climate variables, fused these with IPCC-class projections of future climate change, and projected population abundance (step 3; Jenouvrier 2013). We omitted studies that used population models to project long-run dynamics under a hypothetical stationary future climate (van de Pol et al. 2010; Dybala et al. 2013; Pardo et al. 2017), though such studies are extremely useful for understanding the pathways through which climate affects overall population dynamics (step 2, Jenouvrier 2013; Jenouvrier et al. 2018b).

In total, we found 18 studies that met our criteria for inclusion (Table 12.1). These studies projected the future abundance of 24 avian species. Of these, only three species were projected to increase in abundance under future climate change: Amsterdam albatross (*Diomedea amsterdamensis*; though this population is increasing under current climate conditions), greater snow goose (though forecasts were highly uncertain by 2050), and white-throated dipper (*Cinclus cinclus*). Conversely, declines in abundance were projected for at least 15 species under mid- to high-emissions scenarios. For the remaining species, populations are projected to remain approximately stable (e.g., American wigeon, *Mareca americana*; greater and lesser scaup *Aythya marila* and *A. affinis*, respectively) or forecasts are equivocal owing to different projections among studies (e.g., mallard, *Anas platyrhynchos*) or high forecast uncertainty (wood thrush, *Hylocichla mustelina*).

12.4.2 Roadmap for future research

We focus the remainder of this chapter on the strengths and commonalities among existing avian population projections, while highlighting ongoing challenges and future opportunities. Our goal is not to criticize previous studies, but rather to contextualize existing forecasts and provide a roadmap for future work. Forecasting is an inherently iterative process, benefiting from continuous model assessment and refinement (Dietze et al. 2018).

Age-structure and breeding stage was commonly accounted for in population forecasts (14/18 studies), often by using matrix modelling approaches (Caswell 2001). This reflects the strong body of age- and stage-structured demographic information in ornithology. Spatial structure was considered in eight studies, either by allowing for spatial differences in the amount of future climate change (Jenouvrier et al. 2014), explicitly incorporating dispersal among sub-populations or habitats (Aiello-Lammens et al. 2011; Harris et al. 2012; Bonnot et al. 2017; Jenouvrier et al. 2017), or less frequently, by estimating spatial differences in population responses to climate (Drever et al. 2012; Peery et al. 2012; Zhao et al. 2016). Future research will benefit from examining the effects of individual heterogeneity on climate sensitivity (e.g., owing to differences in body size, individual behaviours, or latent factors). In particular, one promising avenue is the use of finite mixture models to account for unobserved individual heterogeneity in vital rates (Hamel et al. 2017; Jenouvrier et al. 2018a), and to capture latent spatial variation among individuals (Guéry et al. 2017).

Studies varied widely in the sources of climate forecast uncertainty they included (Table 12.1; Figure 12.5), indicating a need for increased consistency in future studies. Internal variability in climate models was rarely incorporated, either by initiating the same climate models with multiple initial conditions, or by generating stochastic climate forecasts from statistical summaries of a single climate model. Conversely,

Table 12.1 List of studies that forecasted avian responses to climate change by explicitly coupling IPCC-class climate simulations with population models. Study species correspond to AOU codes (where applicable). Lead time describes time horizon of forecast. Uncertainty abbreviations are: internal climate variability (CIn), climate model uncertainty (CMo), climate scenario uncertainty (CSc), demographic parameter uncertainty (DPar), demographic process variance (DProc), uncertainty in initial population abundance (PAb), uncertainty in initial population stage structure (PStr), and uncertainty due to demographic stochasticity (DSto). **Bold/underline** indicates source of uncertainty was explicitly included in model.

	Study	Species	Lead time	Spatial extent	Climate variables	Population structure	Uncertainties included	Projected effect of climate change
1	Aiello-Lammens et al. 2011	SNPL	2010–2100	Florida, USA	Sea Level	Stage, Sex, Spatial (with dispersal)	CIn, CMo, **CSc**, DPar, **DProc**, PAb, PStr, **DSto**	Increased risk of population decline and quasi-extinction under 2 m sea level rise
2	Ballerini et al. 2015	ADPE	2010–2100	Ross Sea, Antarctica	Sea Ice Extent	Stage	**CIn**, **CMo**, **CSc**, DPar, **DProc**, PAb, PStr, DSto	Declines and likely local extinction by 2050 in the absence of immigration, under mid-range emissions scenario (A1B) and other scenarios of increased frequency of extreme events
3	Barbraud et al. 2011	BBAL, AMAL, SNPE	2007–2057	Indian Ocean, Australia, Antarctica	Sea Surface Temp, Sea Ice Extent	Stage	CIn, **CMo**, **CSc**, DPar, **DProc**, PAb, PStr, **DSto**	Likely extinction of BBAL. Likely increase of AMAL. Declines of SNPE, depending on scenario of climate change
4	Bonnot et al. 2017	WOTH, PRAW	2000–2100	Central USA	Temp, Precip, Wind Speed	Stage Spatial (with dispersal)	CIn, CMo, **CSc**, **DPar**, **DProc**, PAb, PStr, **DSto**	Slight declines for wood thrush. Severe declines for prairie warbler
5	Drever et al. 2012	MALL, AMWI, Scaup spp., Scoter spp.	2020, 2050, 2080	Boreal forest, North America	Snow cover (spring)	Spatial	**CIn**, CMo, CSc, DPar, DProc, PAb, PStr, DSto	Possible increase in MALL abundance. Possible decrease in AMWI and scaup abundance. Likely decrease in scoter abundance
6	Gamelon et al. 2017	White-throated dipper	2013–2050	Southern Norway	Temp (winter)	Age Stage	**CIn**, CMo, CSc, **DPar**, **DProc**, **PAb**, **PStr**, **DSto**	Probable increase in abundance
7	Gauthier et al. 2016	GSGO	2012–2050	North America (species range)	Temp (spring and summer)	Age Stage	**CIn**, **CMo**, **CSc**, **DPar**, **DProc**, PAb, PStr, DSto	Probable increases under future warming, but low signal-to-noise ratio
8	Harris et al. 2012	Black-backed cockatoo	2000–2100	Kangaroo Island, Australia	Temp (Jan and July), Precip (annual)	Age Stage Spatial (with dispersal)	CIn, **CMo**, **CSc**, DPar, **DProc**, **PAb**, PStr, **DSto**	Population increase under current climate. Severe population declines under high emissions
9	Jenouvrier et al. 2009	EMPE	1960–2100	Terre Adélie (Antarctica)	Sea Ice Extent	Age Stage	**CIn**, **CMo**, CSc, **DPar**, DProc, PAb, PStr, DSto	Decline in abundance, high risk of quasi-extinction by 2100
10	Jenouvrier et al. 2012	EMPE	2010–2100	Terre Adélie (Antarctica)	Sea Ice Conc	Age Stage Sex	**CIn**, **CMo**, CSc, **DPar**, **DProc**, PAb, PStr, DSto	High probability of decline in abundance by 2100 under mid-range emissions scenario

Continued

Table 12.1 Continued

	Study	Species	Lead time	Spatial extent	Climate variables	Population structure	Uncertainties included	Projected effect of climate change
11	Jenouvrier et al. 2014	EMPE	2010–2100	All known emperor penguin colonies	Sea Ice Conc	Age Stage Sex	**CIn**, **CMo**, CSc, **DPar**, **DProc**, PAb, PStr, DSto	Declining population trends for all populations by 2100 under mid-range emissions scenario
12	Jenouvrier et al. 2017	EMPE	2010–2100	All known emperor penguin colonies	Sea Ice Conc	Spatial (dispersal)	**CIn**, **CMo**, CSc, **DPar**, **DProc**, PAb, PStr, DSto	All populations declining by 2100, regardless of dispersal scenario
13	Liao et al. 2015	Amakihi, Apapane, Iiwi	2010–2100	Hawaiian islands	Temp, Precip	Spatial, Disease-structure	CIn, **CMo**, **CSc**, DPar, DProc, PAb, PStr, DSto	Increased warmness and reduced rainfall caused malaria infection of birds, especially at high elevation, leading to decreased population abundance
14	Nur et al. 2012	SOSP	2010–2060	California, USA	Temp, Precip, Sea level, Tide height	Stage	CIn, CMo, CSc, DPar, **DProc**, PAb, PStr, DSto	Increasing abundance if sea level rise is low-med. Decreasing abundance if sea level rise is high
15	Peery et al. 2012	SPOW	2000–2100	Southwest USA	Temp, Precip	Age Stage Spatial (with immigration)	**CIn**, **CMo**, **CSc**, DPar, **DProc**, PAb, PStr, DSto	Severe declines in Arizona and New Mexico populations due to climate change
16	Thomson et al. 2015	SHAL	2100	Tasmanian islands	Daily Max Temp, Sea Surface Height, Rainfall	Age Stage Sex	CIn, **CMo**, **CSc**, **DPar**, **DProc**, PAb, PStr, DSto	Decrease in population abundance under climate warming, especially under hotter A2 scenario
17	Wolf et al. 2010	CAAU	2100	California, USA	Sea Surface Temp, Upwelling	Stage	CIn, CMo, CSc, **DPar**, **DProc**, PAb, PStr, DSto	Declines in population growth, owing to negative relationship between SST and breeding/survival, and projected increases in SST (and potential decreases in upwelling)
18	Zhao et al. 2016	MALL	2100	Praire pothole region, North America	Temp, Precip, Pond Density	Spatial	CIn, CMo, CSc, **DPar**, **DProc**, PAb, PStr, DSto	Declines in Mallard density under future warming (4°C)

climate model uncertainty was commonly considered in forecasts, either by using model-averaged climate projections, or less commonly, by conducting population forecasts using multiple climate models (though only the latter approach provides a stochastic climate-dependent projection and allows for uncertainty due to internal climate variability to be fully propagated to forecasts). Scenario uncertainty was considered in about half the studies, by evaluating population responses under multiple emissions scenarios. When not included, studies invariably chose a mid-range emissions scenario for climate projection. We urge ecologists to take advantage of the free availability of climate forecasts supervised by the Coupled Model Intercomparison Project (https://www.wcrp-climate.org/wgcm-cmip/wgcm-cmip6), allowing a full integration of all three sources of climate forecast uncertainty. The most recently completed phase of the project (CMIP5) includes more climate models and output variables than previous phases, and importantly, includes several runs of the same AOGCMs and experiment. This allows for uncertainty due to internal variability in climate models to be more fully incorporated into ecological projections.

Studies also varied widely in the sources of demographic uncertainty they included (Table 12.1; Figure 12.5). Uncertainty in demographic parameters (including vital rate responses to climate) was explicitly propagated to forecasts in less than half of the studies. We flag this as an important area of improvement for future research. Parameter uncertainty compounds over time in forecasts and is often the dominant source of forecast uncertainty at long timescales (Gauthier et al. 2016; Dietze 2017b). Failure to account for this source of uncertainty results in highly overconfident end-of-century forecasts. It is therefore likely that many of the forecasted population declines in our literature search are less certain than reported.

Demographic process variation (i.e., residual variation in vital rates not explained by climate) was included in almost all studies, likely because vital rate (co)variation and its consequences have a rich tradition in the study of wildlife population ecology (Tuljapurkar 1990; Caswell 2001; Doak et al. 2005). Demographic stochasticity was only included in five studies; most forecasted populations were large enough that the effects of demographic stochasticity are negligible and were thus explicitly omitted (Jenouvrier et al. 2009, 2012, 2014; Barbraud et al. 2011; Gauthier et al. 2016). Yet, the role of demographic stochasticity increases as population abundance declines, which could compound the negative effects of climate change documented by several studies. The inclusion of demographic stochasticity in forecasts may, therefore,

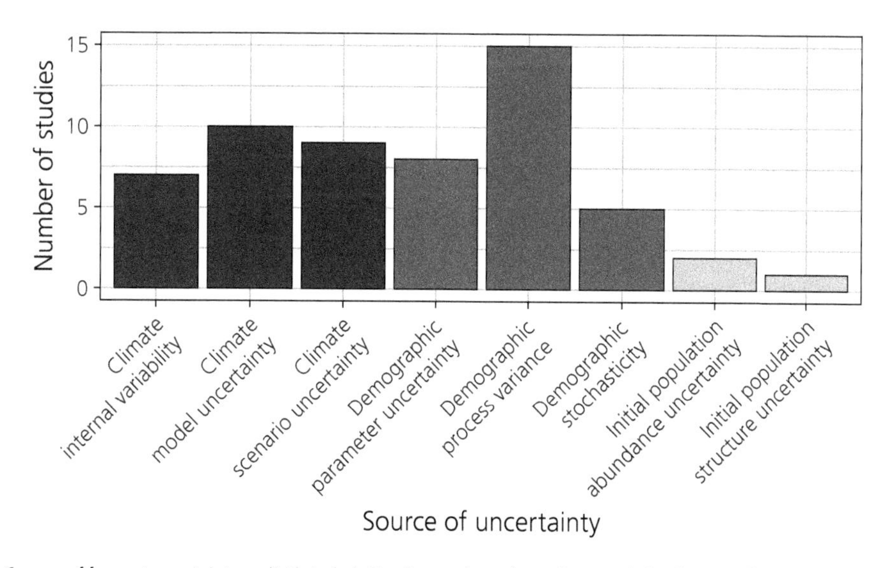

Figure 12.5 Sources of forecast uncertainty explicitly included in climate-dependent avian population forecasts (based on Table 12.1).

be especially important for populations that experience strongly negative effects of projected climate change.

The effect of uncertainty in initial population size was only considered in two studies (Harris et al. 2012; Gamelon et al. 2017), and uncertainty in initial (st)age structure was considered in one (Gamelon et al. 2017). Uncertainty in the initial sex or spatial structure of populations was not considered in any forecasts. Instead, forecasts were often initiated at the long-term stable (st)age distribution. For end-of-century forecasts, these sources of uncertainty are likely negligible relative to climate and parameter uncertainty. However, initial condition uncertainty (in both climate and population state) dominate near-term forecast uncertainty. Near-term forecasting is emerging as an important paradigm with high relevance to dynamic management and decision support (Harris et al. 2018; Humphries et al. 2017; Dietze et al. 2018). While all of the studies in our literature search focused on long-term responses to climate change (i.e., by 2100), we expect that near-term forecasts will become increasingly common over the next decade.

Methods to fully incorporate and partition forecast uncertainties are more accessible than ever (Dietze 2017a, b). Comparative studies will be invaluable for elucidating the dominant sources of uncertainty in population forecasts and how these vary across spatial and temporal scales, among life histories, and between study systems. Accordingly, comparative uncertainty analysis will be critical for guiding management and prioritizing monitoring efforts under climate change.

Other challenges also remain. Many species and populations are considered too data-deficient to assign IUCN Red List status, indicating a critical lack of basic knowledge about population size, trends, and distribution (Butchart and Bird 2010). For a larger number of species, a better understanding of the full life cycle and how it is affected by climate is needed. Well-studied populations will primarily benefit from work aimed at identifying better predictors of population responses that are matched with the scale of climate projections. In parallel, the spatial resolution of climate models is unlikely to improve substantially in the near term. Improved methods for downscaling coarse-grained climate projections may thus be important for generating local-scale predictions of climate, especially in cases where climate at larger spatial scales is not ecologically relevant (Snover et al. 2013).

Finally, eco-evolutionary processes (Chapters 7 and 11) and species interactions (Chapter 15) are critical determinants of future population responses to climate change, though including these processes in population forecasts remains a substantial challenge. Recent studies have identified considerable latent differences among individuals within a population (i.e., unobserved individual heterogeneity), both in terms of life history traits (Jenouvrier et al. 2015; Jenouvrier et al. 2018a) and responses to climate (Guéry et al. 2017). If these differences are genetically determined, individual heterogeneity forms the basis for evolutionary change. Yet, pedigrees are poorly resolved or non-existent in most populations, resulting in large uncertainties in trait heritability. Species interactions can also strongly modify demographic responses to climate (e.g., Ockendon et al. 2014; Liao et al. 2015; Iles et al. 2018). Multi-species population forecasts, especially across trophic levels, are thus an important area of continued research.

12.5 Conclusions

Population forecasting is a nascent discipline compared to the large body of literature on avian natural history, climate impacts, and demography. Despite the large body of literature on the impact of climate on vital rates, there are surprisingly few studies that have developed climate-dependent population models, and even fewer studies linking these population models to climate forecasts from IPCC-class models (to date, only 17 studies to our knowledge). Yet, the increasing sophistication and number of climate models (Chapter 2) and numerous studies of climate impacts on bird demography (Chapters 7 and 11) now allow for building population forecasts under climate change with uncertainties fully specified. Because forecasting is a process that requires iterative model building, performance assessment, and model correction (Dietze et al. 2018), we strongly encourage ecologists to take advantage of the tools and data currently available for this task.

Acknowledgements

We thank Marika Holland and Hal Caswell for productive discussions. This research was funded by NASA (grant NNX14AH74G) and U.S. National Science Foundation (grant OPP- 1341558 and OPP-1246407).

References

Ådahl, E., Lundberg, P., and Jonzén, N. (2006). From climate change to population change: the need to consider annual life cycles. *Global Change Biology*, 12, 1627–33. doi:10.1111/j.1365-2486.2006.01196.x

Aiello-Lammens, M.E., Chu-Agor, M.L., Convertino, M., Fischer, R.A., Linkov, I., and Akçakaya, H.R. (2011). The impact of sea-level rise on Snowy Plovers in Florida: integrating geomorphological, habitat, and metapopulation models. *Global Change Biology*, 17, 3644–54. doi:10.1111/j.1365-2486.2011.02497.x

Ballerini, T., Tavecchia, G., Pezzo, F., Jenouvrier, S., and Olmastroni, S. (2015). Predicting responses of the Adélie penguin population of Edmonson Point to future sea ice changes in the Ross Sea. *Frontiers in Ecology and Evolution*, 3, 1–8. doi:10.3389/fevo.2015.00008

Barbraud, C., Rivalan, P., Inchausti, P., Nevoux, M., Rolland, V., and Weimerskirch, H. (2011). Contrasted demographic responses facing future climate change in Southern Ocean seabirds. *Journal of Animal Ecology*, 80, 89–100. doi:10.1111/j.1365-2656.2010.01752.x

Barbraud, C., and Weimerskirch, H. (2001). Emperor penguins and climate change. *Nature*, 411, 183–6. doi:10.1038/35075554

Blanckenhorn, W.U. (2000). The evolution of body size: what keeps organisms small? *The Quarterly Review of Biology*, 75, 385–407. doi:10.1086/393620

Bonnot, T.W., Thompson, F.R., and Millspaugh, J.J. (2017). Dynamic-landscape metapopulation models predict complex response of wildlife populations to climate and landscape change. *Ecosphere*, 8, e01890. doi:10.1002/ecs2.1890

Butchart, S.H.M., and Bird, J.P. (2010). Data Deficient birds on the IUCN Red List: What don't we know and why does it matter? *Biological Conservation*, 143, 239–47. doi:10.1016/j.biocon.2009.10.008

Carey, C. (2009). The impacts of climate change on the annual cycles of birds. *Philosophical Transactions of the Royal Society of London Series B: Biological Sciences*, 364, 3321–30. doi:10.1098/rstb.2009.0182

Caswell, H. (2001). *Matrix population models: construction, analysis and interpretation*, 2nd edition. Sinauer Associates Inc., Sunderland, Massachusetts.

Che-Castaldo, C., Jenouvrier, S., Youngflesh, C., Shoemaker, K.T., Humphries, G., McDowall, P., Landrum, L., Holland, M.M., Li, Y., and Ji, R. (2017). Pan-Antarctic analysis aggregating spatial estimates of Adélie penguin abundance reveals robust dynamics despite stochastic noise. *Nature Communications*, 8, 832. doi:10.1038/s41467-017-00890-0

Clark, J.S., Carpenter, S.R., Barber, M., Collins, S., Dobson, A., Foley, J.A., Lodge, D.M., Pascual, M., Pielke, R., Pizer, W., Pringle, C., Reid, W.V., Rose, K.A., Sala, O., Schlesinger, W.H., Wall, D.H., and Wear, D. (2001). Ecological forecasts: an emerging imperative. *Science*, 293, 657–60. doi:10.1126/science.293.5530.657

Clutton-Brock, T.H. (1986). Sex ratio variation in birds. *Ibis*, 128, 317–29. doi:10.1111/j.1474-919X.1986.tb02682.x

Cooch, E.G., Lank, D.B., Dzubin, A., Rockwell, R.F., and Cooke, F. (1991a). Body size variation in lesser snow geese: environmental plasticity in gosling growth rates. *Ecology*, 72, 503–12. doi:10.2307/2937191

Cooch, E.G., Lank, D.B., Rockwell, R.F., and Cooke, F. (1991b). Long-term decline in body size in a snow goose population: evidence of environmental degradation? *Journal of Animal Ecology*, 60, 483–96. doi:10.2307/5293

Dietze, M.C. (2017a). *Ecological forecasting*. Princeton University Press, Princeton, NJ.

Dietze, M.C. (2017b). Prediction in ecology: a first-principles framework. *Ecological Applications*, 27, 2048–60. doi:10.1002/eap.1589

Dietze, M.C., Fox, A., Beck-Johnson, L.M., Betancourt, J.L., Hooten, M.B., Jarnevich, C.S., Keitt, T.H., Kenney, M.A., Laney, C.M., and Larsen, L.G. (2018). Iterative near-term ecological forecasting: Needs, opportunities, and challenges. *Proceedings of the National Academy of Sciences of the United States of America*, 115, 1424–32. doi:10.1073/pnas.1710231115

Doak, D.F., Morris, W.F., Pfister, C., Kendall, B.E., and Bruna, E.M. (2005). Correctly estimating how environmental stochasticity influences fitness and population growth. *American Naturalist*, 166, E14–E21. doi:10.1086/430642

Donald, P.F. (2007). Adult sex ratios in wild bird populations. *Ibis*, 149, 671–92. doi:10.1111/j.1474-919X.2007.00724.x

Drever, M.C., Clark, R.G., Derksen, C., Slattery, S.M., Toose, P., and Nudds, T.D. (2012). Population vulnerability to climate change linked to timing of breeding in boreal ducks. *Global Change Biology*, 18, 480–92. doi:10.1111/j.1365-2486.2011.02541.x

Dybala, K.E., Eadie, J.M., Gardali, T., Seavy, N.E., and Herzog, M.P. (2013). Projecting demographic responses to climate change: adult and juvenile survival respond differently to direct and indirect effects of weather in a passerine population. *Global Change Biology*, 19, 2688–97. doi:10.1111/gcb.12228

Ehrlén, J., and Morris, W. F. (2015). Predicting changes in the distribution and abundance of species under environmental change. *Ecology Letters*, 18, 303–14. doi:10.1111/ele.12410

Gamelon, M., Grøtan, V., Nilsson, A.L., Engen, S., Hurrell, J.W., Jerstad, K., Phillips, A.S., Røstad, O.W., Slagsvold, T., Walseng, B., Stenseth, N., and Sæther, B.-E. (2017). Interactions between demography and environmental effects are important determinants of population dynamics. *Scientific Advances*, 3, e1602298. doi: 10.1126/sciadv.1602298

Gardner, J.L., Peters, A., Kearney, M.R., Joseph, L., and Heinsohn, R. (2011). Declining body size: a third universal response to warming? *Trends in Ecology & Evolution*, 26, 285–91. doi:10.1016/j.tree.2011.03.005

Gauthier, G., Péron, G., Lebreton, J.-D., Grenier, P., and van Oudenhove, L. (2016). Partitioning prediction uncertainty in climate-dependent population models. *Proceedings of the Royal Society of London Series B: Biological Sciences*, 283, 20162353. doi:10.1098/rspb.2016.2353

Guéry, L., Descamps, S., Pradel, R., Hanssen, S.A., Erikstad, K.E., Gabrielsen, G.W., Gilchrist, H.G., Bêty, J., and Gaillard, J. (2017). Hidden survival heterogeneity of three Common eider populations in response to climate fluctuations. *Journal of Animal Ecology*, 86, 683–93. doi:10.1111/1365-2656.12643

Hallett, T.B., Coulson, T., Pilkington, J.G., Clutton-Brock, T.H., Pemberton, J.M., and Grenfell, B.T. (2004). Why large-scale climate indices seem to predict ecological processes better than local weather. *Nature*, 430, 71–5. doi:10.1038/nature02708

Hamel, S., Yoccoz, N.G., and Gaillard, J.-M. (2017). Assessing variation in life-history tactics within a population using mixture regression models: a practical guide for evolutionary ecologists. *Biological Reviews*, 92, 754–75. doi:10.1111/brv.12254

Harris, D.J., Taylor, S.D., and White, E.P. (2018). Forecasting biodiversity in breeding birds using best practices. *PeerJ*, 6, e4278. doi:10.7717/peerj.4278

Harris, J.B.C., Fordham, D.A., Mooney, P.A., Pedler, L.P., Araújo, M.B., Paton, D.C., Stead, M.G., Watts, M.J., Akçakaya, H.R., and Brook, B.W. (2012). Managing the long-term persistence of a rare cockatoo under climate change. *Journal of Applied Ecology*, 49, 785–94. doi:10.1111/j.1365–2664.2012.02163.x

Hawkins, E., and Sutton, R. (2009). The potential to narrow uncertainty in regional climate predictions. *Bulletin of the American Meteorological Society*, 90, 1095–107. doi:10.1175/2009BAMS2607.1

Humphries, G.R.W., Naveen, R., Schwaller, M., Che-Castaldo, C., McDowall, P., Schrimpf, M., and Lynch, H.J. (2017). Mapping Application for Penguin Populations and Projected Dynamics (MAPPPD): data and tools for dynamic management and decision support. *Polar Record*, 53, 160–6. doi:10.1017/S0032247417000055

Iles, D.T., Salguero-Gómez, R., Adler, P.B., and Koons, D.N. (2016). Linking transient dynamics and life history to biological invasion success. *Journal of Ecology*, 104, 399–408. doi:10.1111/1365-2745.12516

Iles, D.T., Rockwell, R.F., and Koons, D.N. (2018). Reproductive success of a keystone herbivore is more variable and responsive to climate in habitats with lower resource diversity. *Journal of Animal Ecology*, 87, 1182–91. doi: 10.1111/1365-2656.12837

IPCC (2013). *Climate change 2013: The physical science basis*. Cambridge University Press, Cambridge, UK.

Jenouvrier, S. (2013). Impacts of climate change on avian populations. *Global Change Biology*, 19, 2036–57. doi:10.1111/gcb.12195

Jenouvrier, S., Aubry, L., Barbraud, C., Weimerskirch, H., and Caswell, H. (2018a). Interacting effects of unobserved heterogeneity and individual stochasticity in the life history of the southern fulmar. *Journal of Animal Ecology*, 87, 212–22. doi:10.1111/1365-2656.12752

Jenouvrier, S., Barbraud, C., Cazelles, B., and Weimerskirch, H. (2005a). Modelling population dynamics of seabirds: importance of the effects of climate fluctuations on breeding proportions. *Oikos*, 108, 511–22. doi:10.1111/j.0030–1299.2005.13351.x

Jenouvrier, S., Barbraud, C., and Weimerskirch, H. (2005b). Long-term contrasted responses to climate of two Antarctic seabird species. *Ecology*, 86, 2889–903. doi:10.1890/05-0514

Jenouvrier, S., Caswell, H., Barbraud, C., Holland, M., Stroeve, J., and Weimerskirch, H. (2009). Demographic models and IPCC climate projections predict the decline of an emperor penguin population. *Proceedings of the National Academy of Sciences of the United States of America*, 106, 1844–7. doi:10.1073/pnas.0806638106

Jenouvrier, S., Caswell, H., Barbraud, C., and Weimerskirch, H. (2010). Mating behavior, population growth, and the operational sex ratio: A periodic two-sex model approach. *American Naturalist*, 175, 739–52. doi:10.1086/652436

Jenouvrier, S., Desprez, M., Fay, R., Barbraud, C., Weimerskirch, H., Delord, K., and Caswell, H. (2018b). Climate change and functional traits affect population dynamics of a long-lived seabird. *Journal of Animal Ecology*, 87, 906–20. doi:10.1111/1365-2656.12827

Jenouvrier, S., Garnier, J., Patout, F., and Desvillettes, L. (2017). Influence of dispersal processes on the global dynamics of emperor penguin, a species threatened by climate change. *Biological Conservation*, 212, 63–73. doi:10.1016/j.biocon.2017.05.017

Jenouvrier, S., Holland, M., Stroeve, J., Barbraud, C., Weimerskirch, H., Serreze, M., and Caswell, H. (2012). Effects of climate change on an emperor penguin population: analysis of coupled demographic and climate models. *Global Change Biology*, 18, 2756–70. doi:10.1111/j.1365–2486.2012.02744.x

Jenouvrier, S., Holland, M., Stroeve, J., Serreze, M., Barbraud, C., Weimerskirch, H., and Caswell, H. (2014). Projected continent-wide declines of the emperor penguin under climate change. *Nature Climate Change*, 4, 715–18. doi:10.1038/nclimate2280

Jenouvrier, S., Peron, C., and Weimerskirch, H. (2015). Extreme climate events and individual heterogeneity shape life history traits and population dynamics. *Ecological Monographs*, 85, 605–24. doi:10.1890/14-1834.1

Kéry, M., and Schaub, M. (2012). *Bayesian population analysis using WinBUGS: a hierarchical perspective*. Academic Press, London, UK.

Knutti, R., Masson, D., and Gettelman, A. (2013). Climate model genealogy: Generation CMIP5 and how we got there. *Geophysical Research Letters*, 40, 1194–9. doi: 10.1002/grl.50256

Lambrechts, M.M., Blondel, J., Maistre, M., and Perret, P. (1997). A single response mechanism is responsible for evolutionary adaptive variation in a bird's laying date. *Proceedings of the National Academy of Sciences of the United States of America*, 94, 5153–5. doi:10.1073/pnas.94.10.5153

Lande, R. (1993). Risks of population extinction from demographic and environmental stochasticity and random catastrophes. *American Naturalist*, 142, 911–27. doi:10.1086/285580

Lande, R., Engen, S., and Sæther, B.-E. (2003). *Stochastic population dynamics in ecology and conservation*. Oxford University Press, Oxford, UK.

Liao, W., Timm O.E., Zhang, C., Atkinson, C.T., LaPointe, D.A., and Samuel, M.D. (2015). Will a warmer and wetter future cause extinction of native Hawaiian forest birds? *Global Change Biology*, 21, 4342–52. doi: 10.1111/gcb.13005

Lindström, J. (1999). Early development and fitness in birds and mammals. *Trends in Ecology & Evolution*, 14, 343–8. doi:10.1016/S0169-5347(99)01639-0

McLean, N., Lawson, C.R., Leech, D.I., and van de Pol, M. (2016). Predicting when climate-driven phenotypic change affects population dynamics. *Ecology Letters*, 19, 595–608. doi:10.1111/ele.12599

Nur, N., Salas, L., Veloz, S., Wood, J., Liu, L., and Ballard, G. (2012). *Assessing vulnerability of tidal marsh birds to climate change through the analysis of population dynamics and viability*. Unpublished Report to the California Landscape Conservation Cooperative, version 1.

Ockendon, N., Baker, D.J., Carr, J.A., White, E.C., Almond, R.E. A., Amano, T., Bertram, E., Bradbury, R.B., Bradley, C., Butchart, S.H.M., Doswald, N., Foden, W., Gill, D.J.C., Green, R.E., Sutherland, W.J., Tanner, E.V.J., and Pearce-Higgins, J.W. (2014). Mechanisms underpinning climatic impacts on natural populations: altered species interactions are more important than direct effects. *Global Change Biology*, 20, 2221–9. doi:10.1111/gcb.12559

Oro, D., Torres, R., Rodríguez, C., and Drummond, H. (2010). Climatic influence on demographic parameters of a tropical seabird varies with age and sex. *Ecology*, 91, 1205–14. doi:10.1890/09-0939.1

Pardo, D., Jenouvrier, S., Weimerskirch, H., and Barbraud, C. (2017). Effect of extreme sea surface temperature events on the demography of an age-structured albatross population. *Philosophical Transactions of the Royal Society of London Series B: Biological Sciences*, 372, 20160143. doi:10.1098/rstb.2016.0143

Peery, M.Z., Gutiérrez, R.J., Kirby, R., LeDee, O.E., and LaHaye, W. (2012). Climate change and spotted owls: potentially contrasting responses in the Southwestern United States. *Global Change Biology*, 18, 865–80. doi:10.1111/j.1365–2486.2011.02564.x

Sæther, B.-E., and Bakke, Ø. (2000). Avian life history variation and contribution of demographic traits to the population growth rate. *Ecology*, 81, 642–53. doi:10.1890/0012-9658(2000)081[0642:ALHVAC]2.0.CO;2

Sheridan, J.A., and Bickford, D. (2011). Shrinking body size as an ecological response to climate change. *Nature Climate Change*, 1, 401–6. doi:10.1038/nclimate1259

Sibly, R.M., Witt, C.C., Wright, N.A., Venditti, C., Jetz, W., and Brown, J.H. (2012). Energetics, lifestyle, and reproduction in birds. *Proceedings of the National Academy of Sciences of the United States of America*, 109, 10937–41. doi:10.1073/pnas.1206512109

Snover, A.K., Mantua, N.J., Littell, J.S., Alexander, M.A., McClure, M.M., and Nye, J. (2013). Choosing and using climate change scenarios for ecological-impact assessments and conservation decisions. *Conservation Biology*, 27, 1147–57. doi:10.1111/cobi.12163

Stearns, S.C. (1992). *The evolution of life histories*. Oxford University Press, New York, NY.

Stott, I., Townley, S., and Hodgson, D.J. (2011). A framework for studying transient dynamics of population projection matrix models. *Ecology Letters*, 14, 959–70. doi:10.1111/j.1461–0248.2011.01659.x

Tavecchia, G., Tenan, S., Pradel, R., Igual, J.-M., Genovart, M., and Oro, D. (2016). Climate-driven vital rates do not always mean climate-driven population. *Global Change Biology*, 22, 3960–6. doi:10.1111/gcb.13330

Thomson, R.B., Alderman, R.L., Tuck, G.N., and Hobday, A.J. (2015). Effects of climate change and fisheries bycatch on shy albatross (*Thalassarche cauta*) in southern Australia. *PLoS One*, e0127006. doi: 10.1371/journal.pone.0127006

Travis, J.M.J., Mustin, K., Bartoń, K.A., Benton, T.G., Clobert, J., Delgado, M.M., Dytham, C., Hovestadt, T., Palmer, S.C.F., Van Dyck, H., and Bonte, D. (2012).

Modelling dispersal: an eco-evolutionary framework incorporating emigration, movement, settlement behaviour and the multiple costs involved. *Methods in Ecology and Evolution*, 3, 628–41. doi:10.1111/j.2041–210X.2012.00193.x

Tuljapurkar, S. (1990). *Population dynamics in variable environments*. Springer, New York, NY.

van de Pol, M., Brouwer, L., Brooker, L.C., Brooker, M.G., Colombelli-Négrel, D., Hall, M.L., Langmore, N.E., Peters, A., Pruett-Jones, S., and Russell, E.M. (2013). Problems with using large-scale oceanic climate indices to compare climatic sensitivities across populations and species. *Ecography*, 36, 249–55. doi:10.1111/j.1600–0587.2012.00143.x

van de Pol, M., Vindenes, Y., Sæther, B.-E., Engen, S., Ens, B.J., Oosterbeek, K., and Tinbergen, J.M. (2010). Effects of climate change and variability on population dynamics in a long-lived shorebird. *Ecology*, 91, 1192–204. doi:10.1890/09-0410.1

van Vuuren, D.P., Edmonds, J., Kainuma, M., Riahi, K., Thomson, A., Hibbard, K., Hurtt, G.C., Kram, T., Krey, V., Lamarque, J.-F., Masui, T., Meinshausen, M., Nakicenovic, N., Smith, S.J., and Rose, S.K. (2011). The representative concentration pathways: an overview. *Climatic Change*, 109, 5. doi:10.1007/s10584-011-0148-z

Visser, M.E. (2008). Keeping up with a warming world; assessing the rate of adaptation to climate change. *Proceedings of the Royal Society of London Series B: Biological Sciences*, 275, 649–59. doi:10.1098/rspb.2007.0997

Weimerskirch, H., Lallemand, J., and Martin, J. (2005). Population sex ratio variation in a monogamous long-lived bird, the wandering albatross. *Journal of Animal Ecology*, 74, 285–91. doi:10.1111/j.1365–2656.2005.00922.x

Wolf, S.G., Snyder, M.A., Sydeman, W.J., Doak, D.F., and Croll, D.A. (2010). Predicting population consequences of ocean climate change for an ecosystem sentinel, the seabird Cassin's auklet. *Global Change Biology*, 16, 1923–35. doi: 10.1111/j.1365-2486.2010.02194.x

Zhao, Q., Silverman, E., Fleming, K., and Boomer, G.S. (2016). Forecasting waterfowl population dynamics under climate change—Does the spatial variation of density dependence and environmental effects matter? *Biological Conservation*, 194, 80–8. doi:10.1016/j.biocon.2015.12.006

CHAPTER 13

Consequences of climatic change for distributions

Brian Huntley

13.1 Introduction

Although often considered in isolation from one another, individual species' spatio-temporal distributions, their populations, and community composition, form the vertices of a triangle of inter-dependence (Figure 13.1). Climatic conditions act upon all three vertices, and because interactions between the vertices are bi-directional, climatic changes have both direct and indirect effects upon those vertices. Thus, for example, a temperature change may cause a boundary of a species' distribution to shift either through a direct effect related to the species' physiological tolerance (e.g., of freezing temperatures), through a direct effect on some aspect of the species' demography (e.g., its breeding success), or indirectly because it changes the outcome of a biotic interaction (e.g., competition for a limited resource).

Notwithstanding these inter-dependencies, this chapter focuses upon species' distributions and begins by describing the primary ways in which climatic change can affect these distributions. Interactions of distribution changes with populations and communities are then outlined. Thereafter, evidence relating to the effects upon distributions, and their interactions with populations and communities, is discussed using a chronological framework, beginning with the Quaternary fossil and palaeoenvironmental record, then considering historical and recent observations, followed by future projections. The chapter ends with conclusions reached on the basis of these and related lines of evidence.

13.2 Species' distributions and climatic change

Under the climatic conditions prevailing at a given time, some fraction of the overall range of combinations of climatic conditions under which a given terrestrial species can survive, its fundamental climatic niche *sensu* Hutchinson (1957, reprinted 1991), may be represented and may intersect with one or more continental land mass(es). However, the fraction of the fundamental climatic niche that is represented will change as climatic conditions change, not all suitable combinations of climatic conditions being represented at any time. Thus, when the climatic conditions available at different times are compared, we find combinations of climatic conditions present at some times without an analogue at others, leading to the concepts of 'novel' and 'disappearing' climates (Williams et al. 2007). It is convenient to refer to that fraction of the fundamental climatic niche available at any time as the potentially realizable climatic niche at that time. Biotic interactions and/or biogeographical history will then determine which part(s) of this potentially realizable climatic niche the species actually occupies, this being referred to as its realized climatic

Huntley, B., *Consequences of climatic change for distributions*. In: *Effects of Climate Change on Birds*. Second Edition. Edited by Peter O. Dunn and Anders Pape Møller: Oxford University Press (2019). © Oxford University Press.
DOI: 10.1093/oso/9780198824268.003.0013

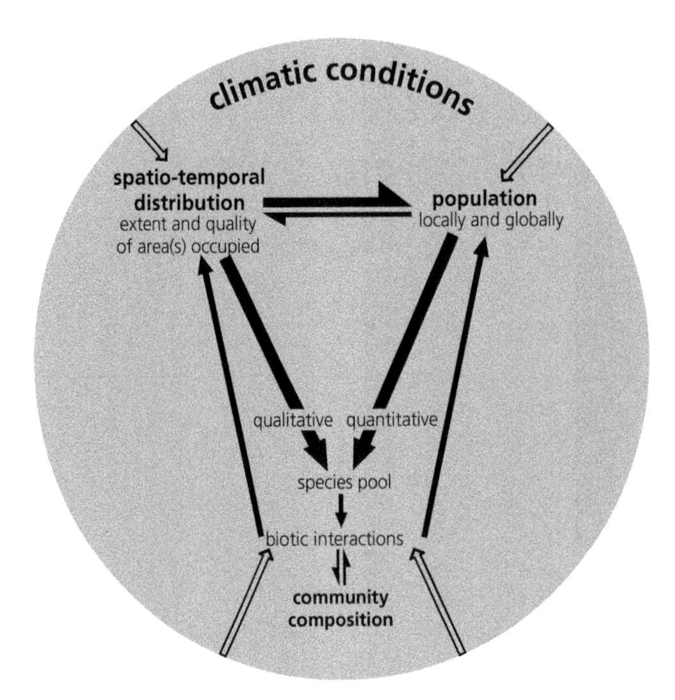

Figure 13.1 A triangle of inter-dependence. Filled arrows indicate the directions and strengths of the interactions between the vertices of the triangle; unfilled arrows indicate where climatic conditions primarily act.

niche *sensu* Hutchinson (1957, reprinted 1991) (Figure 13.2).

When we observe species' distributions it is their realized distributions, and hence their realized climatic niche, that we observe. This has important implications when models fitted to describe relationships between species' present distributions and present climatic conditions are applied to predict distributions for a different climatic regime that may include climatic conditions without a present analogue. When extrapolated into such no-analogue regions of climatic space models fitted using different approaches behave differently, and sometimes unpredictably. Figure 13.2 illustrates an extreme case in which a species' realized climatic niches at two times have no overlap, the species being excluded by biotic interactions from the intervening area of climatic space that is potentially realizable at both times. A model fitted to the species' realized distribution at either time generally would fail to predict its distribution at the other.

The discussion below focuses on the overall geographical distributions of species, but it is important to note that distributions can be examined at a

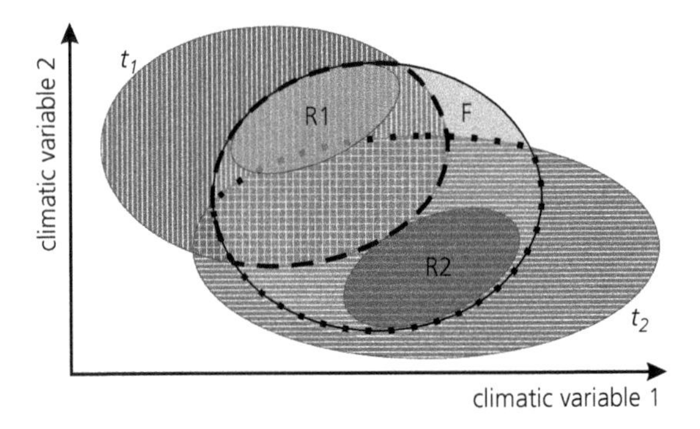

Figure 13.2 Climatic niches of a species. For simplicity only two climatic axes are shown. The large pale grey ellipse is the species' fundamental climatic niche (F) with respect to these axes. Combinations of values of the two variables available on the continents at times t_1 and t_2 are indicated by the vertically and horizontally striped ellipses respectively, their cross-hatched intersection indicating combinations available at both times. The heavy dashes outline that fraction of the species' fundamental niche represented at time t_1, i.e., its potentially realizable niche at that time, whilst the heavy dots outline its potentially realizable niche at time t_2. The darker grey ellipses show the species' realized niches at times t_1 (R1) and t_2 (R2), biotic interactions excluding the species from the central part of its fundamental climatic niche, even though this is part of the potentially realizable niche at both times. Although the lack of overlap between R1 and R2 might be taken to indicate an evolved change in the species' climatic niche, in reality its fundamental climatic niche is conserved and the different realized climatic niches reflect the different potentially realizable niches.

wide range of spatial extents (e.g., individual woodland, county, country, sub-continent) and grains (1 m^2 – 10 000 km^2). The various potential impacts described below, and illustrated in Figure 13.3, are relevant at all scales, albeit to somewhat different degrees.

Perhaps the most obvious impact of climatic change upon a species' distribution, and that which has received most attention, is a spatial shift in the potential distribution, i.e., a shift in its location (Figure 13.3 (i)). A familiar example of such an impact is the general poleward shift of temperate species' distributions when climate warms (Hickling et al. 2006; Mason et al. 2015). Shifts in location are often associated with a second impact, namely a change in extent (Figure 13.3 (ii)). For example, warming will lead to a reduced extent of the potential distribution of northern Eurasian species whose ranges already occupy the poleward part of the land mass. In contrast, other species may experience potential increases in the extent of their distributions. For example, species occupying regions of Europe with a sub-Mediterranean climate are projected to have larger ranges by the end of this century (see Figure 4.16 in Huntley et al. 2007), because the area with such climatic conditions is projected to increase. Species occupying isolated mountains or mountain ranges may experience little, if any, extensive geographical shift in distribution as a result of climatic warming, most simply

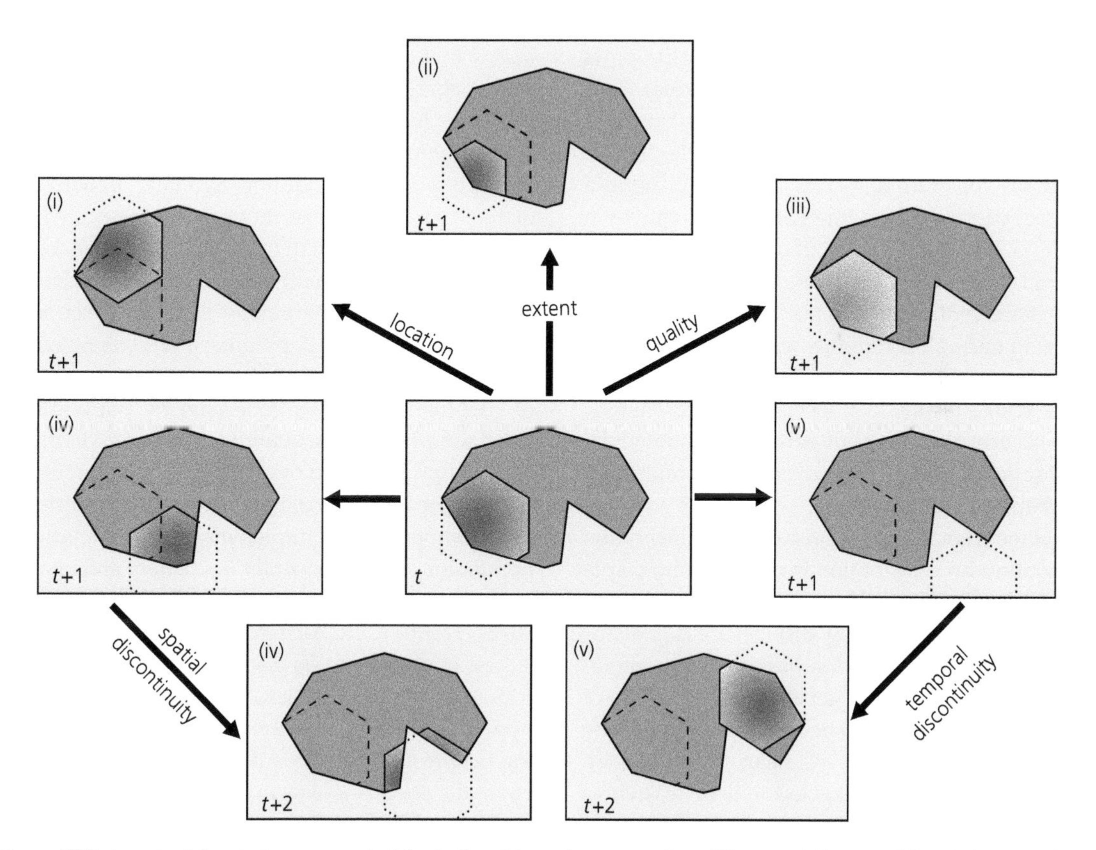

Figure 13.3 Impacts of climatic change on species' distributions. Schematic representations of five potential impacts of climatic change on the distribution of a terrestrial species. Changes in the location (i), extent (ii), or quality (iii) of the realized range, plus the potential for spatial discontinuities to arise (iv) or for a temporal discontinuity in the availability of suitable areas (v), are illustrated. On each panel the blue background represents ocean, whereas the green polygon represents a continental land-mass. The solid-outlined hexagon represents the area where climatic conditions are suitable for the species, whilst the brown-shaded polygon(s) represent(s) the intersection between this climatically suitable area and the land-mass. Intensity of the brown shading indicates suitability, with darkest shading indicating most suitable areas. The species' initial distribution at time *t*, shown on the central panel, is indicated by the dashed polygon on the other panels for times *t*+1 and *t*+2. The unit of time is an interval of sufficient duration for both climatic conditions and the species' distribution to have changed; generally this would be at least several decades.

shifting their distribution upslope. Species already occupying only the higher elevations, however, will generally experience a reduced potential range extent, because the area available decreases, ultimately to zero, as the maximum elevation of a mountain or mountain range is approached (see, e.g., Brambilla et al. 2016).

An often overlooked impact of climatic change on species' distributions is a potential change in quality of the area occupied (Figure 13.3 (iii)). Climatic change may alter the suitability of areas within a species' range such that, even if its distribution neither shifts nor changes in extent, its population size may change. In general, highest local population densities are near the centre of a species' distribution (Gaston 2003), implying that this is where conditions are most suitable, although there are exceptions: For example, the highest quality parts of a species' distribution may be adjacent to a physical barrier, such as a coastline, that truncates the distribution. Less obviously, if available climatic conditions constrain a species' realized niche to a marginal part of its fundamental niche, as shown in Figure 13.2, then the quality of its realized distribution will generally be less than if it was able to realize a more central part of its fundamental niche where, in general, conditions will be more suitable than towards the margins of that niche. Thus, if climatic change results in a species' realized climatic niche occupying a different part of its fundamental climatic niche, its realized distribution often will also change in quality.

Climatic change can also result in fragmentation of a continuous distribution into two or more spatially discontinuous areas, or conversely can lead to coalescence of previously disjunct areas of distribution. Either fragmentation or coalescence generally will impact upon the genetics and evolution of the species, and often will be associated either with evolution by divergence of sister taxa, or with development of hybrid zones or taxa of hybrid origin. Climatic change can also lead to newly suitable areas that are spatially separated from any area occupied by the species. In some cases these newly suitable areas will be geographically remote from the area occupied, or separated from it by a physical barrier across which dispersal of the species is unlikely (Figure 13.3 (iv)), rendering colonization by the species unlikely. Areas newly suitable as a consequence of climatic change can also be temporally discontinuous from the previously occupied area, leading to a period of time when no climatically suitable areas are available to the species (Figure 13.3 (v)). Species may survive such a temporal discontinuity of relatively short duration (e.g., if climatic conditions are not fatal to mature individuals), but longer temporal discontinuities may result in regional extirpation or even global extinction of the species (Huntley 1998).

13.2.1 Interactions between distributions and populations under a changing climate

As discussed above, climatic change can lead to changes in extent and/or quality of a species' distribution that may impact upon its overall population size. A change in a species' distribution extent will generally lead to a change in its population size, unless there is a balancing change in the climatic suitability of the distribution. Similarly, a change in quality of a species' distribution will generally lead to change in its population. In this latter case, however, the change in quality may also lead to a change in range extent if marginally suitable areas, able only to sustain sink populations, are colonized or abandoned as a result of a change in the number of surplus individuals generated by populations occupying the higher quality parts of the distribution. Large changes in overall population can result if changes in range extent and quality reinforce one another; for example, marked reduction in population size will occur if range extent and quality both decrease substantially (see e.g., Huntley et al. 2012a; Huntley and Barnard 2012).

Less obviously, shifts in location of a species' distribution may lead, at least transiently, to marked population reductions. For example, climatic warming may result in newly suitable areas towards and beyond a species' poleward range boundary, while at the same time leading to decreased quality of areas towards its equatorward limit. The population of such a species will decrease if it is unable rapidly to occupy the newly suitable areas and is suffering population declines in those parts of its range that are decreasing in quality. Such population reductions may be magnified if physical barriers limit or prevent colonization of newly suitable areas that are spatially discontinuous from previously occupied areas, or if development of suitable habitat lags behind the availability of suitable climatic conditions (e.g., range shift of Kirtland's warbler (*Setophaga kirtlandii*) requires shifts of its habitat, jack pine (*Pinus banksiana*) forest; Botkin et al. 1991). In extreme cases population reductions resulting from climatic change can result in severe bottlenecks, leading to loss of genetic diversity and of the ability to adapt to further changes in conditions. Temporal discontinuities in the availability of climatically suitable areas are likely to have the most extreme impacts upon species' populations, potentially leading to regional extirpation or even global extinction if the discontinuity exceeds the longevity of mature individuals and conditions prevent species' regeneration.

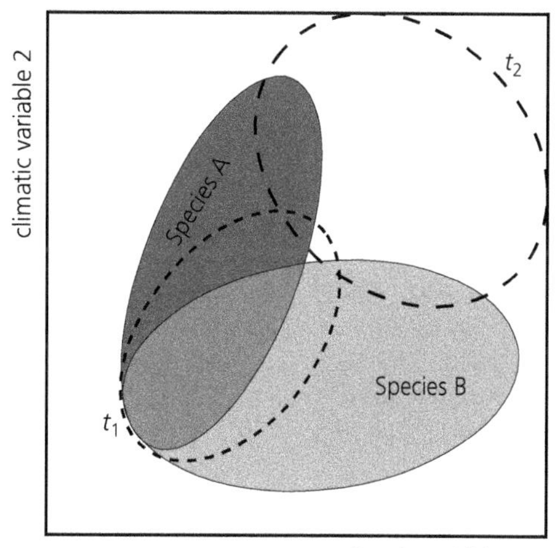

Figure 13.4 Species' individualistic responses. Schematic illustration of one way in which species that initially co-occur can become dissociated following a climatic change. The square represents a bioclimatic space defined by two variables. The pale grey and dark grey ellipses show the fundamental niches of two species, A and B, with respect to these two variables; the mid-grey area highlights where these niches overlap. The dotted and dashed ellipses enclose the combinations of the two bioclimatic variables that are available at times t_1 and t_2, between which a climatic change occurs. At time t_1 the available combinations include that part of the bioclimatic space where the species' niches overlap, as well as parts of both species' niches that do not overlap. Hence, at time t_1, species A and B each potentially occurs alone in some areas but in many areas they may co-occur. At time t_2 the available combinations no longer include those where the species' niches overlap, although both can potentially occur alone in some areas. Hence species whose distributions largely overlap with one another at time t_1 do not anywhere co-occur at time t_2. (Redrawn from Huntley et al. 2007, Figure 1.17.)

13.2.2 Community consequences of climatic change impacts on species' distributions

As a result of the unique character of every species' niche, *sensu* Hutchinson (1957, reprinted 1991), species' responses to climatic change, whether in terms of changes in their distributions or in their population sizes, are individualistic (Huntley 1991) (Figure 13.4). These individualistic responses lead to changes in the species pool at any given location that in turn lead to changes in community composition. Climatic change can also alter community composition, qualitatively and/or quantitatively, without a change in the species pool, because climatic conditions often modulate the outcomes of biotic interactions (Woodward 1987). However, unless the climatic change either is of very small magnitude, or is rapidly reversed, such impacts on biotic interactions will swiftly be overtaken by changes in the species pool as species' distributions respond to the climatic change. As a result of species' individualistic responses to climatic change, communities are only temporary assemblages of species and are in continuous flux as climate continuously changes (Huntley 1991, 1996). Climatic changes that result in some combinations of climatic conditions 'disappearing' and other 'novel' combinations arising lead to some communities disappearing whilst others appear that are without an analogue amongst those initially present (Williams and Jackson 2007).

13.3 Evidence and insights from the Quaternary

13.3.1 Quaternary climatic history

The Quaternary geological period (the past *ca.* 2.6 Ma) comprises the Pleistocene and Holocene (the past 11.7 ka) epochs (Cohen et al. 2013; see also http://www.stratigraphy.org/upload/QuaternaryChart.pdf and http://www.stratigraphy.org/upload/QuaternaryChartExplanation.pdf). Alternating glacial and interglacial climatic conditions superimposed upon an overall global cooling trend characterize the Quaternary, the base of which is defined by the first evidence of cooling of the magnitude associated with glacial stages. During the early Pleistocene the dominant periodicity of the climatic alternations was *ca.* 40 ka, but since *ca.* 1 Ma BP their dominant periodicity has been *ca.* 100 ka and their amplitude has been greater, mainly reflecting a trend towards progressively colder glacial climates. Glacial terminations, the transitions from glacial to interglacial conditions, are the largest magnitude Quaternary climatic changes and also the most rapid global climatic changes associated with the orbital forcing that drives the glacial–interglacial alternations (Hays et al. 1976; Imbrie and Imbrie 1979). During the most recent termination, *ca.* 22 – 8 ka BP, global mean temperature increased by an estimated 4.9°C (Shakun and Carlson 2010). Although extremely rapid in a geological context, the overall mean rate of this global warming (*ca.* 0.35×10^{-3} °C yr^{-1}) is at the very least an order of magnitude slower than that projected by Collins et al. (2013) for the twenty-first century (0.3×10^{-2} °C yr^{-1} being the 5th percentile estimate for RPC 2.6) and potentially as much as *ca.* 140 times slower (0.48×10^{-1} °C yr^{-1} being the 95th percentile estimate for RPC 8.5).

More frequent, rapid climatic fluctuations up to half the magnitude of glacial–interglacial alternations also occurred, especially during glacial stages (Dansgaard–Oeschger cycles, see e.g., Wolff et al. 2010). Although these fluctuations were of global extent, the associated climatic changes were not uniform but varied regionally; most strikingly, high southern latitudes warmed when higher northern latitudes cooled and vice versa (Blunier and Brook 2001). These more frequent climatic fluctuations, with rates and magnitudes more comparable to current and projected climatic changes, had important impacts upon ecosystems (Huntley et al. 2013) and left a legacy in regional biogeographic patterns (Huntley et al. 2016).

13.3.2 Quaternary distribution changes of birds

The extensive Quaternary fossil record of birds (see, e.g. Tyrberg, 1998, updated at http://w1.115.telia.com/~u11502098/pleistocene.html) provides abundant evidence that many species' distributions changed markedly in response to Quaternary climatic changes. Fossils of Arctic species (e.g., snowy owl, *Bubo scandiacus*) occur as far south in Europe as Italy and Spain in last glacial stage deposits. During previous interglacials, especially the last (Eemian) interglacial that is characterized in Europe by summer temperatures warmer than those during the pre-industrial Holocene, the distributions of southern European species extended further north, some as far as the United Kingdom (e.g., Alpine swift *Tachymarptis melba*, booted/Bonelli's eagle *Hieraaetus* sp.). Molecular genetic studies provide similar evidence of substantial range changes, and also allow elucidation of both glacial distributions and patterns of Holocene range expansion of temperate species (e.g., Liukkonen-Anttila et al. 2002; Brito 2005; Pellegrino et al. 2015). Such studies provide evidence that western Eurasian species using habitats such as tundra (Ruokonen et al. 2005) and steppe (Garcia et al. 2011), that were regionally more extensive during glacial stages (Allen et al. 2010), had more extensive distributions and larger populations at such times. They also provide evidence of population fragmentation under glacial conditions (Younger et al. 2015b), marked population increases in species favoured by the glacial to interglacial transition (Younger et al. 2015a), and population bottlenecks when climatic conditions resulted in restricted distributions (Saitoh et al. 2010).

The more extensive fossil records of other taxonomic groups provide complementary evidence of fragmentation or coalescence of species' distributions in response to Quaternary climatic changes, as well as of the individualism of these responses and the consequences for the composition and impermanence of communities (Huntley and Birks 1983; Huntley 1991, 1996; FAUNMAP Working Group 1996; Graham 1997). They also document past communities without any contemporary analogue (Huntley 1990b; Overpeck et al. 1992; Williams et al. 2001; Jackson and Williams 2004), sometimes reflecting very extensive changes in species' distributions (see, e.g., Coope 1973, 1994). They also record changing abundance patterns (Huntley and Birks 1983; Huntley, 1990a) and allow estimation of the rates at which species responded to past climatic changes (mean maximum Holocene rate of range boundary shift for 16 European tree taxa was

0.7375 km yr^{-1}, individual taxon values being 0.2 – 2.0 km yr^{-1} (Huntley and Birks 1983; Huntley 1991)). Finally, they provide clear evidence of species extinctions, whilst ancient DNA studies reveal 'cryptic' extinctions of intraspecific clades, leading to genetic impoverishment and potentially to niche restriction (e.g., saiga antelope, *Saiga tatarica* (Campos et al. 2010), woolly mammoth, *Mammuthus primigenius* (Barnes et al. 2007)). These extinctions are associated with rapid, large-magnitude climatic changes and reflect either temporal or spatial discontinuities in the availability of suitable climatic conditions, or an inability of species to shift their distributions sufficiently rapidly (Huntley 1998). It is almost certain that birds have experienced similar impacts, for which their relatively poorer fossil record has yet to provide evidence.

13.4 Lessons from historical distribution changes

13.4.1 Extinction can threaten widespread as well as restricted-range species

The Quaternary fossil record shows that once widespread species have rapidly become extinct, but much of the concern about potential future extinctions focuses on restricted-range species that are generally considered to be at greater risk (e.g., Foden et al. 2013). This ignores the possibility that widespread species may rapidly become restricted in range, or even extinct, if their required climatic conditions become rare or absent. To date most observed examples of rapid decreases in range extent reflect negative impacts of human activities, especially habitat loss and persecution. Nonetheless, they demonstrate the potential for similarly rapid responses to climatic change. For example, the western range boundary of yellow-breasted bunting (*Emberiza aureola*) has retreated eastwards *ca.* 5000 km in just 33 years (Kamp et al. 2015). Such examples demonstrate that widespread species can rapidly decline towards extinction and emphasize the need for vigilance. Species whose range rapidly becomes extremely restricted also suffer enhanced risk of a stochastic event leading to, or at least contributing to, their extinction (e.g., great auk, *Pinguinus impennis*; Birkhead 1994). One of the most immediate and well-documented threats posed by climatic change is sea-level rise. This potentially can negatively affect species endemic to low-lying islands, especially if they have only small populations. The reality of this threat is illustrated by the recent extinction of the Bramble Cay melomys (*Melomys rubicola*), a small rodent endemic to an island in the Torres Strait that was inundated as a result of sea-level rise (Gynther et al. 2016; Waller et al. 2017).

13.4.2 Environmental change can lead range-restricted species rapidly to become widespread

Just as widespread species' distributions can rapidly decrease, so restricted-range species' distributions can rapidly increase if climatic change increases the area offering suitable conditions. Once again, although the fossil record documents rapid large increases in species' distributions in response to Quaternary climatic changes, recent examples often result primarily from human activities. Northern fulmar (*Fulmarus glacialis*), for example, has substantially expanded its North Atlantic breeding distribution since the mid-eighteenth century, when it was known to breed only on the islands of Grimsey, Iceland, and St Kilda, Outer Hebrides, Scotland (Cramp and Simmons 1977; Burg et al. 2003). It now extends south to Brittany, east to Norway, and west to Labrador and Newfoundland. Fisher (1952, 1966) argued that this distribution increase resulted primarily from increasing availability of offal and discards, initially from whaling and later from commercial fishing fleets. More recently, Grosbois and Thompson (2005) showed that adult over-winter survival of northern fulmar correlates with the winter North Atlantic oscillation, both inter-annually and multi-decadally. Aebischer et al. (1990) reported similar evidence of the influence of the North Atlantic oscillation on North Sea populations of black-legged kittiwake (*Rissa tridactyla*). Thus, whilst the northern fulmar might have benefited from commercial fishing in recent years (Phillips et al. 1999), it is likely that changing climatic conditions were at least as important in causing the species' historical range expansion. Whatever the cause, however, this example shows how a regionally restricted-range species can rapidly become widespread.

13.4.3 'Invasions' can be very rapid

Most species are excluded from some parts of their potential global distribution by dispersal barriers in the form of extensive areas of unsuitable conditions. Climatic or other environmental changes, however, can in some cases reduce or remove such barriers, or chance events enable species to overcome them, for example cattle egret (*Bubulcus ibis*) crossing the Atlantic from Africa to South America (Massa et al. 2014). In either case, species can then 'invade' the newly accessible area of suitable conditions, and may do so extremely rapidly. Eurasian collared dove (*Streptopelia decaocto*), for example, was historically restricted to southern Asia, but then expanded its range across the Middle East and Turkey, subsequently invading Europe from the south east during the twentieth century. Although the environmental changes that enabled it to cross what had previously been unsuitable areas of the Middle East and Turkey almost certainly resulted from human activities, once it reached Europe it encountered continent-wide suitable conditions. Its distribution expanded northwards through the Balkans prior to 1930 before increasing extremely rapidly, reaching the British Isles, Belgium, and Scandinavia by the 1950s. By the 1970s its range extended west to Portugal, north to Iceland, and north-east to Finland (Cramp 1985). In just 40 years (1932–1972) its range expanded by *ca.* 2.5×10^6 km^2; its range boundary moved at a mean rate of 60–80 km yr^{-1}, areas up to 3250 km from its pre-1930 distribution being occupied. This rate of range expansion considerably exceeds those seen at the onset of the Holocene, and is more than an order of magnitude faster than rates of range boundary shift observed in response to recent climatic change (Hickling et al. 2006; Chen et al. 2011).

Species distribution models can predict species' potential global ranges, including climatically suitable areas remote from the species' realized range. For example, extensive areas of Scandinavia are predicted as suitable for the white-tailed ptarmigan (*Lagopus leucura*) (Huntley and Green 2011), the realized distribution of which is restricted to western North America. However, these models cannot predict chance events or human actions that may enable species to cross the barriers isolating these areas.

13.5 Recent distribution changes

Climatic conditions globally have changed at accelerating rates over the past two centuries (Stocker et al. 2013). Many studies have investigated whether species' distributions have shown changes consistent with expected responses to these climatic changes. Early studies used repeat survey data from localities for which historical data were available (Grabherr et al. 1994; Parmesan 1996). Subsequently, data from repeated regional distribution mapping projects (e.g., British Trust for Ornithology atlases of bird distributions in the British Isles (Sharrock 1976; Lack 1986; Gibbons et al. 1993; Holloway 1996; Balmer et al. 2013), Atlas of Southern African Birds projects (Harrison et al. 1997, http://sabap2.adu.org.za/)) have been used (e.g., Thomas and Lennon 1999; Brommer and Møller 2010; Hockey et al. 2011; Gillings et al. 2015).

Most studies have taken a simplistic view of both climatic change and the climatic factors limiting distributions, seeking evidence for poleward distribution shifts considered consistent with 'global warming'. Thus, Thomas and Lennon (1999) examined southern species' northern limits, and vice versa, in Great Britain in two successive distribution atlases (Sharrock 1976; Gibbons et al. 1993), finding a mean northward shift by the former of *ca.* 18 km (0.9 km yr^{-1}), but no consistent pattern for the latter. Given regional climatic warming during the 20-year interval, and correlations of timing and success of reproduction with temperature, they considered response to climatic warming to be the most parsimonious explanation of the range changes. Brommer and Møller (2010) applied a standardized approach to repeat the analysis of Thomas and Lennon (1999), as well as those of Brommer (2004, Finland), Hitch and Leberg (2007, North America), and Zuckerberg et al. (2009, New York State), all of which adopted a similar approach. Brommer and Møller's (2010) approach standardized with respect to the interval between surveys and the correction of range margin change for regional range size change, but maintained the focus on assessing poleward shifts in response to warming. They obtained mean rates of poleward shift of southern species' northern range limits of 0.64–2.52 km yr^{-1}. Other studies report similar rates: northern range boundary shifts of 1.5 km yr^{-1} for 22 bird species in Great Britain (Hickling et al. 2006) and 0.775–1.43 km yr^{-1} from five studies in various parts of Europe and North America (Chen et al. 2011), the latter authors also reporting range centroid shifts of 0.18–1.84 km yr^{-1}.

That this focus on poleward shift is simplistic and inappropriate, however, is apparent from analyses of potential future range changes of European birds presented by Huntley et al. (2008). Not only does none of the scenarios examined project a mean shift of species' potential ranges in a strictly northward direction, they also emphasize the individualism of species' responses, individual species' potential ranges being projected to shift in a wide range of directions. The recent analysis by Gillings et al. (2015) of atlas data for 1988–91 (Gibbons et al. 1993) and 2007–11 (Balmer et al. 2013) for 122 bird species breeding in Great Britain explored the direction in which each species' range boundary and centroid had shifted. Their analysis of range boundary shifts in any one of 24 directions showed northern range boundaries advancing mostly either north-westwards or north-eastwards, rather than north-wards. Shifts were consequently up to 30 per cent greater in magnitude than inferred from examining northward shifts. They also demonstrated the proportion of species showing range margin retreats is similar to that showing range margin advances, that margins are retreating principally from the south-west, and that the magnitude of these retreats is up to double that inferred by examining retreats from the south. Furthermore, striking correlations between the directions species' climatic spaces and ranges had shifted allowed robust attribution of the observed range changes as responses to climatic change. Nonetheless, without more mechanistic studies, attribution of all recent species' range shifts as responses to climatic change is questionable, and in some cases is unlikely to be the case. Hockey et al. (2011), for example, suggested southern African species exhibiting westward range shifts had responded to increased availability of 'artificial' habitat patches, notably woodlands and wetlands, in western South Africa resulting from land-use changes, arguing that neither climatic nor land-use changes alone can account for all observed range changes.

13.6 The problem of attribution

Few observational studies have addressed how key bioclimatic variables act to determine the distribution limits of species, in terms of their effects on critical demographic processes and/or life stages, or of how they relate to inherent physiological limits (but see, e.g., Robinson et al. 2007; Oswald et al. 2008; Wright et al. 2009; du Plessis et al. 2012; Cunningham et al. 2013a, 2013b). Nonetheless, such studies provide the only firm basis for directly attributing observed changes in species' distributions to climatic change. The paucity of such studies, however, has led to a predominance of indirect attribution based on correlations between climatic changes and populations or community composition changes, as exemplified below. Given the direct and indirect linkages between species' distributions, species' populations, and community composition (Figure 13.1), such attribution of changes in populations or communities to climatic change provides strong, albeit indirect, support for the attribution of distribution changes also to climatic change.

13.6.1 Population responses to climatic change

Green et al. (2008) correlated trends in the numbers of pairs of 42 regionally rare bird species breeding annually in the British Isles with the trend in mean climatic suitability of the region estimated using species distribution models (Huntley et al. 2007). Over a 25-year period, trends in numbers of breeding pairs were well predicted by trends in climatic suitability, supporting attribution of the population trends as responses to climatic change. This study also demonstrated the robustness of the species' distribution models used to assess climatic suitability, as well as providing strong support for attribution of the distribution changes underlying the population trends as responses to climatic change. Gregory et al. (2009) explored whether recent population trends of common European bird species were consistent with their potential range changes in response to projected late twenty-first century climatic changes (Huntley et al. 2007, 2008), finding an excellent correspondence. Subsequently, Stephens et al. (2016) performed similar analyses for United States bird species, as well as for updated European data, but extended their analyses to explore intraspecific variability in expected population trends between states/countries. They showed patterns in population trends were consistent with climatic suitability changes, both regionally within a species' range and interspecifically, demonstrating that bird populations in both regions have responded as expected to recent rapid climatic changes.

13.6.2 Community responses to climatic change

Evidence of recent quantitative and qualitative changes in avian community composition comes from repeated surveys of breeding bird communities. Devictor et al. (2008) used French Breeding Bird Survey (FBBS) point count data for 1989–2006 for the 105 most common terrestrial species recorded. For each species a species temperature index (STI) was calculated as the mean breeding season (March–August) temperature of 50 km grid squares from which the species was recorded in The EBCC Atlas of European Breeding Birds (Hagemeijer and Blair 1997). An annual community temperature index (CTI) for each locality was then computed as the weighted mean STI of species recorded at the locality, weights being the number of individuals of each species recorded. Their results showed substantial changes in bird community composition during the study period, and a steady, linear increase in CTI paralleling an increase in March–August mean temperature. However, whereas bird communities with a given CTI were located on average 91 km further north in 2006 than in 1989, the March–August mean temperature increase represented an average shift northwards of equivalent temperatures by 273 km. Thus, whilst the correlation of increasing CTI with increasing breeding season temperatures provides support for attribution of the underlying distribution changes as responses to climatic change, these results also show species' responses are occurring at only one-third of the rate of the climatic change.

Similar results have subsequently been reported from Sweden (Lindström et al. 2013) and Finland (Virkkala and Rajasärkkä 2011), although both analyses concluded that quantitative changes in community composition predominated. This is the expected initial response of communities, because it reflects differential local demographic impacts of the climatic change and does not depend upon species' range shifts. The relatively slow rate of distribution changes also likely largely accounts for the lagged response of CTI to temperature. More recently, using FBBS data for 2001–2012 for the 122 most common species, Gaüzère et al. (2015) showed close correspondence between spatial variations in CTI changes and in spring temperature changes, thus strengthening further the basis for attributing observed changes as responses to climatic change.

13.7 Projections of potential future distribution changes

Numerous studies have used climate envelope models and future climate projections to predict bird species' potential future distributions (Zurell and Engler, this volume). Most of these studies examine species' breeding ranges, very many fewer examining non-breeding ranges (e.g., Barbet-Massin et al. 2009) or both breeding and non-breeding ranges (e.g., Doswald et al. 2009). Although some examine a single species (e.g., Walther et al. 2007), most consider a number of species, either members of a taxonomic unit (e.g., Peterson et al. 2001; Barbet-Massin et al. 2009; Doswald et al. 2009), occupying one or more habitat(s) (e.g., Huntley and Barnard 2012; Brambilla et al. 2016), or found in a country or region (e.g., Peterson 2003; Virkkala et al. 2008). A minority of studies model most or all species occurring in a sub-continental or continental region (e.g., Huntley et al. 2007; Hole et al. 2009; Langham et al. 2015; National Audubon Society 2015). Unfortunately, a lack of consistency between studies with respect to quantification and reporting of predicted range changes limits the scope for synthesis.

13.7.1 Changes of location

Modelling studies consistently predict substantial shifts in species' potential distributions. Using results from HadCM2 general circulation model (GCM) experiments for 'conservative' and 'liberal' emissions scenarios, Peterson (2003) simulated modal potential range centroid shifts of 300–350 km by 2055 for Rocky Mountain and Great Plains birds. Huntley et al. (2008) simulated mean potential range centroid shifts by 2085 for European breeding birds of

258–680 km for the B2 and of 361–882 km for the A2 emissions scenarios. Langham et al. (2015), using results for the A2 emissions scenario derived from an ensemble of GCMs, predicted mean northward shift by 322 km of North American birds' breeding ranges, and by 116 km of their non-breeding ranges, by 2085. Such 'average' values obscure considerable interspecific variation, however, with extreme shifts predicted for some (600–700 km by 2055 for Great Plains birds (Peterson 2003); up to 3578 km by 2085 for European breeding birds (Huntley et al. 2008)), and more modest shifts for others (100–150 km for Great Plains birds; 20–25 km for European breeding birds). Nonetheless, most species' potential range boundaries are consistently predicted to shift at rates of at least 3.6–6.5 km yr^{-1}, with rates of >24 km yr^{-1} for some species. Whether even rates predicted for most species are achievable is questionable given recent and Quaternary range change rates.

In order to survive Quaternary climatic fluctuations, species needed to achieve range shift rates that enabled them to maintain an overlap between their realized and potentially realizable ranges as the latter shifted geographical location in response to climatic changes, particularly the rapid, large magnitude changes characteristic of glacial terminations. Given, however, that warming during the last termination is estimated have been 10× slower than twentieth century warming (Jansen et al. 2007), it is unsurprising that most studies find species' distribution shifts lagging climatic changes. The prospect for the future, when rates of projected climatic change are several times faster than twentieth century changes, is far from rosy.

13.7.2 Changes of extent

Species' potential future range extents, when reported, show a consistent pattern of change, being predicted to be smaller than their present range extent for the majority of species, but to be larger for a minority. Thus, Huntley et al. (2008), using results from experiments using three GCMs and two emissions scenarios, predicted mean potential range extents in 2085 for European breeding birds of 80–81 per cent of the species' present range extent for the moderate B2 emissions scenario and 72–89 per cent for the nearer to 'business as usual' A2 scenario. Of 431 species modelled, 4–22 (mean 11; 2.55 per cent) were predicted to have no potential European range in 2085, whilst a further 1–25 (mean 11; 2.55 per cent) were predicted to have a 2085 potential range extent <10 per cent of their present range extent. For North America, Langham et al. (2015) predicted potential range extents in 2085 less than half present range extent for 21.4 per cent of 588 species modelled. For 94 southern African bird species of the Fynbos and Grassland biomes modelled by Huntley and Barnard (2012), mean predicted potential range extent reductions were 15 per cent, 27 per cent, and 34 per cent by 2025, 2055, and 2085 respectively, although for 12 species of conservation concern mean decrease by 2085 was 58 per cent. Bagchi et al. (2013) projected potential future occurrences in Important Bird Areas (IBAs) in south-east Asia of 370 bird species of conservation concern, predicting reduced potential range extent by 2085 for 88 per cent of species, suitable climate being lost from an average 29 per cent of IBAs offering suitable conditions at present. After accounting for uncertainties in their modelling, 45 per cent of species were predicted to be 'extremely likely' to decline across the IBA network as a whole, whereas only 2 per cent were predicted to be 'extremely likely' to find an increased extent of suitable climatic conditions in the IBA network.

These predictions, however, assume species will fully occupy their potential ranges. Given evidence that range boundary advances are lagging climatic changes, complete dispersal is often contrasted with dispersal failure, the proportional overlap of the potential future and present ranges representing the worst-case outcome of the latter. This approach has the merit that it does not require estimates of likely dispersal rates that are lacking for most species. Huntley et al. (2008) predict mean overlaps of 31.2–47.5 per cent for the A2 and 37.5–52.8 per cent for the B2 emissions scenario. Zero overlap was predicted for 13–59 species (mean 36; 8.3 per cent) and <10 per cent overlap for a further 14–78 species (mean 49; 11.4 per cent). The predicted overlap for >30 per cent of species examined by Langham et al. (2015) was <10 per cent, and for *ca.* 80 per cent of species overlap was <30 per cent. Hole et al. (2009), using models for 1608 bird species of sub-Saharan Africa, reported median overlap of

31.5 per cent by 2085 for 815 'priority' species (i.e., those whose presence triggers IBA designation; Fishpool and Evans 2001) compared to 56.3 per cent for 793 other species.

13.7.3 Changes of quality

Climatic suitability varies across species' distributions; greatest abundance of a species is generally associated with the highest quality areas, i.e., those of greatest climatic suitability for the species. Where abundance data are available, Howard et al. (2014) show models fitted to such data out-perform models fitted to qualitative (presence/absence) data for the same species and region, even when used only to predict species' distributions. Despite availability of extensive datasets for birds from several regions recording some measure of abundance (e.g., Price et al. 1995; Hagemeijer and Blair 1997; Harrison et al. 1997; Robertson et al. 2007), studies fitting models to abundance data are rare, and very few have predicted potential future range quality, and hence abundance, of birds. Huntley et al. (2012a), however, fitted quantitative response surface models for 78 southern African bird species using reporting rates from Harrison et al. (1997) as an abundance proxy, applying the models to predict reporting rates for climate projections for the B2 emissions scenario from three GCMs. Predicted reporting rates were transformed to obtain values proportional to species' density that were summed across grid squares. The ratio of the predicted future to present summed densities provided a measure of the change in range quality, as well as a relative population estimate for projected future climatic conditions. Whereas a median decrease in range extent of *ca.* 30 per cent by 2085 was predicted, the median decrease in summed density was *ca.* 50 per cent, indicating a predicted marked decrease in quality of the climatically suitable area, and hence a more marked decrease in population size than in range extent. Predicted change of summed density was greater than that of range extent for 82 per cent of species. Huntley and Barnard (2012) used the same approach to explore potential climatic change impacts on 94 birds of the Fynbos and Grassland biomes of southern Africa. The mean predicted range extent in 2085 was 66 per cent of present extent, whilst mean predicted reporting rate for grid cells predicted to be occupied in 2085 was 86 per cent of mean reporting rate for cells occupied at present. Once again, not only was range extent predicted to decrease, but the future range was predicted to be less suitable on average than the present range.

13.7.4 Assemblage changes resulting from distribution changes

Given overwhelming evidence of individualistic species' responses to past climatic changes (Huntley 1991), projected extensive changes in species distributions will lead to substantial changes in species assemblages. Langham et al. (2015) assessed predicted North American avian assemblage changes, finding strong spatial patterns with greatest dissimilarities between present and potential future breeding season assemblages in the Boreal forest zone and western mountains, and for non-breeding season assemblages in the eastern Boreal forests and Great Lakes region. Hole et al. (2009) computed predicted species turnover for African IBAs, finding median turnover of 20–26 per cent by 2085 for all species, but 35–45 per cent for 'priority' species. Huntley et al. (2012b) reported similar results for European IBAs where median turnover by 2085 was 33 per cent for all 487 species modelled but 57 per cent for 156 European Union Birds Directive Annex 1 species.

Predicted range quality changes are also expected to result in species richness changes. Huntley et al. (2007) predicted mean species richness decrease of 6.8–14.2 per cent for European breeding birds, assuming complete dispersal; assuming dispersal failure substantially increased the mean decrease to 27–48 per cent. Huntley and Barnard (2012) predicted median richness decreases by 2085 of 36 per cent for Fynbos biome avian assemblages and of 39 per cent for Grassland assemblages. Both studies also predicted decreases in extent and shifts in location of the areas of greatest species richness.

Huntley et al. (2007) used a combined classification of species assemblages predicted by their models for present climate and for 2085 to identify 22 avian zoogeographic regions in Europe, each characterized by its species assemblage. Mapping these regions showed some very extensive predicted shifts in location and/or changes in extent by 2085. The Sub-Mediterranean region, for example, was predicted to shift northwards and double in extent to include southern parts of England and Sweden. In contrast, the Nemoral region, which was initially the most extensive, was predicted to decrease in extent by 45 per cent and to shift north-eastward into areas of southern Fennoscandia and the Baltic States at present forming parts of the South Boreal–Montane or Continental Nemoral regions. Overall, the avian assemblage of most grid squares was predicted to change sufficiently by 2085 for the grid cell to be classified in a different faunal region. Nonetheless, an important but easily overlooked result of this analysis is that all 22 faunal regions were represented both at present and in 2085; none of them had 'disappeared' or was 'novel' *sensu* Williams et al. (2007). Although these results must be interpreted with some caution, because the analysis assumed complete dispersal, they are paralleled by those of two other studies. Hole et al. (2009) showed that, given complete dispersal, >99 per cent of 'priority' bird species present today would retain suitable climatic space somewhere in the IBA network of sub-Saharan Africa at the end of the century. Similarly, Bagchi et al. (2013) reported that complete loss of suitable climatic space from the IBA network was not 'extremely likely' for any of the 370 species they modelled.

13.8 Conclusions

Evidence from both the Quaternary period and the recent past shows species respond individualistically to climatic changes. The predominant response when climatic changes are of large magnitude and relatively rapid is a change in location, and also often in extent, of species' realized geographical distributions. The quality (i.e., mean climatic suitability) of that distribution will also often change, and this, along with changed extent, results in both local abundance and overall population changes. Considerable uncertainties in predicting these changes arise, however, when a species' potentially realizable climatic niche changes as climate changes (Figure 13.3). In such cases the individualistic response of species results in a changed assemblage of species with which it may interact. Most models used to predict species' potential future distributions relate species' present realized distributions to present climate. Hence, they implicitly take into account indirect climatic determinants of species' distributions that operate by modulating the outcomes of biotic interactions under the currently prevailing range of climatic conditions, as well as direct climatic controls that operate on the species itself. Under changed climatic conditions, however, at least some regions are likely to have combinations of conditions not found anywhere today. Predicting the realized climatic niche, and hence the distribution, of species under such no-analogue conditions is a difficult modelling challenge. Firstly, knowledge of species' fundamental climatic niches is needed in order to determine the location and extent of their potentially realizable climatic niches under the new climatic conditions. Secondly, knowledge is required of the outcomes of biotic interactions under novel climatic conditions and among assemblages of species, the distributions of many of which may not overlap anywhere today. Only with such knowledge can the location and extent of species' realized climatic niches under the no-analogue conditions be reliably predicted. Unfortunately, most species' fundamental climatic niches are unknown, and our knowledge of how to predict the outcomes of novel biotic interactions is extremely limited. Notwithstanding these issues, however, useful predictions can be made (Pearson and Dawson 2003), firstly by using models that make sensible and appropriate assumptions, and behave predictably when extrapolated, and secondly, because most biotic interactions are generalist, their outcomes are mainly predictable so long as present and future species' assemblages contain functionally equivalent species, even if taxonomically the species assemblage changes.

Neither the Quaternary nor the recent past can provide analogues for what is to come, however, principally because other human activities are causing unprecedented levels of habitat fragmentation and loss, species persecution, and species introductions to biogeographic regions from which they were previously absent. Together these generate a 'landscape' in which species must now react to climatic change that is quite different from that in which they responded to past climatic changes.

We are also generating a no-analogue global environment in which atmospheric carbon dioxide levels, as well as global mean temperatures, are set to reach levels unprecedented for millions of years. However, whilst future conditions are without a recent geological analogue, the nature and mechanisms of species' responses to the climatic and other environmental changes almost certainly will mirror those seen in response to Quaternary and recent climatic changes. Changes of distribution will be the dominant response, but other consequences seen in the past, including species extinctions (Lister and Stuart 2008), must also be considered likely.

A key area of uncertainty relates to predicting the consequences of most species' apparent inability to shift their geographical distributions sufficiently rapidly to maintain 'dynamic equilibrium' with the extremely rapid, large-magnitude climatic changes projected by 2100. Furthermore, species' failure to keep pace with climatic change is likely to become more marked if, as is currently projected for the RPC 4.5 and RPC 8.5 scenarios (Collins et al. 2013), the rate of climatic change accelerates. Most authors to date have considered the worst case scenario to be one of 'dispersal failure', i.e., a species failing to disperse into and colonize newly climatically suitable areas, whilst being extirpated from areas that are no longer climatically suitable, and thus persisting only in any overlap between its present and potential future ranges. However, because climatic conditions will generally have changed, often substantially, within that area of overlap, persistence there for most species will only be possible if one of two alternative situations prevails.

Firstly, the adaptive genetic variation that enables the species to occupy the range of climatic conditions under which it currently occurs may be present throughout its range, thus enabling the population remaining in the overlap to adapt to the new climatic conditions there. Alternatively, although adaptive genetic variation may be spatially structured, rates of gene flow may be sufficient for alleles associated with adaptation to different climatic conditions to move through the population quickly enough to maintain local adaptation to the changing climate. Given that many species exhibit local adaptation to climatic conditions throughout their range, and that many others exhibit geographical clines in genetically based adaptations to gradients in climatic conditions across their ranges, the likelihood that the first situation generally prevails must be seriously questioned (Bradshaw and McNeilly 1991). It is also questionable whether the second situation prevails for many species, especially in those, like birds, in which gene flow depends on dispersal of individuals. Indeed, given that many bird species exhibit a high degree of philopatry, the rate of gene flow through their overall population is likely to be quite severely limited. That for at least some species neither situation prevails is shown by evidence of genomic variation associated with climate across the breeding range of the American yellow warbler (*Setophaga petechia*), despite the wide range and high dispersal ability of this migratory species (Bay et al. 2018). Furthermore, populations with the greatest mismatch between current and predicted future genomic variation, hence requiring the greatest shifts in allele frequencies to keep pace with future climate change, have experienced the largest population declines over the past 50 years, indicating that failure to adapt to changing climate has already had negative effects on populations.

In extreme cases a species may have considerable overall genetic variability, enabling it to occupy a wide range of climatic conditions, but have this genetic variability geographically structured such that as climatic conditions change in the area of overlap between its present and future distributions, the population there becomes progressively more maladapted as a result of limited gene flow. If this is combined with rapid extirpation of populations that contain alleles that would enable persistence of the species in the overlap area, because the areas occupied by those populations are no longer climatically suitable, then the population in the area of overlap may be unable to persist. In such circumstances a species may suffer extinction even though its present and future ranges overlap, and it currently occupies areas with climatic conditions analogous to future conditions in that overlap. Assessing the magnitude of this potential risk requires efforts to determine the degree of geographic structuring of the adaptive genetic variability enabling species to occupy a range of climatic conditions, as well as measurements of potential rates of gene flow through species' populations. Using such data, a new generation of dynamic species distribution models needs to be developed that incorporate not just dispersal and demographic processes (Huntley et al. 2010), but also the dynamics of gene flow and adaptation. Only with such models can the true magnitude of species' future extinction risks be adequately assessed.

References

Aebischer, N.J., Coulson, J.C., and Colebrook, J.M. (1990). Parallel long-term trends across four marine trophic levels and weather. *Nature*, 347, 753–5.

Allen, J.R.M., Hickler, T., Singarayer, J.S., Sykes, M.T., Valdes, P.J., and Huntley, B. (2010). Last glacial vegetation of northern Eurasia. *Quaternary Science Reviews*, 29, 2604–18.

Bagchi, R., Crosby, M., Huntley, B., et al. (2013). Evaluating the effectiveness of conservation site networks under climate change: Accounting for uncertainty. *Global Change Biology*, 19, 1236–48.

Balmer, D.E., Gillings, S., Caffrey, B.J., Swann, R.L., Downie, I.S., and Fuller, R.J. (2013). *Bird Atlas 2007–11: The breeding and wintering birds of Britain and Ireland*, 720 pp. British Trust for Ornithology Books, Thetford, UK.

Barbet-Massin, M., Walther, B.A., Thuiller, W., Rahbek, C., and Jiguet, F. (2009). Potential impacts of climate change on the winter distribution of Afro-Palaearctic migrant passerines. *Biology Letters*, 5, 248–51.

Barnes, I., Shapiro, B., Lister, A., et al. (2007). Genetic structure and extinction of the woolly mammoth, *Mammuthus primigenius*. *Current Biology*, 17, 1072–5.

Bay, R.A., Harrigan, R.J., Underwood, V.L., Gibbs, H.L., Smith, T.B., and Ruegg, K. (2018). Genomic signals of selection predict climate-driven population declines in a migratory bird. *Science*, 359, 83–6.

Birkhead, T. (1994). How collectors killed: One hundred and fifty years ago next week the last two great auks ever seen were killed at their breeding colony on a tiny island off the coast of Iceland. *New Scientist*, 142, 24–7.

Blunier, T., and Brook, E.J. (2001). Timing of millennial-scale climate change in Antarctica and Greenland during the last glacial period. *Science*, 291, 109–12.

Botkin, D.B., Woodby, D.A., and Nisbet, R.A. (1991). Kirtland's warbler habitats: A possible early indicator of climatic warming. *Biological Conservation*, 56, 63–78.

Bradshaw, A.D., and McNeilly, T. (1991). Evolutionary response to global climatic change. *Annals of Botany*, 67, 5–14.

Brambilla, M., Pedrini, P., Rolando, A., and Chamberlain, D.E. (2016). Climate change will increase the potential conflict between skiing and high-elevation bird species in the Alps. *Journal of Biogeography*, 43, 2299–309.

Brito, P.H. (2005). The influence of Pleistocene glacial refugia on tawny owl genetic diversity and phylogeography in western Europe. *Molecular Ecology*, 14, 3077–94.

Brommer, J.E. (2004). The range margins of northern birds shift polewards. *Annales Zoologici Fennici*, 41, 391–7.

Brommer, J.E., and Møller, A.P. (2010). Range margins, climate change, and ecology. In: A.P. Møller, W. Fiedler, and P. Berthold (eds), *Effects of Climate Change on Birds*, pp. 249–74. Oxford University Press, Oxford, UK.

Burg, T.M., Lomax, J., Almond, R., Brooke, M.D., and Amos, W. (2003). Unravelling dispersal patterns in an expanding population of a highly mobile seabird, the northern fulmar (*Fulmarus glacialis*). *Proceedings of the Royal Society of London Series B: Biological Sciences*, 270, 979–84.

Campos, P.F., Kristensen, T., Orlando, L., et al. (2010). Ancient DNA sequences point to a large loss of mitochondrial genetic diversity in the saiga antelope (*Saiga tatarica*) since the Pleistocene. *Molecular Ecology*, 19, 4863–75.

Chen, I.C., Hill, J.K., Ohlemuller, R., Roy, D.B., and Thomas, C.D. (2011). Rapid range shifts of species associated with high levels of climate warming. *Science*, 333, 1024–6.

Cohen, K.M., Finney, S.C., Gibbard, P.L., and Fan, J.X. (2013). The ICS international chronostratigraphic chart. *Episodes*, 36, 199–204.

Collins, M., Knutti, R., Arblaster, J., et al. (2013). Long-term climate change: Projections, commitments and irreversibility. In: T.F. Stocker, D. Qin, G.-K. Plattner, et al. (eds), *Climate change 2013: The physical science basis. Contribution of Working Group I to the Fifth Assessment Report of the Intergovernmental Panel on Climate Change*, pp. 1029–136. Cambridge University Press, Cambridge, UK, and New York, NY, USA.

Coope, G.R. (1973). Tibetan species of dung beetle from late Pleistocene deposits in England. *Nature*, 245, 335–6.

Coope, G.R. (1994). The response of insect faunas to glacial–interglacial climatic fluctuations. *Philosophical Transactions of the Royal Society of London Series B: Biological Sciences*, 344, 19–26.

Cramp, S. (ed.) (1985). *The birds of the western Palearctic, Vol. IV: Terns to Woodpeckers*, 936 pp. Oxford University Press, Oxford, UK.

Cramp, S., and Simmons, K.E.L. (eds) (1977). *The birds of the western Palearctic, Vol. I: Ostrich to Ducks*, 722 pp. Oxford University Press, Oxford, UK.

Cunningham, S.J., Kruger, A.C., Nxumalo, M.P., and Hockey, P.A.R. (2013a). Identifying biologically meaningful hot-weather events using threshold temperatures that affect life-history. *PLoS One*, 8, e82492.

Cunningham, S.J., Martin, R.O., Hojem, C.L., and Hockey, P.A.R. (2013b). Temperatures in excess of critical thresholds threaten nestling growth and survival in a rapidly-warming arid savanna: A study of common fiscals. *PLoS One*, 8, e74613.

Devictor, V., Julliard, R., Couvet, D., and Jiguet, F. (2008). Birds are tracking climate warming, but not fast enough. *Proceedings of the Royal Society of London Series B: Biological Sciences*, 275, 2743–8.

Doswald, N., Willis, S.G., Collingham, Y.C., Pain, D.J., Green, R.E., and Huntley, B. (2009). Potential impacts of climatic change on the breeding and non-breeding ranges and migration distance of European *Sylvia* warblers. *Journal of Biogeography*, 36, 1194–208.

du Plessis, K.L., Martin, R.O., Hockey, P.A.R., Cunningham, S.J., and Ridley, A.R. (2012). The costs of keeping cool in a warming world: implications of high temperatures for foraging, thermoregulation and body condition of an arid-zone bird. *Global Change Biology*, 18, 3063–70.

FAUNMAP Working Group (1996). Spatial response of mammals to late Quaternary environmental fluctuations. *Science*, 272, 1601–6.

Fisher, J. (1952). A history of the Fulmar *Fulmarus* and its population problems. *Ibis*, 94, 334–54.

Fisher, J. (1966). Fulmar population of Britain and Ireland, 1959. *Bird Study*, 13, 5–76.

Fishpool, L.D.C., and Evans, M.I. (eds) (2001). *Important bird areas in Africa and associated islands: Priority sites for conservation*, 1000 pp. Pisces Publications/Birdlife International, Newbury/Cambridge, UK.

Foden, W.B., Butchart, S.H.M., Stuart, S.N., et al. (2013). Identifying the world's most climate change vulnerable species: A systematic trait-based assessment of all birds, amphibians and corals. *PLoS One*, 8, e65427.

Garcia, J.T., Manosa, S., Morales, M.B., et al. (2011). Genetic consequences of interglacial isolation in a steppe bird. *Molecular Phylogenetics and Evolution*, 61, 671–6.

Gaston, K.J. (2003). *The structure and dynamics of geographic ranges*, 266 pp. Oxford University Press, Oxford, UK.

Gaüzère, P., Jiguet, F., and Devictor, V. (2015). Rapid adjustment of bird community compositions to local climatic variations and its functional consequences. *Global Change Biology*, 21, 3367–78.

Gibbons, D.W., Reid, J.B., and Chapman, R.A. (1993). *The new atlas of breeding birds in Britain and Ireland: 1988–1991*, 520 pp. T. & A.D. Poyser, London.

Gillings, S., Balmer, D.E., and Fuller, R.J. (2015). Directionality of recent bird distribution shifts and climate change in Great Britain. *Global Change Biology*, 21, 2155–68.

Grabherr, G., Gottfried, M., and Pauli, H. (1994). Climate effects on mountain plants. *Nature*, 369, 448.

Graham, R.W. (1997). The spatial response of mammals to Quaternary climate changes. In: B. Huntley, W. Cramer, A.V. Morgan, H.C. Prentice, and J.R.M. Allen (eds), *Past and future rapid environmental changes: The spatial and evolutionary responses of terrestrial biota*, pp. 153–62. Springer-Verlag, Berlin.

Green, R.E., Collingham, Y.C., Willis, S.G., Gregory, R.D., Smith, K.W., and Huntley, B. (2008). Performance of climate envelope models in retrodicting recent changes in bird population size from observed climatic change. *Biology Letters*, 4, 599–602.

Gregory, R.D., Willis, S.G., Jiguet, F., et al. (2009). An indicator of the impact of climatic change on European bird populations. *PLoS One*, 4, e4678.

Grosbois, V., and Thompson, P.M. (2005). North Atlantic climate variation influences survival in adult fulmars. *Oikos*, 109, 273–90.

Gynther, I., Waller, N., and Leung, L.K.-P. (2016). *Confirmation of the extinction of the Bramble Cay melomys* Melomys rubicola *on Bramble Cay, Torres Strait: results and conclusions from a comprehensive survey in August–September 2014*. Department of Environment and Heritage Protection, Queensland Government, Brisbane.

Hagemeijer, E.J.M., and Blair, M.J. (eds) (1997). *The EBCC Atlas of European Breeding Birds: Their distribution and abundance*, 903 pp. T. & A.D. Poyser, London.

Harrison, J.A., Allan, D.G., Underhill, L.G., et al. (eds) (1997). *The atlas of southern African birds, Vol. 1: Non-passerines, Vol. 2: Passerines*, 785 & 732 pp. BirdLife South Africa, Johannesburg.

Hays, J.D., Imbrie, J., and Shackleton, N. (1976). Variations in the Earth's orbit: Pacemaker of the ice ages. *Science*, 194, 1121–32.

Hickling, R., Roy, D.B., Hill, J.K., Fox, R., and Thomas, C.D. (2006). The distributions of a wide range of taxonomic groups are expanding polewards. *Global Change Biology*, 12, 450–5.

Hitch, A.T., and Leberg, P.L. (2007). Breeding distributions of north American bird species moving north as a result of climate change. *Conservation Biology*, 21, 534–9.

Hockey, P.A.R., Sirami, C., Ridley, A.R., Midgley, G.F., and Babiker, H.A. (2011). Interrogating recent range changes in South African birds: confounding signals from land use and climate change present a challenge for attribution. *Diversity and Distributions*, 17, 254–61.

Hole, D.G., Willis, S.G., Pain, D.J., et al. (2009). Projected impacts of climate change on a continent-wide protected area network. *Ecology Letters*, 12, 420–31.

Holloway, S. (1996). *The historical atlas of breeding birds in Britain and Ireland 1875–1900*, 476 pp. T. & A.D. Poyser, London.

Howard, C., Stephens, P.A., Pearce-Higgins, J.W., Gregory, R.D., and Willis, S.G. (2014). Improving species distribution models: the value of data on abundance. *Methods in Ecology and Evolution*, 5, 506–13.

Huntley, B. (1990a). European post-glacial forests: compositional changes in response to climatic change. *Journal of Vegetation Science*, 1, 507–18.

Huntley, B. (1990b). Dissimilarity mapping between fossil and contemporary pollen spectra in Europe for the past 13,000 years. *Quaternary Research*, 33, 360–76.

Huntley, B. (1991). How plants respond to climate change: migration rates, individualism and the consequences for plant communities. *Annals of Botany*, 67, 15–22.

Huntley, B. (1996). Quaternary palaeoecology and ecology. *Quaternary Science Reviews*, 15, 591–606.

Huntley, B. (1998). The dynamic response of plants to environmental change and the resulting risks of extinction. In: G.M. Mace, A. Balmford, and J.R. Ginsberg (eds), *Conservation in a changing world*, pp. 69–85. Cambridge University Press, Cambridge, UK.

Huntley, B., Allen, J.R.M., Collingham, Y.C., et al. (2013). Millennial climatic fluctuations are key to the structure of last glacial ecosystems. *PLoS One*, 8, e61963.

Huntley, B., Altwegg, R., Barnard, P., Collingham, Y.C., and Hole, D.G. (2012a). Modelling relationships between species' spatial abundance patterns and climate. *Global Ecology and Biogeography*, 21, 668–81.

Huntley, B., and Barnard, P. (2012). Potential impacts of climatic change on southern African birds of Fynbos and Grassland biodiversity hotspots. *Diversity and Distributions*, 18, 769–81.

Huntley, B., Barnard, P., Altwegg, R., et al. (2010). Beyond bioclimatic envelopes: Dynamic species' range and abundance modelling in the context of climatic change. *Ecography*, 33, 621–6.

Huntley, B., and Birks, H.J.B. (1983). *An atlas of past and present pollen maps for Europe: 0–13000 B.P.*, 667 pp. Cambridge University Press, Cambridge, UK.

Huntley, B., Collingham, Y., Willis, S., and Hole, D. (2012b). Protected areas and climatic change in Europe: Introduction. In: G. Ellwanger, A. Ssymank, and C. Paulsch (eds), *Natura 2000 and Climate Change—a Challenge*, pp. 7–27. Federal Agency for Nature Conservation, Bonn, Germany.

Huntley, B., Collingham, Y.C., Singarayer, J.S., et al. (2016). Explaining patterns of avian diversity and endemicity: Climate and biomes of southern Africa over the last 140,000 years. *Journal of Biogeography*, 43, 874–86.

Huntley, B., Collingham, Y.C., Willis, S.G., and Green, R.E. (2008). Potential impacts of climatic change on European breeding birds. *PLoS One*, 3, e1439.

Huntley, B., and Green, R.E. (2011). Bioclimatic models of the distributions of gyrfalcons and ptarmigan. In: R.T. Watson, T.J. Cade, M. Fuller, G. Hunt, and E. Potapov (eds), *Gyrfalcons and Ptarmigan in a Changing World*, pp. 329–38. The Peregrine Fund, Boise, Idaho, USA.

Huntley, B., Green, R.E., Collingham, Y.C., and Willis, S.G. (2007). *A climatic atlas of European breeding birds*, 521 pp. Lynx Edicions, Barcelona, Spain.

Hutchinson, G.E. (1957). Concluding remarks. *Cold Spring Harbor Symposia on Quantitative Biology*, 22, 415–27.

Hutchinson, G.E. (1991). Population studies: Animal ecology and demography. *Bulletin of Mathematical Biology*, 53, 193–213.

Imbrie, J., and Imbrie, K.P. (1979). *Ice ages: Solving the mystery*, 224 pp. Macmillan, London.

Jackson, S.T., and Williams, J.W. (2004). Modern analogs in Quaternary paleoecology: Here today, gone yesterday, gone tomorrow? *Annual Review of Earth and Planetary Sciences*, 32, 495–537.

Jansen, E., Overpeck, J., Briffa, K.R., et al. (2007). Palaeoclimate. In: S. Solomon, D. Qin, M. Manning, et al. (eds), *Climate change 2007: The physical science basis. Contribution of Working Group I to the Fourth Assessment Report of the Intergovernmental Panel on Climate Change*, pp. 433–97. Cambridge University Press, Cambridge, UK and New York, NY, USA.

Kamp, J., Oppel, S., Ananin, A.A., et al. (2015). Global population collapse in a superabundant migratory bird and illegal trapping in China. *Conservation Biology*, 29, 1684–94.

Lack, P. (1986). *The atlas of wintering birds in Britain and Ireland*, 447 pp. T. & A.D. Poyser, Calton, Staffs, UK.

Langham, G.M., Schuetz, J.G., Distler, T., Soykan, C.U., and Wilsey, C. (2015). Conservation status of North American birds in the face of future climate change. *PLoS One*, 10, e135350.

Lindström, A., Green, M., Paulson, G., Smith, H.G., and Devictor, V. (2013). Rapid changes in bird community composition at multiple temporal and spatial scales in response to recent climate change. *Ecography*, 36, 313–22.

Lister, A.M., and Stuart, A.J. (2008). The impact of climate change on large mammal distribution and extinction: Evidence from the last glacial/interglacial transition. *Comptes Rendus Geosciences*, 340, 615–20.

Liukkonen-Anttila, T., Uimaniemi, L., Orell, M., and Lumme, J. (2002). Mitochondrial DNA variation and the phylogeography of the grey partridge (*Perdix perdix*) in Europe: From Pleistocene history to present day populations. *Journal of Evolutionary Biology*, 15, 971–82.

Mason, S.C., Palmer, G., Fox, R., et al. (2015). Geographical range margins of many taxonomic groups continue to shift polewards. *Biological Journal of the Linnean Society*, 115, 586–97.

Massa, C., Doyle, M., and Fortunato, R.C. (2014). On how cattle egret (*Bubulcus ibis*) spread to the Americas: Meteorological tools to assess probable colonization trajectories. *International Journal of Biometeorology*, 58, 1879–91.

National Audubon Society (2015). *Audubon's Birds and Climate Change Report: A Primer for Practitioners*. National Audubon Society, New York, NY.

Oswald, S., Bearhop, S., Furness, R.W., Huntley, B., and Hamer, K.C. (2008). Heat-stress in a high-latitude seabird: Effects of temperature and food supply on bathing and nest attendance of great skuas *Catharacta skua*. *Journal of Avian Biology*, 39, 163–9.

Overpeck, J.T., Webb, R.S., and Webb, T., III (1992). Mapping eastern North American vegetation change of the past 18 ka: No-analogs and the future. *Geology*, 20, 1071–4.

Parmesan, C. (1996). Climate and species' range. *Nature*, 382, 765–6.

Pearson, R.G., and Dawson, T.P. (2003). Predicting the impacts of climate change on the distribution of species: Are bioclimate envelope models useful? *Global Ecology and Biogeography*, 12, 361–71.

Pellegrino, I., Negri, A., Boano, G., et al. (2015). Evidence for strong genetic structure in European populations of the little owl *Athene noctua*. *Journal of Avian Biology*, 46, 462–75.

Peterson, A.T. (2003). Projected climate change effects on Rocky Mountain and Great Plains birds: Generalities of biodiversity consequences. *Global Change Biology*, 9, 647–55.

Peterson, A.T., Sanchez-Cordero, V., Soberon, J., Bartley, J., Buddemeier, R.W., and Navarro-Siguenza, A.G. (2001). Effects of global climate change on geographic distributions of Mexican Cracidae. *Ecological Modelling*, 144, 21–30.

Phillips, R.A., Petersen, M.K., Lilliendahl, K., et al. (1999). Diet of the northern fulmar *Fulmarus glacialis*: Reliance on commercial fisheries? *Marine Biology*, 135, 159–70.

Price, J., Droege, S., and Price, A. (1995). *The summer atlas of North American birds*, 364 pp. Academic Press, London.

Robertson, C.J.R., Hyvönen, P., Fraser, M.J., and Pickard, C.R. (2007). *Atlas of bird distribution in New Zealand 1999–2004*, 533 pp. The Ornithological Society of New Zealand, Inc., Wellington, New Zealand.

Robinson, R.A., Baillie, S.R., and Crick, H.Q.P. (2007). Weather-dependent survival: Implications of climate change for passerine population processes. *Ibis*, 149, 357–64.

Ruokonen, M., Aarvak, T., and Madsen, J. (2005). Colonization history of the high-Arctic pink-footed goose *Anser brachyrhynchus*. *Molecular Ecology*, 14, 171–8.

Saitoh, T., Alstrom, P., Nishiumi, I., et al. (2010). Old divergences in a boreal bird supports long-term survival through the ice ages. *BMC Evolutionary Biology*, 10, 13.

Shakun, J.D., and Carlson, A.E. (2010). A global perspective on Last Glacial Maximum to Holocene climate change. *Quaternary Science Reviews*, 29, 1801–16.

Sharrock, J.T.R. (1976). *The Atlas of Breeding Birds in Britain and Ireland*, 477 pp. British Trust for Ornithology, Tring, UK.

Stephens, P.A., Mason, L.R., Green, R.E., et al. (2016). Consistent response of bird populations to climate change on two continents. *Science*, 352, 84–7.

Stocker, T.F., Qin, D., Plattner, G.-K., et al. (eds) (2013). *Climate change 2013: The physical science basis. Contribution of Working Group I to the Fifth Assessment Report of the Intergovernmental Panel on Climate Change*, 1535 pp. Cambridge University Press, Cambridge, UK, and New York, NY, USA.

Thomas, C.D., and Lennon, J.J. (1999). Birds extend their ranges northwards. *Nature*, 399, 213–13.

Tyrberg, T. (1998). *Pleistocene birds of the Palearctic: A catalogue*, 720 pp. Nuttall Ornithological Club, Cambridge, Massachusetts.

Virkkala, R., Heikkinen, R.K., Leikola, N., and Luoto, M. (2008). Projected large-scale range reductions of northern-boreal land bird species due to climate change. *Biological Conservation*, 141, 1343–53.

Virkkala, R., and Rajasärkkä, A. (2011). Northward density shift of bird species in boreal protected areas due to climate change. *Boreal Environmental Research*, 16 (suppl. B), 2–13.

Waller, N.L., Gynther, I.C., Freeman, A.B., Lavery, T.H., and Leung, L.K.-P. (2017). The Bramble Cay melomys *Melomys rubicola* (Rodentia: Muridae): A first mammalian extinction caused by human-induced climate change? *Wildlife Research*, 44, 9–21.

Walther, B.A., Schaffer, N., Van Niekerk, A., Thuiller, W., Rahbek, C., and Chown, S.L. (2007). Modelling the winter distribution of a rare and endangered migrant, the Aquatic Warbler *Acrocephalus paludicola*. *Ibis*, 149, 701–14.

Williams, J.W., and Jackson, S.T. (2007). Novel climates, no-analog communities, and ecological surprises. *Frontiers in Ecology and the Environment*, 5, 475–82.

Williams, J.W., Jackson, S.T., and Kutzbach, J.E. (2007). Projected distributions of novel and disappearing climates by 2100 AD. *Proceedings of the National Academy of Sciences of the United States of America*, 104, 5738–42.

Williams, J.W., Shuman, B.N., and Webb, T. (2001). Dissimilarity analyses of late-Quaternary vegetation and climate in eastern North America. *Ecology*, 82, 3346–62.

Wolff, E.W., Chappellaz, J., Blunier, T., Rasmussen, S.O., and Svensson, A. (2010). Millennial-scale variability during the last glacial: The ice core record. *Quaternary Science Reviews*, 29, 2828–38.

Woodward, F.I. (1987). *Climate and plant distribution*, 174 pp. Cambridge University Press, Cambridge, UK.

Wright, L.J., Hoblyn, R.A., Green, R.E., et al. (2009). Importance of climatic and environmental change in the demography of a multi-brooded passerine, the woodlark *Lullula arborea*. *Journal of Animal Ecology*, 78, 1191–202.

Younger, J., Emmerson, L., Southwell, C., Lelliott, P., and Miller, K. (2015a). Proliferation of East Antarctic Adelie penguins in response to historical deglaciation. *BMC Evolutionary Biology*, 15, 11.

Younger, J.L., Clucas, G.V., Kooyman, G., et al. (2015b). Too much of a good thing: Sea ice extent may have forced emperor penguins into refugia during the last glacial maximum. *Global Change Biology*, 21, 2215–26.

Zuckerberg, B., Woods, A.M., and Porter, W.F. (2009). Poleward shifts in breeding bird distributions in New York State. *Global Change Biology*, 15, 1866–83.

SECTION 4

Interspecific effects of climate change

CHAPTER 14

Host–parasite interactions and climate change

Santiago Merino

14.1 Introduction

Several models predict that climate change will modify the distribution of several diseases. However, although an increasing number of examples of host–parasite interactions modified by climate change exist in the literature, empirical data on this issue are still lacking. Here, I review published findings from the last ten years since the previous edition of this book was published (Merino and Møller 2010). We not only incorporate data from papers published since 2010, but also data from several papers published prior to 2010 that were not included in the previous edition. In addition, we include studies on parasites that do not directly affect birds but have similar life cycles to those that do. Although we attempted a comprehensive literature search, some papers may have been overlooked. However, we are confident that the most relevant studies related to this topic have been included.

The differences between the current scenario of climate change and the one presented in 2010 are not considerable, and although some models have been refined and modified, the predictions have not changed substantially (see Rohr et al. 2011). Thus, researchers studying the effect of climate change on bird–parasite interactions still expect an expansion of the geographical range of several diseases and a potential decrease in others, depending on a variety of factors and the idiosyncrasies of each disease (Lafferty 2009). Furthermore, if the global climate becomes warmer and wetter, parasites in general are still expected to achieve greater transmission over longer periods, with changes in virulence as a likely consequence (Harvell et al. 2002; Kutz et al. 2005; Polley and Thompson 2009).

However, we must again stress that both parasites and hosts are living beings with the capacity to evolve and adapt, thus making it difficult to predict their exact response to environmental changes (Martínez and Merino 2011). For instance, many parasites spend a part of their life cycle independent of their vertebrate hosts and, therefore, are highly affected by environmental conditions during that period. In these cases, climate change may severely affect the development and survival of a parasite and thus its transmission to a vertebrate host. In other cases, climate can affect the bird hosts, resulting in changes in population size, behaviour, and physiology that make it easier or more difficult for parasites to complete their life cycles (Morley and Lewis 2014). Parasites can also be brought to new areas by their hosts or switch hosts in response to a reduction in original host abundance and an increase in the abundance of other hosts (Phillips et al. 2010). Introduced host species present a special case. Since they are often free of parasites or resistant to endemic parasites, they can generally outcompete endemic host species. Introduced species

Merino, S., *Host–parasite interactions and climate change*. In: *Effects of Climate Change on Birds*. Second Edition.
Edited by Peter O. Dunn and Anders Pape Møller: Oxford University Press (2019).
DOI: 10.1093/oso/9780198824268.003.0014

can also introduce their own parasites, which may greatly impact endemic host species (Marzal et al. 2011; Lymbery et al. 2014). In any event, we cannot expect to observe a linear relationship between environmental change and parasite populations due to the varying influence of different factors such as environmental thresholds, host resistance, interactions between parasites and between parasites and their predators, or interactions between several of these variables (Sutherst 2001; Lafferty 2009).

14.2 Climate change and changes in phenology of parasites and their hosts

Climate and climate change can potentially affect many events in the lives of both hosts and parasites, including their interactions with each other. For example, the breeding season can be adjusted by temperature and rainfall directly or indirectly by their effect on food availability. Changes in the synchrony between breeding season and food availability may have important consequences for host or parasite fitness. Warmer winters or earlier springs due to climate change may alter the magnitude and timing of peak parasite transmission, thus affecting host health (Polley and Thompson 2009). As a result, climate change is expected to affect the phenology of parasites and hosts, ultimately altering their interactions.

One of the best examples of how phenological changes in response to climate change can influence bird–parasite interactions is the study of cuckoos *Cuculus canorus* and their hosts by Saino et al. (2009). Cuckoos that migrated to Europe for the breeding season arrived late relative to their hosts that were short-distance migrants. There was less of a mismatch between the arrival times of cuckoos and their hosts that were long-distance migrants. Saino et al. (2009) predicted that cuckoo races that specialize on short-distance migrants will disappear, as they will have fewer opportunities for successful parasitism. To further support their hypothesis, they compared areas across Europe that have different degrees of increase in spring temperature, expecting to find less parasitism of residents or short-distance migrants in areas that experience greater increases in temperature. Specifically, they compared the relative frequency of parasitism of two host categories in 23 European countries before and after 1990 and found that parasitism of residents and short-distance migrants decreased, and it was positively related to increases in spring temperature. In conclusion, their data support the hypothesis that the change in host use by the common cuckoo is at least partially driven by climate change (Møller et al. 2011).

We should interpret these types of results with caution because a study examining the effects of changes in breeding phenology of the great spotted cuckoo *Clamator glandarius* and its magpie host *Pica pica* in Europe found different results in cross-sectional versus longitudinal analyses (Avilés et al. 2014). Cross-sectional analyses, which assess correlations between climate conditions and phenological mismatch between host and parasite at the population level, revealed that climate affects laying date and cuckoos were able to adjust their breeding phenology to that of their magpie hosts. In contrast, longitudinal analyses, which examined within-individual variation in breeding phenology, revealed high individual consistency in magpie host phenology and a low influence of climate. These findings suggest that the phenological mismatch between cuckoos and magpies at the population level is not explained by the host's response to climatic conditions (Avilés et al. 2014).

Other examples that demonstrate changes in phenology of bird parasites linked to climate change are scarce. However, several studies of different tick species that can use birds as hosts have shown or predicted changes in phenology associated with environmental conditions. For example, *Ixodes scapularis* is the principal tick vector of the Lyme borreliosis agent *Borrelia burgdorferi* and other tick-borne zoonoses in northeastern North America. Although the life cycle of this parasite is linked primarily to small mammals and humans, recent research has shown that birds also act as reservoirs, thus contributing to the propagation of these tick-borne diseases (Newman et al. 2015). The seasonal phenology of these ectoparasites is partly controlled by ambient temperature. Therefore, climatic changes may affect the degree of seasonal synchrony of the different tick stages. Furthermore, seasonal synchrony among tick nymphs and larvae may affect the type of diseases they transmit. Asynchrony between nymphs and larvae will favour the transmission of persistent pathogens by allowing infected nymphs to inoculate a population of naive hosts that can subsequently transmit the pathogen to larvae and complete the transmission cycle. However, synchrony between nymphs and larvae favours transmission of pathogens that produce short-lived infections in hosts (Levi et al. 2015).

In another study, Ogden and collaborators (2008) mathematically simulated the phenology of *I. scapularis* for a site in southern Ontario, Canada under monthly temperatures projected under emissions scenario A2 of the Intergovernmental Panel on Climate Change. Following this, they simulated pathogen transmission in a population of white-footed mice (*Peromyscus leucopus*) using a susceptible–infected–recovered (SIR) model. Their results showed that the fitness of pathogens transmitted by ticks differed according to the seasonal pattern of immature tick activity, which was different for each predicted temperature scenario. Although, in general, long-lived, highly transmissible, and low virulence pathogens were the fittest, under some scenarios of seasonal tick activity, the fitness of shorter-lived, less efficiently transmissible, and more virulent pathogens increased. Thus, Ogden and co-authors concluded that climate change may affect the evolutionary processes of pathogens transmitted by the tick *I. scapularis* via effects of temperature on tick seasonality.

Using data gathered over 19 years, Levi et al. (2015) analysed the seasonal synchrony of nymph and larval stages of *I. scapularis*. These authors showed that the phenology of nymphal and larval ticks advanced almost three weeks during years in which May and August were warmer compared with years in which these months were colder. However, increased synchrony was not observed. Warmer weather in October was associated with fewer larvae feeding concurrently with nymphs during the following spring. Based on these data and projected warming by 2050, Levi et al. (2015) hypothesized that the timing of nymphal and larval activity will advance on average by 8–11 and 10–14 days, respectively. Thus, climate warming should maintain or increase transmission of persistent pathogens, but may inhibit pathogens that do not produce long-lasting infections (Levi et al. 2015).

14.3 Climate change and changes in prevalence and intensity of parasitism

Increases in temperatures can accelerate the emergence of some parasitic stages, thus increasing their prevalence and intensity in hosts. For instance, both cercarial production within snails, the first intermediate host of trematodes, and their emergence from these hosts is accelerated in response to temperature increases (Poulin 2006). Thus, global warming may increase the cercarial output of trematodes, potentially impacting hosts, including bird hosts. However, this impact may be mitigated, to some extent, as higher temperatures also decrease the lifespan of cercariae (Poulin 2006). These parasites can indirectly affect birds by reducing the amphipod populations on which they prey, as well as directly affecting birds through schistosomatid infections, which are associated with changes in climatic conditions (Mas-Coma et al. 2009). These particular trematode parasites have been studied primarily for the cercarial dermatitis ('swimmers' itch') they cause in humans swimming in lakes. This disease causes an itchy skin rash but is not contagious (Horak et al. 2000). The increasing incidence of this disease is associated with climate warming as bird host populations are spending more time in lakes, thus changing their migratory behaviour (Guillemain et al. 2013). The increased presence of birds, along with the accelerated production of cercariae with milder temperatures, has increased the prevalence and intensity of these parasites in lakes that previously never or rarely produced the disease. In Europe, cercarial dermatitis is caused by cercariae of the genera *Trichobilharzia*, *Bilharziella*, and *Gigantobilharzia*, all of which are known to parasitize birds (Mas-Coma et al. 2009). These parasites also cause severe diseases in their bird hosts (Lévine et al. 1956; Horak et al. 2000).

The nematode *Heterakis gallinarum* presents another potential example that may provide insight into the rise of incidence of parasite infection associated with increases in temperature. This parasite of chickens and related bird species is transmitted after larval development has occurred within eggs previously deposited in the environment. Increases in temperature cause accelerated development of larvae (Mas-Coma et al. 2008), and thus, a potential increase in the prevalence of infection by this parasite. As the climate changes, *Heterakis* should be closely monitored as it may have commercial implications for the poultry industry.

Several studies have linked environmental conditions and infections by avian blood parasites (for a summary, see Sehgal 2015). For example, Sehgal et al. (2011) collected blood samples from olive sunbirds (*Cyanomitra olivacea*) from 28 different localities in central and western Africa and found that temperature was the most important predictor of *Plasmodium* prevalence whereas humidity appeared to be more important for trypanosomes. In another study, Berthelot's pipits (*Anthus berthelotii*) that inhabited locations with higher minimum temperatures on Tenerife (Canary Islands) were more likely to be infected by malaria (González-Quevedo et al. 2014). Some mathematical models have indicated that the prevalence of *Leucocytozoon*, a malaria-like blood parasite of birds, is influenced by seasonal changes in temperature by affecting the emergence of the insect vector, black flies, in the Colorado Rocky Mountains (Murdock et al. 2013). Climate warming is expected to affect the distribution and the incidence of insect vectors, increasing their altitudinal and latitudinal ranges and, as a consequence, affecting the prevalence of the blood parasites transmitted by them. For example, *Plasmodium* prevalence has increased with increasing global temperatures, especially during the last two decades, as shown by Garamszegi (2011; see Figure 14.1). In this study, the author compiled publications on avian malaria infections over the last seven decades and analysed variation in parasite prevalence in more than 3000 bird species. The prevalence of other blood parasites has also been positively linked to temperature in a bird community of the Australian Wet Tropics along an elevation gradient (Zamora-Vilchis et al. 2012). This effect has been explained by both the increase in vector abundance with temperature and the accelerated development of blood parasites within vectors at elevated temperatures. Based on the regression of parasite prevalence on temperature found in this study, the authors predicted a 10 per cent increase in prevalence of parasites for each 1°C increment in temperature (Zamora-Vilchis et al. 2012). A similar relationship between blood parasites and altitudinal

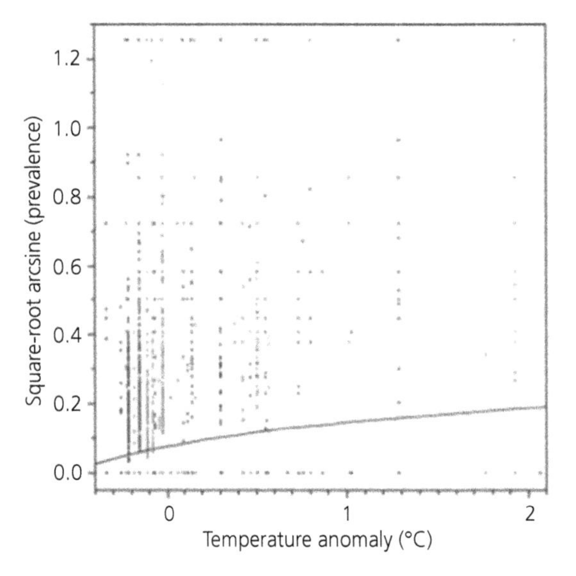

Figure 14.1 The relationship between climate change and the prevalence of avian malaria due to *Plasmodium*. The regression line was fitted to the appropriately transformed data (for a transparent illustration, the Y-axis is left transformed). The data points are species observations from different studies and are coloured based on the geographic origin (red, Africa; green, Europe; yellow, Asia; blue, North America; orange, South America). Taken from Garamszegi (2011) with permission (©John Wiley and Sons).

gradient has also been reported in New Zealand (Niebuhr et al. 2016).

In the case of ectoparasites, things could be different, because they depend more on ambient temperatures for development and completion of their life cycles. Consequently they may be more directly affected by climate change than endoparasites, which once in their bird host are less affected by environmental climatic conditions. However, some ectoparasites live in tight contact with their hosts and it could be expected that direct effects of climate barely affect those parasites. Nonetheless, some studies show how, even in those cases, climate affects ectoparasites. For example, the intensity of feather mites was positively related to temperature across an elevation gradient in five of six species of passerines in Spain (Meléndez et al. 2014). Thus, the effect of climate change along this altitudinal gradient may be affecting mite intensities on their hosts.

Several studies have linked the incidence and the prevalence of the parasitic diptera *Philornis downsi* with climatic events. Although the adult flies feed on fruit, they deposit their eggs in bird nests, and the resulting larvae feed on the blood and tissues of nestlings (O'Connor et al. 2010a), which causes high mortality in some bird species (O'Connor et al. 2010b). Dudaniec et al. (2007) found higher intensities of infection by this parasite in nests of Darwin's finches in a high rainfall year yet lower numbers of parasites in a dry year. However, the authors concluded that this effect is more likely due to an increase in clutch size during years with higher rainfall and not a direct effect of climate on the parasite. This conclusion is in agreement with the results by Koop et al. (2013), who reported that neither the prevalence nor the abundance of *P. downsi* decreased significantly in a dry year compared with an earlier wet year, thus concluding that this species is capable of withstanding the extreme climatic fluctuations characteristic of the Galápagos Islands. However, Antoniazzi et al. (2011) reported that high average maximum temperature and increased rainfall were significantly and positively correlated with mean *Philornis torquans* intensity in Argentinian birds.

In another study, Møller and collaborators (2013) analysed paired information on 89 parasite populations of 24 bird host species using data collected some years ago and again in 2010 with an average interval of 10 years between collection dates. The parasite taxa studied included protozoa, feather parasites, diptera, ticks, mites, and fleas. The study finds that non-dipteran parasites (blood-sucking fleas, ticks, and mites) were the only group that tended to decrease in prevalence between sampling years although differences with other groups were not statistically significant. However, a significant increase in temperature over time was detected, as was an increase in parasite prevalence and abundance. Nevertheless, these variables were not directly related to each other. This finding, and the fact that temperature can significantly affect host variables such as clutch size and body condition, which are also related to parasite infection, led the authors to conclude that the apparent effect of temperature on parasites was mediated through an effect on hosts.

In a recent study, Castaño et al. (2018) experimentally increased the temperature inside nestboxes of a small passerine, the blue tit *Cyanistes caeruleus*, and assessed the effects on both nest-dwelling ectoparasites and flying insect vectors. An average increase of 3°C produced an average reduction of 5.2 per cent in relative humidity within nestboxes. Although populations of the blow fly *Protocalliphora azurea* (based on pupae) and the mite *Dermanyssus gallinoides* were significantly reduced, other parasite populations were unaffected. The effect of temperature on humidity might have mediated these effects given that several arthropods are highly sensitive to water loss (Figure 14.2; Castaño et al. 2018).

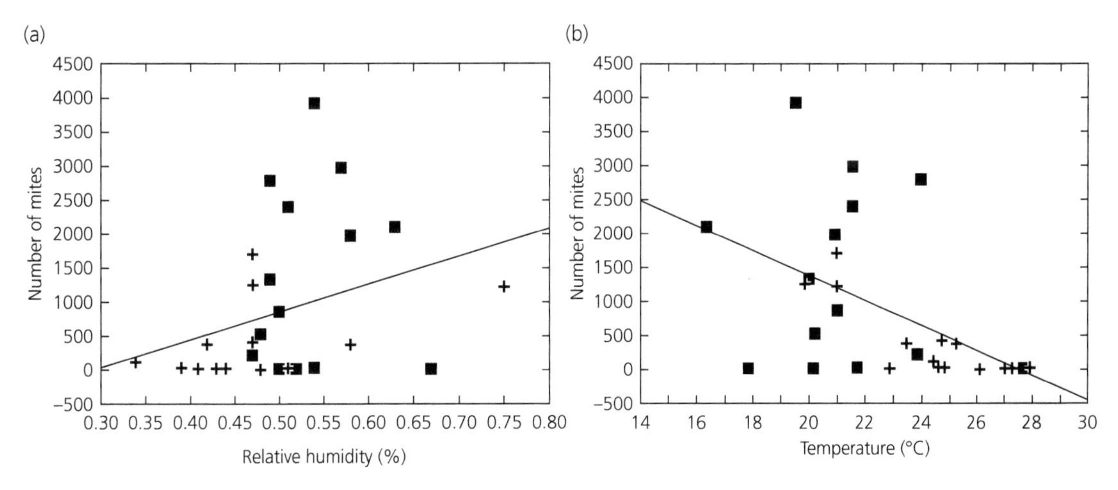

Figure 14.2 Mite numbers were related to temperature (r_s = −0.564, n = 28, P = 0.001) and relative humidity (r_s = 0.372, n = 28, P = 0.050) in blue tit (*Cyanistes caeruleus*) nests following a treatment that increased the temperature inside nests. Data from control (squares) and treated nests (crosses) are shown together. Data from Castaño et al. (2018).

In general, an increase in temperature does not necessarily increase the prevalence and the abundance of parasites. Rather, this is dependent on the association between temperature and other climatic characteristics such as rainfall and the type of parasite and its particular life cycle requirements. In addition, the effects of increased temperatures on hosts may prove to be more important for parasites than any direct effects of climate.

14.4 Range expansion and host–parasite interactions

One of the main predictions of models linking host–parasite interactions with climate change relates to range expansion. Based on a meta-analysis, Chen et al. (2011) estimated that terrestrial vertebrates are expanding their ranges between two and three times faster than previously reported. This study also estimated a distributional change to higher elevations at a median rate of 11.0 metres per decade and to higher latitudes at a median rate of 16.9 kilometres per decade following changes in climate. However, these changes varied considerably among species (Chen et al. 2011). Although parasites and pathogens typically follow hosts in their expansions, there is usually a lag in timing given their absence from host invasion fronts prior to expansion. This absence is likely due to serial founder events in low-density frontal host populations that produce local extinctions and hamper pathogen transmission. Thus, hosts at invasion fronts may temporarily reduce their investment in immune responses, which may facilitate their expansion into new territories. However, the parasites eventually catch up with their hosts in the recently colonized areas (Phillips et al. 2010). The case of birds is especially relevant for range expansions because birds are highly mobile and thus able to expand their range more easily compared to other vertebrates. Moreover, species with migratory behaviour and those introduced by humans are more likely to expand at a fast rate following climate change.

Some studies have directly or indirectly studied range expansions of bird hosts and their association with climate change. For example, in general, latitudinal and altitudinal changes in the range of avian malaria are expected. Avian malaria is already effectively transmitted in Alaska, and it is expected that the expansion of warmer temperatures due to climate change to more northern regions will extend the northern limit of its transmission, which will also allow the development of other vectors and parasites in those areas (Loiseau et al. 2012). Similar results were found for malaria transmission in France (Loiseau et al. 2013). Increases in temperature and rainfall are also expected to increase transmission of avian malaria at higher elevations, as has been predicted for Australia and Hawaii (Zamora-Vilchis et al. 2012; Liao et al. 2015).

A good example of birds potentially expanding the range of disease is the case of *Ixodes scapularis* ticks, the primary tick vector of the Lyme disease pathogen *Borrelia burgdorferi* in North America (as noted in section 14.2). The range expansion of *I. scapularis* in Canada has led to the emergence of Lyme disease in that country. Although temperature appears to be the most important determinant for tick population establishment, long-distance dispersal of ticks by migratory birds and local dispersal by resident hosts appear to be the main factors determining tick range expansion (Leighton et al. 2012). By analysing stable hydrogen isotopes of rectrices from birds carrying ticks, Ogden and collaborators (2015) were able to determine where birds breed and how far north they could migrate. Based on these findings, they concluded that migratory birds have the ability to extend the range of the ticks they carry farther north, even to regions that are presently climatically unsuitable for the establishment of ticks. Thus, under current climate change scenarios, it is expected that ticks will be able to extend their range to more northern regions in the future. In Europe, migrant passerines are known to transport *Ixodes ricinus* ticks infected by a variety of *Borrelia* species to more northern localities (Hasle et al. 2011). Moreover, a range expansion of this tick in some areas of Europe has been predicted on the basis of an ecological niche model under a climate change scenario. However, contractions in other areas have also been predicted (Boeckmann and Joyner 2014).

Birds are also considered the primary hosts and reservoirs responsible for spreading West Nile virus (WNV). High summer temperatures and heavy rainfall are also associated with WNV transmission. Since the introduction of this virus into the United States in 1999, WNV has been expanding rapidly throughout the country, and new associations with native vectors and birds have been developing (Darsie and Ward 2005; Tabachnick 2010). Furthermore, the summer droughts facilitate contact between birds and mosquitoes around the few remaining sources of water, with later rainfalls facilitating the spread of infected birds and mosquitoes. A subsequent period of drought again forces contact between hosts and vectors, thus amplifying the virus. Later rainfalls then allow transmission from infected vectors to humans and other host animals. Although the entry and establishment of WNV in the United States does not appear to be related to climate change, it could have a potential effect on the transmission of WNV (Tabachnick 2010). Other pathogens affecting migratory birds are expected to increase their geographic ranges as a consequence of climate change, although the mechanisms linking these effects are still far from clear, and more ecological studies are clearly needed (Fuller et al. 2012). Multiple factors can affect host-switching during range expansion and they merit further research. Distributional ecology (Estrada-Peña et al. 2014), ecological fitting (Agosta et al. 2010), and the biodiversity-buffers-disease paradigm (Civitello et al. 2015) are among those that still need to be explored in depth.

14.5 Climate change and changes in virulence

Virulence is considered the net effect of parasites on their hosts and, therefore, is a factor in host–parasite interactions (Toft and Karter 1990; Poulin and Combes 1999). In fact, a parasite can show different levels of virulence in different hosts. Thus, climate change may affect virulence when new host–parasite interactions emerge during range expansions. Climate warming may increase the virulence of parasites infecting newly arrived hosts in the expanded areas, or new parasites may be introduced to native hosts of these areas. In both cases, an increase in virulence is expected if warmer temperatures allow parasites to complete their life cycles more rapidly, thus increasing parasite density (Marcogliese 2001; see also section 16.3). Similar processes may also affect introduced species and their parasites (Rahel and Olden 2008). However, there are very few examples of increased virulence mediated by climate change in birds. As noted above, Møller et al. (2013) compared different populations of birds and their parasites across Europe at two different times and found that clutch and brood sizes and body condition were associated with changes in both temperature and parasitism. These authors suggest that these parasites, perhaps mediated through the indirect effects of temperature, are affecting the fecundity and body condition of their hosts.

Increased virulence can also be observed for West Nile virus (WNV) in migratory bird hosts (and mammals), in which new highly virulent strains have evolved following range expansion (Kilpatrick 2011). The virulence of these strains may be related to climate, as in the case of the WN02 strain, which evolved in North America and is more efficiently transmitted by mosquitoes. The difference in transmission of this strain increases with temperature under laboratory conditions, as expected if this strain replicates at a higher rate compared with other strains (Kilpatrick 2011; see section 14.4).

However, in some cases, lower virulence is observed as increases in temperature reduce parasite populations or their activity. For example, ticks find their hosts by climbing vegetation and adopting a sit-and-wait tactic, and this questing activity is reduced at elevated temperatures (Randolph 2009; Dantas-Torres and Otranto 2013).

14.6 Climate change and changes in anti-parasite defences of hosts

Adaptations by birds may emerge following changes in bird–parasite interactions due to the effects of climate change. Both increases and decreases in body temperature are known to affect immune responses against pathogens. However, prior acclimation of individuals to thermal conditions removes the immunosuppressive effect of stress (Martínez and Merino 2011). In any case, the association between thermal stress and immunosuppression leading to a disease episode or an increase in susceptibility has rarely been reported, although several studies have linked the onset of epidemic outbreaks to extreme fluctuations of heat or cold (Travers et al. 2008; Wegner et al. 2008; McClanahan et al. 2009).

One example in birds is a study of the effect of temperature along an altitudinal gradient on both blood parasite infections and major histocompatibility complex (MHC) variation (Zamora-Vilchis et al. 2013). Although parasite intensity was related to both temperature and MHC variability, the authors did not find a relationship between these two factors, concluding that the relationship between immunity and temperature is indirect and mediated by parasite abundance. Birds have defensive behaviours against parasites that may increase or decrease in function, depending on the level of parasite pressure, a factor that can be affected by climate change. For example, female grooming and nest sanitation activities decrease when ectoparasite infestation was experimentally reduced in nests of pied flycatchers (*Ficedula hypoleuca*) (Cantarero et al. 2013). Thus, increases in the abundance of parasites due to climate change will likely affect the amount of time birds devote to hygienic activities.

14.7 Climate change and changes in coevolutionary interactions

Changes in host–parasite interactions may result from processes in which parasites 'missing the boat' offer some advantages to hosts that expand into new areas (MacLeod et al. 2010). However, some diseases adapt easily to new conditions, thus establishing new host–parasite interactions that allow propagation of diseases. An example of this is the aforementioned case of West Nile virus, which has infected new mosquito species in North America and adapted to bird species in that continent (Tabachnick 2010). Other potential examples of parasites adapting to new conditions are migratory cuckoos in Europe that adapt to species available for nest parasitism (Møller et al. 2011) or the effect of *Philornis* parasitism on birds in the Galápagos Islands (Dudaniec et al. 2007). In this respect, generalist parasites introduced into new areas/hosts favoured by climate change may have an advantage, although ecological fitting, the process whereby organisms (in this case parasites) form novel associations with other species (in this case hosts) as a result of the suites of traits that they carry at the time they encounter the novel condition, should also be taken into account (Agosta et al. 2010). Furthermore, the capacity of parasites to adapt to different temperatures may also be the difference between failure and success in establishing a new host (see also Carlson et al. 2017).

14.8 Effects of climate change on bird–parasite interactions: future directions

Birds attract our attention due to their colourful plumages and songs and the ease by which they can be observed. Some of these reasons also contribute to why they are objects of intense research. The contrary can be said of most parasitic species; however, diseases that affect beautiful birds would also be expected to attract great scientific attention. In this chapter, I offer a summary of research advances on bird–parasite interactions related to climate change since the publication by Merino and Møller in 2010. Much of the research attempting to understand how climate change will affect birds and their parasites has centred on certain groups of pathogens, especially zoonotic species that can affect humans. Thus, the majority of the studies that I found that link or potentially link climate change with migratory birds relate to the tick vectors of Lyme disease or the spread of the West Nile virus. Another disease attracting great attention is avian malaria, probably for its parallels with human malaria. Even for lesser studied parasites, their potential direct effects on humans are the main cause for their study as, for example, in the case of cercarial infections in waterbirds.

It is logical that all of these studies attract our attention. However, more studies on the effect of climate change on the expansion, prevalence, intensity, virulence, and coevolutionary relationships of bird parasites are required. Such studies will also be important for humans because birds and parasites are major components of ecosystems (Kuris et al. 2008).

Most studies use data collected in the field to help generate models, including those that make predictions of how changes in temperature may alter future bird–parasite interactions. However, there is still a lack of experimental approaches that produce empirical data to improve these models. The possibility to study bird–parasite interactions under different climatic scenarios also exists, which may provide a better understanding of how these interactions vary with climate effects. The introduction of other variables along with climatic ones in models of bird–parasite relationships has clearly improved our understanding and has shown that even closely related parasites can respond in different ways to the same environmental factors. For example, Pérez-Rodríguez et al. (2013) studied avian blood parasites of different genera in several Iberian populations of blackcap *Sylvia atricapilla* and assessed the relative importance of several potential determinants of variation in the diversity of these parasites, including climate, landscape features, and host population migration. They found that prevalence and richness of parasites were predominantly related to climate, but also to landscape features and host migration. In addition, they found that each parasite genus was affected in different ways by different variables even when infecting the same host species.

Another promising field of study for bird–parasite–climate research in coming years will revolve around invasive species, including not only those introduced by humans but also those that have extended their ranges. There are very few studies of the potential effect of climate and parasitism on the success of invasive species. However, there is little knowledge about how the expansion of non-native species under a climate change scenario may be affected if these species were freed from their common parasites, or if the potential introduction of pathogens of invasive species may be favoured by climate change. The potential response of bird hosts to new infections is also difficult to predict. Therefore, future research should also focus on whether birds can adapt to new challenges produced by the arrival and the adaptation of new pathogens due to climate change. To date, very few studies have investigated the development of resistance in birds challenged by new diseases (e.g. Atkinson et al. 2013). However, the survival of several bird species will likely depend on their capacity to adapt to the presence of novel pathogens.

Acknowledgements

I want to thank Robert Poulin for his comments and suggestions, which considerably improved this chapter. My work is supported by project CGL2015-67789-C2-1-P from the Spanish Ministry of Economy, Industry, and Competitiveness.

References

Agosta, S.J., Niklas, J., and Brooks, D.R. (2010). How specialists can be generalists: Resolving the 'Parasite Paradox' and implications for emerging infectious disease. *Zoologia*, 27, 151–62.

Antoniazzi, L.R., Manzoli, D.E., Rohrmann, D., Saravia, M.J., Silvestri, L., and Beldomenico, P.M. (2011). Climate variability affects the impact of parasitic flies on Argentinean forest birds. *Journal of Zoology*, 283, 126–34.

Atkinson, C.T., Saili, K.S., Utzurrum, R.B., and Jarvi, S.I. (2013). Experimental evidence for evolved tolerance to avian malaria in a wild population of low elevation Hawai'i 'amakihi (*Hemignathus virens*). *EcoHealth*, 10, 366–75. doi: 10.1007/s10393-013-0899-2

Avilés, J.M., Molina-Morales, M., and Martínez, J.G. (2014). Climatic effects and phenological mismatch in cuckoo–host interactions: a role for host phenotypic plasticity in laying date? *Oikos*, 123, 993–1002. doi: 10.1111/oik.01124

Boeckmann, M., and Joyner, T.A. (2014). Old health risks in new places? An ecological niche model for *I. ricinus* tick distribution in Europe under a changing climate. *Health Place*, 30, 70–7.

Cantarero, A., López-Arrabé, J., Alberto, J., Redondo, A.J., and Moreno, J. (2013). Behavioural responses to ectoparasites in pied flycatchers *Ficedula hypoleuca*: an experimental study. *Journal of Avian Biology*, 44, 591–9.

Carlson, C.J., Burgio, K.R., Dougherty, E.R., Phillips, A.J., Bueno, V.M., Clements, C.F., Castaldo, G., et al. (2017). Parasite biodiversity faces extinction and redistribution in a changing climate. *Science Advances*, 3, e1602422. doi: 10.1126/sciadv.1602422

Castaño, F., Martínez, J., Lozano, M., and Merino, S. (2018). Experimental manipulation of temperature reduces ectoparasites in nests of blue tits (*Cyanistes caeruleus*). *Journal of Avian Biology*, 49, e01695. doi: 10.1111/jav.01695

Chen, I.-C., Hill, J.K., Ohlemüller, R., Roy, D.B., and Thomas, C.D. (2011). Rapid range shifts of species associated with high levels of climate warming. *Science*, 133, 1024–6. doi: 10.1126/science.1206432

Civitello, D.J., Cohen, J., Fatima, H., Halstead, N., Liriano, J., McMahon, T.A., Ortega, C.N., et al. (2015). Biodiversity inhibits parasites: broad evidence for the dilution effect. *Proceedings of the National Academy of Sciences of the United States of America*, 112, 8667–71.

Dantas-Torres, F., and Otranto, D. (2013). Seasonal dynamics of *Ixodes ricinus* on ground level and higher vegetation in a preserved wooded area in southern Europe. *Veterinary Parasitology*, 192, 253–8.

Darsie, R.F., Jr., and Ward, R.A. (2005). *Identification and geographical distribution of the mosquitoes of North America, north of Mexico*. University Press of Florida State University System, Gainesville, FL, USA.

Dudaniec, R., Fessl, B., and Kleindorfer, S. (2007). Interannual and interspecific variation in intensity of the parasitic fly, *Philornis downsi*, in Darwin's finches. *Biological Conservation*, 139, 325–32.

Estrada-Peña, A., Ostfeld, R.S., Peterson, A.T., Poulin, R., and de la Fuente, J. (2014). Effects of environmental change on zoonotic disease risk: an ecological primer. *Trends in Parasitology*, 30, 205–14.

Fuller, T., Bensch, S., Müller, I., Novembre, J., Pérez-Tris, J., Ricklefs, R.E., Smith, T.B., and Waldenström, J. (2012). The ecology of emerging infectious diseases in migratory birds: an assessment of the role of climate change and priorities for future research. *Ecohealth*, 9, 80–8. doi: 10.1007/s10393-012-0750-1

Garamszegi, L.Z. (2011). Climate change increases the risk of malaria in birds. *Global Change Biology*, 17, 1751–9.

González-Quevedo, C., Davies, R.G., and Richardson, D.S. (2014). Predictors of malaria infection in a wild bird population: landscape-level analyses reveal climatic and anthropogenic factors. *Journal of Animal Ecology*, 83, 1091–102.

Guillemain, M., Pöysä, H., Fox, A.D., Arzel, C., Dessborn, L., Ekroos, J., Gunnarsson, G., et al. (2013). Effects of climate change on European ducks: what do we know and what do we need to know? *Wildlife Biology*, 19, 404–19.

Harvell, C.D., Mitchell, C.E., Ward, J.R., Altizer, S., Dobson, A.P., Ostfeld, R.S., and Samuel, M.D. (2002). Climate warming and disease risks for terrestrial and marine biota. *Science*, 296, 2158–62.

Hasle, G., Bjune, G.A., Midthjell, L., Røed, K.H., and Leinaas, H.P. (2011). Transport of *Ixodes ricinus* infected with *Borrelia* species to Norway by northward-migrating passerine birds. *Ticks and Tick Borne Diseases*, 2, 37–43. doi: 10.1016/j.ttbdis.2010.10.004

Horak, P., Kolarova, L., and Adema, C. (2000). Biology of the schistosome genus *Trichobilharzia*. *Advances in Parasitology*, 52, 155–233.

Kilpatrick, A.M. (2011). Globalization, land use, and the invasion of West Nile Virus. *Science*, 334, 323–7. doi: 10.1126/science.1201010

Koop, J.A.H., Le Bohec, C., and Clayton, D.H. (2013). Dry year does not reduce invasive parasitic fly prevalence or abundance in Darwin's finch nests. *Reports in Parasitology*, 3, 11–17.

Kuris, A.M., Hechinger, R.F., Shaw, J.C., Whitney, K.L., Aguirre-Macedo, L., Boch, C.A., Dobson, A.P., et al. (2008). Ecosystem energetic implications of parasite and free-living biomass in three estuaries. *Nature*, 454, 515–18. doi: 10.1038/nature06970

Kutz, S.J., Hoberg, E.P., Polley, L., and Jenkins, E.J. (2005). Global warming is changing the dynamics of Arctic host–parasite systems. *Proceedings of the Royal Society of London Series B: Biological Sciences*, 272, 2571–6.

Lafferty, K.D. (2009). The ecology of climate change and infectious diseases. *Ecology*, 90, 888–900.

Leighton, P.A., Koffi, J.K., Pelcat, Y., Lindsay, L.R. and Ogden, N.H. (2012). Predicting the speed of tick invasion: an empirical model of range expansion for the Lyme disease vector *Ixodes scapularis* in Canada. *Journal of Applied Ecology*, 49, 457–64. doi: 10.1111/j.1365–2664.2012.02112.x

Levi, T., Keesing, F., Oggenfuss, K., and Ostfeld, R.S. (2015). Accelerated phenology of blacklegged ticks under climate warming. *Philosophical Transactions of the Royal Society of London Series B: Biological Sciences*, 370, 20130556. http://dx.doi.org/10.1098/rstb.2013.0556

Lévine, N.D., Clark, D.T., and Hanon, L.E. (1956). Encephalitis in a swan due to *Dendritobilharzia* sp. (Trematoda; Schistosomatidae). *Journal of Parasitology*, 42, 496–500.

Liao, W., Timm, O.E, Zhang, C., Atkinson, C.T., LaPointe, D.A., and Samuel, M.D. (2015). Will a warmer and wetter future cause extinction of native Hawaiian Forest birds? *Global Change Biology*, 21, 4342–52. doi: 10.1111/gcb.13005

Loiseau, C., Harrigan, R.J., Bichet, C., Julliard, R., Garnier, S., Lendvai, A.Z., Chastel, O., and Sorci, G. (2013). Predictions of avian *Plasmodium* expansion under climate change. *Scientific Reports*, 3, 1126.

Loiseau, C., Harrigan, R.J., Cornel, A.J., Guers, S.L., Dodge, M., Marzec, T., Carlson, J.S., et al. (2012). First evidence and predictions of *Plasmodium* transmission in Alaskan bird populations. *PLoS One*, 7, e44729. doi:10.1371/journal.pone.0044729

Lymbery, A.J., Morine, M., Kanani, H.G., Beatty, S.J., and Morgan, D.L. (2014). Co-invaders: The effects of alien parasites on native hosts. *International Journal for Parasitology: Parasites and Wildlife*, 3, 171–7.

Marcogliese, D.J. (2001). Implications of climate change for parasitism of animals in the aquatic environment. *Canadian Journal of Zoology*, 79, 1331–52.

MacLeod, C.J., Paterson, A.M., Tompkins, D.M., and Duncan, R.P. (2010). Parasites lost—do invaders miss the boat or drown on arrival? *Ecology Letters*, 13, 516–27. doi: 10.1111/j.1461–0248.2010.01446.x

Martínez, J., and Merino, S. (2011). Host–parasite interactions under extreme climatic conditions. *Current Zoology*, 57, 390–405.

Marzal, A., Ricklefs, R.E., Valkiunas, G., Albayrak, T., Arrier, E., et al. (2011). Diversity, loss, and gain of malaria parasites

in a globally invasive bird. *PLoS One*, 6, e21905. doi:10.1371/journal.pone.0021905
Mas-Coma, S., Valero, M.A., and Bargues, M.D. (2008). Effects of climate change on animal and zoonotic helminthiases. *Revue scientifique et technique (International Office of Epizootics)*, 27, 443–52.
Mas-Coma, S., Valero, M.A., and Bargues, M.D. (2009). Climate change effects on trematodiases, with emphasis on zoonotic fascioliasis and schistosomiasis. *Veterinary Parasitology*, 163, 264–80.
McClanahan, T.R., Weil, E., and Maina, J. (2009). Strong relationship between coral bleaching and growth anomalies in massive Porites. *Global Change Biology*, 15, 1804–16.
Meléndez, L., Laiolo, P., Mironov, S., García, M., Magaña, O., and Jovani, R. (2014). Climate-driven variation in the intensity of a host–symbiont animal interaction along a broad elevation gradient. *PLoS One*, 9, e101942. doi:10.1371/journal.pone.0101942
Merino, S., and Møller, A.P. (2010). Host–parasite interactions and climate change. In: A.P. Møller, W. Fiedler, and P. Berthold (eds), *Effects of Climate Change on Birds*, Chapter 15. Oxford University Press, Oxford, United Kingdom.
Møller, A.P., Merino, S., Soler, J.J., Antonov, A., Badás, E.P., Calero-Torralbo, M.A., De Lope, F., et al. (2013). Assessing the effects of climate on host–parasite interactions: A comparative study of European birds and their parasites. *PLoS One*, 8, e82886. doi:10.1371/journal.pone.0082886
Møller, A.P., Saino, N., Adamík, P., Ambrosini, R., Antonov, A., Campobello, D., Stokke, B.G., et al. (2011). Rapid change in host use of the common cuckoo *Cuculus canorus* linked to climate change. *Proceedings of the Royal Society of London B: Biological Sciences*, 278, 733–8. doi:10.1098/rspb.2010.1592
Morley, N.J., and Lewis, J.W. (2014). Temperature stress and parasitism of endothermic hosts under climate change. *Trends in Parasitology*, 30, 221–7. doi: 10.1016/j.pt.2014.01.007
Murdock, C.C., Foufopoulos, J., and Simon, C.P. (2013). A transmission model for the ecology of an avian blood parasite in a temperate ecosystem. *PLoS One*, 8, e76126.
Newman, E.A., Eisen, L., Eisen, R.J., Fedorova, N., Hasty, J.M., Vaughn, C., and Lane, R.S. (2015). *Borrelia burgdorferi* sensu lato spirochetes in wild birds in Northwestern California: Associations with ecological factors, bird behavior and tick infestation. *PLoS One*, 10, e0118146. doi:10.1371/journal.pone.0118146
Niebuhr, C.N., Poulin, R., and Tompkins, D.M. (2016). Is avian malaria playing a role in native bird declines in New Zealand? Testing hypotheses along an elevational gradient. *PLoS One*, 11, e0165918.
O'Connor, J.A., Robertson, J., and Kleindorfer, S. (2010a). Video analysis of host–parasite interactions in nests of Darwin's finches. *Oryx*, 44, 588–94. doi:10.1017/S0030605310000086
O'Connor, J.A., Sulloway, F.J., Robertson, J., and Kleindorfer, S. (2010b). *Philornis downsi* parasitism is the primary cause of nestling mortality in the critically endangered Darwin's medium tree finch (*Camarhynchus pauper*). *Biodiversity and Conservation*, 19, 853–66.
Ogden, N.H., Bigras-Poulin, M., Hanincová, K., Maarouf, A., O'Callaghan, C.J., and Kurtenbach, K. (2008). Projected effects of climate change on tick phenology and fitness of pathogens transmitted by the North American tick *Ixodes scapularis*. *Journal of Theoretical Biology*, 254, 621–32.
Ogden, N.H., Barker, I.K., Francis, C.M., Heagy, A., Lindsay, L.R., and Hobson, K.A. (2015). How far north are migrant birds transporting the tick *Ixodes scapularis* in Canada? Insights from stable hydrogen isotope analyses of feathers. *Ticks and Tick Borne Diseases*, 6,715–20. doi:10.1016/j.ttbdis.2015.06.004
Pérez-Rodríguez, A., Fernández-González, S., de la Hera, I., and Pérez-Tris, J. (2013). Finding the appropriate variables to model the distribution of vector-borne parasites with different environment preferences: climate is not enough. *Global Change Biology*, 19, 3245–53.
Phillips, B.L., Kelehear, C., Pizzatto, L., Brown, G.P., Barton, D., and Shine, R. (2010). Parasites and pathogens lag behind their host during periods of host range advance. *Ecology*, 91, 872–81.
Polley, L., and Thompson, R.C.A. (2009). Parasite zoonoses and climate change: molecular tools for tracking shifting boundaries. *Trends in Parasitology*, 25, 285–91.
Poulin, R. (2006). Global warming and temperature-mediated increases in cercarial emergence in trematode parasites. *Parasitology*, 132, 143–51.
Poulin, R., and Combes, C. (1999). The concept of virulence: interpretations and implications. *Parasitology Today*, 15, 474–5.
Rahel, F.J., and Olden, J.D. (2008). Assessing the effects of climate change on aquatic invasive species. *Conservation Biology*, 22, 521–33. doi: 10.1111/j.1523–1739.2008.00950.x
Randolph, S. (2009). Epidemiological consequences of the ecological physiology of ticks. *Advances in Insect Physiology*, 37, 297–339.
Rohr, J.R., Dobson, A.P., Johnson, P.T., Kilpatrick, A.M., Paull, S.H., Raffel, T.R., Ruiz-Moreno, D., and Thomas, M.B. (2011). Frontiers in climate change-disease research. *Trends in Ecology & Evolution*, 26, 270–7. doi:10.1016/j.tree.2011.03.002
Saino, N., Rubolini, D., Lehikoinen, E., Sokolov, L.V., Bonisoli-Alquati, A., Ambrosini, R., Boncoraglio, G., and Møller, A.P. (2009). Climate change effects on migration

phenology may mismatch brood parasitic cuckoos and their hosts. *Biology Letters*, 5, 539–41. doi: 10.1098/rsbl.2009.0312

Sehgal, R. (2015). Manifold habitat effects on the prevalence and diversity of avian blood parasites. *International Journal for Parasitology: Parasites and Wildlife*, 4, 421–30.

Sehgal, R., Buermann, W., Harrigan, R., Bonneaud, C., Loiseau, C., Chasar, A., Sepil, I., et al. (2011). Spatially explicit predictions of blood parasites in a widely distributed African rainforest bird. *Proceedings of the Royal Society of London B: Biological Sciences*, 278, 1025e1033.

Sutherst, R.W. (2001). The vulnerability of animal and human health to parasites under global change. *International Journal of Parasitology*, 31, 933–48.

Tabachnick, W.J. (2010). Challenges in predicting climate and environmental effects on vector-borne disease episystems in a changing world. *Journal of Experimental Biology*, 213, 946–54. doi: 10.1242/jeb.037564

Toft, C.A., and Karter, A.J. (1990). Parasite–host coevolution. *Trends in Ecology and Evolution*, 5, 326–9. doi: 10.1016/0169-5347(90)90179-H

Travers, M.A., Le Goïc, N., Huchette, S., Koken, M., and Paillard, C. (2008). Summer immune depression associated with increased susceptibility of the European abalone *Haliotis tuberculata* to *Vibrio harveyi* infection. *Fish and Shellfish Immunology*, 25, 800–8.

Wegner, K.M., Kalbe, M., Milinski, M., and Reusch, T.B.H. (2008). Mortality selection during the 2003 European heat wave in three-spined sticklebacks: Effects of parasites and MHC genotype. *BMC Evolutionary Biology*, 8, 1–12.

Zamora-Vilchis, I., Esparza-Salas, R., Johnson, C.N., Williams, S.E., and Endler, J.A. (2013). Parasite-mediated diversity and selection of MHC genes of birds distributed along an altitudinal gradient: implications for disease impact in a warming climate. In: I. Zamora-Vilchis, *Host–parasite interactions: bird immune genes, blood parasites and climate change implications.* PhD thesis, James Cook University, Queensland, Australia.

Zamora-Vilchis, I., Williams, S.E., and Johnson C.N. (2012). Environmental temperature affects prevalence of blood parasites of birds on an elevation gradient: implications for disease in a warming climate. *PLoS One*, 7, e39208. doi:10.1371/journal.pone.0039208

CHAPTER 15

Predator–prey interactions and climate change

Vincent Bretagnolle and Julien Terraube

15.1 Introduction: Climate change and trophic networks

Impacts of climate change are observed broadly across genes, species, and ecosystems (Bellard et al. 2012; Scheffers et al. 2016; MacLean and Beissinger 2017). Global temperatures have already increased by up to 1.2°C since preindustrial times. Most predictions qualitatively agree that global warming will cause species extinctions and increase disease transmission (Harvell et al. 2002; Chapters 15, 18). Birds have advanced their breeding or migration phenology (Visser et al. 2004; Kristensen et al. 2015; Møller et al. 2006; Chapter 11), while northward density-shifts in birds have been observed in Fennoscandia (Lehikoinen and Virkkala 2016). Climate change is likely to impact all trophic levels (Chambers et al. 2005), although the response of communities and ecosystems to climate change has only recently received its deserved attention (Lurgi et al. 2012; Pearce-Higgins et al. 2015; Møller et al. 2018; Beaugrand and Kirby 2018). Multispecies interaction networks including predation, parasitism, and pollination (Seibold et al. 2018) cannot be assumed to be linear interactions. Species interact with many others, while these interactions may differ in sign, being either positive (i.e., facilitation) or negative (inhibition). Interaction strength is also functionally important with regard to ecosystem processes, because it encompasses transfer of energy through an ecosystem, a particularly important feature for community stability (Paine 1980). Climate change is further expected to affect the magnitude of species interactions themselves (Tylianakis et al. 2008; Cahill et al. 2013; Rosenblatt and Schmitz 2016). Biotic interactions are known to play a major role in the maintenance of ecosystems (Bascompte et al., 2006). Climate change will affect species more through shifts in ecosystem functioning (e.g., pollination, mutualism, parasitism) than by its direct effect (Rand and Tscharntke, 2007; Tylianakis et al. 2008). Furthermore, complex networks of biotic interactions include compensatory mechanisms (Brown et al. 2001) that may buffer the effects of climate change on species (Suttle et al. 2007), but their complexity may also amplify the effects of climate change.

Analysing a single variable (often temperature) for a single life stage of a few selected species may not be satisfactory, therefore, for understanding and predicting the consequences of climate change on the abundance and distribution of organisms (Gilman et al. 2010; Iknayan and Beissinger 2018). A first caveat was revealed by the simple fact that documented advances in breeding phenology of predators may not be in phase with those of their prey (Visser et al. 2004). Examples of differences in phase among trophic levels come from birds feeding on insects (Visser et al. 1998; Pearce-Higgins et al. 2005; see also Devictor et al. 2012), from zooplankton and phytoplankton abundances that have changed at

Bretagnolle, V., and Terraube, J., *Predator–prey interactions and climate change*. In: *Effects of Climate Change on Birds*. Second Edition. Edited by Peter O. Dunn and Anders Pape Møller: Oxford University Press (2019). © Oxford University Press.
DOI: 10.1093/oso/9780198824268.003.0015

different rates (Winder and Schindler 2004), and from animals and plants that have responded heterogeneously to climate change (Nielsen and Møller 2006; Both et al. 2009; Renner and Zohner 2018). A second caveat consists in viewing community response to climate change as the sum of the responses of each individual species (Böhning-Gaese and Lemoine 2004; but see Post and Forchhammer 2002, Devictor et al. 2008), thus considering communities as a 'super organism' with its own climate envelopes (Hole et al. 2009; Wilmers et al. 2007; Chapter 16). Indeed, changes in geographic ranges in response to climate change are known to vary in magnitude between taxa belonging to different trophic levels, or to different taxonomic ranks (Huntley et al. 2004), as species ranges result not only from the direct physiological effects of climate (Porter et al. 2000), but also from indirect effects (Pincebourde et al. 2008). It is reasonable to assume that climate change will affect the temporal and spatial association between interacting species at different trophic levels (Böhning-Gaese and Lemoine 2004), because species often show their own response to temperature (climatic niche envelopes). Lastly, range shifts may have cascading effects on community structure and the functioning of ecosystems (Lovejoy and Hannah 2005; Terraube and Bretagnolle 2018). It is thus extremely likely that the impacts of species range shifts will go far beyond those arising from just adding species to or subtracting species from ecosystems. Such changes may produce trophic cascades or lead to ecological catastrophes such as ecosystem phase shifts (Ainley et al. 2015; Ripple et al. 2014). How such changes will resonate at the ecosystem level is particularly difficult to predict and surely represents a challenge for ecologists. Therefore, it is becoming clear that the response at the community level can be expected to differ from species level responses, and the response of one species can hardly be used to predict the response of another, in particular if the latter belongs to another trophic level.

If we are to understand and predict community or species assemblage responses to environmental variability, we must explicitly use a mechanistic approach including species interactions (Voigt et al. 2007). Understanding how climate change will influence the structure of communities and ecosystems has thus become a major preoccupation for both scientists and managers (Garcia et al. 2014). At the population level, climate change may have important consequences for population regulation, which raises the old question of the respective roles of biotic versus abiotic factors in shaping the regulation of populations and the structure of ecosystems (Martin 2001). The observed effects of climate change on distributions of species has led to the suggestion that biotic factors may be less important than abiotic factors (e.g., temperature), although perhaps more in plants than in animals (Austin 2002).

In this chapter, we first summarize why and how climate change could affect predator–prey interactions. Second, we review the literature about the impact of climate change on predator–prey relationships in birds. A final section will provide prospects for future studies.

15.2 Why and how climate change may affect predator–prey interactions

15.2.1 Predator–prey relationships and other species interactions

The effects of climate change on host–parasite interactions (Møller 2009; Chapter 15) or plant–herbivore interactions appear to be heterogeneous (McCluney et al. 2012, review in Tylianakis et al. 2008). Climate change often disrupts the synchrony of host–parasitoid phenologies and distributions, phenologically isolating emerging parasitoids from host eggs, which, in the absence of alternative hosts, could lead to localized extinctions, potentially releasing host species from parasitism (Wetherington et al. 2017). Among species interactions, predator–prey interactions are of paramount importance both at population and community levels, because predation is a major cause of mortality in animals, and it is thus a key process in animal population dynamics (Murdoch et al. 2003) and the evolution of life history traits (Doligez and Clobert 2003). Climate change may affect predator–prey interactions through changes in prey or predator abundances, the process of predation itself (including defence against predators), and at the community level through trophic cascades and regime shifts. Parasites, interacting with climate change, may also have significant impacts on predator–prey interactions. For example, increased temperatures and parasite infection triggered higher prey consumption rates in invasive amphipods, highlighting the potential synergy between parasite infection and temperature and the increased ecological impact inflicted on native communities (Laverty et al. 2017). Much like plant–herbivore relationships, predator–prey interactions can be altered by phenological changes (Visser et al. 2004; Thackeray et al. 2010). It is further common that these two latter interactions actually interact themselves, and often the responses to rising temperature for plants, herbivores, and predators have effects on phenology (either positive or negative), resulting in strong effects on both herbivore and predator population sizes (Durant et al. 2007). However, predators at higher trophic levels are usually disproportionately affected by environmental perturbation compared to plant–herbivore interactions, be it climate change, competition from invasive species, or habitat modification in general (Voigt et al. 2007). Climate change could affect all species interactions in a community, however; given the key functional role of predators, the effects of climate change on community structure and functioning may be driven by the effects of climate change on predators.

15.2.2 Processes and mechanisms by which global warming may affect predator–prey interactions

Climate parameters have been shown to affect predator–prey relationships: examples include the Moran effect (synchronizing the pattern of population fluctuations in space; Ranta et al. 1999), or large scale climatic fluctuations such as ENSO or NAO (Stenseth et al. 2002). For example, the interactions between snowshoe hare (*Lepus americanus*) and lynx (*Lynx canadensis*) vary geographically, with regions spatially defined by the influence of NAO (Stenseth et al. 1999). Therefore, it is not surprising that as climate changes, predator–prey interactions are also affected (Bastille-Rousseau et al. 2018). Climate may affect species directly either through a change in life history parameters (such as adult survival or fecundity) or a shift in geographic range or phenology, or indirectly through the food (prey) of the predator (affecting its abundance: Both et al. 2006), or through the control exerted by the predator on prey (Stevens et al. 2002; Peach et al. 2004). Disentangling the relative contributions of trophic (indirect) and direct climate effects is critical if we are to understand, and more importantly to predict, climate-driven effects at the community (or species interaction) level.

Typically, predator–prey dynamics are described by quantifying changes in the abundance of prey populations as a consequence of direct consumption by predators (the functional response), and the resulting changes in predator abundance as a consequence of energy transfer from prey to predator breeding (the numerical response; Abrams and Ginzburg 2000). The modulating effect of climate on this dynamic has recently received considerable attention. In particular, climate affects consumption rates (Stenseth et al. 2005; Vucic-Pestic et al. 2011), and thus the dynamics of predator–prey interactions (Post and Stenseth 1999; Stenseth et al. 1999). Climatic effects may be as important as top-down or bottom-up effects in shaping predator–prey dynamics (Vucetich and Peterson 2004). In addition, more subtle changes may be expected, as predators also have non-lethal effects on their prey (Lima 1998; Peckarsky et al. 2008) that may also be affected by climate change.

Expected changes on prey or predators may include the following:

(1) **Changes in distribution**. Changes in distribution may have complex consequences on encounter rate, because both prey and predator, or only one of these, may show a shift in distribution range. The spatial shift may, or may not, be associated with a temporal shift (latitudinal or altitudinal gradient).
(2) **Changes in phenology**. Depending on whether, and to which extent, temperature is a direct cue of processes affecting the population dynamics of the prey and/or the predator, climate change may result in either a temporal mismatch, or a closer match between predator and prey. Possibly as a consequence of differential changes in species' geographic distribution, or through a change in environmental conditions, several processes could alter the timing of predation

events (e.g., prey and predator encounters), which could have strong direct ecological implications, or indirect behavioural consequences (through non-lethal effects). Because temporal and spatial shifts, though being caused by different climate proxies, will result in patterns of predator–prey interaction that are similar, they will be dealt with simultaneously.

(3) **Changes in population density**. Population sizes and densities of either predator or prey may change as a consequence of the direct effects of environmental gradual change (temperature rising, changes in precipitation patterns) or extreme environments (catastrophic events). These changes will occur through climate impacts on vital rates (i.e. survival, breeding success, and dispersal rates). In response, population sizes may change as predicted by classical deterministic models (Murdoch et al. 2003). The direction and magnitude of these changes may differ depending on the shapes of functional and numerical responses (Abrams and Ginzburg 2000), thus varying with the degree of specialization by the predator and the efficiency of energy transfer. Therefore, this will affect the ratio between numbers of predators and prey. Yet, this ratio is of considerable interest here, as a ratio-dependent functional response has been suggested to be more sensitive to the interaction than the Holling functional response (Abrams and Ginzburg, 2000).

(4) **Changes in behaviour, morphology, or physiology**. These are more subtle changes that are nevertheless expected to be induced by climate change, either through direct effects of climate on predator foraging behaviour or predator avoidance behaviour in prey species, phenotypic plasticity, or microevolution. For instance, diurnal activity patterns of predation behaviour may be affected by climate change, with predators or prey having to cool themselves in particular habitats, or through energetic consequences for individuals. As environmental conditions change, populations may respond by progressively changing their morphological characters (e.g., body size), either due to energetic constraints or physiology. Below we detail these mechanisms.

15.2.3 Spatio-temporal effects of climate change on predator–prey interactions

Climate change has the potential to affect the spatial and temporal coincidence of organisms, both potentially leading to disrupted synchrony between predators and prey. Overall, there has been a rapidly growing body of literature exploring how disrupted synchrony affects plant–herbivore interactions (e.g., Bale et al. 2002) and predator–prey interactions (e.g., Logan et al. 2006). While changes in prey reproduction could potentially change predator reproduction, studies have indicated that predators (mainly those preying on insects) become increasingly mistimed in terms of their reproduction relative to the timing of reproduction of their prey (Nielsen and Møller 2006; Both et al. 2009; Chapter 11), resulting in predator breeding cycles becoming mismatched with seasonal peaks in prey availability, with potential consequences in terms of fitness. Although the evidence remains scarce, climate change affects, more or less rapidly, plants, insects, and birds, because of their differential reaction to photoperiod and temperature (reviews in Both et al. 2009; Devictor et al. 2012; Chapters 9, 10). Both et al. (2009) found in a three-trophic level study (caterpillar, four species of passerines that prey upon caterpillars, and a raptor that preys upon the four passerines), that caterpillar phenological response to temperature rising through budburst was on average delayed only 0.25 days, while passerines lagged 0.5 days after the peak in caterpillar abundance. The Eurasian sparrowhawk (*Accipiter nisus*, hereafter, sparrowhawk) had the largest mismatch of all species considered, which doubled over the 20-year study period.

Despite many studies having investigated how climate change has affected and will affect species distribution ranges, very few of them have specifically examined how differential shifts in range may affect predator–prey interactions (but see Peers et al. 2014). It remains unknown whether distribution shifts are more pronounced in predators than in prey, although this is unlikely since many birds (such as passerines in temperate communities) are both predators (of insects) and prey (of ornithophagous raptors). However, it has often been suggested that body size may affect the shift in range because larger-bodied species may be more resistant to adverse climate and thus less prone to move (Brommer 2008), although some studies maintained that larger species may have a poorer ability to respond to environmental changes than smaller species (Stevenson and Bryant 2000). Raptors eating birds, reptiles, and small mammals are, on average, much larger than their prey, so one may expect that raptors are less susceptible than passerines to being affected by climate change. While most studies on birds have focused on small-bodied species (see section 15.3), recent evidence from raptors suggests that large-bodied species are as affected by climate change as smaller ones, with delays in breeding events or movement shifts towards the north similar to those found for passerines (Lehikoinen et al. 2009). For example, rough-legged buzzards (*Buteo lagopus*) have responded to ongoing climate change by advancing their laying dates as a consequence of earlier snowmelt in subarctic areas of Finland and Norway (Terraube et al. 2015).

Although the number of studies available still remains scarce, a general finding is a mismatch between prey and predator phenologies due to climate change. Most studies have focused on specialist consumers failing to synchronize with their prey during the breeding period, but little is known about how generalist consumers respond to phenological shifts across multiple food resources and if this could alter food webs through a mechanism other than trophic mismatch. Deacy et al. (2017) showed that, in Alaska, warmer than usual spring periods induced phenological synchrony between two important food sources for Kodiak brown bears (*Ursus arctos middendorffi*). The bears switched from capturing salmon in shallow streams to foraging for berries on the surrounding hills, attenuating a trophic linkage with disproportionate ecological significance.

15.2.4 Climate change effects on population dynamics of prey and predator

15.2.4.1 Effects on numbers

Predators have long been identified as being limited or regulated by their food supply (e.g., Sinclair and Krebs 2002). Climate change is likely to impact directly upon the abundance of key invertebrate prey for bird predators (Bale et al. 2002). Food availability is a classical limiting factor for the productivity of insectivorous birds (Marshall et al. 2002), and thus its shortage due to global warming may have significant impacts on bird populations, although this has seldom been studied (Pearce-Higgins et al. 2010). However, recent research in Britain indicated that the negative effects of climate-driven asynchrony on annual productivity did not drive long-term population trends in 21 species of insectivorous birds, which suggested that the relationship between asynchrony and population trends is driven by a wider sensitivity of those species to other environmental pressures (Franks et al. 2018). There are still too few long-term studies that examine the effects of climate change on population growth rates (and not on a single vital rate) of a predator species and its main prey (but see Millon et al. 2014). Nonetheless, these types of studies are indispensable for fully understanding how demographic compensation could affect variation in population growth rates and, ultimately, shifts in species ranges in response to climate change (Villellas et al. 2015).

Both the mean and among-year variance in climate variables such as temperature and precipitation are predicted to change. However, the potential impact of changing climatic variability on the fate of populations has remained largely unexamined. In an analysis of 36 plant and animal species spanning a broad range of life histories and environments, Morris et al. (2008) examined how sensitive long-term stochastic population growth rates were likely to be affected by changes in means and standard deviations of vital rates in response to changing climate. They found that short-lived species (insects and annual plants and algae) were more negatively affected by increasing variability in vital rate relative to longer-lived species (perennial plants, birds, ungulates), and there was no additional effect of taxonomic group. However, in a global assessment of the impact of rapid climate warming and anthropogenic land use conversion on 987 populations of 481 species of terrestrial birds and mammals since 1950, Spooner et al. (2018) found that declines in population abundance for both birds and mammals are greater in areas where mean temperature has increased more rapidly, and that this effect is more pronounced for birds. Therefore, recent studies strongly suggest that vertebrate predators are being strongly impacted by climate change in combination with other anthropogenic stressors.

15.2.4.2 The complex interplay between density dependence and climate change

It is becoming increasingly evident that population growth of predators is not purely determined by the rate of prey consumption, as usually considered in classic models of predator–prey interactions (Lotka–Volterra or logistic models), but that other factors including social interaction, interference, and territoriality also come into play. More recently, models incorporating ratio-dependent functional responses have been proposed (Berryman 1992). A key factor in modelling the dynamics of predators and their prey is density-dependent regulation, which may occur both in the prey and the predator. For instance, predation rates decrease when predator densities reach their carrying capacity (e.g. Lande et al. 2003). For prey, inverse density dependence, or the Allee effect, is apparently not rare (Stephens and Sutherland 1999). Climate change may affect predator numbers, prey numbers, density dependence in predator numbers, density dependence in prey numbers, and density dependence in the prey–predator interaction itself, or, of course, any combination thereof. In addition, nonlinearity (Henden et al. 2009) and the interaction between intrinsic density dependence and extrinsic environmental variation remains a major challenge in predator–prey interactions (Turchin 1995).

The interplay between density-dependent processes (both acting at the prey and the predator levels) and external factors, such as climate, is a classical problem in population ecology, which is further complicated by the contrasting effects of different types of predator–prey interactions (i.e., specialist or generalist predators, see section 15.2.5.2) and atypical dynamics (such as cycles). Also, the directional trend of climate change (both in average and variance) adds complexity to the analyses. When these factors and their interactions are not taken into account, they can produce misleading conclusions for the processes at work. A good example is provided by elk (*Cervus canadensis*) population trends in Yellowstone National Park, where a decline in elk followed the reintroduction of wolves (*Canis lupus*), leading to the idea that wolf predation was the key driver of elk decline. However, a more careful analysis (by modelling) concluded that elk decline was influenced by other factors, especially lower than average annual rainfall, rather than wolf predation, which appeared to primarily be compensatory (Vucetich et al. 2005).

Although models and theoretical predictions have been used repeatedly, empirical data (not to mention experimental evidence) are scarce (Wright et al. 2009). While many authors have analysed predator–prey interactions by incorporating density dependence (Sutherland 2006), predation (Evans 2004), or climate (Kausrud et al. 2008), fewer have analysed them together (but see Vucetich et al. 2005, Millon et al. 2014). Theoretical models of predator–prey interactions have been tested with observational data mainly using organisms with short generation times and rapid dynamics, especially invertebrates from aquatic ecosystems, while terrestrial studies come mainly from cyclic dynamics (e.g. Capuccino and Price 1995; Tyson and Lutscher 2016). Climate change is known to be able to drive population dynamics from stable to cyclic dynamics (Coulson et al. 2001), and is also suspected to be the cause of the recent dampening of cycles of small mammals or moths, as well as their predators (review in Ims et al. 2008; Cornulier et al. 2013). Precise quantification of links between climate and cycles (both prey and predator) often rely on limited empirical data. Analysing population models of consumer–resource systems suggested that direct density dependence is primarily related to intratrophic interactions, whereas delayed density dependence in time series may be related to biological interactions. Therefore, numerical changes in delayed density dependence (i.e. the second order autoregressive coefficient) of a predator–prey interaction due to climate change may reflect changes in the strength of predator–prey interaction (Stenseth et al. 2002).

Using large-scale indices of climatic variability as proxies of climate has allowed much progress in understanding the role of environmental factors on population dynamics (review in Stenseth et al. 2002). For instance, Gamelon et al. (2017) have shown that it is essential to integrate density-dependent feedback into predictive models when investigating the effects of climate change on population dynamics. The variation in the North Atlantic Oscillation (NAO) was also found to interfere with density dependence and predation (Vucetich and Peterson 2004; Wilmers et al. 2006) on the population dynamics of mammalian herbivores. In addition, the same factors acted on driving synchrony (Post and Forchhammer 2004, 2006). Although density dependence, climate, and predation together determine population dynamics, it is clear that mechanistic models are required to understand more thoroughly how these factors interact (Stenseth et al. 2002). Wilmers et al. (2007) have developed a modelling framework to explore the effects of different predation strategies on the response of age-structured prey populations under climate change. They showed that predation acts in opposition to temporal correlation in climatic conditions to suppress prey population fluctuations. However, in some cases predation pressure and climate change appear to interact synergistically to affect negatively the population dynamics of prey species (Pokallus and Pauli 2015).

15.2.5. Climate change and the nature of predator–prey interaction

15.2.5.1 Top-down versus bottom-up control

A general and still debated issue concerns whether predators control prey populations (review in Murdoch et al. 2003). In particular the question whether density-dependent or density-independent (in other words, biotic versus abiotic) factors control prey or predator populations has been revived. Whether predators control, or even dampen, prey population fluctuations have mostly been evaluated with models, and to a lesser extent with experiments. Loss of apex predators has been linked to prey release (Soulé et al. 1988) and can lead to alternative ecosystem states (Estes et al. 2011). More generally, prey populations exist at lower densities when exposed to predators (Mech and Peterson 2002). A meta-analysis of experimental studies suggested that predation usually results in trophic cascades (Schmitz et al. 2000; see section 15.2.5.4). However, the issue of top-down versus bottom-up control of prey populations is still hotly debated (Ripple et al. 2014), and evidence so far suggests that prey control the system more often than the reverse (Vucetich et al. 2005), particularly when anthropogenic pressure induces decreases in apex predators below the density where they carry out structurally important top-down functions (Pasanen-Mortensen et al. 2017).

Regulating processes (either top-down or bottom-up) may change due to climate change (i.e., predators are not regulating prey currently, but could become a regulating factor due to climate change). A nice example is provided by mammalian predator–prey interactions. Although the abundance of migratory caribou (*Rangifer tarandus*) is not controlled by predation (Vors and Boyce 2009), wolves can negatively affect prey abundance when caribou are feeding on resources that are insufficient to maintain high densities. As climate change induces loss of lichen as a winter-food resource (Cornelissen et al. 2001), migratory caribou may face increased predation pressure in the future. But a more complex scenario appears when elk is added to the caribou–wolf interaction, because one of the prey is preferred by wolf, but supports higher predation pressure than the other, the 'predation pit' hypothesis (Vors and Boyce 2009). A similar scenario is suggested by Millon et al. (2009) with regard to blackbird (*Turdus merula*) and song thrush (*T. philomelos*), both prey of sparrowhawk, possibly leading to reduced thrush populations. In the case of a mismatch in phenology between prey and predator, it is further conceivable that prey are more prone to respond to temperature increases than the predator, because this gives the prey a new way to escape from predation, and hence a higher selective pressure for early breeding (Both et al. 2009).

15.2.5.2 Specialist versus generalist predators

While specialist predators are mainly dependent on a specific prey species, generalist predators are theoretically able to switch among alternative prey according to their current abundance or profitability

(Andersson and Erlinge 1977). Generalist predation in particular may either stabilize (Erlinge et al. 1988) or destabilize the community (Bonsall and Hassell 1997), depending on ecological conditions. Generalist predators are theoretically expected to maintain relatively constant vital rates because they can shift to alternative prey (i.e., they respond mainly functionally to variation in the abundance of their preferred prey), and therefore should display more stable populations than specialist predators (Redpath and Thirgood 1999). Thus, they are expected to cope better with global changes than specialist predators, which are generally believed to be more sensitive to environmental change than generalists. However, there are very few studies that have shown that specialist predators are more affected by climate change than generalist predators (but see Rand and Tscharntke 2007).

Most studies of predator–prey interactions have involved specialist predators and one or sometimes a few prey species (Korpimäki and Hakkarainen 1991; Nielsen 1999). In addition, most studies involved cyclic predator–prey dynamics of voles and their raptor predators in Fennoscandia (Korpimäki et al. 2003; Lehikoinen et al. 2009). The question thus arises whether the findings can be generalized to more complex food webs (Millon et al. 2009) in other areas. In particular, how generalist predators will respond to changes in their prey community is not predicted by current theoretical models (but see Baudrot et al. 2016), despite generalist predators probably constituting the majority of avian predators (review in Valkama et al. 2005). However, a generalist predator may be partly sensitive to changes in diversity or composition of prey communities. In addition, if generalist predators shift from one prey to another in the community, this may affect interactions among species, both at the level of competing prey and at the level of other predators of the trophic network (Hoy et al. 2017). Very few empirical studies are however currently available, mainly because it is far more difficult to deal with a generalist predator that preys upon dozens of prey species and has to be monitored over many years in order to analyse functional responses (but see Redpath and Thirgood 1999; Rutz and Bijlsma 2006; Millon et al. 2009).

15.2.5.3 Functional versus numerical responses

The functional response of a predator largely determines the effect of a predator on the prey population (Abrams and Ginzburg 2000). In addition, the shapes of both numerical and functional responses of predators have strong effects for prey community stability and composition (Jaksic et al. 1992). Predation plays a major role in shaping the structure and dynamics of ecological communities, and the functional response of a predator is of crucial importance to the dynamics of any predator–prey system by linking the trophic levels. However, quantitative and precise studies of functional response are scarce, in particular in the ornithological literature (review in Valkama et al. 2005). In addition there is a great deal of debate on the way we measure functional response and how to 'scale up' from local (individual) measurements to a population-level function (Englund and Leonardsson 2008), and a major difficulty remains to measure functional response in the field at (very) low prey density. Finally, both for mammal and bird predators, very few studies have attempted to document variation in the functional response according to season, social status, or sex. Whether climate modulates the functional response remains poorly understood. A few examples have found contrasting effects of abiotic climatic factors (NAO or ENSO) on the functional response of large carnivore species (Sinclair et al. 2013; Bowler et al. 2014).

15.2.5.4 Trophic cascades and regime shifts

If climate change impacts upon several trophic levels simultaneously (which is very likely), then wholesale community changes may become evident, and constitute a 'regime shift' (Rodionov 2004). Climate change, while affecting predator–prey interactions, may also affect community structure through a trophic cascade (Ripple et al., 2014). In terrestrial ecosystems, trophic cascades result for instance from the indirect effect of predators on plants mediated by herbivores (Paine 1980; Polis et al. 2000). Trophic cascades partly also result from the nonlinear (and thus, non-trivial) nature of species interactions (McCann 2007). They provide examples of how indirect effects propagate in communities via consumption of prey by predators.

Trophic cascades have, however, been more documented in aquatic ecosystems than in terrestrial ecosystems, the reason being unclear, but perhaps related to simplified interactions within each trophic level in the former. A terrestrial example comes from a long-term study of a three-trophic level system including grey wolf, moose, and their winter food resource, balsam fir (*Abies balsamea*; Wilmers et al. 2006). The balsam fir increases in abundance with winter snow (in relation to NAO), and influences wolf kill rates of moose, with cascading effects on balsam fir growth (Post and Stenseth 1999). A virus outbreak in the wolf population allowed testing the transient effects of reduction of predation pressure and climatic variation on the dynamics of this three-level food chain. When wolf numbers dampened, there was a switch from top-down to bottom-up regulation of the moose population, with a stronger influence of climate on moose population growth rate, underlining synergistic interactions between predators (mediated by pathogens) and climate in trophic control. Similarly, a controlled experiment tested how cascading trophic interactions initiated by arthropod predators were affected by changes in rainfall (mimicking climate change) and resulted in leaf litter decomposition changes (Lensing and Wise 2006). More recently, a detailed trophic cascade was analysed in relation to the presence of dingoes (*Canis lupus dingo*; Letnic et al. 2012; Gordon et al. 2016; Morris and Letnic 2017).

15.2.6 Adaptation and selection

While most birds apparently have responded to climate change through phenotypic plasticity (e.g., Charmantier et al. 2008), evolutionary responses have also been documented (Parmesan and Yohe 2003; Root et al. 2003; Møller et al. 2004). For example, changes in body size of birds during a period of only 50 years have been detected (Yom-Tov 2001), as well as changes in the proportion of colour morphs in avian predators (Karell et al. 2011). Such morphological changes probably reflect the impact of abiotic conditions and subsequent selection on body size, and this may result in parallel effects of Bergmann's rule on latitudinal trends in body size (Yom-Tov 2001). However, predators are often long-lived, or at least longer lived than their prey, and, therefore, their biological responses to environmental change may be lagged by one or more generations with respect to their prey (Sæther et al. 2005). Predators are also on average larger than their prey, particularly in aquatic ecosystems as well as in birds (but this is not necessarily true for mammalian predators). Stevenson and Bryant (2000) have suggested with a simple energetic model that small-bodied species will be able to advance breeding dates more easily than large-bodied ones as a response to increasing temperature. These two factors may suggest that larger predators (or longer-lived, which to a large extent is correlated) may take more time to respond evolutionarily to their changing environment than their prey, and therefore may be more vulnerable to climate change in the long term. These differential responses across trophic levels may impact ecosystem functioning, because predators at higher trophic levels may decline more strongly because of the asynchrony with the phenology of their prey (Both et al. 2009).

Finally, it should be borne in mind that predators can also have direct non-consumptive effects on prey and subordinate intraguild predator populations by causing changes in traits such as behaviour and, particularly, habitat selection patterns, growth, and development (Schmitz et al. 2004; Terraube and Bretagnolle 2018). Non-consumptive effects may be equally or more important than consumption for predator–prey population and community dynamics (review in Peckarsky et al. 2008), having indirect effects on other organisms in the community and on ecosystem function (Suraci et al. 2016). The recognized contribution of non-consumptive effects of predators on prey population dynamics may also be affected by climate change, as recently pointed out (Veselý et al. 2017; Lord et al. 2017).

15.3 Climate change and predator–prey relationships in birds: the evidence so far

15.3.1 Taxonomic bias

More work on the effects of climate change on predator–prey interactions is available for mammals than for birds. In addition, although there is a large literature on small mammal cycles, sometimes involving avian predators (raptors, owls, skuas), most often mammalian predators are thought to play the key role and have therefore been more studied than their avian counterparts. This is unfortunate because there are fundamental differences between bird and mammalian predators. (1) Raptors are central place foragers, while mammals are not necessarily or only rarely so. Central place foraging leads to differences in the predator–prey relationship, in particular the spatial aspects of predator–prey encounters (Orians and Pearson 1979), energetic constraints (Orians and Pearson 1979), or travel costs. (2) Raptors, like most birds, are socially monogamous, while mammals are mainly polygynous (Caizergues and Lambrechts 1999). As the breeding system may affect population dynamics, and more generally, other life history traits (Bennett and Owens 2002), it is likely that climate change will affect predator–prey interactions in different ways for mammalian and avian predators. In addition, despite the growing evidence that climate change will affect species interactions, most studies in birds have dealt with small-sized passerines (Møller 2002; Peach et al. 2004; Tylianakis et al. 2008), usually cavity nesters. Fewer studies have addressed bird predators, though numbers are increasing (Rutz and Bijlsma 2006; Millon et al. 2008; Anctil et al. 2014; Terraube et al. 2015; Robinson et al. 2017; Terraube et al. 2017). In this section, we choose to separate studies on terrestrial and marine ecosystems because of differences in the availability of data. Indeed, in marine ecosystems data on prey are difficult to collect and are thus scarce, in contrast to terrestrial ecosystems where climate effects are well documented (Beaugrand and Kirby 2018).

15.3.2 Trophic interactions in marine ecosystems

Several studies have focused on the impact of climate change on seabirds (see many references in Jenouvrier 2013), which are often top predators in marine pelagic ecosystems. Because of this trophic position, they have repeatedly been used as indicators of the state of marine ecosystems (Wanless et al. 2007; Ainley et al. 2015). Most seabird studies on the effects of climate change have concentrated on Antarctic and arctic seabirds (e.g. Møller et al. 2006), with very few concerning tropical (Precheur et al. 2016; Nicoll et al. 2017) or subtropical/temperate seabirds (Sydeman et al. 2015). During recent decades, because of climate change, there have been more El Ninõ than La Ninã events, leading to a generally lower than average Southern Oscillation Index (Murphy et al. 2007; Chapter 2) and, consequently, the Antarctic peninsula is experiencing one of the most rapid warming in the world (Ducklow et al. 2007; Cresswell et al. 2008). The more positive Southern Oscillation Index (Trathan et al. 2007) acts through an increase in sea surface temperature and sea ice season duration and a decreased sea ice extent (Jenouvrier et al. 2003; Ducklow et al. 2007). This may have strong consequences for the entire food web in the southern oceans (Trathan et al. 2007), because these phenomena affect primary production and cause a decline in the abundance and a change in distribution of krill (Euphausiacea).

15.3.2.1 Spatio-temporal effects of changes in prey abundance and distribution on top marine predators

Using a long-term dataset (55 years) of dates of first arrival and laying for the entire community of Antarctic seabirds in East Antarctica, Barbraud et al. (2008) showed that, in contrast to the Northern hemisphere, arrival and laying dates of Antarctic birds are delayed as a consequence of reduced krill abundance and delayed access to colonies resulting from late sea ice breakup. King penguins (*Aptenodytes patagonicus*) forage over the polar front and dive to the thermocline to feed on myctophid fish (Péron et al. 2012). Here, climate change affects the position of both the polar front and the thermocline, showing that climate change in polar marine ecosystems not only impacts horizontal, but also vertical, prey distribution through modifications of sea temperature and sea ice extent. In both studies, the distance to foraging areas is affected and, if increased, may have a negative impact on breeding success (Durant et al. 2007) because of longer foraging trips during the breeding season and thus reduced parent and chick body conditions (Trathan et al. 2007). Climate change has already increased water temperature in the Norwegian Sea and thus modified currents and shifted the herring stock to the north, far away from breeding grounds of puffin

(*Fratercula arctica*), thus creating spatial mismatch between prey and predator (Durant et al. 2007). Although a warming climate may change the distribution of prey for little auks (*Alle alle*), they may not be negatively affected, because their energy expenditure will decrease with increasing temperature (Amélineau et al. 2018). Further research is urgently needed to understand how different anthropogenic stressors like overfishing, pollutants, and climate change interact to affect prey availability and distribution and how they reverberate into seabird demography. This is key to assess seabird resilience to global change and to prioritize conservation efforts (Oro 2014).

15.3.2.2 Effect of climate change on top marine predator life-history traits and predator density and demography

In the Southern Ocean, some marine predator populations show periodicity in population and breeding performance, driven by physical forcing from, e.g., the El Niño-Southern Oscillation, ENSO (Jenouvrier et al. 2009; Ducklow et al. 2007). During the late 1970s, change in population periodicity and sudden variation in population trends have been recorded, indicating a regime shift in the Southern Ocean (Chapter 2), potentially driven by climate change. Furthermore, during warm anomalies, birds skip breeding probably because the food availability was low and limiting for the highly energy demanding reproductive activities (Jenouvrier et al. 2003). Cresswell et al. (2008) showed that rapid changes in the mean supply and the patchiness of krill could have an effect on female and chick condition and thus on breeding success of macaroni penguins (*Eudyptes chrysolophus*). Few studies have reported a decline in avian prey quality due to climate change in marine ecosystems. Wanless et al. (2007) reported that climate change impacts the breeding success of black-legged kittiwakes (*Rissa tridactyla*) by affecting the quality of prey, i.e. lower condition of the sandlance (*Ammodytes marinus*). Climate change has also been shown to decrease adult survival in the little auk in the Arctic through a change in prey quality or quality (Hovinen et al. 2014). Hilton et al. (2006) studied the effects of climate change on the rockhopper penguin (*Eudyptes chrysocome*), which has experienced a marked population decline throughout most of its circumpolar breeding distribution. Using stable isotope analyses and feather samples dating back to 1861, they found evidence for a decreased signature in primary productivity over time. This decline was associated with annual variations in sea surface temperature, and may reflect a reduced carrying capacity for penguin populations.

Changes in prey abundance, distribution, and quality due to climate change have affected the life history traits of seabirds, and, overall, there is a spatial mismatch between the distributions of seabirds and their prey leading to longer foraging trips. A recent meta-analysis of 209 phenological time series from 145 breeding populations evidenced that, between 1952 and 2015, seabird populations worldwide have not adjusted their breeding seasons over time or in response to sea surface temperature (Keogan et al. 2018). This limited temperature-mediated plasticity of reproductive timing highlights the high vulnerability of marine top predators to future mismatch with lower-trophic-level resources.

15.3.3 Bird predator–prey interactions in terrestrial ecosystems

15.3.3.1 Changes in phenology and their consequences for predators

Many studies of the impact of climate change on terrestrial birds have revealed a mismatch between advancing food abundance peaks and the timing of highest energy requirements for the nestlings (Both et al. 2009; Burgess et al. 2018), although only 37 per cent report a negative effect on reproductive success (Chapter 9). When early spring temperatures are high, European insectivorous passerines tend to lay their first egg earlier to match peak food abundance (Cresswell and McCleery 2003). Climate change also has an effect on the timing of migration (Chapter 8). In North America, southward migration of sandpipers and falcons were expected to be strongly related to timing of snowmelt (Niehaus and Ydenberg 2006). However, sandpipers seem to respond less strongly than their falcon predators to variation in timing of snowmelt, leading sandpipers (adults and juveniles) to encounter more predators on their stopover sites. These different effects of earlier snowmelt indicate that climate change could alter the ecological dynamics of predator–prey systems (Niehaus and Ydenberg 2006).

15.3.3.2 Changes in prey accessibility and availability, and effects on predator foraging success

Birds feeding on soil-dwelling invertebrate prey have to forage in dry surface soil as climate change generates high temperature and low rainfall. Indeed, dry surface soil makes it more difficult for birds to probe, and soil invertebrates tend to bury themselves deeper in the ground and thus become inaccessible to predators (Peach et al. 2004). The induced food shortage has consequences for bird condition and thus for breeding success (Peach et al. 2004; Green 1988). Additionally, so far few studies have investigated the effects of climate on bird foraging success during the non-breeding period. Terraube et al. (2017) showed that, in boreal ecosystems, increased frequency of rainy days in autumn influences the foraging success of pygmy owls (*Glaucidium passerinum*), potentially through reduced vulnerability of small mammals to predation in rainy weather. In this case, climate change is making prey inaccessible, or at least less accessible, to avian predators. In contrast, other authors have reported potential benefits of global warming for avian predators, such as increased prey availability in aquatic ecosystems. Stevens et al. (2002) described a possible improvement of foraging success by snail kites (*Rostrhamus sociabilis*) in an increasing temperature scenario in Florida wetlands. The snail kite is a specialist predator of apple snails (*Pomacea paludosa*). At low water temperature, apple snails become inactive and tend to bury themselves, while at higher temperature apple snails are active and consequently become accessible. Nevertheless, severe drying events render apple snails unavailable for kites (Stevens et al. 2002). If water temperature increases due to global warming, it is easy to imagine a shift in snail kite distribution to the north, advancement in laying, or an increase in breeding success due to an improvement in apple snail availability and in foraging success of snail kites. In crag martins (*Ptyonoprogne rupestris*), high temperatures during brood care lower breeding success, because drought decreases availability of aquatic insects for offspring (Acquarone et al. 2003).

15.3.3.3 Generalists versus specialist bird predators

Generalist predators are expected to cope with climate change better than specialists (see section 15.2.5.2). However, the few studies available so far indicate that generalist avian predators are not necessarily able to shift to other prey species (e.g. sparrowhawks in Nielsen and Møller 2006). Two studies involved raptor generalist predators. In Denmark, a long-term study of sparrowhawks, one of the most common bird-eating predators of the Palearctic, showed they have experienced important changes in the composition of their prey community during the last 40 years (Millon et al. 2009). Contrary to expectations for a generalist predator, sparrowhawks seemed to be predominantly sensitive to changes in the abundance of only two main prey species (skylark, *Alauda arvensis*, and blackbird, *Turdus merula*). In another generalist predator, the common buzzard (*Buteo buteo*), Lehikoinen et al. (2009) found that climate had a much stronger effect than vole abundance on timing of breeding in the raptor. The buzzard, though breeding at the northernmost limit of its range in Finland, should have benefited from increased temperatures during the breeding season. Interestingly that was not the case, as the Finnish breeding population has crashed, possibly partly due to asymmetrical climate change, i.e. an increase in winter and early spring temperatures on the one hand (leading to earlier breeding), but no such warming in late spring and summer temperatures on the other hand (leading to poor meteorological conditions during chick rearing and ultimately lower productivity). Gilg et al. (2009) reported a study about the impact of climate change on cyclic predator–prey population dynamics in the high Arctic. In this case, the long-tailed skua (*Stercorarius longicaudus*) and the snowy owl (*Bubo scandiacus*) feed almost exclusively on lemmings (*Dicrostonyx groenlandicus*), and so they could be classified as specialist predators. Until 2000, lemmings displayed regular 4-year cyclic dynamics, but afterward, the density of lemmings has remained at a low level (Gilg et al. 2009). As a consequence, the snowy owl has been absent since 2000. Gilg et al.

(2009) ran models with different climate change scenarios and showed that in all cases climate change will lead to an increase in the duration of the lemming population cycle and a decrease in the maximum population densities, which may ultimately lead to local extinction of the owl (Schmidt et al. 2012). Overall, contrary to predictions, studies showed that generalist predators may not cope better with climate change than specialists.

15.3.4 Concluding remarks on the impact of climate change on bird predator–prey interactions

Studies of the effects of global warming on predator–prey interactions in marine ecosystems have investigated almost exclusively bottom-up processes. Through an effect on prey, global warming indirectly affects many life history traits and demographic parameters of seabird predators. In terrestrial ecosystems, studies mainly focused on the match–mismatch processes in insectivorous birds of Europe or North America, although recently several studies involved raptors. Both in marine and terrestrial ecosystems climate change has started to affect avian predators and has already had significant effects on several aspects of avian predator–prey interactions. Few studies have detailed the effects of climate change on both prey and their avian predators, with the exception of tits and caterpillars (Both et al. 2009; see also Nielsen and Møller 2006), and few have identified by which mechanisms such effects have occurred. However, Terraube et al. (2015) showed that climate change impacts rough-legged buzzards in Lapland indirectly through a climate-driven decrease in the abundance of their main prey (voles) and not through direct negative effects of adverse weather on nestling survival and breeding success. A similar conclusion arose from a study focusing on tawny owl population dynamics (Millon et al. 2014).

15.4 Future prospects

Despite the recognized and acknowledged importance of the effects of climate change on trophic interactions, there are too few studies currently available to suggest any clear general trend (Gilman et al. 2010). In addition, given that the detected effects are sometimes in opposite directions, and that changes observed in one species may indirectly affect other parts of the community through competition or predation, it is currently almost impossible to predict the future effects of global change at the community level using predator–prey interactions. Below we list areas that we believe should be tackled in the near future.

15.4.1 Long-term studies

We need more long-term field studies of communities (Jenouvrier 2013). Although there are several models and theoretical approaches that predict effects of climate change on communities, few empirical data support these models. Therefore, it is essential to emphasize the need for maintenance of long-term biological datasets to validate predictions. In addition, new studies should be started and carried out in poorly studied although important ecosystems, such as the tropics, where climate change is nevertheless acting (e.g., Chamaille-Jammes et al. 2008). Recent research revealed strong interactive effects of climate and land-use change on ecological communities in tropical grasslands and savannahs (Newbold 2018), highlighting the need to better assess the consequences of these modifications in community structure in terms of trophic cascades and ecosystem shifts. The relationship between climate and population cycles suggests a causal relationship between climate change and cycle dampening for small mammals and their predators, as well as insect cycles (Jepsen et al. 2008). Again, long-term studies may reveal unexpected patterns. Since the mid 1990s, small mammal cycles have dampened in Fennoscandia (Ims et al. 2008), but more recent data suggest that cycles may restart again, at least in the boreal zone (Brommer et al. 2010; Cornulier et al. 2013). Long-term empirical studies could help shed light on other poorly understood aspects like how individual variability in predator behaviour may modify the effects of climate change on predator–prey interactions. This would improve our ability to predict the demographic response of both predators and prey species to environmental change (Pettorelli et al. 2015). Finally, we need a better integration between observational, experimental, and modelling studies on mechanisms of species interactions along environmental gradients.

15.4.2 Modelling and experimental studies

Above all, we need more studies that apply a mechanistic approach, because it allows us to gain understanding of underlying causes (Sutherst et al. 2007). To date, there have been few modelling attempts to predict the effects of climate change on predator–prey relationships. Such models may help to address complex issues such as trophic cascades, hyper predation processes, and compensation, so far poorly studied in the context of global warming (but see Emmerson et al. 2005). While the relative importance of deterministic and stochastic factors has been a central tenet in population ecology for at least five decades (see Coulson et al. 2004 for a summary), the theoretical debate in predator–prey ecology moved from deterministic models to individual-based models, stochastic and numerical models (van der Meer and Smallegange 2009; Bocedi et al. 2014). Indeed, even simple individual-based models including predator–prey interactions may help in understanding how climate-driven changes in distribution or breeding phenology of prey may have population consequences for predators (Peers et al. 2014). In addition, modelling studies would help answer questions about how climate change affects the synchronization of breeding phenologies, and how it will affect population dynamics when individual heterogeneity (and variance in fitness) is taken into account. Recent advances in evolutionary models will help disentangle these effects. Similarly, recent methodological advances integrating principles from consumer–resource analyses, resource selection theory, and species distribution modelling, will enhance quantitative prediction of shifts in species range (Pellissier et al. 2013; Trainor and Schmitz 2014).

15.4.3 Evolutionary questions and conservation issues

Given that predator–prey interactions are often viewed as an evolutionary arms race, the effects of global warming on predator–prey interactions beg for evaluating its longer-term, evolutionary implications. For instance, the mismatch in phenology between prey and predator may lead to counterintuitive selective pressures. For example, it remains to be studied whether predators are phenotypically plastic with regard to prey choice, whether prey are able to cope rapidly with a changing predator community, and at which rate microevolutionary change could allow either prey or predator to cope with its changing environment. Thus we need better analyses of the effects of temporal and spatial climatic variation, and quantification of species traits in species interactions. With regard to the interaction itself, apart from being specialist or generalist predators, there is also a gradient in how predators acquire and use food for reproduction (i.e., the income vs capital breeder dichotomy), which has strong effects on life history traits of predators. Very few studies of birds have studied whether this difference in processing and using energy is affected by climate change, although it is likely.

Shifts in predator–prey dynamics can trigger trophic cascades and affect communities at large scale with conservation implications worldwide. For example, in the Arctic, polar bears (*Ursus arctos*) have recently been recorded shifting to foraging on nesting seabirds because of sea-ice loss and, consequently, lower access to ringed seals (Iverson et al. 2014). This has potential consequences for large-scale population dynamics of seabirds and the whole Arctic food web. Similarly, the dynamics of small rodent species, such as voles and lemmings, that were previously characterized by large amplitude regular cycles has recently changed more or less simultaneously across Europe (Ims et al. 2008). Thus, climate change could affect predators indirectly through modifications in population dynamics of voles (Solonen 2006). As a consequence, specialist predators of cyclic rodents have already declined (Millon and Bretagnolle 2008). Changes in small herbivore dynamics have the potential to lead to a regime shift, thus representing a new challenge for the conservation of biodiversity. The life histories of predators are seemingly adapted to these interactions (Ims and Fuglei 2005), but it is unknown to what extent their populations can be sustained under different dynamics. Worryingly, models in Gilg et al. (2009) showed that prey–predator communities will be severely impacted by climate change. Further research is also needed to understand the effects of conservation actions (e.g., protected areas establishment and management) on the resilience of predator–prey interactions under climate change and how this reverberates at the community and ecosystem scales. Most of the available evidence comes from marine reserves (Ling and Johnson 2012), but the effects of protected areas on predator–prey interactions in terrestrial ecosystems and their potential buffer effect against climate change remains virtually unknown. Does the maintenance of complex food webs in protected areas help buffering ecosystems against regime shifts?

As a consequence of their life history traits, predators are supposed to be more affected by the adverse effects of climate change than lower trophic levels. Indeed, many emblematic predators, such as raptors, are already of conservation concern (see, e.g., Bennett and Owens 2002 for a comparative analysis on birds; McClure et al. 2018). These concerns, and the ongoing climate-driven shifts in trophic interactions, may have important consequences for population dynamics, community structure, and ecosystem resilience, posing a challenge for conservation in the near future.

Acknowledgments

We would like to thank Haneke Gillis for her important help in the processing of the first version of this chapter in 2009, and Anders Møller and Peter Dunn for their careful reading and comments on successive drafts.

References

Abrams, P., and Ginzburg, L. (2000). The nature of predation: prey dependent, ratio dependent or neither? *Trends in Ecology and Evolution*, 15, 337–41.

Acquarone, D., Cucco, M., and Malarcarne, G. (2003). Reproduction of the crag martin (*Ptyonoprogne rupestris*) in relation to weather and colony size. *Ornis Fennica*, 80, 79–85.

Ainley, D.G., Ballard, G., Jones, R.M., Jongsomjit, D., Pierce, S.D., Smith, W.O., Jr, and Veloz, S. (2015). Trophic cascades in the western Ross Sea, Antarctica: revisited. *Marine Ecology Progress Series*, 534, 1–16.

Amélineau, F., Fort, J., Mathewson, P.D., Speirs, D.C., Courbin, N., Perret, S., Porter, W.P., Wilson, R.J., and Grémillet, D. (2018). Energyscapes and prey fields shape a North Atlantic seabird wintering hotspot under climate change. *Royal Society Open Science*, 5, 171883.

Anctil, A., Franke, A., and Bêty, J. (2014). Heavy rainfall increases nestling mortality of an Arctic top predator: experimental evidence and long-term trend in peregrine falcons. *Oecologia*, 174, 1033–43.

Andersson, M., and Erlinge, S. (1977). Influence of predation on rodent populations. *Oikos*, 29, 591–7.

Austin, M. (2002). Spatial prediction of species distribution: An interface between ecological theory and statistical modelling. *Ecological Modelling*, 157, 101–18.

Bale, J., Masters, G., Hodkinson, I., Awmack, C., Bezemer, T., Brown, V., Butterfield, J., Buse, A., Coulson, J., and Farrar, J. (2002). Herbivory in global climate change research: Direct effects of rising temperature on insect herbivores. *Global Change Biology*, 8, 1–16.

Barbraud, C., Marteau, C., Ridoux, V., Delord, K., and Weimerskirch, H. (2008). Demographic response of a population of white-chinned petrels *Procellaria aequinoctialis* to climate and longline fishery bycatch. *Journal of Applied Ecology*, 45, 1460–7.

Bascompte, J., Jordano, P., and Olesen, J. (2006). Asymmetric coevolutionary networks facilitate biodiversity maintenance. *Science*, 312, 431–3.

Bastille-Rousseau, G., Schaefer, J.A., Peers, M.J.L., Ellington, E.H., Mumma, M.A., Rayl, N.D., Mahoney, S.P., and Murray, D.L. (2018). Climate change can alter predator–prey dynamics and population viability of prey. *Oecologia*, 186,141–50.

Baudrot, V., Perasso, A., Fritsch, C., Giraudoux, P., and Raoul, F. (2016). The adaptation of generalist predators' diet in a multi-prey context: insights from new functional responses. *Ecology*, 97, 1832–41.

Beaugrand, G., and Kirby, R.R. (2018). How do marine pelagic species respond to climate change? Theories and observations. *Annual Review of Marine Science*, 10, 169–97.

Bellard, C., Bertelsmeier, C., Leadley, P., Thuiller, W., and Courchamp, F. (2012). Impacts of climate change on the future of biodiversity. *Ecology Letters*, 15, 365–77.

Bennett, P., and Owens, I. (2002). *Evolutionary ecology of birds: Life histories, mating systems and extinction*. Oxford University Press, Oxford, UK.

Berryman, A. (1992). The origins and evolution of predator–prey theory. *Ecology*, 73, 1530–5.

Bocedi, G., Palmer, S.C.F., Pe'er, G., Heikkinen, R.K., Matsinos, Y.G., Watts, K., and Travis, J.M.J. (2014). RangeShifter: a platform for modelling spatial eco-evolutionary dynamics and species' responses to environmental changes. *Methods in Ecology and Evolution*, 5, 388–96.

Böhning-Gaese, K., and Lemoine, N. (2004). Importance of climate change for the ranges, communities and conservation of birds. *Advances in Ecological Research*, 35, 211–36.

Bonsall, M., and Hassell, M. (1997). Apparent competition structures ecological assemblages. *Nature*, 388, 371–3.

Both, C., Bouwhuis, S., Lessells, C., and Visser, M. (2006). Climate change and population declines in a long-distance migratory bird. *Nature*, 441, 81–3.

Both, C., van Asch, M., Bijlsma, R., van den Burg, A., and Visser, M. (2009). Climate change and unequal phenological changes across four trophic levels: Constraints or adaptations? *Journal of Animal Ecology*, 78, 73–83.

Bowler, B., Krebs, C., O'Donoghue, M., and Hone, J. (2014). Climatic amplification of the numerical response of a predator population to its prey. *Ecology*, 95, 1153–61.

Brommer, J. (2008). Extent of recent polewards range margin shifts in Finnish birds depends on their body mass and feeding ecology. *Ornis Fennica*, 85, 109–17.

Brommer, J., Pietiäinen, H., Ahola, K., Karell, P., Karstinen, T., and Kolunen, H. (2010). The return of the vole cycle in southern Finland refutes the generality of the loss of cycles through 'climatic forcing'. *Global Change Biology*, 16, 577–86.

Brown, J., Whitham, T., Morgan Ernest, S., and Gehring, C. (2001). Complex species interactions and the dynamics of ecological systems: Long-term experiments. *Science*, 293, 643–50.

Burgess, M.D., Smith, K.W., Evans, K.L., Leech, D., Pearce-Higgins, J.W., Branston, C.J., Briggs, K., Clark, J.R., du Feu, C.R., Lewthwaite, K., Nager, R.G., Sheldon, B.C., Smith, J.A., Whytock, R.C., Willis, S.G., and Phillimore, A.B. (2018). Tritrophic phenological match-mismatch in space and time. *Nature Ecology and Evolution*, 2, 970–5.

Cahill, A.E., Aiello-Lammens, M.E., Fisher-Reid, M.C., Hua, X., Karanewsky, C.J., Yeong Ryu, H., Sbeglia, G.C., Spagnolo, F., Waldron, J.B., Warsi, O., and Wiens, J.J. (2013). How does climate change cause extinction? *Proceedings of the Royal Society of London Series B: Biological Sciences*, 280, 20121890.

Caizergues, A., and Lambrechts, M. (1999). Male 'macho' mammals exploiting females versus male 'Don Juan' birds exploited by females: The opposite-sex exploitation (osex) theory. *Ecology Letters*, 2, 204–6.

Capuccino, N., and Price, P. (1995). *Population dynamics: New approaches and synthesis.* Academic Press, London, UK.

Chamaille-Jammes, S., Fritz, H., Valeix, M., Murindagomo, F., and Clobert, J. (2008). Resource variability, aggregation and direct density dependence in an open context: The local regulation of an African elephant population. *Journal of Animal Ecology*, 77, 135–44.

Chambers, L., Hughes, L., and Weston, M. (2005). Climate change and its impact on Australia's avifauna. *Emu*, 105, 1–20.

Charmantier, A., McCleery, R., Cole, L., Perrins, C., Kruuk, L., and Sheldon, B. (2008). Adaptive phenotypic plasticity in response to climate change in a wild bird population. *Science*, 320, 800–3.

Cornelissen, J., Callaghan, T., Alatalo, J., Michelsen, A., Graglia, E., Hartley, A., Hik, D., Hobbie, S., Press, M., Robinson, C., Henry, G., Shaver, G., Phoenix, G., Jones, D., Jonasson, S., Chapin, F., Molau, U., Neill, C., Lee, J., Melillo, J., Sveinbjornsson, B., and Aerts, R. (2001). Global change and Arctic ecosystems: Is lichen decline a function of increases in vascular plant biomass? *Journal of Ecology*, 89, 984–94.

Cornulier, T., Yoccoz, N.G., Bretagnolle, V., Brommer, J.E., Butet, A., Ecke, F., Elston, D.A., Framstad, E., Henttonen, H., Hörnfeldt, B., Huitu, O., Imholt, C., Ims, R.A., Jacob, J., Jedrzejewska, B., Millon, A., Petty, S.J., Pietiäinen, H., Tkadlec, E., Zub, K., and Lambin, X. (2013). Europe-wide dampening of population cycles in keystone herbivores. *Science*, 340, 63–6.

Coulson, T., Catchpole, E., Albon, S.D., Morgan, B.J.T., Pemberton, J., Clutton-Brock, T., Crawley, M., and Grenfell, B. (2001). Age, sex, density, winter weather, and population crashes in Soay sheep. *Science*, 292, 1528–31.

Coulson, T., Rohani, P., and Pascual, M. (2004). Skeletons, noise and population growth: The end of an old debate? *Trends in Ecology and Evolution*, 19, 359–64.

Cresswell, K.A., Wiedenmann, J., and Mangel, M. (2008). Can macaroni penguins keep up with climate- and fishing-induced changes in krill? *Polar Biology*, 31, 641–9.

Cresswell, W., and McCleery, R. (2003). How great tits maintain synchronization of their hatch date with food supply in response to long-term variability in temperature. *Journal of Animal Ecology*, 72, 356–66.

Deacy, W., Armstrong, J., Leacock, W., Robbins, C., Gustine, D., Ward, E., Erlenbach, J., and Stanford, J. (2017). Phenological synchronization disrupts trophic interactions between Kodiak brown bears and salmon. *Proceedings of the National Academy of Sciences of the United States of America*, 114, 10432–7.

Devictor, V., Julliard, R., Couvet, D., and Jiguet, F. (2008). Birds are tracking climate warming, but not fast enough. *Proceedings of the Royal Society of London Series B: Biological Sciences*, 275, 2743–8.

Devictor, V., van Swaay, C., Brereton, T., Brotons, L., Chamberlain, D., Heliola, J., Herrando, S., Julliard, R., Kuussaari, M., Lindstrom, A., Reif, J., Roy, D.B., Schweiger, O., Settele, J., Stefanescu, C., Van Strien, A., Van Turnhout, C., Vermouzek, Z., WallisDeVries, M., Wynhoff, I., and Jiguet, F. (2012). Differences in the climatic debts of birds and butterflies at a continental scale. *Nature Climate Change*, 2, 121–4.

Doligez, B., and Clobert, J. (2003). Clutch size reduction as a response to increased nest predation rate in the collared flycatcher. *Ecology*, 84, 2582–8.

Ducklow, H., Baker, K., Martinson, D., Quetin, L., Ross, R., Smith, R., Stammerjohn, S., Vernet, M., and Fraser, W. (2007). Marine pelagic ecosystems: The west Antarctic peninsula. *Philosophical Transactions of the Royal Society of London Series B: Biological Sciences*, 362, 67–94.

Durant, J.M., Hjermann, D., Ottersen, G., and Stenseth, N.C. (2007). Climate and the match or mismatch between predator requirements and resource availability. *Climate Research*, 33, 271–83.

Emmerson, M., Bezemer, M., Hunter, M., and Jones, T. (2005). Global change alters the stability of food webs. *Global Change Biology*, 11, 490–501.

Englund, G., and Leonardsson, K. (2008). Scaling up the functional response for spatially heterogeneous systems. *Ecology Letters*, 11, 440–9.

Erlinge, S., Göransson, G., Högstedt, G., Jansson, G., Liberg, O., Loman, J., Nilsson, I., von Schantz, T., and Sylvén, M. (1988). More thoughts on vertebrate predator regulation of prey. *American Naturalist*, 132, 148–54.

Estes, J.A., Terborgh, J., Brashares, J.S., Power, M.E., Berger, J., Bond, W.J., Carpenter, S.R., Essington, T.E., Holt, R.D., Jackson, J.B.C., Marquis, R.J., Oksanen, L., Oksanen, T., Paine, R.T., Pikitch, E.K., Ripple, W.J., Sandin, S.A., Scheffer, M., Schoener, T.W., Shurin, J.B., Sinclair, A.R.E., Soule, M.E., Virtanen, R., and Wardle, D.A. (2011). Trophic downgrading of planet Earth. *Science*, 333, 301–6.

Evans, K. (2004). The potential for interactions between predation and habitat change to cause population declines of farmland birds. *Ibis*, 146, 1–13.

Franks, S.E., Pearce-Higgins, J.W., Atkinson, S., Bell, J.R., Botham, M.S., Brereton, T.M., Harrington, R., and Leech, D.I. (2018). The sensitivity of breeding songbirds to changes in seasonal timing is linked to population change but cannot be directly attributed to the effects of trophic asynchrony on productivity. *Global Change Biology*, 24, 957–71.

Gamelon, M., Grøtan, V., Nilsson, A.L., Engen, S., Hurrell, J.W., Jerstad, K., Phillips, A.S., Røstad, O.W., Slagsvold, T., and Walseng, B. (2017). Interactions between demography and environmental effects are important determinants of population dynamics. *Science Advances*, 3, e1602298.

Garcia, R.A., Cabeza, M., Rahbek, C., and Araújo, M.B. (2014). Multiple dimensions of climate change and their implications for biodiversity. *Science*, 344, 1247579.

Gilg, O., Sittler, B., and Hanski, I. (2009). Climate change and cyclic predator–prey population dynamics in the high-Arctic. *Global Change Biology*, 15, 2634–52.

Gilman, S.E., Urban, M.C., Tewksbury, J., Gilchrist, G.W., and Holt, R.D. (2010). A framework for community interactions under climate change. *Trends in Ecology and Evolution*, 15, 2634–52.

Gordon, C.E., Eldridge, D.J., Ripple, W.J., Crowther, M.S., Moore, B.D., and Letnic, M. (2016). Shrub encroachment is linked to extirpation of an apex predator. *Journal of Animal Ecology*, 86, 147–57.

Green, R.E. (1988). Effects of environmental factors on the timing and success of breeding of common snipe *Gallinago gallinago* (Aves, Scolopacidae). *Journal of Applied Ecology*, 25, 79–93.

Harvell, C., Mitchell, C., Ward, J., Altizer, S., Dobson, A., Ostfeld, R., and Samuel, M. (2002). Climate warming and disease risks for terrestrial and marine biota. *Science*, 296, 2158–62.

Henden, J., Ims, R., and Yoccoz, N. (2009). Nonstationary spatio-temporal small rodent dynamics: Evidence from long-term Norwegian fox bounty data. *Journal of Animal Ecology*, 78, 636–45.

Hilton, G., Thompson, D., Sagar, P., Cuthbert, R., Cherel, Y., and Bury, S. (2006). A stable isotopic investigation into the causes of decline in a sub-Antarctic predator, the rockhopper penguin *Eudyptes chrysocome*. *Global Change Biology*, 12, 611–25.

Hole, D., Willis, S., Pain, D., Fishpool, L., Butchart, S., Collingham, Y., Rahbek, C., and Huntley, B. (2009). Projected impacts of climate change on a continent-wide protected area network. *Ecology Letters*, 12, 420–31.

Hovinen, J.E.H., Welcker, J., Descamps, S., Strøm, H., Jerstad, K., Berge, J., et al. (2014). Climate warming decreases the survival of the little auk (*Alle alle*), a high Arctic avian predator. *Ecology and Evolution*, 4, 3127–38.

Hoy, S.R., Petty, S.J., Millon, A., Whitfield, D.P., Marquiss, M., Anderson, D.I.K., Davison, M., and Lambin, X. (2017). Density-dependent increase in superpredation linked to food limitation in a recovering population of northern goshawks *Accipiter gentilis*. *Journal of Avian Biology*, 48, 1205–15.

Huntley, B., Green, R., Collingham, Y., Hill, J., Willis, S., Bartlein, P., Cramer, W., Hagemeijer, W., and Thomas, C. (2004). The performance of models relating species geographical distributions to climate is independent of trophic level. *Ecology Letters*, 7, 417–26.

Iknayan, K.J., and Beissinger S.R. (2018). Collapse of a desert bird community over the past century driven by climate change. *Proceedings of the National Academy of Sciences of the United States of America*, 115, 8597–602.

Ims, R., and Fuglei, E. (2005). Trophic interaction cycles in tundra ecosystems and the impact of climate change. *BioScience*, 55, 311–22.

Ims, R., Henden, J., and Killengreen, S. (2008). Collapsing population cycles. *Trends in Ecology and Evolution*, 23, 79–86.

Iverson, S.A., Gilchrist, H.G., Smith, P.A., Gaston, A.J., and Forbes, M.R. (2014). Longer ice-free seasons increase the risk of nest depredation by polar bears for colonial breeding birds in the Canadian Arctic. *Proceedings of the Royal Society of London Series B: Biological Sciences*, 281: 20133128. doi: 10.1098/rspb.2013.3128

Jaksic, F., Jimenez, J., Castro, S., and Feinsinger, P. (1992). Numerical and functional response of predators to a long-term decline in mammalian prey at a semiarid Neotropical site. *Oecologia*, 89, 90–101.

Jenouvrier, S. (2013). Impacts of climate change on avian populations. *Global Change Biology*, 19, 2036–57.

Jenouvrier, S., Barbraud, C., and Weimerskirch, H. (2003). Effects of climate variability on the temporal population dynamics of southern fulmars. *Journal of Applied Ecology*, 72, 576–87.

Jenouvrier, S., Thibault, J., Viallefont, A., Vidal, P., Ristow, D., Mougin, J., Brichetti, P., Borg, J., and Bretagnolle, V. (2009). Global climate patterns explain range-wide synchronicity in survival of a migratory seabird. *Global Change Biology*, 15, 268–79.

Jepsen, J., Hagen, S., Ims, R., and Yoccoz, N. (2008). Climate change and outbreaks of the geometrids *Operophtera brumata* and *Epirrita autumnata* in subarctic birch forest: Evidence of a recent outbreak range expansion. *Journal of Animal Ecology*, 77, 257–64.

Karell, P., Ahola, K., Karstinen, T., Valkama, J., and Brommer, J.E. (2011). Climate change drives microevolution in a wild bird. *Nature Communications*, 2, 208.

Kausrud, K., Mysterud, A., Steen, H., Vik, J.O., Østbye, E., Cazelles, B., Framstad, E., Eikeset, A., Mysterud, I., Solhøy, T., and Stenseth, N.C. (2008). Linking climate change to lemming cycles. *Nature*, 456, 93–8.

Keogan, K., Daunt, F., Wanless, S., et al. (2018). Global phenological insensitivity to shifting ocean temperatures among seabirds. *Nature Climate Change*, 8, 313.

Korpimäki, E., and Hakkarainen, H. (1991). Fluctuating food supply affects the clutch size of Tengmalm's owl independent of laying date. *Oecologia*, 85, 543–52.

Korpimäki, E., Klemola, T., Norrdahl, K., Oksanen, L., Oksanen, T., Banks, P., Batzli, G., and Henttonen, H. (2003). Vole cycles and predation. *Trends in Ecology and Evolution*, 18, 494–5.

Kristensen, N.P., Johansson, J., Ripa, J., and Jonzen, N. (2015). Phenology of two interdependent traits in migratory birds in response to climate change. *Proceedings of the Royal Society of London Series B: Biological Sciences*, 282, 20150288.

Lande, R., Engen, S., and Saether, B.-E. (2003). *Dynamics in ecology and conservation*. Oxford University Press, Oxford, UK.

Laverty, C., Brenner, D., McIlwaine, C., Lennon, J.J., Dick, J.T.A., Lucy, F.E., and Christian, K.A. (2017). Temperature rise and parasitic infection interact to increase the impact of an invasive species. *International Journal of Parasitology*, 47, 291–6.

Lehikoinen, A., and Virkkala, R. (2016). North by northwest: climate change and directions of density shifts in birds. *Global Change Biology*, 22, 1121–9.

Lehikoinen, A., Byholm, P., Ranta, E., Saurola, P., Valkama, J., Korpimäki, E., Pietiainen, H., and Henttonen, H. (2009). Reproduction of the common buzzard at its northern range margin under climatic change. *Oikos*, 118, 829–36.

Lensing, J., and Wise, D. (2006). Predicted climate change alters the indirect effect of predators on an ecosystem process. *Proceedings of the National Academy of Sciences of the United States of America*, 103, 15502–5.

Letnic, M., Ritchie, E.G., and Dickman, C.R. (2012). Top predators as biodiversity regulators: the dingo *Canis lupus dingo* as a case study. *Biological Reviews*, 87, 390–413.

Lima, S.L. (1998). Nonlethal effects in the ecology of predator–prey interactions. What are the ecological effects of anti-predator decision-making? *Bioscience*, 48, 25–34.

Ling, S., and Johnson, C. (2012). Marine reserves reduce risk of climate-driven phase shift by reinstating size- and habitat-specific trophic interactions. *Ecological Applications*, 22, 1232–45.

Logan, J., Wolesensky, W., and Joern, A. (2006). Temperature-dependent phenology and predation in arthropod systems. *Ecological Modelling*, 196, 471–82.

Lord, J.P., Barry, J.P., and Graves, D. (2017). Impact of climate change on direct and indirect species interactions. *Marine Ecology Progress Series*, 571, 1–11.

Lovejoy, T., and Hannah, L (2005). *Climate change and biodiversity*. Yale University Press, New Haven, CT.

Lurgi, M., López, B.C., and Montoya, JM. (2012). Novel communities from climate change. *Philosophical Transactions of the Royal Society of London Series B: Biological Sciences*, 367, 2913–22.

MacLean, S.A., and Beissinger, S.R. (2017). Species' traits as predictors of range shifts under contemporary climate change: A review and meta-analysis. *Global Change Biology*, 23, 4094–105.

Marshall, M., Cooper, R., DeCecco, J., Strazanac, J., and Butler, L. (2002). Effects of experimentally reduced prey abundance on the breeding ecology of the red-eyed vireo. *Ecological Applications*, 12, 261–80.

Martin, T. (2001). Abiotic vs. biotic influences on habitat selection of coexisting species: Climate change impacts? *Ecology*, 82, 175–88.

McCann, K. (2007). Protecting biostructure. *Nature*, 446, 29.

McCluney, K.E., Belnap, J., Collins, S.L., Gonzalez, A.L., Hagen, E.M., Holland, J.N., Kotler, B.P., Maestre, F.T., Smith, S.D., and Wolf, B.O. (2012). Shifting species interactions in terrestrial dryland ecosystems under altered water availability and climate change. *Biological Reviews*, 87, 563–82.

McClure, C.J.W., Westrip, J.R.S., Johnson, J.A., Schulwitz, S.E., Virani, M.Z., Davies, R, Symes, A., Wheatley, H.,

Thorstrom, R., Amar, A., Buij, R., Jones, V.J., Williams, N.P., Buechley, E.R., and Butchart, S.H.M. (2018). State of the world's raptors: distributions, threats, and conservation recommendations. *Biological Conservation*, 227, 390–402. doi: 0.1016/j.biocon.2018.08.012

Mech, L., and Peterson, R. (2002). Wolf-prey relationships. In: L. Mech and L. Boitani (eds), *Wolves: Behavior, ecology, conservation*, pp. 131–160. University of Chicago Press, Chicago, IL.

Millon, A., and Bretagnolle, V. (2008). Predator population dynamics under a cyclic prey regime: Numerical responses, demographic parameters and growth rates. *Oikos*, 117, 1500–10.

Millon, A., Arroyo, B., and Bretagnolle, V. (2008). Variable but predictable prey availability affects predator breeding success: Natural versus experimental evidence. *Journal of Zoology*, 275, 349–58.

Millon, A., Nielsen, J., Bretagnolle, V., and Møller, A.P (2009). Predator–prey relationships in a changing environment: the case of the sparrowhawk on its avian prey community in a rural area. *Journal of Animal Ecology*, 78, 1086–95.

Millon, A., Petty, S.J., Little, B., Gimenez, O., Cornulier, T., and Lambin, X. (2014). Dampening prey cycle overrides the impact of climate change on predator population dynamics: a long-term demographic study on tawny owls. *Global Change Biology*, 20, 1770–81.

Møller, A.P. (2002). North Atlantic Oscillation (NAO) effects of climate on the relative importance of first and second clutches in a migratory passerine bird. *Journal of Animal Ecology*, 71, 201–10.

Møller, A.P. (2009). Host–parasite interactions and vectors in the barn swallow in relation to climate change. *Global Change Biology*, 16, 1158–70.

Møller, A.P., Flensted-Jensen, E., and Mardal, W. (2006). Rapidly advancing laying date in a seabird and the changing advantage of early reproduction. *Journal of Animal Ecology*, 75, 657–65.

Møller, A.P., Fiedler, W., and Berthold, P. (2004). *Birds and climate change*. Academic Press, London, UK.

Møller, A.P., Thorup, O., and Laursen, K. (2018). Predation and nutrients drive population declines in breeding waders. *Ecological Applications*, 28, 1292–130.

Morris, T., and Letnic, M. (2017). Removal of an apex predator initiates a trophic cascade that extends from herbivores to vegetation and the soil nutrient pool. *Proceedings of the Royal Society of London Series B: Biological Sciences*, 284, 20170111.

Morris, W., Pfister, C., Tuljapurkar, S., Haridas, C., Boggs, C., Boyce, M., Bruna, E., Church, D., Coulson, T., and Doak, D. (2008). Longevity can buffer plant and animal populations against changing climatic variability. *Ecology*, 89, 19–25.

Murdoch, W., Briggs, C., and Nisbet, R. (2003). *Consumer–resource dynamics*. Princeton University Press, Princeton, NJ.

Murphy, E., Trathan, P., Watkins, J., Reid, K., Meredith, M., Forcada, J., Thorpe, S., Johnston, N., and Rothery, P. (2007). Climatically driven fluctuations in Southern Ocean ecosystems. *Proceedings of the Royal Society of London Series B: Biological Sciences*, 274, 3057–67.

Newbold, T. (2018). Future effects of climate and land-use change on terrestrial vertebrate community diversity under different scenarios. *Proceedings of the Royal.Society of London Series B: Biological Sciences*, 285, 20180792.

Nicoll, M.A.C., Nevoux, M., Jones, C.G., Ratcliffe, N., Ruhomaun, K., Tatayah, V., and Norris, K. (2017). Contrasting effects of tropical cyclones on the annual survival of a pelagic seabird in the Indian Ocean. *Global Change Biology*, 23, 550–565.

Niehaus, A., and Ydenberg, R. (2006). Ecological factors associated with the breeding and migratory phenology of high-latitude breeding western sandpipers. *Polar Biology*, 30, 11–17.

Nielsen, J.T., and Møller, A.P. (2006). Effects of food abundance, density and climate change on reproduction in the sparrowhawk *Accipiter nisus*. *Oecologia*, 149, 505–18.

Nielsen, O. (1999). Gyrfalcon predation on ptarmigan: Numerical and functional responses. *Journal of Animal Ecology*, 68, 1034–50.

Orians, G., and Pearson, N. (1979). On the theory of central place foraging. In: D.J. Horn, G.R. Staris, and R.D. Mitchell (eds), *Analysis of ecological systems*, pp. 153–77. Ohio State University Press, Columbus, OH.

Oro, D. (2014). Seabird and climate: knowledge, pitfalls, and opportunities. *Frontiers in Ecology and Evolution*, 79, 1 12.

Paine, R. (1980). Food webs: Linkage, interaction strength and community infrastructure. *Journal of Animal Ecology*, 49, 667–85.

Parmesan, C., and Yohe, G. (2003). A globally coherent fingerprint of climate change impacts across natural systems. *Nature*, 421, 37–42.

Pasanen-Mortensen, M., Elmhagen, B., Lindén, H., Bergström, R., Wallgren, M., van der Velde, Y., and Cousins, S.A.O. (2017). The changing contribution of top-down and bottom-up limitation of mesopredators during 220 years of land use and climate change. *Journal of Animal Ecology*, 86, 566–76.

Peach, W., Robinson, R., and Murray, K. (2004). Demographic and environmental causes of the decline of rural song thrushes *Turdus philomelos* in lowland Britain. *Ibis*, 146, 50–9.

Pearce-Higgins, J., Yalden, D., and Whittingham, M. (2005). Warmer springs advance the breeding phenology of golden plovers *Pluvialis apricaria* and their prey (Tipulidae). *Oecologia*, 143, 470–6.

Pearce-Higgins, J.W., Dennis, P., Whittingham, M.J., and Yalden, D.W. (2010). Impacts of climate on prey abundance account for fluctuations in a population of a northern wader at the southern edge of its range. *Global Change Biology*, 16, 12–23.

Pearce-Higgins, J.W., Eglington, S.M., Martay, B., and Chamberlain, D.E. (2015). Drivers of climate change impacts on bird communities. *Journal of Animal Ecology*, 84, 943–54.

Peckarsky, B., Abrams, P., Bolnick, D., Dill, L., Grabowski, J., Luttbeg, B., Orrock, J., Peacor, S., Preisser, E., and Schmitz, O. (2008). Revisiting the classics: Considering non-consumptive effects in textbook examples of predator–prey interactions. *Ecology*, 89, 2416–25.

Peers, M.J.L, Wehtje, M., Thornton, D.H., and Murray, D.L. (2014). Prey switching as a means of enhancing persistence in predators at the trailing southern edge. *Global Change Biology*, 20, 1126–35.

Pellissier, L., Rohr, R.P., Ndiribe, C., Pradervand, J.-N., Salamin, N., Guisan, A., and Wisz, M. (2013). Combining food web and species distribution models for improved community projections. *Ecology and Evolution*, 3, 4572–83.

Péron, C., Weimerskirch, H., and Bost, C.A. (2012). Projected poleward shift of king penguins' (*Aptenodytes patagonicus*) foraging range at the Crozet Islands, southern Indian Ocean. *Proceedings of the Royal Society of London Series B: Biological Sciences*, 279, 2515–23.

Pettorelli, N., Hilborn, A., Duncan, C., and Durant, S.M. (2015). Individual variability: the missing component to our understanding of predator–prey interactions. *Advances in Ecological Research*, 52, 19–44.

Pincebourde, S., Sanford, E., and Helmuth, B. (2008). Body temperature during low tide alters the feeding performance of a top intertidal predator. *Limnology and Oceanography*, 53, 1562–73.

Pokallus, J.W., and Pauli, J.N. (2015). Population dynamics of a northern-adapted mammal: disentangling the influence of predation and climate change. *Ecological Applications*, 25, 1546–56.

Polis, G., Sears, A., Huxel, G., Strong, D., and Maron, J. (2000). When is a trophic cascade a trophic cascade? *Trends in Ecology and Evolution*, 15, 473–5.

Porter, W., Budaraju, S., Stewart, W., and Ramankutty, N. (2000). Calculating climate effects on birds and mammals: Impacts on biodiversity, conservation, population parameters, and global community structure. *Integrative and Comparative Biology*, 40, 597–630.

Post, E., and Forchhammer, M. (2002). Synchronization of animal population dynamics by large-scale climate. *Nature*, 420, 168–71.

Post, E., and Forchhammer, M. (2004). Spatial synchrony of local populations has increased in association with the recent northern hemisphere climate trend. *Proceedings of the National Academy of Sciences of the United States of America*, 101, 9286–90.

Post, E., and Forchhammer, M. (2006). Spatially synchronous population dynamics: An indicator of Pleistocene faunal response to large-scale environmental change in the Holocene. *Quaternary International*, 151, 99–105.

Post, E., and Stenseth, N.C. (1999). Climatic variability, plant phenology, and northern ungulates. *Ecology*, 80, 1322–39.

Precheur, C., Barbraud, C., Martail, F., Mian, M., Nicolas, J.-C., Brithmer, R., Belfan, D., Conde, B., and Bretagnolle, V. (2016). Some like it hot: effect of environment on population dynamics of a small tropical seabird in the Caribbean region. *Ecosphere*, 7, e01461.

Rand, T., and Tscharntke, T. (2007). Contrasting effects of natural habitat loss on generalist and specialist aphid natural enemies. *Oikos*, 116, 1353–62.

Ranta, N., Kaitala, V., and Lindström, J. (1999). Spatially autocorrelated disturbances and patterns in population synchrony. *Proceedings of the Royal Society of London Series B: Biological Sciences*, 266, 1851–6.

Redpath, S., and Thirgood, S. (1999). Numerical and functional responses in generalist predators: Hen harriers and peregrines on Scottish grouse moors. *Journal of Animal Ecology*, 68, 879–92.

Renner, S.S., and Zohner, C.M. (2018). Climate change and phenological mismatch in trophic interactions among plants, insects, and vertebrates. *Annual Review of Ecology, Evolution and Systematics*, 49, 165–82.

Ripple, W.J., Estes, J.A., Beschta, R.L., Wilmers, C.C., Ritchie, E.G., Hebblewhite, M., Berger, J., Elmhagen, B., Letnic, M., Nelson, M.P., Schmitz, O.J., Smith, D.W., Wallach, A.D., and Wirsing, A.J. (2014). Status and ecological effects of the world's largest carnivores. *Science*, 343, 1241484.

Robinson, B.G., Franke, A., and Derocher, A.E. (2017). Weather-mediated decline in prey delivery rates causes food-limitation in a top avian predator. *Journal of Avian Biology*, 48, 748–58.

Rodionov, S. (2004). A sequential algorithm for testing climate regime shifts. *Geophysical Research Letters*, 31, L09204.

Root, T., Price, J., Hall, K., Schneider, S., Rosenzweig, C., and Pounds, J. (2003). Fingerprints of global warming on wild animals and plants. *Nature*, 421, 57–60.

Rosenblatt, A.E., and Schmitz, O.J. (2016). Climate change, nutrition, and bottom-up and top-down food web processes. *Trends in Ecology and Evolution*, 31, 965–75.

Rutz, C., and Bijlsma, R. (2006). Food-limitation in a generalist predator. *Proceedings of the Royal Society of London Series B: Biological Sciences*, 273, 2069–76.

Sæther, B.-E., Engen, S., Møller, A.P., Visser, M., Matthysen, E., Fiedler, W., Lambrechts, M., Becker, P., Brommer, J., and Dickinson, J. (2005). Time to extinction of bird populations. *Ecology*, 86, 693–700.

Scheffers, B.R., De Meester, L., Bridge, T.C.L., Hoffmann, A.A., Pandolfi, J.M., Corlett, R.T., Butchart, S.H.M., Pearce-Kelly, P., Kovacs, K.M., Dudgeon, D., Pacifici, M., Rondinini, C., Foden, W.B., Martin, T.G., Mora, C., Bickford, D., and Watson, J.E.M. (2016). The broad footprint of climate change from genes to biomes to people. *Science*, 354, aaf7671.

Schmidt, N.M., Ims, R.A., Høye, T.T., Gilg, O., Hansen, L.H., Hansen, J., Lund, M., Fuglei, E., Forchhammer, M.C., and Sittler, B. (2012). Response of an Arctic predator guild to collapsing lemming cycles. *Proceedings of the Royal Society of London Series B: Biological Sciences*, 279, 4417–22.

Schmitz, O., Hamback, P., Beckerman, A., and Leibold, M. (2000). Trophic cascades in terrestrial systems: a review of the effects of carnivore removals on plants. *American Naturalist*, 155, 141–53.

Schmitz, O., Krivan, V., and Ovadia, O. (2004). Trophic cascades: The primacy of trait-mediated indirect interactions. *Ecology Letters*, 7, 153–63.

Seibold, S., Cadotte, M.W., MacIvor, J.S., Thorn, S., and Müller, J. (2018). The necessity of multitrophic approaches in community ecology. *Trends in Ecology and Evolution*, 33, 754–64.

Sinclair, A., and Krebs, C. (2002). Complex numerical responses to top-down and bottom-up processes in vertebrate populations. *Philosophical Transactions of the Royal Society of London Series B: Biological Sciences*, 357, 1221–32.

Sinclair, A.R.E., Metzger, K.L., Fryxell, J.M., Packer, C., Byrom, A.E., Craft, M.E., Hampson, K., Lembo, T., Durant, S.M., Forrester, G.J., Bukombe, J., Mchetto, J., Dempewolf, J., Hilborn, R., Cleaveland, S., Nkwabi, A., Mosser, A., and Mduma, S.A.R. (2013). Asynchronous food-web pathways could buffer the response of Serengeti predators to El Niño southern oscillation. *Ecology*, 94,1123–30.

Solonen, T. (2006). Overwinter population change of small mammals in southern Finland. *Annales Zoologici Fennici*, 43, 295–302.

Soulé, M.E., Bolger, D.T. Alberts, A.C., Wright, J., Sorice, M., and Hill, S. (1988). Reconstructed dynamics of rapid extinctions of chaparral-requiring birds in urban habitat islands. *Conservation Biology*, 2, 75–91.

Spooner, F.E.B., Pearson, R.G., and Freeman, R. (2018). Rapid warming is associated with population decline among terrestrial birds and mammals globally. *Global Change Biology*, 24, 4521–31.

Stenseth, N.C., Chan, K., Tong, H., Boonstra, R., Boutin, S., Krebs, C., Post, E., O'Donoghue, M., Yoccoz, N., and Forchhammer, M. (1999). Common dynamic structure of Canada lynx populations within three climatic regions. *Science*, 285, 1071–3.

Stenseth, N.C., Mysterud, A., Ottersen, G., Hurrell, J., Chan, K., and Lima, M. (2002). Ecological effects of climate fluctuations. *Science*, 297, 1292–6.

Stenseth, N.C., Ottersen, G., Hurrell, J., and Belgrano, A. (eds) (2005). *Marine ecosystems and climate variation: The North Atlantic: A comparative perspective*. Oxford University Press, Oxford, UK.

Stephens, P., and Sutherland, W.J. (1999). Consequences of the Allee effect for behaviour, ecology and conservation. *Trends in Ecology and Evolution*, 14, 401–5.

Stevens, A., Welch, Z., Darby, P., and Parcival, H. (2002). Temperature effects on Florida apple snail activity: Implications for snail kite foraging success and distribution. *Wildlife Society Bulletin*, 30, 75–81.

Stevenson, I., and Bryant, D. (2000). Avian phenology: Climate change and constraints on breeding. *Nature*, 406, 366–7.

Sutherland, W.J. (2006). *Ecological census techniques*. Cambridge University Press, Cambridge, UK.

Suraci, J.P., Clinchy, M., Dill, L.M, Roberts, D., and Zanette, L.Y. (2016). Fear of large carnivores causes a trophic cascade. *Nature Communications*, 7, 10698.

Sutherst, R., Maywald, G., and Bourne, A. (2007). Including species interactions in risk assessments for global change. *Global Change Biology*, 13, 1843–59.

Suttle, K., Thomsen, M., and Power, M. (2007). Species interactions reverse grassland responses to changing climate. *Science*, 315, 640–2.

Sydeman, W.J., Thompson, S.A., Santora, J.A., Koslow, J.A., Goericke, R., and Ohman, M.D. (2015). Climate–ecosystem change off southern California: Time dependent seabird predator–prey numerical responses. *Deep-Sea Research II*, 112, 158–70.

Terraube, J., Villers, A., Poudré, L., Varjonen, R., and Korpimäki, E. (2017). Increased autumn rainfall disrupts predator–prey interactions in fragmented boreal forests. *Global Change Biology*, 23, 1361–73.

Terraube, J., and Bretagnolle, V. (2018). Top-down limitation of mesopredators by avian top predators: a call for research on cascading effects at the community and ecosystem scale. *Ibis*, 160, 693–702.

Terraube, J., Villers, A., Ruffino, L., Iso-Iivari, L., Henttonen, H., Oksanen, T., and Korpimäki, E. (2015). Coping with fast climate change in northern ecosystems: Mechanisms underlying the population-level response of a specialist avian predator. *Ecography*, 38, 690–9.

Thackeray, S.J., Sparks, T.H., Frederiksen, M., Burthe, S., Bacon, P.J., Bell, J.R., Botham, M.S., Brereton, T.M., Bright, P.W., Carvalho, L., Clutton-Brock, T., Dawson, A., Edwards, M., Elliott, J.M., Harrington, R., Johns, D.,

Jones, I.D., Jones, J.T., Leech, D.I., Roy, D.B., Scott, W.A., Smith, M., Smithers, R.J., Winfield I.J., and Wanless, S. (2010). Trophic level asynchrony in rates of phenological change for marine, freshwater and terrestrial environments. *Global Change Biology*, 16, 3304–13.

Trainor, A.M., and Schmitz, O.J. (2014). Infusing considerations of trophic dependencies into species distribution modelling. *Ecology Letters*, 17, 1507–17.

Trathan, P., Forcada, J., and Murphy, E. (2007). Environmental forcing and southern ocean marine predator populations: Effects of climate change and variability. *Philosophical Transactions of the Royal Society of London Series B: Biological Sciences*, 362, 2351–65.

Turchin, P. (1995). Population regulation: Old arguments and a new synthesis. In N. Cappucino and P.W. Price (eds), *Population dynamics: New approaches and synthesis*, pp. 19–40. Academic Press, New York, NY.

Tylianakis, J., Didham, R., Bascompte, J., and Wardle, D. (2008). Global change and species interactions in terrestrial ecosystems. *Ecology Letters*, 11, 1351–63.

Tyson, R., and Lutscher, F. (2016). Seasonally varying predation behavior and climate shifts are predicted to affect predator–prey cycles. *American Naturalist*, 188, 539–53.

Valkama, J., Korpimäki, E., Arroyo, B., Beja, P., Bretagnolle, V., Bro, E., Kenward, R., Manosa, S., Redpath, S., Thirgood, S., and Viñuela, J. (2005). Birds of prey as limiting factors of gamebird populations in Europe: A review. *Biological Reviews*, 80, 171–203.

van der Meer, J., and Smallegange, I. (2009). A stochastic version of the Beddington–DeAngelis functional response: Modelling interference for a finite number of predators. *Journal of Animal Ecology*, 78, 134–42.

Veselý, L., Boukal, D.S., Buřic, M., Kozák, P., Kouba, A., and Sentis, A. (2017). Effects of prey density, temperature and predator diversity on nonconsumptive predator-driven mortality in a freshwater food web. *Scientific Reports*, 7, 18075.

Villellas, J., Doak, D.F., Garcıa, M.B., and Morris, W.F. (2015). Demographic compensation among populations: what is it, how does it arise and what are its implications? *Ecology Letters*, 18, 1139–52.

Visser, M., Both, C., and Lambrechts, M. (2004). Global climate change leads to mistimed avian reproduction. *Advances in Ecological Research*, 35, 89–110.

Visser, M., Van Noordwijk, A., Tinbergen, J., and Lessells, C. (1998). Warmer springs lead to mistimed reproduction in great tits (*Parus major*). *Proceedings of the Royal Society of London Series B: Biological Sciences*, 265, 1867–70.

Voigt, W., Perner, J., and Jones, T.H. (2007). Using functional groups to investigate community response to environmental changes: Two grassland case studies. *Global Change Biology*, 13, 1710–21.

Vors, L., and Boyce, M. (2009). Global declines of caribou and reindeer. *Global Change Biology*, 15, 1365–2486.

Vucetich, J., and Peterson, R. (2004). The influence of top-down, bottom-up and abiotic factors on the moose (*Alces alces*) population of Isle Royale. *Proceedings of the Royal Society of London, Series B: Biological Sciences*, 271, 183–9.

Vucetich, J., Smith, D., and Stahler, D. (2005). Influence of harvest, climate and wolf predation on Yellowstone elk, 1961–2004. *Oikos*, 111, 259–70.

Vucic-Pestic, O., Ehnes, R.B., Rall, B.C., and Brose, U. (2011). Warming up the system: higher predator feeding rates but lower energetic efficiencies. *Global Change Biology*, 17, 1301–10.

Wanless, S., Frederiksen, M., Daunt, F., Scott, B., and Harris, M. (2007). Black-legged kittiwakes as indicators of environmental change in the North Sea: Evidence from long-term studies. *Progress in Oceanography*, 72, 30–8.

Wetherington, M.T., Jennings, D.E., Shrewsbury, P.M., and Duan, J.J. (2017). Climate variation alters the synchrony of host–parasitoid interactions. *Ecology and Evolution*, 7, 8578–87.

Wilmers, C., Post, E., and Hastings, A. (2007). The anatomy of predator–prey dynamics in a changing climate. *Journal of Animal Ecology*, 76, 1037–44.

Wilmers, C., Post, E., Peterson, R., and Vucetich, J. (2006). Predator disease out-break modulates top-down, bottom-up and climatic effects on herbivore population dynamics. *Ecology Letters*, 9, 383–9.

Winder, M., and Schindler, D. (2004). Climate change uncouples trophic interactions in an aquatic ecosystem. *Ecology*, 85, 2100–6.

Wright, L., Hoblyn, R., Green, R., Bowden, C., Mallord, J., Sutherland, W.J., and Dolman, P. (2009). Importance of climatic and environmental change in the demography of a multi-brooded passerine, the woodlark *Lullula arborea*. *Journal of Animal Ecology*, 78, 1191–202.

Yom-Tov, Y. (2001). Global warming and body mass decline in Israeli passerine birds. *Proceedings of the Royal Society of London Series B: Biological Sciences*, 268, 947–52.

CHAPTER 16

Bird communities and climate change

Lluís Brotons, Sergi Herrando, Frédéric Jiguet, and Aleksi Lehikoinen

16.1 General context

Climate is an important factor shaping the structure and composition of plant and animal communities. Climate drives many aspects of the ecology of species directly or indirectly through changes in habitat type and structure (Berthold 1998), so that climate gradients are thought to be a key determinant of communities at large spatial scales (Forsman and Mönkkönen 2003; Holm and Svenning 2014). Changes in climatic conditions are suggested to have been the main causes of species extinctions and large-scale community changes and re-organization in the geological past (Mayhew et al. 2008). Since current rates of change in climatic conditions are faster than those previously reported (Chapter 2), species extinction rates and community structure changes are expected to increase accordingly (Pimm 2009).

Disruptions of current climatic regimes may lead to the loss of a large fraction of the planet's biological diversity even if we only consider a simple mechanism such as the shift of species ranges when temperatures rise (Krosby et al. 2015; Thomas et al. 2004). However, climate change is much more complex in both space and time, and the shift of species ranges in latitude and altitude is not the only factor affecting ecological communities. First, climate change is likely to be complex and differ regionally in intensity and directionality leading to idiosyncratic impacts on species ecology. Second, climate affects a great number of basic ecological mechanisms and therefore the number of impacted species at different complexity levels is large (Walther 2007). Therefore, climate will differentially change the distribution of resources at different sites, creating mismatches in resource use and availability in different species (Pimm 2009). Such mismatches will eventually lead to cascade effects and asynchronies between the responses of different trophic groups, enhancing the negative and positive cumulative impacts of climate change on ecological communities (Brown et al. 1997; Martin and Maron 2012). How will these additive effects—across species, space, and time—affect present and future bird communities?

The community approach can assess the summed effects of different factors of change on different species, trophic levels, and spatial scales, ideally integrating co-occurrence structure to consider species interactions. In this review, we use this perspective to analyse the effects of climate change on bird communities. Here, we build from previous reviews of this recent research field (Brotons and Jiguet 2010; Møller et al. 2004) and assess the present state of the art to identify key points mining our capability to anticipate impacts of climate change on bird communities. Below, we first discuss current methods used to anticipate potential changes in bird communities derived from climate changes and then report on the responses observed so far at the community level for birds.

Brotons, L., Herrando, S., Jiguet, F., and Lehikoinen, A., *Bird communities and climate change*. In: *Effects of Climate Change on Birds*. Second Edition. Edited by Peter O. Dunn and Anders Pape Møller: Oxford University Press (2019).
 DOI: 10.1093/oso/9780198824268.003.0016

16.2 Methods to predict the impacts of climate change on bird communities

There are two main approaches to predict the impact of climate change on bird communities (Table 16.1). First, predictions for communities may be derived from single-species studies, which assume the response is sufficiently general or consistent among species (Jenouvrier 2013). Second, there are a variety of modelling methods that can be used to make predictions for entire communities. For example, species distribution models combine current species occurrence data and climatic variables to predict the future distribution of species under climate change scenarios (Chapter 8). Under the assumption that a given species will conserve its climatic preferences in the future, we can predict the potential future distribution under different assumptions of species dispersal ability using future climatic scenarios provided by the International Panel on Climate Change (IPCC) (Şekercioğlu et al. 2012; Thomas et al. 2004). These single-species models can then be expanded to make predictions about communities. However, to synthesize multi-species impacts of climate disruption, we must take a next step in considering changes in patterns of species

Table 16.1 Examples of the main predictions of climate change impacts on bird communities derived from species-specific and modelling-based studies. Community derived impacts for different species traits are based on specific findings and their associated references.

Species traits	Community derived impact	Type of study	Finding	References
Breeding	Increases in generalist species	Species based	Advancement in breeding phenology	(Kluen et al. 2017; Winkler et al. 2002)
Breeding	Decreases in species richness and community turnover	Species based	Increases in asynchrony between trophic levels	(Franks et al. 2018; Sanz et al. 2003)
Winter survival	Regional variation in species richness	Species based	Increases in winter survival in temperate climates and decrease in tropical climates	(Barbraud et al. 1999)
Migration	Increases in species richness due to increased residency	Species based	Changes the proportion of migratory individuals in migratory species	(Meller et al. 2016; Potvin et al. 2016; Visser et al. 2009)
Migration	Decreases in species richness due to decreases in populations of long-distance migrants	Species based	Altered migration regimes in long-distance migrants	(Cotton 2003)
Body size	Changes in competition between species leading to species turnover	Species based	Decreasing body size with temperature	(Gardner et al. 2009; Yom-Tov 2001)
Species distribution	Increases in species richness in temperature-limited systems and decreases in precipitation-limited systems	Modelling	Species richness related to ecosystem productivity	(Jetz et al. 2007)
Species distribution	Northern shifts of winter range	Species based	Species distribution matching changes in climatic conditions	(La Sorte et al. 2009; Lehikoinen et al. 2013)
Species distribution	Changes in species richness patterns at different locations	Modelling	Species distribution matching changes in climatic conditions	(Huntley et al. 2007)
Species distribution	Species turnover due to idiosyncratic responses to climate	Modelling	Species distribution matching changes in climatic conditions	(Lawler et al. 2009; Stralberg et al. 2009)
Migratory component	Decreases of abundance of long-distance migrants with increased winter temperatures	Modelling	Association between climatic gradients and proportion of migrants in the community	(Lemoine and Böhning-Gaese 2003)
Thermal composition	Community shifts to more species which like it hot	Modelling	Community thermal index increased locally	(Devictor et al. 2012; Prince and Zuckerberg 2015; Tayleur et al. 2016)

co-occurrence. Joint species distribution models (JSDMs) integrate the relationship between species and their environments and potential biotic interactions, using co-occurrence patterns to identify ecological processes (Pollock et al. 2014). JSDMs estimate distributions of multiple species simultaneously and allow decomposition of species co-occurrence patterns into components describing shared environmental responses and residual patterns of co-occurrence. If one has clear a priori predictions for some species interactions, JSDMs are useful for validating suspected interactions beyond their dependence on environmental factors (competition, predation, parasitism, amongst others).

16.3 Predicted changes in bird communities

We used models and observational studies to summarize predicted impacts on bird communities (Table 16.1). These impacts generally focus on species richness, turnover, and function.

16.3.1 Predicted changes in species richness

Many studies have projected future changes in species distributions and patterns of diversity (species range changes, richness, and extinction rates), with an emphasis on range contractions and potential for species extinctions (Massimino et al. 2017; Stralberg et al. 2015). Two major publications using this approach over large spatial scales are '*A climatic atlas of European breeding birds*' (Huntley et al. 2007), and an assessment of global bird diversity (Jetz et al. 2007). Both studies compiled maps of species richness and potential climate-induced changes in species richness for the twenty-first century. Assessing the potential ability of species to disperse from current to predicted future range is crucial, and assumptions of total or null dispersion can lead to opposite regional trends in species richness, as shown by Barbet-Massin et al. (2009) when studying wintering ranges of Afro-Palearctic migrant passerines. In the case of no dispersal, local species richness is at best stable. However, birds are highly mobile (Paradis et al. 1998) and considering range shifts of a few hundred kilometres during a century does not seem unreasonable for breeding ranges (but see Howard et al. (2018) for climate-related mobility constraints to migratory species).

When combining species distribution models to obtain information on species assemblages, another option is to combine raw data of predicted suitability probability, instead of combining threshold-dependent presence–absence model outputs. This approach was used to study how current reserves in the Cerrado region of Brazil will protect bird diversity in the face of climate-driven changes (Marini et al. 2009). Their study was based on the combination of species distribution models obtained for 38 endemic or rare bird species (Figure 16.1). Comparing current and future (2046–2060) distributions with the current Brazilian reserve system highlighted large gaps in their protection. The cumulative climatic suitability of all species under future climate scenarios was proposed as a useful tool to implement new reserves, especially in the south-eastern part of the Cerrado region and in the mountains of eastern Brazil.

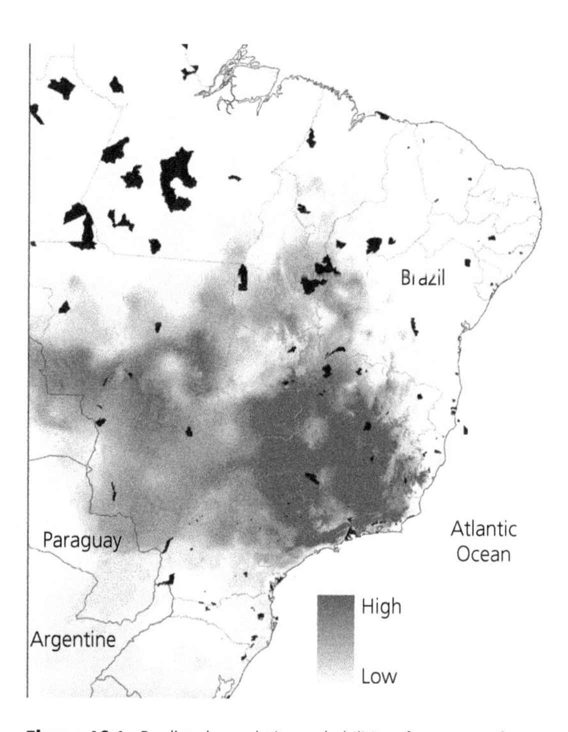

Figure 16.1 Predicted cumulative probabilities of occurrence for the 38 Cerrado bird species in Brazil for the future (2046–2060), as obtained with an ensemble-forecasting approach. The colours in the maps represent the cumulative probability of occurrence (climatic suitability) within a pixel, ranging from 0 to 1 (pale to dark green). Large (> 30 000 ha) reserves are represented by black polygons. Based on Marini et al. (2009).

16.3.2 Predicted changes in species turnover and function

At the community level, range shifts lead to a turnover of species, i.e. community reassembly. In contrast, phenotypic adaptation allows species to persist *in situ*, conserving community composition. So far, community reassembly and adaptation have mostly been studied separately. In nature, however, both processes would occur simultaneously. Schaefer et al. (2008) developed a model aimed to predict the impact of global climate change on migratory bird communities to assess the extent to which reassembly and adaptation may contribute to changes across bird assemblages in Europe. The magnitude of community reassembly was measured using spatial variation in the proportion of potentially migratory species. These spatial relationships were used to make temporal predictions about changes in migratory species under global climate change. According to their models, increasing winter temperature is expected to lead to declines in the proportion of migratory species, whereas increasing spring temperature and decreasing spring precipitation may lead to increases. They concluded that under current climate change forecasts, changes in the proportion of migratory species will be modest, and communities of migratory birds in Europe are projected to change through shifts in the proportion of migratory species rather than species turnover. However, we could also hypothesize that warmer winters would lead to declines in numbers of long-distance migrants if resident birds benefit from warmer winters and impose increasing competitive pressure on migrants.

Although high rates of projected species turnover have been identified for many geographic regions, this does not directly consider the degree to which novel or 'no-analogue' communities may be anticipated. Such turnover patterns would also intuitively be highly scale-dependent, making the topic particularly difficult to study. Entirely unique combinations of species and new interactions that occur among species may lead to even greater rates of local extirpation if species cannot adapt sufficiently quickly (Şekercioğlu et al. 2012). Indeed, climate disruption is predicted to induce independent shifts in species distributions, resulting in rapid development of novel species assemblages that could challenge the capacity of species to co-exist and adapt. For example, a recent study projected that by 2070 over half of California could be occupied by novel bird assemblages that have no analogy with currently observed species assemblages (Stralberg et al. 2015). These changes could have potentially dramatic consequences for species interactions and ecosystem function.

Bird assemblages fulfil a critical set of ecological functions for ecosystems and these have been projected to change highly unevenly across space. A recent study has combined species and community perspectives to assess the consequences of projected geographic range changes for the diverse functional attributes of avian assemblages worldwide (Barbet-Massin and Jetz 2015). Differences in functional structure across space arose from both changes in the number of species and changes in their local functional redundancy or distinctiveness. These changes often resulted in substantial losses of functional diversity that could lead to decreases in ecosystem health. Future range expansions could eventually counter functional losses in high-latitude regions, but offer little room for compensation in many tropical and subtropical biomes.

A more specific example of the functional role of birds in ecosystems is that of seed dispersal and frugivorous birds (Kati and Şekercioğlu 2006). Bird species profiting from seeds and fruits are known to spatially track food resources according to the species requirements imposed by weather conditions. In this context, changes in climate may alter the need for frugivorous species to search for new food sources, heavily impacting on the ecological network and structure of communities (Rivalan et al. 2007), potentially causing cascade effects at other trophic levels. Most frugivorous birds belong to particular orders such as perching birds (Passeriformes), woodpeckers (Piciformes), parrots (Psittaciformes) and pigeons (Columbiformes), and current climate explains more variance in species richness in these functional groups than in any other. Actual evapotranspiration is the best single climatic predictor of avian frugivory (Kissling et al. 2009), which allows prediction of impacts of future climatic change on frugivorous species. At a global level, diet and distribution information for 8472 species has been used to predict changes in guild structure under future climate change scenarios

(Ko et al. 2014). This analysis revealed that dietary guilds vary geographically; for example, insectivores are found more often in the tropics, while carnivores and omnivores are more common at higher latitudes. Under climate change forecasts for 2100, they predict the greatest change in dietary guilds at higher latitudes because of poleward shifts in distribution. However, many of the predictions are specific to particular regions and subject to increasing uncertainty depending on assumptions about dispersal distance.

16.4 Observed community changes: What are the observed responses so far?

In this section, we review the evidence that bird communities are already responding to climate change and to what degree observed changes fit predictions (Table 16.1). Despite the large body of evidence that global climate change has changed the structure of ecological communities, our understanding of the causes and mechanisms by which these changes occur in time and space is still partial.

16.4.1 Large scale changes in distributions matching climate change

Global warming predicts that species should shift their range poleward, everything else being equal (La Sorte and Thompson 2007; Thomas and Lennon 1999). Evidence from large-scale studies suggests that species distributions are generally changing in line with climate change predictions during both breeding and non-breeding seasons (Chen et al. 2011; Santangeli and Lehikoinen 2017), but there are often lags in their responses compared to reported climate change (Devictor et al. 2012; Quintero and Wiens 2013). Data from long-term bird monitoring schemes in Europe suggest that species predicted to expand their range are increasing and species predicted to shrink their distributions are decreasing (Gregory et al. 2009). This pattern is widespread; however, a closer analysis showed the trend in Europe is mainly driven by decreases in cold-sensitive species and in America by warm-expanding species (Figure 16.2; Stephens et al. 2016). These contrasting results support a general widespread effect of climate change, but demonstrate that measured responses are highly idiosyncratic and depend on region, specific climatic constraints, and species-specific traits such as habitat selection, feeding ecology, or migration ecology (Brommer 2008; MacLean and Beissinger 2017).

Continental-scale studies of temperature-related climatic debt show that changes in community composition are rapid, and that northward shifts in temperature in Europe are occurring faster than changes in community composition (Devictor et al. 2012). Both birds and butterflies are not keeping up with temperature increases, and marked north–south gradients in these responses suggest changes in precipitation unrelated to temperature may be behind regional differences (Gillings et al. 2015). The role of aridity and water-related constraints on bird communities is an active area of research (VanDerWal et al. 2012). In fact, in Mediterranean countries, butterfly species from arid environments showed the strongest population decreases, whereas the most severe bird population declines were found for species inhabiting humid habitats (Herrando et al. 2019), contrary to patterns in central and northern Europe (Pearce-Higgins et al. 2015; Santangeli and Lehikoinen 2017). Climate change has caused deserts to become warmer and drier than any other ecoregion in the lower United States in the past 50 years. This was associated with a 43 per cent decline in the number of bird species in the Mojave desert since the early 1900s (Iknayan and Beissinger 2018). However, some recent evidence points to the possibility that novel climate conditions leading to drought impacts on vegetation may be disproportionally stronger in more humid environments (Choat et al. 2012). In tropical habitats, longer dry seasons also lag behind demographic and distribution changes in otherwise protected and pristine habitats (Brawn et al. 2017). Changes have been reported in communities of birds, reptiles, and amphibians as a consequence of changes in the dry-season mist frequency, which has declined dramatically since the mid-1970s and is negatively correlated with sea surface temperature in the equatorial Pacific (Pounds et al. 1999).

16.4.2 Community shifts and thermal ranges

Under the influence of global warming, the geographical location of the climatic niche of a species is expected to shift, which should affect the overall species assemblage, i.e., community composition. Some scientists have attempted to measure community shifts by assessing changes in community

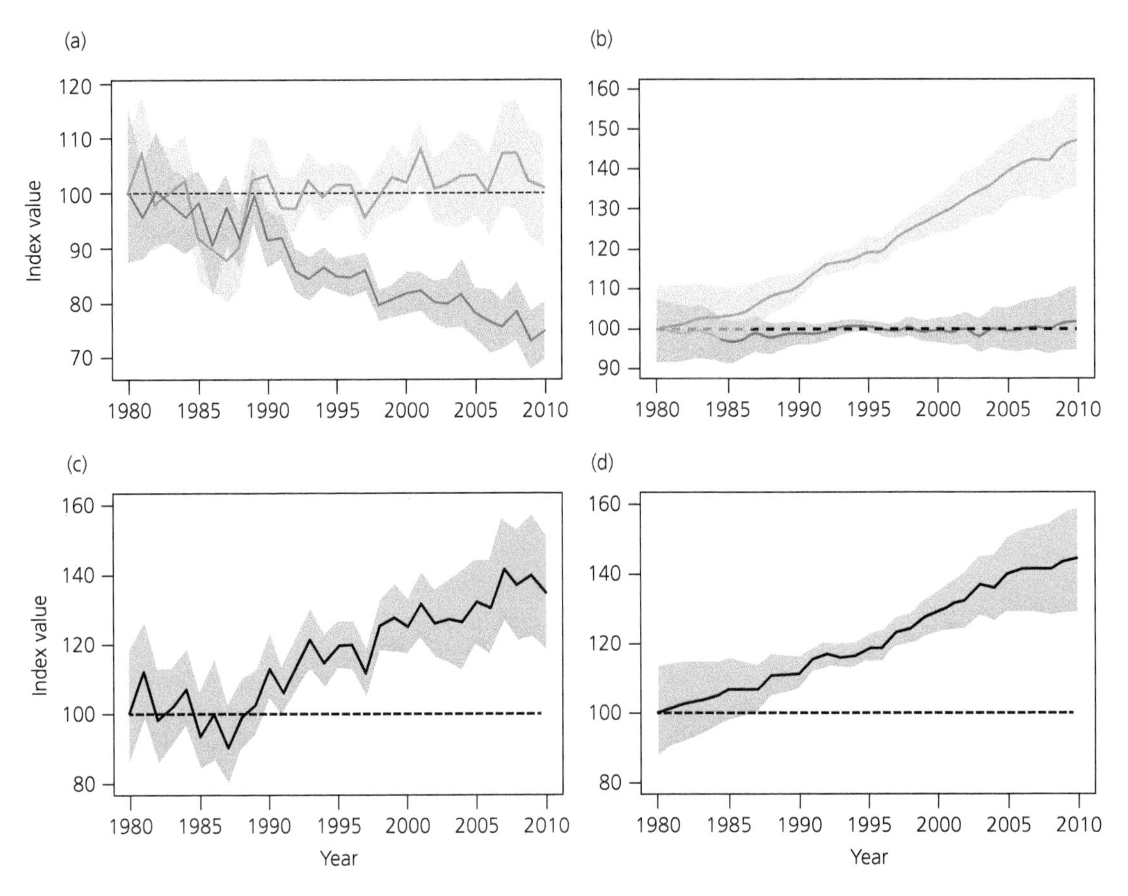

Figure 16.2 Trends in the community thermal indices (CTI) for local bird communities in Europe and the United States for the period 1980–2010. Multispecies population indices (CII) for species expected to respond positively to climate change (CST+, orange lines) and those expected to respond negatively to climate change (CST–, blue lines) were combined across all eligible countries of Europe (A) and states of the United States (B). Shaded polygons in each case indicate 90% confidence intervals. Annual values of the ratio of the CST+ index to the CST– index (Climate Impact Indicator, CII) are shown for Europe (C) and the United States (D). Bird populations in both Europe and the United States show an increasing climate change signal, although the community changes involved are different. From Stephens et al. (2016).

composition, using space-for-time substitution to assess associations with temperature. However, this approach has often been adopted with little empirical support. La Sorte et al. (2009) examined actual changes in three community attributes (species richness, body mass, and occupancy) for 227 terrestrial species wintering in North America from 1975 to 2001. They concluded that in the face of rapid climate change, applying space-for-time substitution as a predictive tool could be problematic, with communities reorganizing and giving origin to novel species combinations. In this context, it should be noticed that temporal data are associated with intrinsic temporal auto-correlation that has rarely been taken into account in the analyses, and that this can potentially lead to misinterpretations of the mechanisms producing changes in biological and community characteristics (Chapter 6).

Changes in species communities in relation to climate have also been investigated using the community temperature index (CTI) approach (Devictor et al. 2008). CTI reflects the composition of the community in relation to dominance of warm- and cold-dwelling species. The species temperature index, STI, aims at capturing species' temperature niche and can be measured using temperature within species' distribution area. CTI is typically calculated using abundance of species and their STIs, and thus it measures the average temperature niche of a community (Devictor et al. 2012; Godet et al. 2011).

Among birds, the first CTI studies have shown that bird communities have changed as expected from climate change with increasing dominance of warm-dwelling species (Devictor et al. 2012; Godet et al. 2011; Lindström et al. 2013). However, these changes in communities have been slower than changes in temperatures. Speed of change in breeding community structure measured as CTI has been 3–6 times slower than the speed of temperature change, and the CTI approach has especially been used during the breeding season, but a few studies have also investigated community changes during the non-breeding season (Godet et al. 2011; Santangeli and Lehikoinen 2017). The Finnish case study showed that land bird communities during winter follow temperature changes even slower than breeding bird communities (Santangeli and Lehikoinen 2017).

A case study from Sweden has shown that breeding CTI tracked patterns of temperature increase, stability, or decrease with a lag period of 1–3 yr (Lindström et al. 2013). On the other hand, a French study found that local species communities remained relatively unaffected by a sudden extreme heat wave (Jiguet et al. 2011). Communities are following temperatures less intensively with increasing elevation, habitat diversity, and the degree to which the landscape is 'natural', according to the French study (Gauzere et al. 2017). This could suggest that complementary effects of local topographic heterogeneity are providing more diverse microclimates and increasing sustainability in natural and diversified landscapes. At the species scale, results also showed that the more a species benefits from protected areas, the less vulnerable it was to temperature changes. Protected areas were also more effective in mitigating the impact of climate change on the less common and northernmost birds (Gauzere et al. 2017). In Finland, in contrast, CTI change has been similar inside and outside Finnish protected areas, but protected areas in general hold colder communities compared to outside areas (Santangeli et al. 2017).

16.4.3 Observed impacts on different functional groups

Climate change is likely to impact particular functional groups of birds more heavily than others; e.g., habitat specialists, migratory species, pollinators, predators, or frugivores. In a recent comprehensive review, MacLean and Beissinger (2017) found that ecological and life history traits had limited success in accounting for variation among species in range shifts over the past century. Only habitat breadth showed significant relationships with range shifts; more generalists had large range shifts, while specialist species occupying higher altitudes or latitudes had smaller range shifts. In general, the current understanding of species traits as predictors of range shifts is limited, and standardized study is needed for traits to be valid indicators of vulnerability in assessments of climate change impacts.

However, more specific evidence for some groups suggests that the functional integrity of bird communities may be compromised in a climate change context. The degree of mismatch may differ among species and be particularly large for migratory species, eventually leading to declining populations and local extinctions. Long-term effects on species richness and composition of ecological communities have been predicted using modelling approaches, but so far has hardly been demonstrated in the field. Lemoine et al. (2007) found that the composition of bird communities corresponded significantly to predicted climate changes, and alternative factors such as changes in land use were of secondary importance, suggesting that global climate change is already significantly influencing species richness and composition of European bird communities in terms of their migratory component. In a previous study, Lemoine and Böhning-Gaese (2003) reported changes in bird communities fitting predictions from spatial relationships between bird community structure and climate in central Europe. In this region, winter temperature increased significantly between two study periods, whereas spring temperature and precipitation did not. The significant declines of long-distance migrants were of a magnitude that could be explained by observed climate change, suggesting that increasingly warmer winters pose a more severe threat to long-distance migrants than to other bird groups (but see La Sorte and Fink 2017).

Another example of functional changes at the community level reported in the literature is driven by both declines of cold-dwelling species as well as increases of warm-dwelling species. A Swedish case study suggested that community changes were influenced by changes of both rare and abundant species.

In eastern North America, winter bird community structure of urban birds has changed with increasing numbers of warm-adapted species. In Sweden the changes in breeding communities were faster in the northern part of the country (Tayleur et al. 2016), whereas in eastern North America change in winter bird communities was strongest in southerly latitudes and driven primarily by local increases in abundance and regional patterns of colonization by southerly distributed birds (Prince and Zuckerberg 2015).

16.5 Indirect climate change effects on bird communities: The complexity of interacting global change components

Climate change is not likely to impact bird communities in isolation, but will act via indirect ecological cascade effects (Ockendon et al. 2014) and interact with a range of other global change impacts, leading to regional specific effects (Figure 16.2). If habitat destruction or population overexploitation is severe, species loss can occur directly and abruptly. Yet the final descent to extinction is often driven by synergistic processes that can be disconnected from the original cause of decline. Brook et al. (2008) have shown that owing to interacting and self-reinforcing processes, estimates of extinction risk for most species are more severe than previously recognized (Hylander and Ehrlén 2013). Therefore, conservation actions that only target single-threat drivers are inadequate due to cascading effects caused by unmanaged synergies. Indeed, works are increasingly focusing on how climate change interacts with and exacerbates other ongoing threats to biodiversity.

Many of these mechanisms have already been described and usually lead to major impacts on community change, but given the potential complexity of these interactions, the exact pathways that these impacts may follow are largely unexplored (Sirami et al. 2017). For example, Martin and Maron (2012) showed that changes in snowfall impacted herbivory by elk (*Cervus canadensis*), which in turn affected habitat quality and productivity of songbird communities. Understanding such indirect impacts is of major importance because they may have major consequences for bird communities and could help interpret current mismatches or lags between predicted impacts of climate change and observed community responses. For example, shifting ranges are inhibited, but not blocked, in landscape zones in which the degree of habitat fragmentation allows persistence (Lawler et al. 2013). In areas where the spatial cohesion of the habitat is below the critical level of metapopulation persistence, expansion of ranges will be blocked (Jarzyna et al. 2015). Here we briefly review recent developments in the unravelling of potential indirect effects of climate change on bird communities and identify this front line as a major challenge for future research.

16.5.1 Interactions between climate change and other components of global change

16.5.1.1 Invasive species

Invasive species are affecting ecological systems worldwide, and their increase may be partly due to climate change (Walther et al. 2009). In the same way that climate change affects the ecology of native species, it might also directly influence the likelihood of alien species becoming invasive. It can also increase the probability of impacting new systems to different degrees from simple space occupancy to complete transformation, where alien species dominate function or richness, leading to reduced diversity of native species. Some invasive species, such as the ring-necked parakeet (*Psittacula krameri*) and lovebirds (*Agapornis* spp.) may be favoured by warming climates in recent years (Jiguet 2009), leading to a direct impact on native communities. In other cases, bird communities may be impacted indirectly by invasive species that transmit disease or influence food availability. For example, avian malaria plays an important role in limiting the distribution and population sizes of many Hawaiian birds, and projected climate change is likely to eliminate most disease-free habitat in Hawaii in the next century (Benning et al. 2002; Chapter 14). Kilpatrick (2006) examined alternative management scenarios for conservation of native Hawaiian birds. The analyses suggested that differences in life history cause some species to be more susceptible to local extinctions from transmission of malaria, but that climate change generally will result in stronger negative impacts on the native bird communities (Garamszegi 2010).

Indirect impacts of invasive species can also come from alteration of the food web. Increases in precipitation due to climate change promotes a wider distribution of invasive Argentine ants and increases colonization rate to new areas (Heller et al. 2008). Recent evidence suggests that reproduction of canopy-foraging bird species that mostly rely on caterpillars for feeding their young could be compromised by invasion of Argentine ants, because they influence the availability of arthropod prey and, hence, the trophic web (Estany-Tigerström et al. 2009). Thus, invasive species have the potential to lead to generalized impacts at the community level (Clavero et al. 2009) by affecting different groups of birds through different mechanisms, including parasitism, predation, and competition (Walther et al. 2009).

16.5.1.2 Land-use changes in the context of climate change

Although changes in land use and climate have an impact on ecological communities, it is unclear which of these factors are currently most important (Harris et al. 2014; Bowler et al. 2018; Yalcin and Leroux 2018). Some studies have sought to determine the influence of land use and climate alteration on changes in abundance (Lemoine et al. 2007), but there is an urgent need to develop effective integrative, community-based approaches to track impacts of climate change on biodiversity independently from other components of global change (Hockey et al. 2011). The complex interaction between climate and land-use changes has been analysed through analyses of land use and climate change at global (Jetz et al. 2007) and regional levels (Bowler et al. 2018; Yalcin and Leroux 2018), as well as with changes in community temperature descriptors such as CTI.

For instance, Mediterranean bird communities in forest habitats had colder-dwelling bird species with more northern distributions compared to farmland, areas burned by wildfire, or urban areas (Clavero et al. 2011a). Clavero et al. (2011b) assessed whether these land-use related changes in CTI overlapped changes related to temperature such as gradients in altitude and latitude. They found that CTI varied as much along gradients of land use as it did along gradients of temperature. Thus it is important to incorporate interactions between both climate change and land use on commonly used biodiversity indicators. This could, for instance, explain why Finnish protected areas, which mainly consist of forest landscapes, have colder communities on average than areas outside protected areas (Santangeli et al. 2017). This highlights that human land-use changes can reverse, hide, or exacerbate the impacts of climate change when measured through community-level climate change indicators (Clavero et al. 2011b). Despite the fact that CTI changes can be influenced by habitat change and quality, habitat-specific changes have rarely been investigated. In the Netherlands, CTI has declined rapidly in farmland communities, likely due to intensification of agriculture rather than climate change (Kampichler et al. 2012). Similar results have been found in England where intensive agriculture has led to the situation that bird communities have not reorganized successfully in response to climate change. Increases in CTI have primarily been attributed to rapid loss of cold-dwelling species due to high-intensity land use, whereas land-use practices have prevented increases of warm-dwelling species (Oliver et al. 2017).

Increasing evidence suggests that the effects of climate change on bird communities should be investigated separately for each type of habitat and land use (Kampichler et al. 2012). For example, in Mediterranean forests northern species have recently increased in abundance and distribution due to widespread effects of forest maturation, thus acting opposite to the predicted impacts of climate change (Gil-Tena et al. 2009). Furthermore, Hockey and Midgley (2009) documented the chronology and habitat use of 18 regionally indigenous bird species that colonized the extreme south-western corner of Africa after the late 1940s. Observations of these colonization events concur with a 'climate change' explanation, assuming extrapolation of Northern Hemisphere results and simplistic application of theory. However, when closely scrutinized, all but one may be more parsimoniously explained by direct anthropogenic changes to the landscape than by indirect effects of climate change. This suggests that observed climate change has not yet been sufficient to trigger extensive shifts in ranges of indigenous birds, or that a priori assumptions are incorrect. Either way, these studies highlight the danger of naïve attribution of range changes to climate change only (Lemoine et al. 2007; Sirami et al. 2017), even if range changes are in accordance with predictions of climate change models.

16.5.2 Changes in disturbance regimes

Sudden changes in community composition may be the result of changes in disturbance regimes, rather than the direct result of gradual climate change (White et al. 2011; Palmer et al. 2017). The magnitude of these projected climate changes and their subsequent impacts on different communities will vary regionally. Coastal wetlands in the south-eastern United States have naturally evolved under a regime of rising sea level and specific patterns of hurricane frequency, intensity, and timing. Michener et al. (1997) reviewed the ecological effects of tropical storms and hurricanes and indicated that storm timing, frequency, and intensity can alter coastal wetland hydrology, geomorphology, biotic structure, energetics, and nutrient cycling. Research conducted to examine impacts of Hurricane Hugo on colonial waterbirds highlighted the importance of long-term studies for identifying complex interactions that may otherwise be dismissed as stochastic processes. Rising sea level and even modest changes in frequency, intensity, timing, and distribution of tropical storms and hurricanes are expected to have substantial impacts on coastal wetland patterns and processes. Persistence of coastal wetlands will be determined by interactions of climate and anthropogenic effects, especially how humans respond to rising sea level and how further human encroachment on coastal wetlands affects resource exploitation, pollution, and water use.

Fire is another widespread natural disturbance agent in many systems. Animal succession and restoration of community structure after fire are related to vegetation regeneration, which in turn is influenced by climate. Species richness and community composition are strongly impacted by fire and later by site-specific successional processes (Clavero et al. 2011a; Fontaine et al. 2009). In areas with an increasing fire frequency, bird communities may shift their relative composition towards early successional stages. In many regions such as the Mediterranean these community shifts are generally predicted from climate warming, with more temperate species being more common in more mature forest habitats. In this context, the two processes may interact but at the community level impacts of climate change on bird communities are very high (Puig-Gironès et al. 2017). Understanding such processes is critical for being able to forecast expected impacts of climate change on bird communities. Because land cover changes result from a variety of processes, it is still unclear how effectively species distribution models capture responses to these changes, including those derived from climate (Titeux et al. 2016; Vallecillo et al. 2009).

16.6 Bird community and climate changes: Perspectives and the way forward

Careful reporting of current community changes and careful confrontation of model projections with hard data on community dynamics are two ways to better identify likely impacts of climate change on bird communities and should guide future research. Confidence in projections of future distributions of species and community descriptors will also require demonstration that recently observed changes could be predicted adequately (Green et al. 2008). At present, there is general agreement about the role of climate warming on changes in species distributions including community shifts fitting predictions.

Long-term datasets have proven essential here, and require special attention because at present climate change research on birds heavily relies on information gathered in citizen science-related programmes, especially for analyses at the community level (Cooper et al. 2014). However, once the complexity of direct and indirect effects of climate change is recognized, assessments and predictions of impacts at the community level should move forward and explicitly recognize the major role of cascade effects, lags in responses to current change between species, and interactions between climate and other global change components (Sirami et al. 2017). Despite different studies showing that bird communities are rapidly changing, the lack of adjustment between the complexity of expected impacts and our current modelling capabilities leads to differences between the observations and the available predictions (Chapter 13; La Sorte et al. 2009).

An essential step forward in our ability to generate predictions about impacts of climate change on bird communities will come from the development of a new generation of models. New advances in mechanistic dynamic species distribution models specifically and explicitly integrate species interactions and ecological constraints such as dispersal, allowing future scenarios of species distribution changes to be ecologically sound. Some examples of these approaches have recently appeared in the literature (Methorst et al. 2017; Willis et al. 2009). These methods use a range of dynamic modelling frameworks combining physiological constraints and climate, but also land-use changes and dispersal abilities, and they are usually able to capture recently observed changes in expanding species accurately (McRae et al. 2008). The ability to integrate bottom-up demographic processes like these with top-down constraints imposed by climate and land use in a dynamic modelling environment is a key advantage of the resulting frameworks. In the future, modelling approaches that combine climate suitability and spatially explicit population models, incorporating demographic variables and habitat availability, are likely to be valuable tools in projecting responses of species to climatic change and hence for anticipating management to facilitate dispersal and persistence of species.

Developing a comprehensive understanding of the ecological ramifications of global change will necessitate close coordination among scientists from multiple disciplines and a balanced mixture of approaches. Cooper et al. (2014) have proposed that insights may be gained through careful design and implementation of broad-scale comparative studies that incorporate salient patterns and processes, including anthropogenic influences. Well-designed, broad-scale comparative studies could serve as the scientific framework for developing relevant and focused long term ecological research, monitoring programmes, experiments, and modelling studies (Lenoir and Svenning 2014). The investigation of impacts of climate change at the community level is affected by the complexity of the process involved. We lack good information on bird community changes and explicit integration of climate change interactions with other global change components. These are currently the greatest challenges for assessing community-scale impacts of climate change on birds. As a final message, we propose the need to use common conceptual frameworks for advancement (Methorst et al. 2017; Willis et al. 2009) and suggest some lines of development in this formidable quest ahead:

(1) Developing hypotheses based on dynamic and joint species modelling of future projections that explicitly incorporate expected interactions and ecological constraints: trophic levels, competition and commensalism, and host–parasite interactions. These future scenarios of community change could be viewed only as a range of null hypotheses indicative of potential future responses of communities to the complexity of mechanisms involved in the predictions, including evolutionary processes.
(2) Explicit integration of the complexity and interactive nature of climate change impacts, including the multidimensional character of climate going beyond temperature changes, with other components of global change such as invasive species and land-use changes, including alterations in perturbation regimes.
(3) Recording of community responses at the landscape and regional levels to allow validation of predictions using hard data on community descriptors. Comparison of observed community changes with the range of changes predicted by different null hypotheses of change will allow identification of prevailing factors influencing community dynamics in specific situations. Only by building on such a structured and hypothesis-based framework will we be able to advance in our ability to understand and anticipate the certainly complex effects of climate change on bird communities.

Acknowledgments

The work was partly supported by Spanish Government Grants CGL2017-89999-C2/BOS, Catalan Government grant SGR2018-531, and Academy of Finland grant 275606.

References

Barbet-Massin, M., and Jetz, W. (2015). The effect of range changes on the functional turnover, structure and diversity of bird assemblages under future climate scenarios. *Global Change Biology*, 21, 2917–28.

Barbet-Massin, M., Walther, B.A., Thuiller, W., Rahbek, C., and Jiguet, F. (2009). Potential impacts of climate change on the winter distribution of Afro-Palaearctic migrant passerines. *Biology Letters*, 5, 248–51.

Barbraud, C., Barbraud, J.C., and Barbraud, M. (1999). Population dynamics of the White Stork *Ciconia ciconia* in western France. *Ibis*, 141, 469–79.

Benning, T.L., Pointe, D., Atkinson, C.T., and Vitousek, P.M. (2002). Interactions of climate change with biological invasions and land use in the Hawaiian Islands: modeling the fate of endemic birds using a geographic information system. *Proceedings of the National Academy of Sciences of the United States of America*, 99, 14246–9.

Berthold, P. (1998). Vogelwelt und Klima: Gegenwärtige veränderungen. *Naturwissenschaftliche Rundschau*, 51, 337–46.

Bowler, D.E., Heldbjerg, H., Fox, A.D., O'Hara, R.B., and Böhning-Gaese, K. (2018). Disentangling the effects of multiple environmental drivers on population changes within communities. *Journal of Animal Ecology*, 87, 1034–45.

Brawn, J.D., Benson, T.J., Stager, M., Sly, N.D., and Tarwater, C.E. (2017). Impacts of changing rainfall regime on the demography of tropical birds. *Nature Climate Change*, 7, 133–6.

Brommer, J.E. (2008). Extent of recent polewards range margin shifts in Finnish birds depends on their body mass and feeding ecology. *Ornis Fennica*, 85, 109–17.

Brook, B.W., Sodhi, N.S., and Bradshaw, C.J.A. (2008). Synergies among extinction drivers under global change. *Trends in Ecology & Evolution*, 23, 453–60.

Brotons, L., and Jiguet, F. (2010). Bird communities and climate change. In: A. P. Møller, W. Fiedler, and P. Berthold (eds), *Birds and climate change*, pp. 275–29. Elsevier, Amsterdam.

Brown, J.H., Valone, T.J., and Curtin, C.G. (1997). Reorganization of an arid ecosystem in response to recent climate change. *Proceedings of the National Academy of Sciences of the United States of America*, 94, 9729–33.

Chen, I.C., Hill, J.K., Ohlemüller, R., Roy, D.B., and Thomas, C.D. (2011). Rapid range shifts of species associated with high levels of climate warming. *Science*, 333, 1024–6.

Choat, B., Jansen, S., Brodribb, T.J., Cochard, H., Delzon, S., Bhaskar, R., et al. (2012). Global convergence in the vulnerability of forests to drought. *Nature*, 491, 752–5.

Clavero, M., Brotons, L., and Herrando, S. (2011a). Bird community specialization, bird conservation and disturbance: the role of wildfires. *Journal of Animal Ecology*, 80, 128–36.

Clavero, M., Brotons, L., Pons, P., and Sol, D. (2009). Prominent role of invasive species in avian biodiversity loss. *Biological Conservation*, 142, 2043–9.

Clavero, M., Villero, D., and Brotons, L. (2011b). Climate change or land use dynamics: Do we know what climate change indicators indicate? *PLoS ONE*, 6, e18581.

Cooper, C.B., Shirk, J., and Zuckerberg, B. (2014). The invisible prevalence of citizen science in global research: Migratory birds and climate change. *PLoS One*, 9, e106508.

Cotton, P.A. (2003). Avian migration phenology and global climate change. *Proceedings of the National Academy of Sciences of the United States of America*, 100(21), 12219–22.

Devictor, V., Julliard, R., Couvet, D., and Jiguet, F. (2008). Birds are tracking climate warming, but not fast enough. *Proceedings of the Royal Society of London Series B: Biological Sciences*, 275, 2743–8.

Devictor, V., van Swaay, C., Brereton, T., Brotons, L., Chamberlain, D., Heliola, J., et al. (2012). Differences in the climatic debts of birds and butterflies at a continental scale. *Nature Climate Change*, 2, 121–4.

Estany-Tigerström, D., Bas, J.M., and Pons, P. (2009). Does Argentine ant invasion affect prey availability for foliage-gleaning birds? *Biological Invasions*, 12, 827–39.

Fontaine, J.B., Donato, D.C., Robinson, W.D., Law, B.E., and Kauffman, J.B. (2009). Bird communities following high-severity fire: Response to single and repeat fires in a mixed-evergreen forest, Oregon, USA. *Forest Ecology and Management*, 257, 1496–504.

Forsman, J.T., and Mönkkönen, M. (2003). The role of climate in limiting European resident bird populations. *Journal of Biogeography*, 30, 55–70.

Franks, S.E., Pearce-Higgins, J.W., Atkinson, S., Bell, J.R., Botham, M.S., Brereton, T.M., et al. (2018). The sensitivity of breeding songbirds to changes in seasonal timing is linked to population change but cannot be directly attributed to the effects of trophic asynchrony on productivity. *Global Change Biology*, 24, 957–71.

Garamszegi, L.Z. (2010). Climate change increases the risk of malaria in birds. *Global Change Biology*, 17, 1751–9.

Gardner, J.L., Heinsohn, R., and Joseph, L. (2009). Shifting latitudinal clines in avian body size correlate with global warming in Australian passerines. *Proceedings of the Royal Society of London Series B: Biological Sciences*, 276, 3845–52.

Gauzere, P., Prince, K., and Devictor, V. (2017). Where do they go? The effects of topography and habitat diversity on reducing climatic debt in birds. *Global Change Biology*, 23, 2218–29.

Gil-Tena, A., Brotons, L., and Saura, S. (2009). Mediterranean forest dynamics and forest bird distribution changes in the late 20th century. *Global Change Biology*, 15, 474–85.

Gillings, S., Balmer, D.E., and Fuller, R.J. (2015). Directionality of recent bird distribution shifts and climate change in Great Britain. *Global Change Biology*, 21, 2155–68.

Godet, L., Jaffre, M., and Devictor, V. (2011). Waders in winter: long-term changes of migratory bird assemblages facing climate change. *Biology Letters*, 7, 714–17.

Green, R.E., Collingham, Y.C., Willis, S.G., Gregory, R.D., Smith, K.W., and Huntley, B. (2008). Performance of climate envelope models in retrodicting recent changes in bird population size from observed climatic change. *Biology Letters*, 4, 599–602.

Gregory, R.D., Willis, S., Jiguet, F., Voříšek, P., Klvaňová, A., van Strien, A., et al. (2009). An indicator of the impact of climatic change on European bird populations. *PLoS One*, 4, e4678.

Harris, J.B.C., Putra, D.D., Gregory, S.D., Brook, B.W., Prawiradilaga, D.M., Sodhi, N.S., et al. (2014). Rapid deforestation threatens mid-elevational endemic birds but climate change is most important at higher elevations. *Diversity and Distributions*, 20, 773–85.

Heller, N.E., Sanders, N.J., Shors, J.W., and Gordon, D.M. (2008). Rainfall facilitates the spread, and time alters the impact, of the invasive Argentine ant. *Oecologia*, 155, 385–95.

Herrando, S., Titeux, N., Brotons, L., Anton, M., Ubach, A., Villero, D., García-Barros, E., Munguira, M.L., Godinho, C. and Stefanescu, C. (2019) Contrasting impacts of precipitation on Mediterranean birds and butterflies. Scientific Reports, 9, 5680.

Hockey, P.A.R., and Midgley, G.F. (2009). Avian range changes and climate change: A cautionary tale from the Cape Peninsula. *Ostrich*, 80, 29–34.

Hockey, P.A.R., Sirami, C., Ridley, A.R., Midgley, G.F., and Babiker, H.A. (2011). Interrogating recent range changes in South African birds: confounding signals from land use and climate change present a challenge for attribution. *Diversity and Distributions*, 17, 254–61.

Holm, S.R., and Svenning, J.C. (2014). 180,000 years of climate change in Europe: avifaunal responses and vegetation implications. *PLoS One*, 9, e94021.

Howard, C., Stephens, P.A., Tobias, J.A., Sheard, C., Butchart, S.H.M., and Willis, S.G. (2018). Flight range, fuel load and the impact of climate change on the journeys of migrant birds. *Proceedings of the Royal Society of London Series B: Biological Sciences*, 285, 20172329–9.

Huntley, B., Green, R.E., Collingham, Y.C., and Willis, S.G. (2007). *A climatic atlas of European breeding birds*. Lynx Edicions, Barcelona, Spain.

Hylander, K., and Ehrlén, J. (2013). The mechanisms causing extinction debts. *Trends in Ecology & Evolution*, 28, 341–6.

Iknayan, K.J., and Beissinger, S.R. (2018). Collapse of a desert bird community over the past century driven by climate change. *Proceedings of the National Academy of Sciences of the United States of America*, 115, 8597–602.

Jarzyna, M.A., Porter, W.F., Maurer, B.A., Zuckerberg, B., and Finley, A.O. (2015). Landscape fragmentation affects responses of avian communities to climate change. *Global Change Biology*, 21, 2942–53.

Jenouvrier, S. (2013). Impacts of climate change on avian populations. *Global Change Biology*, 19, 2036–57.

Jetz, W., Wilcove, D.S., and Dobson, A.P. (2007). Projected impacts of climate and land-use change on the global diversity of birds. *PLoS Biology*, 5, 1211–19.

Jiguet, F. (2009). Inséparable de Fischer *Agapornis fischeri*. Inséparable masqué *Agapornis personatus*. In: A. Flitti, B. Kabouche, Y. Kayser, and G. Olioso (eds), *Atlas des oiseaux nicheurs de Provence-Alpes-Côte d'Azur*. LPO PACA. Delachaux et Niestlé, Paris, France.

Jiguet, F., Brotons, L., and Devictor, V. (2011). Community responses to extreme climatic conditions. *Current Zoology*, 57, 406–13.

Kampichler, C., van Turnhout, C.A.M., Devictor, V., and van der Jeugd, H.P. (2012). Large-scale changes in community composition: determining land use and climate change signals. *PLoS One*, 7. https://doi.org/10.1371/journal.pone.0035272

Kati, V.I., and Şekercioğlu, Ç.H. (2006). Diversity, ecological structure, and conservation of the landbird community of Dadia reserve, Greece. *Diversity and Distributions*, 12, 620–9.

Kilpatrick, A.M. (2006). Facilitating the evolution of resistance to avian malaria in Hawaiian birds. *Biological Conservation*, 128, 475–85.

Kissling, W.D., Böhning-Gaese, K., and Jetz, W. (2009). The global distribution of frugivory in birds. *Global Ecology and Biogeography*, 18, 150–62.

Kluen, E., Nousiainen, R., and Lehikoinen, A. (2017). Breeding phenological response to spring weather conditions in common Finnish birds: resident species respond stronger than migratory species. *Journal of Avian Biology*, 48, 611–19.

Ko, C.Y., Schmitz, O.J., Barbet-Massin, M., and Jetz, W. (2014). Dietary guild composition and disaggregation of avian assemblages under climate change. *Global Change Biology*, 20, 790–802.

Krosby, M., Wilsey, C.B., McGuire, J.L., Duggan, J.M., Nogeire, T.M., Heinrichs, J.A., et al. (2015). Climate-induced range overlap among closely related species. *Nature Climate Change*, 5, 883–6.

La Sorte, F.A., and Fink, D. (2017). Projected changes in prevailing winds for transatlantic migratory birds under global warming. *Journal of Animal Ecology*, 86, 273–84.

La Sorte, F.A., and Thompson, F.R. (2007). Poleward shifts in winter ranges of North American birds. *Ecology*, 88, 1803–12.

La Sorte, F.A., Lee, T.M., Wilman, H., and Jetz, W. (2009). Disparities between observed and predicted impacts of climate change on winter bird assemblages. *Proceedings of the Royal Society of London Series B: Biological Sciences*, 276, 3167–74.

Lawler, J.J., Ruesch, A.S., Olden, J.D., and McRae, B.H. (2013). Projected climate-driven faunal movement routes. *Ecology Letters*, 16, 1014–22.

Lawler, J.J., Shafer, S.L., White, D., Kareiva, P., Maurer, E.P., Blaustein, A.R., and Bartlein, P.J. (2009). Projected climate-induced faunal change in the Western Hemisphere. *Ecology*, 90, 588–97.

Lehikoinen, A., Jaatinen, K., Vähätalo, A.V., Clausen, P., Crowe, O., Deceuninck, B., et al. (2013). Rapid climate driven shifts in wintering distributions of three common waterbird species. *Global Change Biology*, 19, 2071–81.

Lemoine, N., and Böhning-Gaese, K. (2003). Potential impact of global climate change on species richness of long-distance migrants. *Conservation Biology*, 17, 577–86.

Lemoine, N., Bauer, H.G., Peintinger, M., and Böhning-Gaese, K. (2007). Effects of climate and land-use change

on species abundance in a central European bird community. *Conservation Biology*, 21, 495–503.
Lenoir, J., and Svenning, J.C. (2014). Climate-related range shifts—a global multidimensional synthesis and new research directions. *Ecography*, 38, 15–28.
Lindström, A., Green, M., Paulson, G., Smith, H.G., and Devictor, V. (2013). Rapid changes in bird community composition at multiple temporal and spatial scales in response to recent climate change. *Ecography*, 36, 313–22.
MacLean, S.A., and Beissinger, S.R. (2017). Species' traits as predictors of range shifts under contemporary climate change: A review and meta-analysis. *Global Change Biology*, 23, 4094–105.
Marini, M.A., Barbet-Massin, M., Lopes, M.E., and Jiguet, F. (2009). Major current and future gaps of Brazilian reserves to protect Neotropical savanna birds. *Biological Conservation*, 142, 3039–50.
Martin, T.E., and Maron, J.L. (2012). Climate impacts on bird and plant communities from altered animal–plant interactions. *Nature Climate Change*, 2, 195–200.
Massimino, D., Johnston, A., Gillings, S., Jiguet, F., and Pearce-Higgins, J.W. (2017). Projected reductions in climatic suitability for vulnerable British birds. *Climatic Change*, 145, 117–30.
Mayhew, P.J., Jenkins, G.B., and Benton, T.G. (2008). A long-term association between global temperature and biodiversity, origination and extinction in the fossil record. *Proceedings of the Royal Society of London Series B: Biological Sciences*, 275, 47–53.
McRae, B.H., Schumaker, N.H., McKane, R.B., Busing, R.T., Solomon, A.M., and Burdick, C.A. (2008). A multimodel framework for simulating wildlife population response to land-use and climate change. *Ecological Modelling*, 219, 77–91.
Meller, K., Vähätalo, A.V., Hokkanen, T., Rintala, J., Piha, M., and Lehikoinen, A. (2016). Interannual variation and long-term trends in proportions of resident individuals in partially migratory birds. *Journal of Animal Ecology*, 85, 570–80.
Methorst, J., Böhning-Gaese, K., Khaliq, I., and Hof, C. (2017). A framework integrating physiology, dispersal and land-use to project species ranges under climate change. *Journal of Avian Biology*, 48, 1532–48.
Michener, W.K., Blood, E.R., Bildstein, K.L., Brinson, M.M., and Gardner, L.R. (1997). Climate change, hurricanes and tropical storms, and rising sea level in coastal wetlands. *Ecological Applications*, 7, 770–801.
Møller, A.P., Fiedler, W., and Berthold, P. (eds) (2004). *Birds and climate change*. Elsevier, Amsterdam.
Ockendon, N., Baker, D.J., Carr, J.A., White, E.C., Almond, R.E.A., Amano, T., et al. (2014). Mechanisms underpinning climatic impacts on natural populations: altered species interactions are more important than direct effects. *Global Change Biology*, 20, 2221–9.
Oliver, T.H., Gillings, S., Pearce-Higgins, J.W., Brereton, T., Crick, H.Q.P., Duffield, S.J., et al. (2017). Large extents of intensive land use limit community reorganization during climate warming. *Global Change Biology*, 23, 2272–83.
Palmer, G., Platts, P.J., Brereton, T., Chapman, J.W., Dytham, C., Fox, R., Pearce-Higgins, J.W., Roy, D.B., Hill, J.K., and Thomas, C.D. (2017). Climate change, climatic variation and extreme biological responses. *Philosophical Transactions of the Royal Society of London Series B: Biological Sciences*, 372, 20160144.
Paradis, E., Baillie, S.R., Sutherland, W.J., and Gregory, R.D. (1998). Patterns of natal and breeding dispersal in birds. *Journal of Animal Ecology*, 67, 518–36.
Pearce-Higgins, J.W., Eglington, S.M., Martay, B., and Chamberlain, D.E. (2015). Drivers of climate change impacts on bird communities. *Journal of Animal Ecology*, 84, 943–54.
Pimm, S.L. (2009). Climate disruption and biodiversity. *Current Biology*, 19, R595–R601.
Pollock, L.J., Tingley, R., Morris, W.K., Golding, N., O'Hara, R.B., Parris, K.M., et al. (2014). Understanding co-occurrence by modelling species simultaneously with a Joint Species Distribution Model (JSDM). *Methods in Ecology and Evolution*, 5, 397–406.
Potvin, D.A., Välimäki, K., and Lehikoinen, A. (2016). Differences in shifts of wintering and breeding ranges lead to changing migration distances in European birds. *Journal of Avian Biology*, 47, 619–28.
Pounds, J.A., Fogden, M.P.L., and Campbell, J.H. (1999). Biological response to climate change on a tropical mountain. *Nature*, 398, 611–15.
Prince, K., and Zuckerberg, B. (2015). Climate change in our backyards: the reshuffling of North America's winter bird communities. *Global Change Biology*, 21, 572–85.
Puig-Gironès, R., Brotons, L., and Pons, P. (2017). Aridity influences the recovery of vegetation and shrubland birds after wildfire. *PLoS One*, 12, e0173599–17.
Quintero, I., and Wiens, J.J. (2013). Rates of projected climate change dramatically exceed past rates of climatic niche evolution among vertebrate species. *Ecology Letters*, 16, 1095–103.
Rivalan, P., Frederiksen, M., Loïs, G., and Julliard, R. (2007). Contrasting responses of migration strategies in two European thrushes to climate change. *Global Change Biology*, 13, 275–87.
Santangeli, A., and Lehikoinen, A. (2017). Are winter and breeding bird communities able to track rapid climate change? Lessons from the high North. *Diversity and Distributions*, 23, 308–16.
Santangeli, A., Rajasarkka, A.I., and Lehikoinen, A. (2017). Effects of high latitude protected areas on bird communities under rapid climate change. *Global Change Biology*, 23, 2241–9.

Sanz, J.J., Potti, J., Moreno, J., Merino, S., and Frias, O. (2003). Climate change and fitness components of a migratory bird breeding in the Mediterranean region. *Global Change Biology*, 9, 461–72.

Schaefer, H.C., Jetz, W., and Böhning-Gaese, K. (2008). Impact of climate change on migratory birds: community reassembly versus adaptation. *Global Ecology and Biogeography*, 17, 38–49.

Sirami, C., Caplat, P., Popy, S., Clamens, A., Arlettaz, R., Jiguet, F., et al. (2017). Impacts of global change on species distributions: obstacles and solutions to integrate climate and land use. *Global Ecology and Biogeography*, 26, 385–94.

Stephens, P.A., Mason, L.R., Green, R.E., Gregory, R.D., Sauer, J.R., Alison, J., et al. (2016). Consistent response of bird populations to climate change on two continents. *Science*, 352, 84–7.

Stralberg, D., Jongsomjit, D., Howell, C.A., Snyder, M.A., Alexander, J.D., Wiens, J.A., and Root, T.L. (2009). Re-shuffling of species with climate disruption: A No-Analog future for California birds? *PLoS One*, 4, e6825.

Stralberg, D., Matsuoka, S.M., Hamann, A., Bayne, E.M., Solymos, P., Schmiegelow, F.K.A., et al. (2015). Projecting boreal bird responses to climate change: the signal exceeds the noise. *Ecological Applications*, 25, 52–69.

Şekercioğlu, Ç.H., Primack, R.B., and Wormworth, J. (2012). The effects of climate change on tropical birds. *Biological Conservation*, 148, 1–18.

Tayleur, C.M., Devictor, V., Gauzere, P., Jonzen, N., Smith, H.G., and Lindström, A. (2016). Regional variation in climate change winners and losers highlights the rapid loss of cold-dwelling species. *Diversity and Distributions*, 22, 468–80.

Thomas, C.D., and Lennon, J.J. (1999). Birds extend their ranges northwards. *Nature*, 399, 213.

Thomas, C.D., Cameron, A., Green, R.E., Bakkenes, M., Beaumont, L.J., Collingham, Y.C., et al. (2004). Extinction risk from climate change. *Nature*, 427, 145–8.

Titeux, N., Henle, K., Mihoub, J.B., Regos, A., Geijzendorffer, I.R., Cramer, W., et al. (2016). Biodiversity scenarios neglect future land-use changes. *Global Change Biology*, 22, 2505–15.

Vallecillo, S., Brotons, L., and Thuiller, W. (2009). Dangers of predicting bird species distributions in response to land-cover changes. *Ecological Applications*, 19, 538–49.

VanDerWal, J., Murphy, H.T., Kutt, A.S., Perkins, G.C., Bateman, B.L., Perry, J.J., and Reside, A.E. (2012). Focus on poleward shifts in species' distribution underestimates the fingerprint of climate change. *Nature Climate Change*, 2, 1–5.

Visser, M.E., Perdeck, A.C., van Balen, J.H., and Both, C. (2009). Climate change leads to decreasing bird migration distances. *Global Change Biology*, 15, 1859–65.

Walther, G.R. (2007). Tackling ecological complexity in climate impact research. *Science*, 315, 606–7.

Walther, G.R., Roques, A., Hulme, P.E., Sykes, M.T., Pysek, P., Kuhn, I., et al. (2009). Alien species in a warmer world: risks and opportunities. *Trends in Ecology & Evolution*, 24, 686–93.

White, J.D., Gutzwiller, K.J., Barrow, W.C., Johnson-Randall, L., Zygo, L., and Swint, P. (2011). Understanding interaction effects of climate change and fire management on bird distributions through combined process and habitat models. *Conservation Biology*, 25, 536–46.

Willis, S.G., Thomas, C.D., Hill, J.K., Collingham, Y.C., Telfer, M.G., Fox, R., and Huntley, B. (2009). Dynamic distribution modelling: predicting the present from the past. *Ecography*, 32, 5–12.

Winkler, D.W., Dunn, P.O., and McCulloch, C.E. (2002). Predicting the effects of climate change on avian life-history traits. *Proceedings of the National Academy of Sciences of the United States of America*, 99, 13595–9.

Yalcin, S., and Leroux, S.J. (2018). An empirical test of the relative and combined effects of land-cover and climate change on local colonization and extinction. *Global Change Biology*, 24, 3849–61.

Yom-Tov, Y. (2001). Global warming and body mass decline in Israeli passerine birds. *Proceedings of the Royal Society of London Series B: Biological Sciences*, 268, 947–52.

CHAPTER 17

Fitting the lens of climate change on bird conservation in the twenty-first century

Peter P. Marra, Benjamin Zuckerberg, and Christiaan Both

17.1 Introduction

Climate plays a fundamental role in shaping the biology of all species. A fact that has left scientists scrambling for the past several decades to understand how a rapidly *changing* climate will impact ecological and evolutionary phenomena, including species distributions, the timing of life history events, and the probability of species extinctions (e.g., Parmesan and Yohe 2003; Root et al. 2003). Ample evidence shows that human-induced climate change is actively and rapidly impacting habitats and the organisms that depend on those habitats—globally (e.g., McCarty et al. 2017). In addition, because of the rapid rate of climate change, and given what we know about how animals can respond and adapt, considerable evidence now exists that animals, including many birds, are vulnerable to modern climate change (e.g., Jenouvrier et al. 2009; Pacifici et al. 2015; but see Hof et al. 2011).

Understanding and predicting the future ecological impacts of climate change, and how humans could optimally respond and adapt in terms of conservation, still remains mostly a black box for several reasons (Knudsen et al. 2011). First, we still lack a robust understanding of how climate variability (e.g., temperature, precipitation) itself, either directly or indirectly, influences the biology of organisms. In fact, biological traits of organisms have only occasionally been used to understand species' responses to climate change (Buckley and Kingsolver 2012), let alone whether heritable variation is present for selection to act upon. Second, when evidence points to a species being vulnerable to the effects of climate change, there is a lack of specific and timely recommendations for managers to reduce that vulnerability (Mawdsley et al. 2009; Heller and Zavaleta 2009). Taken together, these issues complicate and impede our ability to predict how climate change and its associated uncertainty will impact individuals and populations in the future. Despite this uncertainty, it remains essential that we continue to quantify past and current responses to climate change at population levels and to incorporate this information into modelling approaches to predict future responses and develop robust strategies for conservation.

Birds are perhaps one of the best-studied and monitored classes of vertebrate organisms (Bonnet et al. 2002). Despite this, many of the mechanistic underpinnings between birds and climate remain poorly understood. We do know that bird distributions are linked to climate, and that they may change their underlying distributions to track their shifting niches, the phenology of their life history events, or

Marra, P. P., Zuckerberg, B., and Both, C., *Fitting the lens of climate change on bird conservation in the twenty-first century.* In: *Effects of Climate Change on Birds*. Second Edition. Edited by Peter O. Dunn and Anders Pape Møller: Oxford University Press (2019). © Oxford University Press. DOI: 10.1093/oso/9780198824268.003.0017 both (Chapter 13). Finally, evidence for phenotypic adaptation to a changing thermal environment is also mounting. Specifically, consistent with Bergmann's rule, evidence also exists that body sizes, and other morphological features, are changing in response to a warming climate (Chapter 10; Miller et al. 2018) but there are exceptions. These behavioral and phenotypic responses likely vary between species, depending on their life history. One such contrast exists between resident species, which can track and respond to local changes, and migratory birds, which can move thousands of kilometres between breeding and non-breeding areas to track food resources and avoid climatic extremes. Thus, birds, whether it be a resident northern cardinal (*Cardinalis cardinalis*) or a long-distance migratory Arctic tern (*Sterna paradisaea*), are responding phenotypically to changing environments to the degree to which they are able. Morphology can constrain responses to climate change (Møller et al. 2017). The extent to which humans can actively reduce the impacts of climate change on bird and other animal populations will help these species persist into the future.

Climate change adaptation refers to efforts by humans that reduce the vulnerability of species and ecosystems to the effects of modern climate change. Many government and private agencies involved in bird conservation recognize the need for climate change adaptation in their strategic planning, but with an ever-expanding list of conservation concerns, many managers consider climate change an issue that is too big and complex, controversial, and lacking in real-world case studies. For birds, adaptation strategies often include maintaining or creating habitat, establishing and enhancing protected areas, increasing connectivity among populations, and improved monitoring to reduce extinction risk. However, one of the major obstacles of applying such broad strategies in conservation is the lack of specific, tangible actions that managers can implement on the ground. In essence, *strategies* provide a useful starting point for including climate change

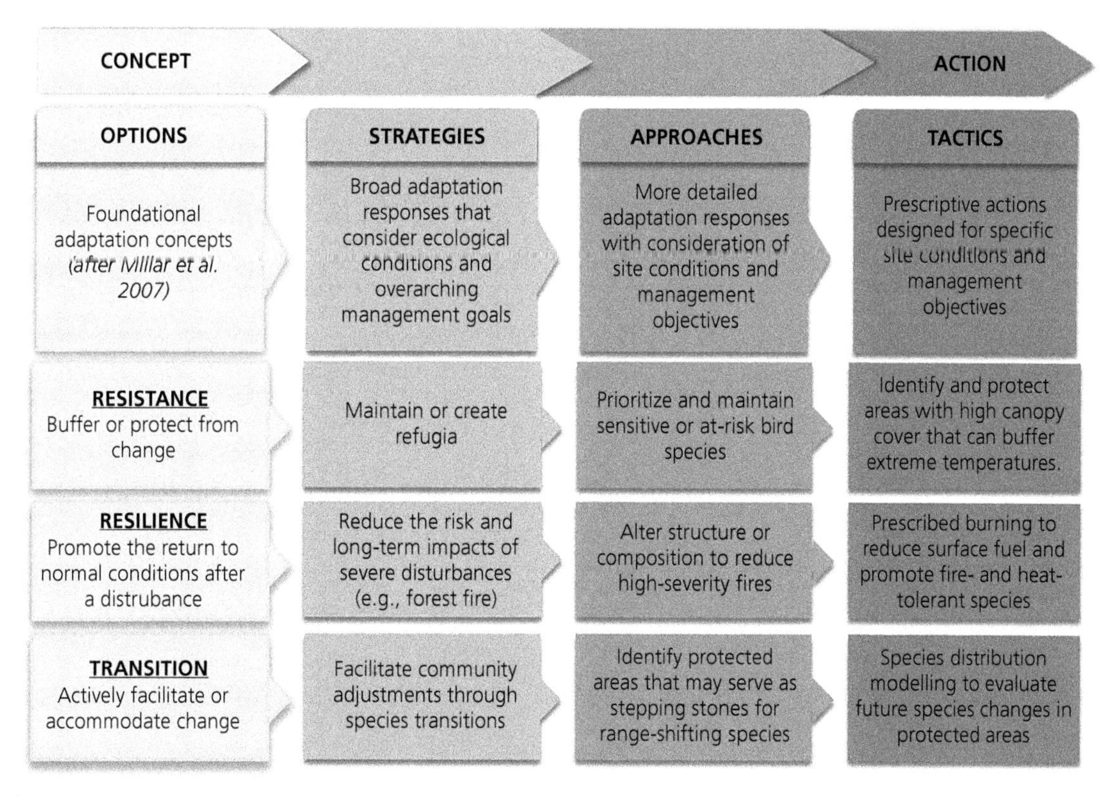

Figure 17.1 Climate change adaptation is a continuum of management actions within broader concepts of resistence, resilience, and transition. Adapation actions range from relatively broad strategies to more detailed tactics that can be implemented by managers and conservation planners. Diagram adapted from Swanston et al. (2016).

adaptation in bird conservation, but *tactics* require prescriptive and specific actions (Figure 17.1). The goal of this chapter is to review how we assess which species are most impacted by climate change and then provide a framework and examples of common strategies and tactics managers can use to incorporate climate change adaptation into bird conservation. In doing so, we present a suite of strategies designed to translate broad conservation concepts into targeted and prescriptive actions for bird conservation (Figure 17.1). The first step prior to the implementation of any adaptation strategy, however, is identifying and predicting which species and populations are most vulnerable to the effects of future climate change.

17.2 Predicting future responses by birds to climate change

In an effort to quantitatively determine how species and their distributions will be impacted in the future, several different types of modelling approaches have been developed and applied at multiple scales to predict changes in species distributions, vulnerability, and population viability (Table 17.1; Dawson et al. 2011).

17.2.1 Species distribution models

Species distribution models (SDMs), also known as ecological niche models or, in the case of specific climate applications, bioclimatic envelope models (see also Chapters 7 and 15) were some of the first approaches used to predict how species would be impacted by future climate change. Species distribution models correlate field data on species' locations (i.e., presence/absence, abundance) with environmental variables (Elith and Leathwick 2009; Franklin 2010).

The bleak picture coming into focus from studies using SDMs (Table 17.1) is that bird species, at the regional, continental, and global levels, will experience

Table 17.1 A selection of studies using species distribution models, climate change vulnerability assessments, and mechanistic population models that project the impacts of future climate change on bird geographic distributions and population viability. All references are available online in a supplemental Reference section (www.oup.co.uk/companion/dunn&moller).

Reference	Region/Habitat	Species	Inputs	Outcomes	Time frame
Species Distribution Models					
Elsen and Tingley 2015	Global/Montane	Montane landbirds	Mountain topography	True mountaintop species represent highest global conservation priority regardless of topography but depending on mountain topography, species distribution, elevation, and direction of range shifts may have either positive or negative outcomes for species.	N/A
Huntley et al. 2008	Europe	All breeding birds (n = 431)	European breeding bird distributions, mean monthly climate data 1961–90, six future climate scenarios, three general circulation models, A2 & B2 emissions scenarios	The potential future range of an average species is expected to shift nearly 550km NE and reduce in size by a fifth by the end of the century. The average current and potential future range overlap is 40% but for some species, there is no future potential range at all. Arctic, sub-Arctic, and some Iberian species are projected to suffer the greatest range loss while some species found only in Europe are likely to increase their risk of extinction. Overall, there will be a general decrease in species richness throughout Europe.	2070–99

Table 17.1 Continued

Reference	Region/Habitat	Species	Inputs	Outcomes	Time frame
Jetz et al. 2007	Global	All 8750 landbird species	Millennium Ecosystem Assessment global scenarios (4), bird distribution data	Mean expected range contraction is 21–26% by 2050 and 29–35% by 2100. Approximately 170–260 species by 2050 and an additional 83–195 species by 2100 are projected to experience greater than 50% range declines.	2050 & 2100
Langham et al. 2015	US & Canada	588 species	North American Breeding Bird Survey, Audubon Christmas Bird Count, 17 bioclimatic variables, 3 emissions scenarios	314 species (53%) are projected to lose more than half of their current geographic range.	By 2100
Lee and Barnard 2016	South Africa Fynbos biome	6 endemics	Density estimates, occupancy, 19 climatic variables, MaxEnt	For all species, climatic space changes are unfavourable under future climate scenarios and more influenced by temperature compared to precipitation.	2100
Peterson et al. 2002	Mexico	All 1179 species	Ecological niche model, HadCM2 general circulation model under a conservative and liberal climate scenario	Species' distributions show increasing tendencies to contract under realistic dispersal scenarios.	2055
Schwartz et al. 2006	Eastern United States	116 species	Breeding Bird Survey data, habitat suitability models, HADCM2SUL and CGCM1 climate models	Extinction vulnerability increases with decreasing range size. Species with small range size are more likely to have future ranges largely or wholly disjunct from current ranges.	
Scridel et al. 2018	Holarctic mountain and upland	2316 species	Meta-analysis of the direct and indirect effects of climate change on mountain birds in the Holarctic region	Climate change is affecting the reproduction, survival, population trends, and distributions of mountain birds both directly and indirectly, yet patterns are highly variable.	N/A
Şekercioĝlu et al. 2008	Global	8459 landbirds	Elevational ranges, 4 Millennium Assessment habitat-loss scenarios, 5 estimates of surface warming, and 3 estimates of shifts in elevational limits	Intermediate scenarios projected 400–550 landbird extinctions while worst-case scenarios projected about 2498 extinctions (30% of all landbirds). An additional 1770–2650 species would be at risk of extinction. Worldwide every degree of warming projected a nonlinear increase in bird extinctions of about 100–500 species.	2090–2100
Şekercioĝlu et al. 2012	Global/tropics	Tropical	Literature review	Tropical mountain birds, species without access to higher elevations, coastal forest birds, and range-restricted species especially vulnerable. Birds that experience limited temperature variation and low basal metabolic rates most prone to temperature warming and heat waves. Some models predict extinctions of 600–900 landbird species, 89% of which occur in the tropics.	By 2100

(continued)

Table 17.1 Continued

Reference	Region/Habitat	Species	Inputs	Outcomes	Time frame
Stralberg et al. 2015	North America/ boreal & southern Arctic	80 boreal breeders	Climate, density, land use, and topography data	Across species, climate change represented the largest component (0.44) of variance in projected abundance change. Of the 80 species modelled, 30 were projected to decline by 2040, 34 species by 2070, and 37 species by 2100. Most species exhibited a northward distributional shift with range centroids shifting an average of 18m upward in elevation, 3 degrees north in latitude, and 3 degrees west in longitude.	2040–2100
Thomas et al. 2004	Global	Mexico n = 186, Europe n=34, Queensland n=13, South Africa n=5	Species–area relationships, 3 climate projections	For mid-range climate scenarios, 5% or 8% (with or without dispersal) of species are expected to become extinct in Mexico and 0 or 51% (with or without dispersal) in South Africa. For maximum expected climate change, 7% or 48% (with or without dispersal) of species are expected to become extinct in Europe and 85% (with dispersal) in Queensland.	2050–2100
Wiens et al. 2009	California	60 breeding landbird species	Bird and vegetation distribution, 8 climatic variables, 2 climate models	Most species are projected to decrease in distribution. Changes in total species richness vary over the state with both projected increases and decreases in areas.	By 2070
Climate Change Vulnerability Assessments					
Barbet-Massin et al. 2012	Western Palearctic	409 European landbird species	7 SDMs, 5 climate change scenarios, 3 emission and land use scenarios	Using both climate and habitat variables, it was predicted that the range of 71% of species would decrease by 2050. Variations in species richness showed large decreases in the southern regions of Europe, as well as increases mainly in Scandinavia and northern Russia.	2050
Burthe et al. 2014	North Sea, Scotland	45 seabird species	Colony-based and at-sea bird data counts, sea surface temperatures, productivity and survival of 25 species with sufficient data. A qualitative approach for remaining species.	53% of species showed negative relationships with SST. Trends in counts and demography were combined with climate vulnerability to give an index of population concern to future climate warming, and 44% of species were classified as high or very high concern.	1980–2011
Byers and Norris 2011	West Virginia	18 species	Climate Change Vulnerability Index. Climate data used an ensemble of 16 global circulation models from Climate Wizard	One species categorized as highly vulnerable, 3 as moderately vulnerable, 6 as presumed stable, 8 as increase likely.	2060
Carr et al. 2014	West Africa	1172 species	IUCN's Climate Change Vulnerability Assessment Framework. Guided by CCVA trait groups, selected taxon-specific biological, ecological, physiological, and environmental traits.	Using optimistic assumptions of missing data a total of 17 (1.5%), 247 (21%), and 309 (26%) bird species were considered climate change-vulnerable by the years 2025, 2055, and 2085 respectively.	2025, 2055, 2085

Table 17.1 Continued

Reference	Region/Habitat	Species	Inputs	Outcomes	Time frame
Case et al. 2015	Northwestern North America	113 species	Scientific literature and expert knowledge (workshops). Sensitivity scores assessed based on nine factors.	Birds had a median sensitivity score of 52, compared to reptiles and amphibians, which were highest (median =76), and plants, which were lowest (median =48). Birds had the largest number of species, and also had the smallest range of scores: 21–71. The 64 bird species with 'at risk' designations had sensitivity scores not significantly different than bird species without the designations.	N/A
Coe et al. 2012	Coronado National Forest	8 species	Literature review. Modelled maps of current vegetation (Year 2005) and projected change (Year 2090). Examined 30 high priority vertebrate species using System for Assessing Vulnerability of Species (SAVS).	Provided support for the likelihood of several key environmental changes on the CNF. Vertebrate assessment showed that most species have vulnerability according to the multiple criteria, and all species had vulnerability in some areas. Birds were the taxonomic group with the highest average overall vulnerability score (+7.4).	2090
Culp et al. 2017	Upper Midwest and Great Lakes Region	46 species	CCVA incorporating the full annual cycle. Methodology included background risk, climate change exposure × climate sensitivity, adaptive capacity to climate change, and indirect effects of climate change.	Study ranked 10 species as highly vulnerable to climate change and two as having low vulnerability. The remaining 34 species were ranked as moderately vulnerable.	N/A
Foden et al. 2013	World	9856 species	Assessed three dimensions of climate change vulnerability using species' biological traits and their modelled exposure to projected climatic changes.	Found that 608–851 (6–9%) of bird species are both highly climate change vulnerable and already threatened with extinction on the IUCN Red List. The remaining highly climate change vulnerable species represent new priorities for conservation.	2050, 2090
Galbraith et al. 2014	Western Hemisphere	52 shorebird taxa, 49 species (three species were split into two populations)	Categorical risk model based on anticipated changes in breeding, migration, and wintering habitat, degree of dependence on ecological synchronicities, migration distance, and degree of specialization on breeding, migration, or wintering habitat.	The majority of taxa, 47 (90%), are predicted to experience an increase in risk of extinction. No species was reclassified into a lower-risk category, although 6 species had at least one risk factor decrease in association with climate change.	2050, 2100
Gardali et al. 2012	California	358 bird taxa	Assessed climate vulnerability by scoring sensitivity and exposure. Using seven exposure and sensitivity criteria.	Using the combined sensitivity and exposure scores as an index, they classified 128 species as vulnerable to climate change. Birds associated with wetlands had the largest representation on the list relative to other habitat groups. Of the 29 state or federally listed taxa, 21 were also classified as climate vulnerable, further raising their conservation concern.	2060, 2099

(continued)

Table 17.1 Continued

Reference	Region/Habitat	Species	Inputs	Outcomes	Time frame
Garnett et al. 2013	Australia	1232 ultrataxa (706 species, 926 subspecies)	Employing risk models that vulnerability arises from a combination of sensitivity and exposure. 18 climate models were used to identify the median area of climate space suitable for each taxon in 2085.	The climate space of 101 Australian terrestrial and inland water bird taxa is likely to be entirely gone by 2085, 16 marine taxa have breeding sites that are predicted to be at least 10% less productive than today, and 55 terrestrial taxa are likely to be exposed to more frequent or intense fires.	2085
Hof et al. 2017	Subarctic and Arctic Norway, Sweden, and Finland (Fennoscandia)	251 regional breeding bird species	Constructed a vulnerability matrix integrating a climatic exposure-based vulnerability index and a natural history trait-based vulnerability index.	Climate change is predicted to significantly reduce the current breeding range of species adapted to cold climates in Subarctic and Arctic Europe. Overall, 15% of the bird species included in this study are predicted to experience a reduction in their breeding range, according to the species distribution models based mostly on climatic variables.	2070
Liebezeit et al. 2012	Arctic Alaska	54 regional breeding bird species	NatureServe Climate Change Vulnerability Index (CCVI). Used data from 5 global circulation models, 2 emissions scenarios, and at 2 spatial resolutions.	The CCVI results ranked 2 species as highly vulnerable, 7 as moderately vulnerable, and 5 as likely to increase.	2050
Magginni et al. 2014	Switzerland	Swiss breeding birds	Five indicators expressing three operational aspects of vulnerability: projected change in distribution, reservoirs for the species, and population trend.	Swiss birds inhabiting coniferous woodlands, alpine habitats, and wetlands have higher vulnerability to climate and land-use change than species in other habitats.	2050–2100
Pearce-Higgins et al. 2017	Great Britain	180 species	Distribution data of target species linked with 4 bioclimate variables. Information from climate change projections, observed changes in species distributions for simplified risk assessment. For a subset of species, additional ecological information was integrated for a more comprehensive risk assessment.	Simplified risk assessment classified approximately 10% of species as being at high risk and 8% at medium risk of range loss while approximately 25% of species were classified as having high and 27% medium opportunity to expand their distribution. Approximately 27% of species are likely to have both risks and opportunities. Overall, the distributions of most species are likely to change.	2070–2099
Reece et al. 2013	Florida	44 birds of conservation concern that were most at risk from sea level rise	Results of an expert opinion-based survey with four modules: vulnerability, lack of adaptive capacity, conservation value, and information availability.	The Florida grasshopper sparrow (*Ammodramus savannarum floridanus*), whooping crane (*Grus americana*), and Western snowy plover (*Charadrius alexandrinus nivosus*) are the three bird species identified as having the highest conservation priority. The Florida grasshopper sparrow was rated as highly likely to be extinct by 2100 due to habitat loss and potentially invasive fire ants and/or disease.	By 2100

Table 17.1 Continued

Reference	Region/Habitat	Species	Inputs	Outcomes	Time frame
Rempel and Hornseth 2017	Terrestrial areas of US and Canadian watersheds of the Great Lakes Basin	Eastern meadowlark (*Sturnella magna*), wood thrush (*Hylocichla mustelina*), and hooded warbler (*Setophaga citrina*).	Climate change severity index in migratory range, moisture, temperature based on mid-level greenhouse gas scenario, measures of indirect exposure to climate change, measures of sensitivity and adaptive capacity, and a species' documented and modelled response to climate change.	Eastern meadowlark and wood thrush are highly vulnerable to climate change, but hooded warbler was less vulnerable.	2041–2100
Siegel et al. 2014	California Sierra Nevadas	168 regional breeding bird species	NatureServe Climate Change Vulnerability Index (CCVI) under two different downscaled climate models with divergent projections about future precipitation.	Only one species received the most vulnerable rank, Extremely Vulnerable. Sixteen species scored as Moderately Vulnerable using one or both climate models.	2040–2069
Zack et al. 2010	Great Plains	20 grassland bird species	NatureServe's Climate Change Vulnerability Index (CCVI) with climate projections and historical climate data.	None of the species analysed ranked as 'highly vulnerable'. Only Lesser Prairie Chicken generated 'moderately vulnerable' index values.	2040–2069
Population Models					
Both et al. 2006	The Netherlands	Ten pied flycatcher populations	Population size, laying date, date of peak caterpillar biomass, percentage of great tits producing second broods (as a proxy for phenological state of each area).	In areas where food availability peaks early and breeding is mistimed populations have declined by 90%. In areas with a late food peak, early-breeding birds still breed at the right time and population declines were weak.	1987–2003
Jenouvrier et al. 2009	Antarctica	Emperor penguins	A long-term demographic dataset and projections of sea ice extent from IPCC models.	The probability of quasi-extinction of this population is at least 36% by 2100. The median population size is projected to decline from around 6000 to around 400 breeding pairs in this time.	By 2100
McCauley et al. 2017	United States	Henslow's Sparrow (*Ammodramus henslowii*)	Demographic meta-analysis from published studies.	Strong climate (summer precipitation/temperature)–demography linkage positively affecting nesting success. Populations of this species will likely move southwesterly due to changes in precipitation.	1981–2050
Jenouvrier et al. 2018	Sub-Antarctic breeding and Tasmanian waters during non-breeding	Black-browed Albatross (*Thalassarche melanophris*)	Population data from about 200 breeding pairs for almost 40 years put into a structured matrix population model.	Population growth rate driven by carry-over effect between multiple seasons and demographic processes. Changes in sea surface temperature in late winter, through its effect on juvenile survival, has the largest impact on population growth rate.	N/A

large geographic range shifts (largely declines at the warm, and potentially expansions at the cooler, margins), likely leading to range contractions or extinctions. Two examples of SDM approaches, one from North America and another a global analysis, illustrate the potential magnitude of climate change on large and varied taxonomic groupings of birds. First, in North America, Langham et al. (2015) used the Breeding Bird Survey (BBS) and the Audubon Christmas Bird Count (CBC) to quantify breeding and non-breeding bird distributions for 588 species. Under a range of future emission scenarios, 314 of the species (53 per cent) were projected to lose more than half of their current geographic range. For 126 of these species, no range expansion was associated with habitat loss, whereas for 188 of the species, colonizing new habitat was possible. Next, in a global analysis of 8750 land bird species, Jetz et al. (2007) projected changes in the range of species driven by climate change and the effects of direct land-use change using the Millennium Ecosystem Assessment scenarios. Under the most conservative scenario, at least 400 species are projected to suffer >50 per cent range reductions by the year 2050 and over 900 species by the year 2100 (and as many as 1800 using the more liberal scenarios). Interestingly, although their results do find that the impacts of climate change at high latitudes will be considerable, land use at tropical latitudes and its impact on highly specialized and endemic bird species is more significant for range sizes.

While the predictions of species distribution models are certainly sobering, it is also important to remember that these models require a robust understanding of species–climate relationships. Like any modelling approach, SDMs are ultimately limited by the quality and quantity of both the biological and environmental data (Beale et al. 2008). As the geographic scale and the number of species increase, the availability and quality of the biological data decreases simply because our knowledge of the species' biology becomes more difficult to quantify. The SDM approach also assumes that species are in equilibrium with their climate space, and it is entirely dependent on the niche concept itself (Guisan and Thuiller 2005; Chapter 13). It is important to recognize that SDMs informed only with climate (i.e., bioclimatic envelope models) produce a representation of *potential* climate space, not necessarily where a species will or will not be in the future, because other non-climate factors may just as important (e.g., dispersal and/or habitat limitations). Equally important is that many correlative SDMs do not incorporate critical biological mechanisms, including demography, dispersal, species interactions, physiology, evolution, and intraspecific trait variation (Chapter 13). Finally, SDMs are often focused on predicting changes in distribution in relation to various climate change scenarios that have many uncertainties and may be overly simplistic (McMahon et al. 2011). Assumptions and limitations aside, SDMs offer a valuable tool for predicting the coarse effects of climate change for multiple species but should be supplemented by approaches on quantifying the underlying mechanisms that inform species' vulnerability to climate change.

17.2.2 Climate change vulnerability assessments

The need to assess the complete vulnerability of a species from impending climate change has led to the development of climate change vulnerability assessments (CCVAs; Figure 17.2; Table 17.1; e.g., Glick et al. 2011; Pacifici et al. 2015). CCVAs are another quantitative approach to help identify species or habitats to target for conservation, prioritize land acquisition decisions, and pinpoint specific factors that contribute to vulnerability (Glick et al. 2011; Culp et al. 2017). Vulnerability itself can be defined in multiple ways, but for the purposes of this chapter we follow the IPCC definition as the 'predisposition to be adversely affected'. Vulnerability is a function of both intrinsic and extrinsic factors (McCarthy et al. 2001) and includes three components: exposure, sensitivity, and adaptive capacity (Figure 17.2; IPCC 2007; Dawson et al. 2011; Moritz and Agudo 2013). *Exposure* refers to the rate and magnitude of climate change experienced at a specific location or by a species, throughout their annual cycle. *Sensitivity* is the degree to which the survival, persistence, fitness, performance, or regeneration of a species or population is dependent on the prevailing climate, particularly on climate variables that are likely to undergo change in the near future. Finally, the *adaptive capacity* of a species refers to the capacity of a species or constituent populations to cope with climate change by persisting *in situ*, by shifting to more suitable local microhabitats, or by migrating to more suitable regions. The concept of adaptive capacity can also reflect the potential for management actions to reduce negative impacts (Klausmeyer and Shaw 2009). Bird species or populations may therefore differ in their vulnerability to climate change due to differences in any or all of these three components.

To date, most North American CCVAs have focused solely on the breeding season (Table 17.1; e.g. Young et al. 2011; Gardali et al. 2012; NatureServe v2.1 https://connect.natureserve.org/science/climate-change/ccvi) and have not considered the complex annual life cycle of migratory animals (Small-Lorenz et al. 2013; Culp et al. 2017). As a result, these assessments have suggested, perhaps incorrectly, that migratory birds are less vulnerable than other taxa—in part due to their high dispersal ability and often relatively broad habitat use. For example, of eight climate change vulnerability assessments that included migratory birds (165 species and subspecies), zero species were classified as extremely vulnerable to climate change, 3 per cent were highly vulnerable, 16 per cent were moderately vulnerable, 50 per cent were not vulnerable and stable, and 21 per cent were not vulnerable and likely to increase (10 per cent were not able to be classified, NatureServe v.2.1, https://connect.natureserve.org/science/climate-change/ccvi). Although, it may be true that these characteristics ameliorate the effects of climate change, it is not possible to assess whether this is true for migratory species without incorporating climate from throughout their full annual cycles. Failing to consider risk throughout the full annual cycle can lead to incorrect conclusions and inefficient allocation of resources, decreasing our ability to conserve habitat where it is most important.

If we hope to understand biological responses and vulnerability to climate change, it is essential to incorporate year-round climate and life history data into assessments for *linked populations* of migratory birds. Breeding and wintering locations are said to

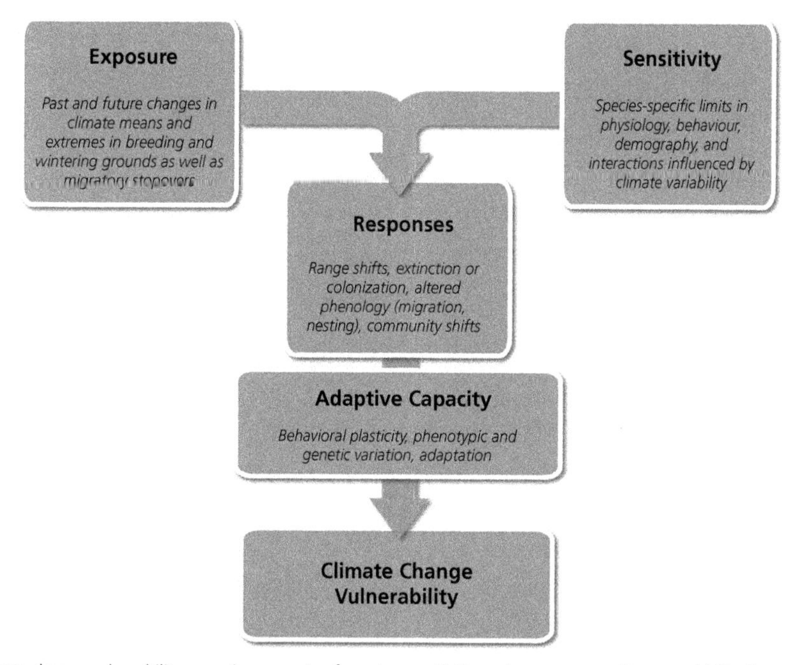

Figure 17.2 Climate change vulnerability comprises aspects of species sensitivity and expsosure to climate variability. Importantly, species and populations can reduce their sensivitity and/or exposure to climate change through their adaptive capacity. Climate change vulnerability assessments seek to identify those species and populations most vulnerable to the effects of modern change as well as identify those mechanisms that drive species' sensitivity and exposure.

be *linked* when individuals from the breeding population are the same as those in the wintering population. For example, the Kirtland's warbler (*Setophaga kirtlandii*) breeds in Michigan's lower peninsula and then migrates to the Bahamas for the non-breeding season. In a recent study of the Kirtland's warbler (Rockwell et al. 2012), there was a strong relationship between winter rainfall in the Bahamas and the timing of spring arrival to breeding areas in Michigan. The same study showed that for every one-inch decline in annual Bahamian rainfall, annual fecundity declined by 0.6 young per warbler pair (Rockwell et al. 2012). In another study that used 26 years of breeding bird survey data, American redstarts (*Setophaga ruticilla*) exhibited a strong positive response in abundance to wetter conditions in the western Caribbean where these populations over-winter (Wilson et al. 2011). Unfortunately, climate change is expected to bring drier conditions to many tropical areas (Neelin et al. 2006), which means that we may see migratory birds like the American redstart delaying spring migration and ultimately arriving late to temperate breeding sites.

These findings illustrate that assessments without year-round climate and life history data are likely to draw inaccurate conclusions about the vulnerability of migratory species in the future. Furthermore, like species distribution models, CCVAs have major components that are challenging to quantify, and incorporate assumptions that likely introduce uncertainty into model output. For example, the ability of a species to physiologically tolerate changes in temperature (thermal sensitivity) or its adaptive capacity remain difficult to parameterize and often require indirect approaches and different methodologies. This has all resulted in many different approaches in the development of CCVAs, ranging from trait-based analyses for hundreds of species to species-specific population models (Table 17.1). Unfortunately, a recent study compared the output of 12 different CCVAs for British butterflies and birds and found that, at least in this case, the different methodologies are not consistent with one another in categorizing species to similar risk groups. Thus, some standardization in CCVA methodology will be needed (Wheatley et al. 2017).

17.2.3. Population modelling

In a few cases, when long-term demographic data are available for populations, more robust quantitative modelling approaches can be used to project the impact of projected climate change on future population dynamics (Table 17.1). For example, Jenouvrier et al. (2009; Chapter 12) parameterized a stochastic population model that combined a long-term demographic dataset (1962–2005) from a breeding colony of emperor penguins (*Aptenodytes forsteri*) with the extent of Arctic sea ice and general climate circulation models to examine future population viability of the species. What their modelling efforts discovered was that there was a 36 per cent probability of complete quasi-extinction (decline of at least 95 per cent) by 2100 and that the population was expected to decline from about 6000 to 400 breeding pairs over this time frame. Emperor penguins will need to adapt to changing environmental conditions if they are to avoid extinction—an unlikely outcome over this time frame of climate change for such a long-lived species. Although this approach is one of the more robust approaches for predicting the future impacts of climate change, it is a single-species study that depends on a unique and rare forty-plus year dataset (see Chapter 12 for more details). Furthermore, it is still subject to the uncertainties associated with the available climate models.

Although there are issues associated with species distribution models, climate change vulnerability assessments, and stochastic population models for predicting the future of bird populations, they remain essential first steps and the best approaches available for prioritizing which species are in need of attention.

17.3 Climate change adaptation in bird conservation

Once a species or population is identified as vulnerable to the effects of climate change, the next steps should focus on how can managers and conservation planners can reduce the sensitivity or exposure of a vulnerable species (i.e., vulnerability). The most effective route for reducing climate change vulnerability is through climate change adaptation, which developed from three broader concepts in ecosystem management (Millar et al. 2007): *resistance*, *resilience*, and *transition* (Figure 17.1). *Resistance* strategies improve the defences or adaptive capacity of a species or population against anticipated changes in climate. *Resilience* strategies accommodate or recognize some degree of change, but encourage a return to pre-disturbance conditions. Finally, strategies of *transition* anticipate drastic changes in climate and promote actions that help facilitate the response of species, populations, and communities to a novel climate space.

17.3.1 Resistance: Identifying and managing climate refugia

The concept of resistance focuses on improving the capacity of a species or population to remain relatively unaffected by climate change over time. Resistance strategies are likely most effective for species with low sensitivity or high adaptive capacity to climate change or in areas buffered from severe climate change impacts. An important adaptation strategy for increasing resistance is maintaining or creating climate change refugia (Figure 17.1). In paleoecology *climate change refugia* are geographic regions that harboured relict populations during eras of rapid climate change (e.g., glaciation) (Cain 1944; Haffer 1969). In modern conservation, the concept of refugia applies to areas characterized by relatively stable local climatic conditions (e.g., microclimates) that persist despite widespread changes at regional and global scales (Ashcroft et al. 2012; Keppel et al. 2012; Morelli et al. 2016). These localized areas of stable climates represent 'microrefugia' that emerge from even slight differences in elevation, shade, aspect, vegetation, or the presence of water or snow (Dobrowksi 2011). Coarse-scale climate models generally fail to capture these microrefugia, but the differences in temperatures resulting from microclimates (e.g., between forests and grasslands) can be of similar magnitude to those predicted from anthropogenic climate change (Suggitt et al. 2011).

Unlike other less mobile taxa, birds have a high capacity to seek out and use microrefugia that could buffer them from unsuitable climate or extreme weather events, such as drought or heat waves. In forests, microclimates can emerge due to differences in stand densities, basal area, and forest edges (Latimer and Zuckerberg 2017), all of which can potentially alter bird behaviour, movement, and abundance. For example, microclimate variability was a stronger predictor than vegetation of forest bird distributions during the summer in the Cascade Mountains of Oregon (Frey et al. 2016). Many species, such as hermit warblers (*Setophaga occidentalis*), avoided warmer sites and settled at cooler sites by the end of the summer season (Figure 17.3). Of 15 forest bird species, almost half were as likely to be associated with warm sites as with cool sites, but despite these species-specific differences, microclimate associations suggest that maintaining microrefugia may be important for increasing the resilience of montane bird communities to rapid climate change (Betts et al. 2018). Managers of vulnerable bird populations should consider important microrefugia, and their associated drivers (e.g., old growth forest, valleys, south-facing slopes) to identify areas that warrant further protection (Morelli et al. 2016). Other resistance actions might include protecting riparian areas, island creation for coastal breeding birds, and controlling invasive plant species. Through these tactics, managers are increasing the resilience of bird populations to the effects of climate change by providing refugia that are essentially relict habitats. Unfortunately, little is known about the minimum number or size of refugia to maintain minimum effective population sizes over the long term. Although resilience may be effective in the short term, it is likely that buffering vulnerable bird populations will require greater resources and intensive management over the long term as the climate shifts further from historical conditions.

17.3.2 Resilience: Accommodating climate-induced disturbance

The concept of resilience encompasses a broader recognition that some degree of climate-induced disturbance occurs within systems, but targeted management allows species and communities to accommodate that change and return to near-prior conditions. For purposes of climate change adaptation, resilience strategies increase the ability of a vulnerable species to recover following disturbances such as droughts, flooding, heat waves, and wildfires

Figure 17.3 Identifying and managing climate change refugia is an example of managing for resistance. In Western Cascade Mountains of Oregon, USA, many forest birds, such as hermit warblers, will settle at cooler sites and shift away from warmer sites throughout the breeding season. These cooler sites, often associated with old growth forests, could serve as important microrefugia buffering forest birds from extreme temperatures. Hermit Warbler photo by Hankyu Kim.

(Figure 17.2). Importantly, these strategies are most appropriate in areas where disturbance is already a natural component of the variability of the ecosystem (e.g., fire-dependent ecosystems, habitat creation through windthrow), and birds have adapted life histories allowing them to bounce back from such disturbances.

An example of the critical importance of resilience is in the management and conservation of forest birds exposed to high severity wildfires in the western United States. Beginning in the mid-1980s, forest wildfires demonstrated a sustained increase throughout most of the western US, a trend attributed to climate change and increasing temperatures, drier summers, drought conditions, and earlier spring snowmelt (Westerling et al. 2006; Littell et al. 2009; Westerling 2016). The California spotted owl (*Strix occidentalis occidentalis*) is a highly vulnerable species of conservation concern due to its declining population numbers and dependency on high-canopy cover forests dominated by large trees (Tempel et al. 2016). There is an increasing recognition that, in addition to other threats such as invading barred owls (*Strix varia*) and habitat loss, climate change and high-severity fire poses an imposing threat to spotted owl populations (Jones et al. 2016a, b). The U.S. Forest Service manages the majority of spotted owl habitat and must balance the objectives of conserving spotted owls and promoting the resilience of Sierra Nevada forests in the face of climate change by restoring historical wildfire regimes without endangering already declining spotted owl populations (Tempel et al. 2014; Peery et al. 2017).

In practice, implementing fire management strategies that do not also pose risks to spotted owls is challenging. Conserving habitat known to be important to spotted owls may lead to dense stands with high fuel loadings that are at risk from high-severity fire, but fuel reduction strategies (e.g., reducing canopy cover, or large-tree density) can potentially exacerbate spotted owl population declines. Despite these challenges, large high-severity fires pose an increasing risk to spotted owls, and these risks will likely increase with climate change (Jones et al. 2016b). Given these challenges, Peery et al. (2017) recommended resilience strategies that should be implemented at scales important to the ecology and behaviour of spotted owls. At the activity-centre scale (defined as areas of long-term nesting and roosting within an owl territory), management should focus on careful, low-intensity vegetation treatments that maintain high-quality nesting habitat (i.e., old forest). At larger home range and landscape scales, management can be more proactive with an emphasis on forest restoration, prescribed burning, and maintaining heterogeneous forests that increase resilience. Conserving and restoring owl habitat (large trees, moderate stem density, and canopy cover) in areas that could support climate-resilient conditions (i.e., lower climatic water deficit rates) may better align the distribution of owl habitat with forest restoration goals (Peery et al. 2017). Maintaining key habitat within activity centres and territories will likely promote population growth in the short term, whereas reducing risk from high-severity fire across the broader landscape could promote population viability in the long term (Peery et al. 2017). Importantly, all efforts in resilience benefit from a robust feedback loop that generates and incorporates new information and learning in an adaptive management framework.

17.3.3 Transition: Protected areas and facilitating change

Unlike resistance and resilience, transition strategies recognize that the effects of climate change on species and populations are inevitable and focus on predicting and facilitating species' responses to those changes (Figure 17.2). Transition strategies can involve approaches such as assisted migration, captive breeding programmes, and reintroduction, but a primary strategy is evaluating and predicting the future use and connectivity of protected areas. Establishing protected areas is one of the most widely used strategies for biodiversity conservation (Gaston et al. 2008; Watson et al. 2014) and is a critical conservation tool for species that might be shifting their ranges in response to modern climate change. Recent climate change has already pushed the climate space of many bird species towards higher latitudes and altitudes, and, given their static nature, shifting ranges pose a unique problem for the role of protected area networks (Chapter 13). Unlike the refugia concept, protected areas serve as potential 'stepping stones' for species as they track their optimal climate space over time (Hannah et al. 2014). For example, during recent range boundary expansions several bird species preferentially colonized protected areas in Great Britain (Thomas et al. 2012); this was especially important for wetland birds that used protected areas as 'landing pads' when arriving and establishing in a new region (Hiley et al. 2013). Similarly, in a study of range shifts for 139 savannah bird species in Tanzania, climate was a primary driver of colonization that occurred disproportionately in protected areas, whereas local extinctions were less impacted by climate and more affected by processes occurring outside protected areas (e.g., habitat loss and agricultural development; Beale et al. 2013). These studies are in agreement that protected areas play a more important role as stepping stones of colonization for warm-adapted birds as opposed to holdouts for cold-adapted birds facing climate-induced extinction (but see Gillingham et al. 2015). However, the value of any given protected area likely varies with species life history characteristics (e.g., habitat specialization) (Thomas et al. 2012) and location with respect to range margins (Hiley et al. 2013; Gillingham et al. 2015).

Protected areas are clearly important for the overall maintenance of bird diversity under climate change (Thomas and Gillingham 2015; Gaüzère et al. 2016), but given future climate projections, there is mounting concern that protected areas may become climatically unsuitable for those species that they were designated to protect. The legal resilience of protected areas is often based on the occurrence and abundance of species of conservation concern, and there is the possibility that future shifts in species and communities could compromise the effectiveness and justification of existing protected area networks. To test this, Johnston et al. (2013) developed statistical models that linked climate to the abundance of internationally important sea and water bird populations in northwestern Europe. Using these models, they estimated that future climate change would cause strong declines for more than half of the bird populations. Despite these future declines, however, most Special Protection Areas in the United Kingdom retained at least 1 per cent of the national population, which is sufficient to maintain the legal status of the areas. Furthermore, sites that currently protect large populations were projected to be important in the future. Within the United States, the national park system confers strong protection for many bird species. To understand how climate change is likely to alter bird communities in national parks, Wu et al. (2018) used species distribution models relating North American Breeding Bird Survey (summer) and Audubon Christmas Bird Count (winter) observations to climate data from the early 2000s and projected to the mid-century under different emissions scenarios. Using climate suitability projections for 513 bird species across 274 national parks, they found that national parks would become increasingly important for birds in the coming decades as colonizations will exceed extirpations in 62–100 per cent of parks. Parks at higher latitudes, such as those in the Midwest and Northeast, are expected to see particularly high rates of turnover in bird communities (Figure 17.4). Within protected areas with high predicted turnover, managers could consider practices that promote colonization such as the use of playbacks or habitat creation. Given the projected change in suitable climate space for many of these bird species, it is likely that future protected area planning will need to accommodate and anticipate these novel community assemblages.

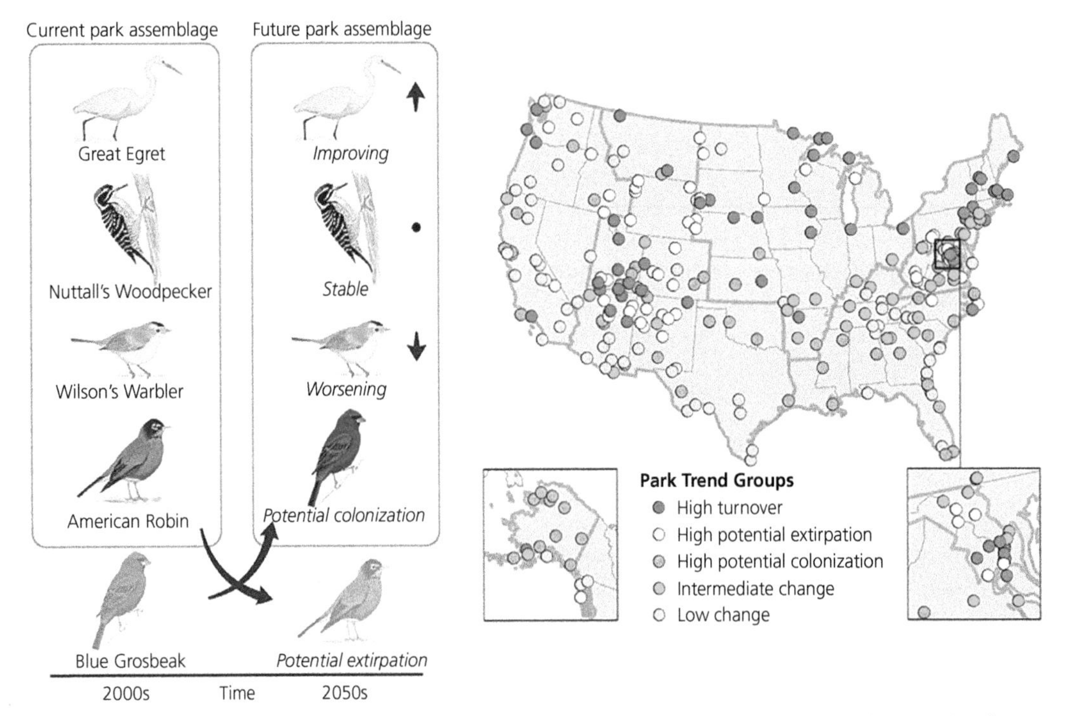

Figure 17.4 Protected areas play a critical role in bird conservation; however, it is not clear how climate change might impact the effectiveness of conservation areas. Scientists use species distribution models to predict the changing composition of bird communities in national parks throughout the U.S. As an example, these models were used to project bird community changes at Golden Gate National Recreation Area. Climate suitability was projected to improve for the great egret (*Ardea alba*), remain stable for the Nuttall's woodpecker (*Picoides nuttallii*), worsen for the Wilson's warbler (*Cardellina pusilla*), and is at risk of disappearing for the American robin (*Turdus migratorius*). On the other hand, the blue grosbeak (*Passerina caerulea*) is not currently found in the park, but climate is projected to become suitable for this species (bird illustrations by Kenn Kaufman). Across 274 U.S. national parks, there will be a wide range of community changes, with some of the highest likelihood of turnover of communities for parks at more northerly latitudes.

17.4 The need for long-term monitoring and research

More than any other area of ecological research, long-term monitoring of bird populations and associated environmental variables is essential for understanding the impacts of climate and climate change. The impacts of climate change will likely manifest themselves over broad spatial and temporal scales, and, as a result, citizen science programmes that monitor bird populations at regional and continental scales have played and will continue to play an important role in detecting the impacts of climate change on birds (Dickinson et al. 2010; Cooper et al. 2014; La Sorte et al. 2018). Programs such the BBS, CBC, eBird, Pan-European Common Birds Monitoring Scheme, and hundreds of bird atlas programmes are essential for tracking bird responses to climate change across the world. Many of these efforts are beginning to incorporate climate change predictions in the implementation of their monitoring efforts. For example, the National Audubon Society launched the Climate Watch Program (www.audubon.org/conservation/climate-watch) to enlist the public in monitoring the occurrence of climate-vulnerable species and as a means of validating and refining models for where these species' ranges will shift in response to a changing climate. The hope is to locate surveys in areas that are *projected* to be climatically suitable for target species in the future. In doing so, Climate Watch serves as an adaptive early-warning system for detecting range shifts and identifying areas of high climatic suitability for vulnerable species and on-the-ground conservation efforts. Although citizen science is an essential component of monitoring bird populations from regional to continental to global scales, many of these programmes are limited to occurrence and abundance data, are constrained to areas that are easily accessible to volunteers (roadsides, residential areas, etc.), and have trouble identifying the underlying mechanisms of species' climate change vulnerability.

17.5 Critical future needs

In this chapter, we have reviewed the best approaches and associated predictions for how climate change may affect bird populations in the future, and how conservationists may counter and adapt to threats to maximize the conservation of bird species globally. Most, but not all, of the projected effects are based on phenomenological pattern descriptions, without a mechanistic understanding of species' climate sensitivity, ecology, or evolutionary potential (but see Population Models in Table 17.1; Visser 2008). Moving forward, we should try and improve our basic understanding of ecological mechanisms driving vulnerability to climate change. It is beyond doubt that more detailed studies focused on gathering demographic data for a mechanistic knowledge of species' adaptive capacity as well as their potential constraints will improve our ability to help species adapt to climate change.

As described above, species distribution models assume that species are inert entities, rather than flexible and evolving organisms that may actually adapt to change. We still lack any understanding of which species will or will not be able to adapt to ongoing climate change. Species likely differ in how plastic they are to environmental change, and how much standing genetic variation is present in populations to evolve. Species with large ranges can show local adaptation to the conditions (see e.g., Pulido et al. 1996; Silverin et al. 1993), and this may facilitate adaptive responses if phenotypes disperse to areas with different climates, or track climate change through dispersal (Berg et al. 2010; Studds et al. 2008; Burger et al. 2013). One of the largest challenges in predicting future range and population trends will be to understand how biotic interactions themselves may change (Van der Putten et al. 2010). Trophic interactions can change because phenological trends may be unequal across trophic levels (Both et al. 2009; Thackeray et al. 2016), but little is known about how this affects food-web structure and dynamics. We are still a long way from building these biological complexities into species distribution and population models, but these processes seem to be essential for maintaining viable populations into the future.

As we expect the most severe consequences of climate change to occur on the margins of a species' range, long-term studies should be established along edges of species' ranges as well as in the interior. Indeed, Cuervo and Møller (2013) show the largest fluctuations do indeed occur at range margins, with European bird species, especially at their southern limits. Also, most long-term studies are concentrated in temperate or Arctic regions, whereas regions with the highest biodiversity lack comprehensive monitoring programmes to establish whether populations adapt to climate change, and how this impacts population viability (Şekercioĝlu et al. 2012). Importantly, longitudinal studies are the only means by which we can study individuals throughout their annual cycles and lifetimes. This should include tracking studies of individual animals and populations to understand how their movements change in response to climate and climate change across generations.

Finally, seasonality is a feature of almost any ecosystem on the globe, and species vary largely in how they deal with these changes. Forecasting the effects of climate change must include not only a within-season component but also how within-season changes fit within the context of the annual cycle and interact with events between seasons, i.e., seasonal interactions (Marra et al. 2015). For example, knowledge of when and where in the annual cycle populations are limited is essential, but then knowing how conditions in that season carry over to interact with other seasons provides a more complete understanding of the biological response. Arctic breeding shorebird populations are limited by food availability on intertidal wintering (Van Gils et al. 2006) and staging areas (Piersma et al. 2016), and not by the vast area of tundra breeding grounds. However, climate change is expected to be most severe in the Arctic (IPCC 2007), reducing and limiting available breeding areas in the future (Wauchope et al. 2017). Impacts are already being detected in red knot (*Calidris canutus*), which are becoming increasingly mismatched with the local insect food peak in the Arctic, resulting in retarded chick growth leading to shorter-billed individuals which have lower survival on the wintering grounds because they cannot reach deeper buried bivalves (van Gils et al. 2016). Subtle, but profound, seasonal interactions such as this are likely common in most systems, but typically we do not have the biological knowledge necessary to decipher these details and include them in projection models.

17.6 Conclusions

Predicting future response and vulnerability of birds with some level of confidence such that decision makers can implement adaptation strategies will be one of the greatest challenges moving forward in conservation. Undoubtedly, there will be winners and losers (i.e., extinctions) that emerge as environments and habitats continue to change over the next 50–100 years. In fact, if we look back over the last 500 years or so, avian species extinctions have been caused by a variety of causes, though few if any were climate induced; the vast majority of vertebrate extinctions have been caused by overharvesting, habitat loss, and alien invasive species such as pathogens, rats, and cats. Although these threats continue to play prominently in the decline of species globally, future climate change, either directly or indirectly, as predicted by the various approaches outlined in this chapter, will likely be a dominant driver of declines and extinctions of species in the future. Ultimately, predicting the magnitude of effects of climate change on birds and identifying strategic prescriptive actions with confidence will require quantitatively disentangling the interacting and confounding effects of habitat loss, invasive species, and other factors on the population dynamics and spatial distribution of species. Despite these challenges and uncertainties, bird conservation efforts must move forward equipped with the best available scientific knowledge and managers should consider the entire suite of climate change adaptation strategies that can be implemented on the ground. There is an ever-growing need for a full spectrum of management actions that can help sustain bird populations in the face of climate change. If the dire predictions become true, given limited resources, a triage approach may become necessary, but let us hope that we can modify human behaviour to minimize the impacts that future warming will have on our ecosystems and their dependent species.

References

Ashcroft, M.B., Gollan, J.R., Warton, D.I., and Ramp, D. (2012). A novel approach to quantify and locate potential microrefugia using topoclimate, climate stability, and isolation from the matrix. *Global Change Biology*, 18, 1866–79.

Beale, C.M., Baker, N.E., Brewer, M.J., and Lennon, J.J. (2013). Protected area networks and savannah bird biodiversity in the face of climate change and land degradation. *Ecology Letters*, 16, 1061–8.

Beale, C.M., Lennon, J.J., and Gimona, A. (2008). Opening the climate envelope reveals no macroscale associations with climate in European birds. *Proceedings of the National Academy of Sciences of the United States of America*, 105, 14908–12.

Berg, M.P., Kiers, E.T., Driessen, G., van der Heijden, M., Kooi, B.W., Kuenen, F., Liefting, M., Verhoef, H.A., and Ellers, J. (2010). Adapt or disperse: Understanding species persistence in a changing world. *Global Change Biology*, 16, 587–98.

Betts, M.G., Phalan, B., Frey, S.J.K., Rousseau, J.S., and Yang, Z. (2018). Old-growth forests buffer climate-sensitive bird populations from warming. *Diversity and Distributions*, 24, 439–47.

Bonnet, X., Shine, R., and Lourdais, O. (2002). Taxonomic chauvinism. *Trends in Ecology & Evolution*, 17, 1–3.

Both, C., Van Asch, M., Bijlsma, R.G., Van Den Burg, A.B., and Visser, M.E. (2009). Climate change and unequal phenological changes across four trophic levels: Constraints or adaptations? *Journal of Animal Ecology*, 78, 73–83.

Buckley, L.B., and Kingsolver, J.G. (2012). Functional and phylogenetic approaches to forecasting species' responses to climate change. *Annual Review of Ecology, Evolution, and Systematics*, 43, 205–26.

Burger, C., Nord, A., Nilsson, J.Å., Gilot-Fromont, E., and Both, C. (2013). Fitness consequences of northward dispersal as possible adaptation to climate change, using experimental translocation of a migratory passerine. *PLoS One*, 8, e83176.

Cain, S. (1944). *Foundations of Plant Geography*. Harper and Brothers, New York, NY.

Cooper, C.B., Shirk, J., and Zuckerberg, B. (2014). The invisible prevalence of citizen science in global research: Migratory birds and climate change. *PLoS One*, 9, e106508.

Cuervo, J.J., and Møller, A.P. (2013). Temporal variation in population size of European bird species: effects of latitude and marginality of distribution. *PLoS One*, 8, e77654.

Culp, L.A., Cohen, E.B., Scarpignato, A.L., Thogmartin, W.E., and Marra, P.P. (2017). Full annual cycle climate change vulnerability assessment for migratory birds. *Ecosphere*, 8, e01565.

Dawson, T.P., Jackson, S.T., House, J.I., Prentice, I.C., and Mace, G.M. (2011). Beyond predictions: Biodiversity conservation in a changing climate. *Science*, 332, 53–8.

Dickinson, J.L., Zuckerberg, B., and Bonter, D.N. (2010) Citizen science as an ecological research tool: Challenges and benefits. *Annual Review of Ecology, Evolution, and Systematics*, 41, 149–72.

Dobrowski, S.Z. (2011). A climatic basis for microrefugia: The influence of terrain on climate. *Global Change Biology*, 17, 1022–35.

Elith, J., and Leathwick, J.R. (2009). Species Distribution Models: Ecological explanation and prediction across space and time. *Annual Review of Ecology, Evolution, and Systematics*, 40, 677–97.

Franklin, J. (2010). *Mapping species distribution: Spatial inference and prediction*. Cambridge University Press, New York, NY.

Frey, S.J.K., Hadley, A.S., and Betts, M.G. (2016). Microclimate predicts within-season distribution dynamics of montane forest birds. *Diversity and Distributions*, 22, 944–59.

Gardali, T., Seavy, N.E., DiGaudio, R.T., and Comrack, L.A. (2012). A climate change vulnerability assessment of California's at-risk birds. *PLoS One*, 7, 331.

Gaston, K.J., Jackson, S.F., Cantú-Salazar, L., and Cruz-Piñón, G. (2008). The ecological performance of protected areas. *Annual Review of Ecology, Evolution, and Systematics*, 39, 93–113.

Gaüzère, P., Jiguet, F., and Devictor, V. (2016). Can protected areas mitigate the impacts of climate change on bird's species and communities? *Diversity and Distributions*, 22, 625–37.

Gillingham, P.K., Bradbury, R.B., Roy, D.B., Anderson, B.J., Baxter, J.M., Bourn, N.A.D. Crick, H.Q.P., Findon, R.A., Fox, R., Franco, A., Hill, J.K., Hodgson, J.A., Holt, A.R., Morecroft, M.D., Hanlon, N.J.O., Oliver, T.O.M.H., Pearce-Higgins, J.W., Procter, D.A., Thomas, J.A., and Thomas, C.D. (2015). The effectiveness of protected areas in the conservation of species with changing geographical ranges. *Biological Journal of the Linnean Society*, 115, 707–17.

Glick, P., Stein, B.A., and Edelson, N.A. (2011). *Scanning the conservation horizon: a guide to climate change vulnerability assessment*. National Wildlife Federation, Washington, DC.

Guisan, A., and Thuiller, W. (2005). Predicting species distribution: Offering more than simple habitat models. *Ecology Letters*, 8, 993–1009.

Haffer, J. (1969). Speciation in Amazonian forest birds. *Science*, 165, 131–7.

Hannah, L., Flint, L., Syphard, A.D., Moritz, M.A., Buckley, L.B., and McCullough, I.M. (2014). Fine-grain modeling of species' response to climate change: Holdouts, stepping-stones, and microrefugia. *Trends in Ecology & Evolution*, 29, 390–7.

Heller, N.E., and Zavaleta, E.S. (2009). Biodiversity management in the face of climate change: A review of 22 years of recommendations. *Biological Conservation*, 142, 14–32.

Hiley, J.R., Bradbury, R.B., Holling, M., and Thomas, C.D. (2013). Protected areas act as establishment centres for species colonizing the UK. *Proceedings of the Royal Society of London Series B: Biological Sciences*, 280, 20122310.

Hof, C., Levinsky, I., Araújo, M.B., and Rahbek, C. (2011). Rethinking species' ability to cope with rapid climate change. *Global Change Biology*, 17, 2987–90.

IPCC (2007). *The Physical Science Basis: Working Group I Contribution to the Fourth Assessment Report of the IPCC. Climate Change 2007*. S. Solomon, D. Qin, M. Manning, Z. Chen, M. Marquis, K.B. Averyt, M.M.B. Tignor, and H.L. Miller (eds). Cambridge University Press, Cambridge, UK.

Jenouvrier, S., Caswell, H., Barbraud, C., Holland, M., Strœve, J., and Weimerskirch, H. (2009). Demographic models and IPCC climate projections predict the decline of an emperor penguin population. *Proceedings of the National Academy of Sciences of the United States of America*, 106, 1844–7.

Jetz, W., Wilcove, D.S., and Dobson, A.P. (2007). Projected Impacts of Climate and Land-Use Change on the Global Diversity of Birds. *PLoS Biology*, 5, e157.

Johnston, A., Ausden, M., Dodd, A.M., Bradbury, R.B., Chamberlain, D.E., Jiguet, F., Thomas, C.D., Cook, A.S.C.P., Newson, S.E., Ockendon, N., Rehfisch, M.M., Roos, S., Thaxter, C.B., Brown, A., Crick, H.Q.P., Douse, A., McCall, R.A., Pontier, H., Stroud, D.A., Cadiou, B., Crowe, O., Deceuninck, B., Hornman, M., and Pearce-Higgins, J.W. (2013). Observed and predicted effects of climate change on species abundance in protected areas. *Nature Climate Change*, 3, 1055–61.

Jones, G.M., Gutiérrez, R.J., Tempel, D.J., Whitmore, S.A., Berigan, W.J., and Peery, M.Z. (2016a). Megafires: an emerging threat to old-forest species. *Frontiers in Ecology and the Environment*, 14, 300–6.

Jones, G.M., Gutiérrez, R.J., Tempel, D.J., Zuckerberg, B., and Peery, M.Z. (2016b). Using dynamic occupancy models to inform climate change adaptation strategies for California spotted owls. *Journal of Applied Ecology*, 53, 895–905.

Keppel, G., Van Niel, K.P., Wardell-Johnson, G.W., Yates, C.J., Byrne, M., Mucina, L., Schut, A.G.T., Hopper, S.D., and Franklin, S.E. (2012). Refugia: identifying and understanding safe havens for biodiversity under climate change. *Global Ecology and Biogeography*, 21, 393–404.

Klausmeyer, K.R., and Shaw, M.R. (2009). Climate change, habitat loss, protected areas and the climate adaptation potential of species in Mediterranean ecosystems worldwide. *PLoS One*, 4, e6392.

Knudsen, E., Lindén, A., Both, C., Jonzén, N., Pulido, F., Saino, N., Sutherland, W.J., Bach, L.A., Coppack, T., Ergon, T., Gienapp, P., Gill, J.A., Gordo, O., Hedenstrom, A., Lehikoinen, E., Marra, P.P., Møller, A.P., Nilsson, A.L.K., Peron, G., Ranta, E., Rubolini, D., Sparks, T.H., Spina, F., Studds, C.E., Sæther, S.A., Tryjanowski, P., and Stenseth, N.C. (2011). Challenging claims in the study of migratory birds and climate change. *Biological Reviews*, 86, 928–46.

La Sorte, F.A., Lepczyk, C.A., Burnett, J.L., Hurlbert, A.H., Tingley, M.W., and Zuckerberg, B. (2018). Opportunities and challenges for big data ornithology. *Condor*, 120, 414–26.

Langham, G.M., Schuetz, J.G., Distler, T., Soykan, C.U., and Wilsey, C. (2015). Conservation status of North American birds in the face of future climate change. *PLoS One*, 10, e0135350.

Latimer, C.E., and Zuckerberg, B. (2017). Forest fragmentation alters winter microclimates and microrefugia in human-modified landscapes. *Ecography*, 40, 158–70.

Littell, J.S., Mckenzie, D., Peterson, D.L., and Westerling, A.L. (2009). Climate and wildfire area burned in western U.S. ecoprovinces, 1916–2003. *Ecological Applications*, 19, 1003–21.

Marra, P.P., Cohen, E.B., Loss, S.R., Rutter, J.E., and Tonra, C.M. (2015). A call for full annual cycle research in animal ecology. *Biology Letters*, 11, 20150552.

Mawdsley, J.R., O'Malley, R., and Ojima, D.S. (2009). A review of climate-change adaptation strategies for wildlife management and biodiversity conservation. *Conservation Biology*, 23, 1080–9.

McCarthy, J.J., Canziani, O.F., Leary, N.A., Dokken, D.J., and White, K.S. (2001). *Climate change 2001: impacts, adaptation, and vulnerability: contribution of Working Group II to the third assessment report*. Intergovernmental Panel on Climate Change, Cambridge University Press, Cambridge, UK.

McCarty, J.P., Wolfenbarger, L.L., and Wilson, J.A. (2017). Biological impacts of climate change. *eLS*, 1–13. https://doi.org/10.1002/9780470015902.a0020480.pub2

McMahon, S.M., Harrison, S.P., Armbruster, W.S., Bartlein, P.J., Beale, C.M., Edwards, M.E., Kattge, J., Midgley, G., Morin, X., and Prentice, I.C. (2011). Improving assessment and modelling of climate change impacts on global terrestrial biodiversity. *Trends in Ecology & Evolution*, 26, 249–59.

Millar, C.I., Stephenson, N.L., and Stephens, S.L. (2007). Climate change and forests of the future: Managing in the face of uncertainty. *Ecological Applications*, 17, 2145–51.

Miller, C.R., Latimer, C.E., and Zuckerberg, B. (2018). Bill size variation in northern cardinals associated with anthropogenic drivers across North America. *Ecology and Evolution*, 8, 4841–51.

Møller, A.P., Rubolini, D., and Saino, N. (2017). Morphological constraints on changing avian migration phenology. *Journal of Evolutionary Biology*, 30, 1177–84.

Morelli, T.L., Daly, C., Dobrowski, S.Z., Dulen, D.M., Ebersole, J.L., Jackson, S.T., Lundquist, J.D., Millar, C.I., Maher, S.P., Monahan, W.B., Nydick, K.R., Redmond,

K.T., Sawyer, S.C., Stock, S., and Beissinger, S.R. (2016). Managing climate change refugia for climate adaptation. *PLoS One*, 11, e0159909.

Moritz, C., and Agudo, R. (2013). The future of species under climate change: resilience or decline? *Science*, 341, 504–8.

Neelin, J.D., Munnich, M., Su, H., Meyerson, J.E., and Holloway, C.E. (2006). Tropical drying trends in global warming models and observations. *Proceedings of the National Academy of Sciences of the United States of America*, 103, 6110–15.

Pacifici, M., Foden, W.B., Visconti, P., Watson, J.E.M., Butchart, S.H.M., Kovacs, K.M., Scheffers, B.R., Hole, D.G., Martin, T.G., Akçakaya, H.R., Corlett, R.T., Huntley, B., Bickford, D., Carr, J.A., Hoffmann, A.A., Midgley, G.F., Pearce-Kelly, P., Pearson, R.G., Williams, S.E., Willis, S.G., Young, B., and Rondinini, C. (2015). Assessing species vulnerability to climate change. *Nature Climate Change*, 5, 215–25.

Parmesan, C., and Yohe, G. (2003). A globally coherent fingerprint of climate change impacts across natural systems. *Nature*, 421, 37–42.

Peery, M.Z., Gutiérrez, R.J., Manley, P.N., Stine, P.A., and North, M.P. (2017). *Synthesis and interpretation of California spotted owl research within the context of public forest management. Bioregional Assessment of the California Spotted Owl.* USDA Forest Service, Albany, CA. PSW-GTR-254, 263–91.

Piersma, T., Lok, T., Chen, Y., Hassell, C.J., Yang, H.Y., Boyle, A., Slaymaker, M., Chan, Y.C., Melville, D.S., Zhang, Z.W., and Ma, Z. (2016). Simultaneous declines in summer survival of three shorebird species signals a flyway at risk. *Journal of Applied Ecology*, 53, 479–90.

Pulido, F., Berthold, P., and van Noordwijk, A.J. (1996). Frequency of migrants and migratory activity are genetically correlated in a bird population: evolutionary implications. *Proceedings of the National Academy of Sciences of the United States of America*, 93, 14642–7.

Rockwell, S.M., Bocetti, C.I., and Marra, P.P. (2012). Carry-over effects of winter climate on spring arrival date and reproductive success in an endangered migratory bird, Kirtland's Warbler (*Setophaga kirtlandii*). *The Auk*, 129, 744–52.

Root, T.L., Price, J.T., Hall, K.R., Schneider, S.H., Rosenzweig, C., and Pounds, J.A. (2003). Fingerprints of global warming on wild animals and plants. *Nature*, 421, 57–60.

Şekercioğlu, C.H., Primack, R.B., and Wormworth, J. (2012). The effects of climate change on tropical birds. *Biological Conservation*, 148, 1–18.

Silverin, B., Massa, R., and Stokkan, K.A. (1993). Photoperiodic adaptation to breeding at different latitudes in great tits. *General and Comparative Endocrinology*, 90, 14–22.

Small-Lorenz, S.L., Culp, L.A., Ryder, T.B., Will, T.C., and Marra, P.P. (2013). A blind spot in climate change vulnerability assessments. *Nature Climate Change*, 3, 91–3.

Studds, C.E., Kyser, T.K., and Marra, P.P. (2008). Natal dispersal driven by environmental conditions interacting across the annual cycle of a migratory songbird. *Proceedings of the National Academy of Sciences of the United States of America*, 105, 2929–33.

Suggitt, A.J., Gillingham, P.K., Hill, J.K., Huntley, B., Kunin, W.E., Roy, D.B., and Thomas, C.D. (2011). Habitat microclimates drive fine-scale variation in extreme temperatures. *Oikos*, 120, 1–8.

Swanston, C.W., Janowiak, M.K., Brandt, L.A., Butler, P.R., Handler, S.D., Shannon, P.D., Derby Lewis, A., Hall, K., Fahey, R.T., Scott, L., Kerber, A., Miesbauer, J.W., and Darling, L. (2016). *Forest Adaptation Resources: climate change tools and approaches for land managers*. 2nd edition. Gen. Tech. Rep. NRS-GTR-87-2. U.S. Department of Agriculture, Forest Service, Northern Research Station, Newtown Square, PA.

Tempel, D.J., Gutiérrez, R.J., Whitmore, S.A., Reetz, M.J., Stoelting, R.E., Berigan, W.J., Seamans, M.E., and Peery, M.Z. (2014). Effects of forest management on California spotted owls: Implications for reducing wildfire risk in fire-prone forests. *Ecological Applications*, 24, 2089–106.

Tempel, D.J., Keane, J.J., Gutiérrez, R.J., Wolfe, J.D., Jones, G.M., Koltunov, A., Ramirez, C.M., Berigan, W.J., Gallagher, C.V., Munton, T.E., Shaklee, P.A., Whitmore, S.A., and Peery, M.Z. (2016). Meta-analysis of California Spotted Owl (*Strix occidentalis occidentalis*) territory occupancy in the Sierra Nevada: Habitat associations and their implications for forest management. *The Condor*, 118, 747–65.

Thackeray, S.J., Henrys, P.A., Hemming, D., Bell, J.R., Botham, M.S., Burthe, S., Helaouet, P., Johns, D.G., Jones, I.D., Leech, D.I., Mackay, E.B., Massimino, D., Atkinson, S., Bacon, P.J., Brereton, T.M., Carvalho, L., Clutton-Brock, T.H., Duck, C., Edwards, M., Elliott, J.M., Hall, S.J., Harrington, R., Pearce-Higgins, J.W., Hoye, T.T., Kruuk, L.E., Pemberton, J.M., Sparks, T.H., Thompson, P.M., White, I., Winfield, I.J., and Wanless, S. (2016). Phenological sensitivity to climate across taxa and trophic levels. *Nature*, 535, 241–5.

Thomas, C.D., and Gillingham, P.K. (2015). The performance of protected areas for biodiversity under climate change. *Biological Journal of the Linnean Society*, 115, 718–30.

Thomas, C.D., Gillingham, P.K., Bradbury, R.B., Roy, D.B., Anderson, B.J., Baxter, J.M., Bourn, N.A.D., Crick, H.Q.P., Findon, R.A., Fox, R., Hodgson, J.A., Holt, A.R., Morecroft, M.D., O'Hanlon, N.J., Oliver, T.H., Pearce-Higgins, J.W., Proctor, D.A., Thomas, J.A., Walker, K.J., Walmsley, C.A., Wilson, R.J., and Hill, J.K. (2012). Protected areas facilitate species' range expansions. *Proceedings of the National Academy of Sciences of the United States of America*, 109, 14063–8.

Van der Putten, W.H., Macel, M., and Visser, M.E. (2010). Predicting species distribution and abundance responses to climate change: Why it is essential to include biotic interactions across trophic levels. *Philosophical Transactions of the Royal Society of London Series B: Biological Sciences*, 365, 2025–34.

Van Gils, J.A., Lisovski, S., Lok, T., Meissner, W., Ozarowska, A., De Fouw, J., Rakhimberdiev, E., Soloviev, M.Y., Piersma, T., and Klaassen, M. (2016). Body shrinkage due to Arctic warming reduces red knot fitness in tropical wintering range. *Science*, 352, 819–21.

Van Gils, J.A., Piersma, T., Dekinga, A., Spaans, B., and Kraan, C. (2006). Shellfish dredging pushes a flexible avian top predator out of a marine protected area. *PLoS Biology*, 4, 2399–404.

Visser, M.E. (2008). Keeping up with a warming world: Assessing the rate of adaptation to climate change. *Proceedings of the Royal Society of London Series B: Biological Sciences*, 275, 649–59.

Watson, J.E.M., Dudley, N., Segan, D.B., and Hockings, M. (2014). The performance and potential of protected areas. *Nature*, 515, 67–73.

Wauchope, H.S., Shaw, J.D., Varpe, Ø. Lappo, E.G., Boertmann, D., Lanctot, R.B., and Fuller, R.A. (2017). Rapid climate-driven loss of breeding habitat for Arctic migratory birds. *Global Change Biology*, 23, 1085–94.

Westerling, A.L., Hidalgo, H.G., Cayan, D.R., and Swetnam, T.W. (2006). Warming and earlier spring increase Western U.S. forest wildfire activity. *Science*, 313, 940–3.

Westerling, A.L. (2016). Increasing western US forest wildfire activity: Sensitivity to changes in the timing of spring. *Philosophical Transactions of the Royal Society of London Series B: Biological Sciences*, 371, 20150178.

Wheatley, C.J., Beale, C.M., Bradbury, R.B., Pearce-Higgins, J.W., Critchlow, R., and Thomas, C.D. (2017). Climate change vulnerability for species: Assessing the assessments. *Global Change Biology*, 23, 3704–15.

Wilson, S., Ladeau, S.L., Tøttrup, A.P., and Marra, P.P. (2011). Range-wide effects of breeding- and nonbreeding-season climate on the abundance of a Neotropical migrant songbird. *Ecology*, 92, 1789–98.

Wu, J.X., Wilsey, C.B., Taylor, L., and Schuurman, G.W. (2018). Projected avifaunal responses to climate change across the U.S. National Park System. *PLoS One*, 13, e0190557.

Young, B., Byers, E., Gravuer, K., Hall, K., Hammerson, G., and Redder, A. (2011). *Guidelines for Using the NatureServe Climate Change Vulnerability Index, version 2.1.* NatureServe, Arlington, VA.

CHAPTER 18

Climate change in other taxa and links to bird studies

David W. Inouye

18.1 Introduction—Links between birds and other organisms

Birds occupy a variety of positions in food webs, serving as herbivores, predators, omnivores, pollinators, hosts, and prey. Given this variety of ecological roles it's not surprising that climate change can affect their interactions with other taxa in myriad ways. The direct effects of climate that can influence birds' interactions with other species include the phenology of migration (Kullberg et al. 2015; Cadahia et al. 2017; Courter 2017; Kolarova et al. 2017; Probst et al. 2017), and changes in distribution (Thomas and Lennon 1999; Hitch and Leberg 2007; Doswald et al. 2009; Zuckerberg et al. 2009; Freeman and Class Freeman 2014; Bateman et al. 2015), including elevational range changes (Tingley et al. 2012). Birds' attempts to remain within their historic climate zones (Tingley et al. 2009), can generate changes in local abundance (Huang et al. 2017) and potentially new kinds of interactions as they interact in reassorted communities.

18.2 What can we learn from studies of other taxa?

Birds are embedded in communities and ecosystems, and depend on other taxa for resources such as food and nest sites. They interact with other species in sometimes extensive networks that include predators, prey, competitors, and parasites, and can serve as mutualists in dispersal or pollination. These networks change over time and space, as birds may rely on different resources at different ages, or different times of the year, and will encounter different ecological communities as they migrate or change latitudinal or altitudinal distributions in response to the changing climate. Thus, a complete understanding of the ecological or conservation status of a bird species will require data on a host of other taxa, all of which are likely to be responding to the changing climate. An understanding of these responses will help set the context for understanding how and why avian species are in turn responding to the changing climate.

18.3 Overview of effects across trophic levels (i.e., what happens when ecological linkages are stretched or broken?)

Whether the bird species are herbivores or carnivores, resident or migratory, their interactions with the species they depend on for nutrition are likely being affected by the ongoing changes in phenology, distribution, and abundance of both plants and animals. The Green Wave Hypothesis is that migrating animals should track or 'surf' high-quality forage at the leading edge of spring green-up (Merkle et al. 2016), and evidence for large mammalian herbivores

Inouye, D.W., *Climate change in other taxa and links to bird studies*. In: *Effects of Climate Change on Birds*. Second Edition. Edited by Peter O. Dunn and Anders Pape Møller: Oxford University Press (2019).
DOI: 10.1093/oso/9780198824268.003.0018

provides some support for it (Bischof et al. 2012). It also has some support from the avian literature. For example, herbivorous waterfowl such as geese that reproduce in the Arctic are faced with different responses by plants along their migratory route that may disrupt the historical pattern of the 'green wave' of spring along the latitudinal gradient. Lameris et al. (2017) used warming experiments to study the effects of warming on plant biomass and nitrogen concentration at three sites along the migration route of barnacle geese (*Branta leucopsis*), and found that the experimental warming increased biomass accumulation but sped up the decline in nitrogen content at the Arctic breeding site but not at the lower-latitude wintering site or stop-over sites. They concluded that warming will likely change the arrival dates in the Arctic and result in a phenological mismatch, and negatively affect reproduction.

Climate change effects on small mammal populations in the Arctic (e.g., lemmings) can cascade into effects on their avian predators (snowy owls, *Bubo scandiacus*; long-tailed skuas, *Stercorarius longicaudus*) (Schmidt et al. 2012). For example, arctic ground squirrels have lower survival in areas where shrubs are encroaching, making them more susceptible to predators (Wheeler et al. 2018). A lack of snow cover for insulating the ground can also influence small mammal populations; in the winter of 2017–18 in Colorado, with near-record low snowfall, the ground froze and appeared to cause the local extinction of pocket gophers, which have to be able to access the ground surface during the winter to deposit dirt from their tunnels as they excavate roots for food (Inouye, pers. obs.).

Temperature can affect the abundance and the timing of migrations of fish (Taylor 2008; Kuczynski et al. 2017), which can have consequences for avian predators. Insect migrations are also changing, with potential consequences for insectivorous birds (Bell et al. 2015). The phenology of emergence from hibernation by mammals (Inouye et al. 2000; Lane et al. 2012; Sheriff et al. 2013), and of abundance of insect species (Ellwood et al. 2012; Boggs 2016; Forrest 2016; Scranton and Amarasekare 2017) are changing in response to climate change, and because these species can be important food resources for birds, there are potential impacts on these predator–prey interactions.

Given the large number of interacting species in ecological communities it's not surprising that they would exhibit disparate responses to climate change, resulting in unequal rates of change in phenology over time, and hence the potential for phenological mismatch as warming progresses (Ovaskainen et al. 2013). There are a few instances where we have information on how prey species are changing their phenology, and how bird species are responding. For example, the onset of breeding by pied flycatchers (*Ficedula hypoleuca*) and great tits (*Parus major*) in Scandinavia is tracking closely the spring phenology of local insects (Valtonen et al. 2017). In another study, however, caterpillar prey phenology was changing more rapidly than that of predatory birds (Visser et al. 2006; Burger et al. 2012), Similarly, hummingbird arrival dates at the end of migration to the breeding ground are changing, although not as quickly as the flowering dates of their floral resources (McKinney et al. 2012). Thus, historical patterns in phenology can change at different rates in different species, creating new patterns of phenological mismatch that can break food web links due to temporal or spatial changes in distribution. These mismatches can have significant consequences for fitness of bird populations (Senner et al. 2017), although if the birds have sufficient flexibility in their foraging behaviour they may be able to compensate (Mallord et al. 2017).

Not all impacts of climate change cascade into negative effects on birds, however. For example, the rapid melting of Arctic glaciers can affect phytoplankton and raise them closer to the surface, making them more accessible to seabirds, facilitating their foraging and reproduction (Stempniewicz et al. 2017).

18.3.1 Birds as hosts (disease, parasites)

Parasites and their vectors are also subject to changes in their distribution and abundance as a consequence of the changing climate. For example, the altitudinal distribution of malaria-bearing mosquitoes has increased in Hawaii, which has had an impact on native Hawaiian birds (Atkinson et al. 2014; Glad and Crampton 2015; Liao et al. 2017). A similar pattern has been observed in the Colorado Rocky Mountains, with disease-transmitting mosquito species moving up in altitude and affecting hosts that previously reproduced at altitudes high enough to avoid them (R. Conover, pers. comm.). Increasing latitudinal distribution of avian malaria is also a concern, as shown by recent studies in Alaska (Loiseau et al. 2012) and France (Loiseau et al. 2013). West Nile virus is another disease that can affect birds, and its distribution is also being affected by climate change (Harrigan et al. 2014). The diverse parasitic taxa that can affect birds, including protozoa, feather parasites, diptera, ticks, mites, and fleas, may in part be buffered from climate change by virtue of living on their avian hosts, but parasite abundance is increasing over time, with consequences for clutch size, brood size, and body condition (Møller et al. 2014). Warming in the Arctic may facilitate trematode life cycles, making it more likely they will parasitize avian hosts (Galaktionov 2016).

18.3.2 Birds as prey

Polar bears have been strongly affected by the warming climate and subsequent loss in the distribution of sea ice, as this affects their ability to harvest seals as prey (Hamilton et al. 2017). An unforeseen consequence has been increasing predation on bird eggs, nestlings, and even adults (Dey et al. 2017; Waters 2017). Some bird species depend on camouflage for protection against predators. Ptarmigan (*Lagopus* spp.) are an example, and they change their plum age colour to white during the winter to provide a match with the snow. But as the temporal pattern of snow cover diminishes, there is the potential for them to turn white before the ground is white, and to remain white after the snow melts, which would greatly increase the ability of predators to spot them. A similar problem faces snowshoe hares (*Lepus americanus*) (Zimova et al. 2014), which can experience strong selection on coat colour phenology; animals mismatched with the background experienced weekly survival decreases up to 7 per cent (Zimova et al. 2016).

18.3.3 Habitat

Most bird species have relatively specific habitat requirements, often determined by the type of vegetation present. For example, woodpeckers need trees, and some birds require alpine tundra. Tree distributions in turn determine woodpecker distributions; a recent survey of Picid distributions found that there was a strong positive relationship between woodpecker species richness and current tree cover and annual precipitation, respectively (Ilsoe et al. 2017). Because we know from climate models and recent empirical evidence that precipitation patterns are changing, and that tree distributions will too, we can infer that woodpecker distributions will have to change too. Modelling studies suggest that many plant species will respond to climate change with altered distributions and abundances. For example, the alpine area in the Italian alps, the suitable habitat for rock ptarmigan (*Lagopus muta*), is projected to decrease by up to 49 per cent, largely because there is insufficient area for that habitat to expand above its current altitudinal distribution (Ferrarini et al. 2017). Piñon-juniper woodlands in the western USA have experienced large-scale mortality in the past few decades, with reduced canopy cover and decreased cone production, and climate models predict these trends will continue (Johnson et al. 2017). This will have a large impact on distribution and abundance of the several bird species, such as pinyon jays (*Gymnorhinus cyanocephalus*), that rely on pine nuts from those trees.

We know that birds can respond to changes in habitat and temperatures by changing their distribution. A study of European species found that their ranges expanded north by 37 km between 1990 and 2008, but the climate envelope shifted much more rapidly, creating a 212 km 'climate gap' (Devictor et al. 2012). By comparison, the gap for butterfly species was only 135 km behind the climate. This is further evidence that ecological communities including birds are in the process of reassorting, and that these changes will continue for a while even if humans are able to slow or halt the ongoing climate change.

18.4 Case studies

18.4.1 *Delphinium* species and climate change effects on hummingbird resources

Broad-tailed hummingbirds (*Selasphorus platycercus*) are a resident breeding species at the Rocky Mountain

Biological Laboratory (RMBL), in Colorado, USA. They depend on a variety of wildflowers for nectar in this montane environment, some of which are specialized for pollination by hummingbirds, and others of which are shared with other pollinators like bumble bees. At this study site, we have long-term records of the weather, snowfall and snowpack, and phenology and abundance of wildflowers. We also have data on arrival dates of males in the spring (McKinney et al. 2012), These data have shown us that the growing season is typically starting earlier now than it used to, reflecting warmer spring temperatures and earlier snowmelt, and hence flowering is trending earlier. The migratory hummingbirds are arriving earlier than they used to, but their phenological advance is not matching that of some of the wildflowers that they rely on, suggesting that if these trends continue the earliest-flowering species visited by hummingbirds may eventually finish flowering before the birds arrive (McKinney et al. 2012).

The changing climate is also having effects on floral abundance, as the frequency of frost damage to flowers has increased (Inouye 2008), and because lower snowpack is correlated with lower flower production of some of the flowers that hummingbirds visit (Inouye and McGuire 1991; Inouye et al. 2002; Saavedra et al. 2003; Miller-Rushing and Inouye 2009). At least at one level, we know that hummingbirds respond to variation in flower abundance; in years with high abundance we caught fewer hummingbirds at feeders (Inouye et al. 1991). We interpret this as showing that birds may travel greater distances to feeders in years with lower flower abundance, when they may not find enough nectar in wildflowers close to nest sites. In addition to the changes in abundance, we also know that there is a community-level reassortment of phenology taking place at RMBL. Some species are advancing significantly while others are not changing, and many are also changing the shape of their flowering curves in ways that alter patterns of co-flowering (CaraDonna et al. 2014).

We also know that variation in floral resources is affecting another important group of pollinators in the montane meadows, bumblebees (Ogilvie et al. 2017). These bees are also moving up in altitude, presumably in response to the changing climate (Pyke et al. 2016). Bumblebees share some floral resources with hummingbirds, so changes in their distributions have the potential to impact the birds. Probably not in response to changing resources, but more likely in response to the warming temperatures, it seems that black-chinned hummingbirds (*Archilochus alexandri*) are moving up in altitude; we observed a male at 9500 feet for the first time in 2017.

In the same habitat other species are also responding to the changing climate. Foxes have increased their altitudinal range, as have moose, mosquitoes, fish, and a ground squirrel species. Yellow-bellied marmots (*Marmota flaviventris*) now emerge from hibernation about three weeks earlier than they did in the 1970s (Inouye et al. 2000), and put on significantly more fat than before, which has resulted in decreased winter mortality (Ozgul et al. 2010). Ground squirrels and chipmunks in the same habitats are also changing their hibernation behaviours, although not in the same way as marmots (B. Barr, pers. comm.). These results are similar to those reported for a red squirrel population in the southwest Yukon, Canada, for which timing of breeding has advanced by 18 days in a 10-year period, prompted by increasing spring temperatures and food supply. The researchers attributed the change to a combination of phenotypic changes within generations and genetic changes among generations, in response to the rapidly changing environment (Réale et al. 2003).

18.4.2 Caterpillars, birds, and trees

Birds that rely on caterpillars to feed nestlings in the spring are at the mercy of two trophic levels, the trees that produce leaves and the Lepidoptera whose caterpillars eat those leaves. Both the plant and herbivore species respond to changes in their phenology as the temperature warms, with earlier leaf-out dates and earlier hatch dates for caterpillars, but their responses are not synchronized. Thus the warming climate is disrupting the historical synchrony, for example of oak trees and winter moths in the UK (Visser and Holleman 2001). In turn, the changing but disparate responses of caterpillars and birds have affected great tit reproduction (Visser et al. 2006). Similar patterns of phenological change have also been implicated in population declines of the trans-Saharan migratory wood warbler *Phylloscopus sibilatrix* (Mallord et al. 2017).

18.4.3 Frost effects in a warming world

A paradoxical consequence of climate warming has been an increase in the incidence of late-spring frost events, which can have ecosystem-wide effects (Gu et al. 2008). This is an outcome when the advance in green-up timing outpaces that of the last spring freeze. For example, at 9500 feet in the Colorado Rocky Mountains the date of the last spring frost was historically about 10 June, and although the date of snowmelt has advanced, sometimes occurring in mid-April on south-facing slopes, the last frost date has not advanced significantly (Inouye 2008). A consequence is that the growing season starts earlier, so that frost-sensitive plant species may have developed buds by the time the frost arrives, sometimes causing close to 100 per cent mortality. If the buds are killed, there are then no floral resources for pollinators, or seeds for the variety of animals that rely on them, and there will be no seedlings the following spring. This phenomenon of 'false spring' warming events that trigger flowering is also having severe consequences for agriculture, such as the loss of fruit following frost events in 2007 and 2012 that caused billions of dollars of losses to mid-west and eastern US agriculture (Peterson and Abatzoglou 2014), and it can have consequences for both insects and small mammals (Inouye 2000).

18.5 Critical needs and future directions

Studies of the ecological effects of climate change are somewhat challenging if you rely on observational data, as those data are only accumulated at the rate of one datum per year, and there can be substantial inter-annual variation that may require decades of observations before clear trends are evident. What was the arrival date of the migratory species on its breeding ground? What is the first date of flowering? Or what is the peak abundance of the resource of interest? Thus, access to long-term historical datasets is valuable because they can document the relationship between the ecological variable of interest and an appropriate climatic variable. Climate projections can then be combined with the historical data to permit forecasting of the effects of future climates on the ecological variables or interactions of interest. In this context, the citizen science efforts of programmes such as the USA National Phenology Network, Bumble Bee Watch, or the variety of programmes supervised by the Cornell Lab of Ornithology, are tremendous opportunities both for interested participants and for the scientists who can benefit from those datasets about birds, their food sources, and other species with which they interact. The data on spatial distributions, and how they are changing (both latitude and altitude), and data on abundances are all opportunities for non-scientists to contribute to our understanding of the ecological effects of climate change.

Another opportunity for data collection is the increasing frequency of extreme weather events, which in turn generate consequences such as heat waves, wildfires, and floods, which can have dramatic consequences for plant and animal species. Responses can be rapid, such as the recent mass mortality of flying foxes in response to record high temperatures in Australia (https://news.nationalgeographic.com/2018/01/australian-heat-wave-flying-fox-deaths-koala-spd/), or slower, such as the bleaching of coral reefs in response to warming ocean temperatures. Climate change vulnerability assessments (Wheatley et al. 2017) are one way to gain insights into how these events may influence distribution and abundance of flora and fauna in the future.

References

Atkinson, C.T., Utzurrum, R.B., Lapointe, D.A., Camp, R.J., Crampton, L.H., Foster, J.T., and Giambelluca, T.W. (2014). Changing climate and the altitudinal range of avian malaria in the Hawaiian islands—an ongoing conservation crisis on the island of Kaua'i. *Global Change Biology*, 20, 2426–36.

Bateman, B.L., Pidgeon, A.M., Radeloff, V.C., Vanderwal, J., Thogmartin, W.E., Vavrus, S.J., and Heglund, P.J. (2015). The pace of past climate change vs. potential bird distributions and land use in the United States. *Global Change Biology*, 22, 1130–44.

Bell, J.R., Alderson, L., Izera, D., Kruger, T., Parker, S., Pickup, J., Shortall, C.R., Taylor, M.S., Verrier, P., and Harrington, R. (2015). Long-term phenological trends, species accumulation rates, aphid traits and climate: Five decades of change in migrating aphids. *Journal of Animal Ecology*, 84, 21–34.

Bischof, R., Loe, L.E., Meisingset, E.L., Zimmermann, B., Moorter, B.V., and Mysterud, A. (2012). A migratory northern ungulate in the pursuit of spring: Jumping or surfing the green wave? *American Naturalist*, 180, 407–24.

Boggs, C.L. (2016). The fingerprints of global climate change on insect populations. *Current Opinion in Insect Science*, 17, 69–73.

Burger, C., Belskii, E., Eeva, T., Laaksonen, T., Mägi, M., Mänd, R., Qvarnström, A., Slagsvold, T., Veen, T., Visser, M.E., Wiebe, K.L., Wiley, C., Wright, J., and Both, C. (2012). Climate change, breeding date and nestling diet: How temperature differentially affects seasonal changes in pied flycatcher diet depending on habitat variation. *Journal of Animal Ecology*, 81, 926–36.

Cadahia, L., Labra, A., Knudsen, E., Nilsson, A., Lampe, H.M., Slagsvold, T., and Stenseth, N.C. (2017). Advancement of spring arrival in a long-term study of a passerine bird: Sex, age and environmental effects. *Oecologia*, 184, 917–29.

CaraDonna, P.J., Iler, A.M., and Inouye, D.W. (2014). Shifts in flowering phenology reshape a subalpine plant community. *Proceedings of the National Academy of Sciences of the United States of America*, 111, 4916–21.

Courter, J.R. (2017). Changes in spring arrival dates of rufous hummingbirds (*Selasphorus rufus*) in western North America in the past century. *Wilson Journal of Ornithology*, 129, 535–44.

Devictor, V., Van Swaay, C., Brereton, T., Brotons, L., Chamberlain, D., Heliola, J., Herrando, S., Julliard, R., Kuussaari, M., Lindstrom, A., Reif, J., Roy, D.B., Schweiger, O., Settele, J., Stefanescu, C., Van Strien, A., Van Turnhout, C., Vermouzek, Z., WallisDeVries, M., Wynhoff, I., and Jiguet, F. (2012). Differences in the climatic debts of birds and butterflies at a continental scale. *Nature Climate Change*, 2, 121–4.

Dey, C.J., Richardson, E., McGeachy, D., Iverson, S.A., Gilchrist, H.G., and Semeniuk, C.A.D. (2017). Increasing nest predation will be insufficient to maintain polar bear body condition in the face of sea ice loss. *Global Change Biology*, 23, 1821–31.

Doswald, N., Willis, S.G., Collingham, Y.C., Pain, D.J., Green, R.E., and Huntley, B. (2009). Potential impacts of climatic change on the breeding and non-breeding ranges and migration distance of European *Sylvia* warblers. *Journal of Biogeography*, 36, 1194–208.

Ellwood, E., Diez, J., Ibáñez, I., Primack, R., Kobori, H., Higuchi, H., and Silander, J. (2012). Disentangling the paradox of insect phenology: Are temporal trends reflecting the response to warming? *Oecologia*, 168, 1161–71.

Ferrarini, A., Alatalo, J.M., and Gustin, M. (2017). Climate change will seriously impact bird species dwelling above the treeline: A prospective study for the Italian Alps. *Science of the Total Environment*, 590, 686–94.

Forrest, J.R. (2016). Complex responses of insect phenology to climate change. *Current Opinion in Insect Science*, 17, 49–54.

Freeman, B.G., and Class Freeman, A.M. (2014). Rapid upslope shifts in New Guinean birds illustrate strong distributional responses of tropical montane species to global warming. *Proceedings of the National Academy of Sciences of the United States of America*, 111, 4490–4.

Galaktionov, K.V. (2016). Transmission of parasites in the coastal waters of the Arctic seas and possible effect of climate change. *Biology Bulletin*, 43, 1129–47.

Glad, A., and Crampton, L.H. (2015). Local prevalence and transmission of avian malaria in the alakai plateau of Kauai, Hawaii, USA. *Journal of Vector Ecology*, 40, 221–9.

Gu, L., Hanson, P.J., Post, W.M., Kaiser, D.P., Yang, B., Nemani, R., Pallardy, S.G., and Meyers, T. (2008). The 2007 eastern US spring freeze: Increased cold damage in a warming world? *BioScience*, 58, 253–62.

Hamilton, C.D., Kovacs, K.M., Ims, R.A., Aars, J., and Lydersen, C. (2017). An Arctic predator-prey system in flux: Climate change impacts on coastal space use by polar bears and ringed seals. *Journal of Animal Ecology*, 86, 1054–64.

Harrigan, R.J., Thomassen, H.A., Buermann, W., and Smith, T.B. (2014). A continental risk assessment of West Nile Virus under climate change. *Global Change Biology*, 20, 2417–25.

Hitch, A.T., and Leberg, P.L. (2007). Breeding distributions of North American bird species moving north as a result of climate change. *Conservation Biology*, 21, 534–9.

Huang, Q.Y., Sauer, J.R., and Dubayah, R.O. (2017). Multidirectional abundance shifts among North American birds and the relative influence of multifaceted climate factors. *Global Change Biology*, 23, 3610–22.

Ilsoe, S.K., Kissling, W.D., Fjeldsa, J., Sandel, B., and Svenning, J.C. (2017). Global variation in woodpecker species richness shaped by tree availability. *Journal of Biogeography*, 44, 1824–35.

Inouye, D.W. (2000). The ecological and evolutionary significance of frost in the context of climate change. *Ecology Letters*, 3, 457–63.

Inouye, D.W. (2008). Effects of climate change on phenology, frost damage, and floral abundance of montane wildflowers. *Ecology*, 89, 353–62.

Inouye, D.W., Barr, B., Armitage, K.B., and Inouye, B.D. (2000). Climate change is affecting altitudinal migrants and hibernating species. *Proceedings of the National Academy of Sciences of the United States of America*, 97, 1630–3.

Inouye, D.W., Calder, W.A., and Waser, N.M. (1991). The effect of floral abundance on feeder censuses of hummingbird abundance. *Condor*, 93, 279–85.

Inouye, D.W., and McGuire, A.D. (1991). Effects of snowpack on timing and abundance of flowering in *Delphinium nelsonii* (Ranunculaceae): Implications for climate change. *American Journal of Botany*, 78, 997–1001.

Inouye, D.W., Morales, M.A., and Dodge, G.J. (2002). Variation in timing and abundance of flowering by *Delphinium barbeyi* Huth (Ranunculaceae): The roles of snowpack, frost, and La Niña, in the context of climate change. *Oecologia*, 139, 543–50.

Johnson, K., Sadoti, G., and Smith, J. (2017). Weather-induced declines in piñon tree condition and response of a declining bird species. *Journal of Arid Environments*, 146, 1–9.

Kolarova, E., Matiu, M., Menzel, A., Nekovar, J., Lumpe, P., and Adamik, P. (2017). Changes in spring arrival dates and temperature sensitivity of migratory birds over two centuries. *International Journal of Biometeorology*, 61, 1279–89.

Kuczynski, L., Chevalier, M., Laffaille, P., Legrand, M., and Grenouillet, G. (2017). Indirect effect of temperature on fish population abundances through phenological changes. *PLoS One*, 12, 13.

Kullberg, C., Fransson, T., Hedlund, J., Jonzén, N., Langvall, O., Nilsson, J., and Bolmgren, K. (2015). Change in spring arrival of migratory birds under an era of climate change, Swedish data from the last 140 years. *Ambio*, 44, 69–77.

Lameris, T.K., Jochems, F., Van Der Graaf, A.J., Andersson, M., Limpens, J., and Nolet, B.A. (2017). Forage plants of an Arctic-nesting herbivore show larger warming response in breeding than wintering grounds, potentially disrupting migration phenology. *Ecology and Evolution*, 7, 2652–60.

Lane, J.E., Kruuk, L.E.B., Charmantier, A., Murie, J.O., and Dobson, F.S. (2012). Delayed phenology and reduced fitness associated with climate change in a wild hibernator. *Nature*, 489, 554–7.

Liao, W., Atkinson, C.T., Lapointe, D.A., and Samuel, M.D. (2017). Mitigating future avian malaria threats to Hawaiian forest birds from climate change. *PLoS One*, 12, 25.

Loiseau, C., Harrigan, R.J., Bichet, C., Julliard, R., Garnier, S., Lendvai, A.Z., Chastel, O., and Sorci, G. (2013). Predictions of avian *Plasmodium* expansion under climate change. *Scientific Reports*, 3, 6.

Loiseau, C., Harrigan, R.J., Cornel, A.J., Guers, S.L., Dodge, M., Marzec, T., Carlson, J.S., Seppi, B., and Sehgal, R.N.M. (2012). First evidence and predictions of *Plasmodium* transmission in Alaskan bird populations. *PLoS One*, 7, 5.

Mallord, J.W., Orsman, C.J., Cristinacce, A., Stowe, T.J., Charman, E.C., and Gregory, R.D. (2017). Diet flexibility in a declining long-distance migrant may allow it to escape the consequences of phenological mismatch with its caterpillar food supply. *Ibis*, 159, 76–90.

McKinney, A.M., CaraDonna, P.J., Inouye, D.W., Barr, B., Bertelsen, C.D., and Waser, N.M. (2012). Asynchronous changes in phenology of migrating broad-tailed hummingbirds and their early-season nectar resources. *Ecology*, 93, 1987–93.

Merkle, J.A., Monteith, K.L., Aikens, E.O., Hayes, M.M., Hersey, K.R., Middleton, A.D., Oates, B.A., Sawyer, H., Scurlock, B.M., and Kauffman, M.J. (2016). Large herbivores surf waves of green-up during spring. *Proceedings of the Royal Society of London Series B: Biological Sciences*, 283: rspb.2016.0456

Miller-Rushing, A.J., and Inouye, D.W. (2009). Variation in the impact of climate change on flowering phenology and abundance: An examination of two pairs of closely related wildflower species. *American Journal of Botany*, 96, 1821–9.

Møller, A.P., Merino, S., Soler, J.J., Antonov, A., Badás, E.P., Calero-Torralbo, M.A., De Lope, F., Eeva, T., Figuerola, J., Flensted-Jensen, E., Garamszegi, L.Z., González-Braojos, S., Gwinner, H., Hanssen, S.A., Heylen, D., Ilmonen, P., Klarborg, K., Korpimäki, E., Martínez, J., Martínez-De La Puente, J., Marzal, A., Matthysen, E., Matyjasiak, P., Molina-Morales, M., Moreno, J., Mousseau, T.A., Nielsen, J.T., Pap, P.L., Rivero-De Aguilar, J., Shurulinkov, P., Slagsvold, T., Szép, T., Szöllősi, E., Török, J., Vaclav, R., Valera, F., and Ziane, N. (2014). Assessing the effects of climate on host-parasite interactions: A comparative study of European birds and their parasites. *PLoS One*, 8, e82886.

Ogilvie, J.E., Griffin, S.R., Gezon, Z.J., Inouye, B.D., Underwood, N., Inouye, D.W., and Irwin, R.E. (2017). Interannual bumble bee abundance is driven by indirect climate effects on floral resource phenology. *Ecology Letters*, 20, 1507–15.

Ovaskainen, O., Skorokhodova, S., Yakovleva, M., Sukhov, A., Kutenkov, A., Kutenkova, N., Shcherbakov, A., Meyke, E., and Delgado, M.D.M. (2013). Community-level phenological response to climate change. *Proceedings of the National Academy of Sciences of the United States of America*, 110, 13434–9.

Ozgul, A., Childs, D.Z., Oli, M.K., Armitage, K.B., Blumstein, D.T., Olson, L.E., Tuljapurkar, S., and Coulson, T. (2010). Coupled dynamics of body mass and population growth in response to environmental change. *Nature*, 466, 482–5.

Peterson, A., and Abatzoglou, J.T. (2014). Observed changes in false springs over the contiguous United States. *Geophysical Research Letters*, 41, 2156–62.

Probst, J.C., Therrien, J.F., Goodrich, L.J., and Bildstein, K.L. (2017). Increase in numbers and potential phenological adjustment of ruby-throated hummingbirds (*Archilochus colubris*) during autumn migration at Hawk Mountain Sanctuary, eastern Pennsylvania, 1990–2014. *Wilson Journal of Ornithology*, 29, 360–4.

Pyke, G.H., Thomson, J.D., Inouye, D.W., and Miller, T.J. (2016). Effects of climate change on phenologies and distributions of bumble bees and the plants they visit. *Ecosphere*, 7, e01267.

Réale, D., Mcadam, A.G., Boutin, S., and Berteaux, D. (2003). Genetic and plastic responses of a northern mammal to climate change. *Proceedings of the Royal Society of London Series B: Biological Sciences*, 270, 591–6.

Saavedra, F., Inouye, D.W., Price, M.V., and Harte, J. (2003). Changes in flowering and abundance of *Delphinium nuttallianum* (Ranunculaceae) in response to a subalpine climate warming experiment. *Global Change Biology*, 9, 885–94.

Schmidt, N.M., Ims, R.A., Høye, T.T., Gilg, O., Hansen, L.H., Hansen, J., Lund, M., Fuglei, E., Forchhammer, M.C., and Sittler, B. (2012). Response of an Arctic predator guild to collapsing lemming cycles. *Proceedings of the Royal Society of London Series B: Biological Sciences*, 279, 4417–22.

Scranton, K., and Amarasekare, P. (2017). Predicting phenological shifts in a changing climate. *Proceedings of the National Academy of Sciences of the United States of America*, 114, 13212–17.

Senner, N.R., Stager, M., and Sandercock, B.K. (2017). Ecological mismatches are moderated by local conditions for two populations of a long-distance migratory bird. *Oikos*, 126, 61–72.

Sheriff, M.J., Richter, M.M., Buck, C.L., and Barnes, B.M. (2013). Changing seasonality and phenological responses of free-living male Arctic ground squirrels: The importance of sex. *Philosophical Transactions of the Royal Society of London Series B: Biological Sciences*, 368, rstb.2012.0480.

Stempniewicz, L., Goc, M., Kidawa, D., Urbanski, J., Hadwiczak, M., and Zwolicki, A. (2017). Marine birds and mammals foraging in the rapidly deglaciating Arctic fjord—numbers, distribution, and habitat preferences. *Climatic Change*, 140, 533–48.

Taylor, S.G. (2008). Climate warming causes phenological shift in pink salmon, *Oncorhynchus gorbuscha*, behavior at Auke Creek, Alaska. *Global Change Biology*, 14, 229–35.

Thomas, C.D., and Lennon, J.J. (1999). Birds extend their ranges northwards. *Nature*, 399, 213.

Tingley, M.W., Koo, M.S., Moritz, C., Rush, A.C., and Beissinger, S.R. (2012). The push and pull of climate change causes heterogeneous shifts in avian elevational ranges. *Global Change Biology*, 18, 3279–90.

Tingley, M.W., Monahan, W.B., Beissinger, S.R., and Moritz, C. (2009). Birds track their Grinnellian niche through a century of climate change. *Proceedings of the National Academy of Sciences of the United States of America*, 106, 19637–43.

Valtonen, A., Latja, R., Leinonen, R., and Pöysä, H. (2017). Arrival and onset of breeding of three passerine birds in eastern Finland tracks climatic variation and phenology of insects. *Journal of Avian Biology*, 48, 785–95.

Visser, M., Holleman, L., and Gienapp, P. (2006). Shifts in caterpillar biomass phenology due to climate change and its impact on the breeding biology of an insectivorous bird. *Oecologia*, 147, 164–72.

Visser, M.E., and Holleman, L.J.M. (2001). Warmer springs disrupt the synchrony of oak and winter moth phenology. *Proceedings of the Royal Society of London Series B: Biological Sciences*, 268, 1–6.

Waters, H. (2017). Can these seabirds adapt fast enough to survive a melting Arctic? *Audubon*, Winter 2017. https://www.audubon.org/magazine/winter-2017/can-these-seabirds-adapt-fast-enough-survive

Wheatley, C.J., Beale, C.M., Bradbury, R.B., Pearce-Higgins, J.W., Critchlow, R., and Thomas, C.D. (2017). Climate change vulnerability for species—assessing the assessments. *Global Change Biology*, 23, 3704–15.

Wheeler, H.C., Høye, T.T., and Svenning, J.C. (2018). Wildlife species benefitting from a greener Arctic are most sensitive to shrub cover at leading range edges. *Global Change Biology*, 24, 212–23.

Zimova, M., Mills, L.S., Lukacs, P.M., and Mitchell, M.S. (2014). Snowshoe hares display limited phenotypic plasticity to mismatch in seasonal camouflage. *Proceedings of the Royal Society of London Series B: Biological Sciences*, 281, rspb.2014.0029.

Zimova, M., Mills, L.S., and Nowak, J.J. (2016). High fitness costs of climate change-induced camouflage mismatch. *Ecology Letters*, 19, 299–307.

Zuckerberg, B., Woods, A.M., and Porter, W.F. (2009). Poleward shifts in breeding bird distributions in New York state. *Global Change Biology*, 15, 1866–83.

CHAPTER 19

Conclusions

Anders Pape Møller and Peter O. Dunn

19.1 Concluding remarks

In the eight years since the first edition of this book we have learned much more about the effects of climate change on birds. However, as noted in the previous edition, we still know much less than we thought we knew. We have made advances, but in some important areas, such as understanding how productivity and population trends respond to climate, we still need to learn much more. For example, increasingly sophisticated and large-scale studies have found evidence that climate change is associated with changes in the distribution of species; however, the causes of these changes is generally unknown, and they could easily be due to land cover changes caused directly by human development, rather than temperature. Unfortunately, the many potential causes of population change are rarely studied directly in conjunction with climate change. This points to one of the main limitations of the field; there are few experimental studies that can make inferences about causation. While there are many experiments on plants or insects, there are few on birds (e.g., Martin and Maron 2012).

Our aim in this second edition has been to provide an up-to-date overview of the effects of climate change on birds, including what has been studied and what still needs more research. In Chapter 2 we began with the overwhelming evidence for climate change and how this is linked to man-made changes in the atmosphere. For ecologists there are many sources of climate data as well as methodological issues pertaining to research on climate change. Chapter 3 reviews some of the most widely used sources of climate data, while Chapter 4 summarizes the many data sources for information on migration, phenology, distribution, and population trends of birds, including new resources, such as eBird and similar databases, which provide detailed resources. Chapter 5 discusses some of the ways to identify the climate variables that birds respond to, and how to compare these responses between populations and species. These data are being used in increasingly more sophisticated distribution models that integrate more aspects of demography, environment, genetics, and interspecific interactions (Chapter 6). Models of the effects of climate on population size are also advancing with incorporation of age structure, temporal changes in density-dependence, and more sophisticated methods of incorporating stochasticity into predicting population trends (Chapter 7).

In terms of the observed responses of birds to climate change, migration has shown some of the strongest responses, but only recently with more sophisticated tracking, such as geolocators, have we started to understand more about changes in migration by individuals and the carry-over effects on populations (Chapter 8). Another area of intense research is the effect of climate change on the timing of reproduction. Although we know much more now about the physiological and environmental basis of the timing of breeding, it is still unclear how this relates to population trends, as the rate of change in breeding date is not associated with population trends (Chapter 9). Numerous physiological responses of

Møller, A.P. and Dunn, P.O., *Conclusions*. In: *Effects of Climate Change on Birds*. Second Edition.
Edited by Peter O. Dunn and Anders Pape Møller: Oxford University Press (2019). © Oxford University Press.
DOI: 10.1093/oso/9780198824268.003.0019

birds to climate change have also been reported, with resulting changes in morphology and behaviour (Chapter 10) that provide a critical link to population and distribution models. How bird populations respond to climate change is a fundamental question. To date, most studies indicate that behavioral plasticity, rather than adaptation, is the most common response, but there is evidence that the extent of plasticity might vary between populations and species with important demographic consequences (Chapter 11). Forecasting the population consequences of climate change is a relatively new field with only 18 studies incorporating IPCC projections, and most of these projections predicted that populations will decline in the future (Chapter 12). Range expansion has often been claimed to result from climate change, but the empirical evidence suggests that there is not a simple northward shift, as many species are also moving east or west. This variation highlights our lack of understanding of the direct causes of these shifts (Chapter 13). Parasites and emerging diseases have appeared as a serious threat from climate change, but so far there has been relatively little effort to study parasites in relation to climate change, despite the potentially important human and domestic animal health consequences (Chapter 14). Predators and their prey are affected differently by climate change, perhaps mainly because generation times differ and there are many potential interactions, just as for parasites and their hosts. To date, however, there have been few studies of predator–prey interactions in birds and how they have been affected by climate change (Chapter 15). This detailed information about species interactions, as well as more integration of data on changes in climate, land use, pollution, and invasive species, is needed to better understand the causes of changes in bird communities (Chapter 16). New spatially explicit models that integrate demography with climate and land use are emerging and likely to become important tools for predicting responses to global change. Despite the lack of knowledge in some areas, conservation biologists need to make recommendations to reduce the vulnerability of species to global change. Chapter 17 reviews how we assess species vulnerability and provides some strategies that can be used to incorporate climate adaptation into conservation. Lastly, Chapter 18 examines the effects of climate change on other taxa to provide some perspective on studies of birds within the larger field of climate change biology.

19.2 Future prospects for research

In the first edition, we compiled a list of areas in serious need of research relating to the impact of climate change on populations of birds (Møller et al. 2010). Although these recommendations still stand, hardly any have been pursued in any serious way. This suggests that there is still considerable scope for innovative and ground-breaking research. While each of the 17 main chapters in this book provides extensive overviews of areas in need of research, we believe that there are five areas that are particularly important and in need of further scientific enquiry.

First, currently most research is based on change in a single weather variable (predominantly temperature) and a single character in a bird species (or other organism). Numerous factors change simultaneously: agriculture, forestry, and fisheries intensify and become ever less sustainable. Such changes in land-use and exploitation are likely to interact with climate, exacerbating the impact of climate change. For example, the California spotted owl (*Strix o. occidentalis*) depends on high-canopy cover forests and is threatened by logging, invasive barred owls (*Strix varia*), and large and more frequent wildfires that are exacerbated by climate change (Chapter 17).

Second, almost the entire literature on climate change relates to changing temperatures, mainly mean temperatures, but sometimes also minimum and maximum temperatures, and their effects on phenology. However, we know very little about the effects of multiple climatic conditions. For example, temperature has changed in many parts of Europe and North Africa, but rainfall has also increased dramatically during recent decades. Wind speeds have decreased over land, but increased over the oceans. Thus weather has become warmer, but also much more humid, and either more or less windy, which can affect aerial foragers. The effects of such multiple climate factors remain to be determined.

Third, the effects of climate change occur at many different trophic levels, complicating our ability to understand the consequences. Take for example mistiming of breeding by tits in relation to food availability. Tits (and other birds) do not only have to time their breeding to take into account maximum food availability, but also parasite abundance, predators, and intra- and interspecific competitors. It is difficult to talk about mistiming without analysing all such interacting species. All co-occurring species cannot simultaneously respond to a similar degree to all these factors. Likewise, any analysis of the effects of climate change on fitness components relies on optimization of each of these components, given the constraints on such factors.

Fourth, we need to integrate studies across the life cycle to better understand the limiting factors that are affecting bird populations. Many researchers focus on one topic (e.g., phenology or modelling) and one part of the annual cycle (e.g., breeding or migration), but such narrow studies can miss important demographic bottlenecks if they occur in areas or at times that we do not study. More collaborative teams are needed to examine the entire annual cycle of species and how it relates to population size. Excellent examples come from studies of black-tailed godwits (*Limosa limosa*) and barn swallows (*Hirundo rustica*) in Europe and Kirtland's warbler (*Setophaga kirtlandii*) in North America.

Fifth, the current climate change literature is filled with promises about predictions of future population changes, and changes in range size and communities of birds and many other organisms (Chapters 7, 8, 14, 17, and 18). Much of this literature is often based on relatively weak and untested assumptions. Therefore, many so-called projections about future developments are unlikely to be fulfilled. The main reasons are that climate change is only one factor, while many others are likely to act simultaneously, and often synergistically, on populations. Given that our current bases of knowledge are mainly derived from studies in optimal habitats under relatively benign conditions (Møller et al. 1998), it remains to be seen whether responses can be extended to these novel and extreme conditions. While there are requests by decision makers, conservationists, and the public to make such predictions, they are often surrounded by high levels of uncertainty that are often not discussed. This issue of uncertainty in our predictions is highlighted in the chapters that focus on population (Chapters 7, 12) and community (Chapter 16) projections. The IPCC uses specific statements about their conclusions and level of confidence. For example, 'very likely' means a 90–100 per cent probability of occurrence. We believe that ecologists would benefit from a more rigorous and forthright presentation of these issues with decision makers so our science is not dismissed by a small number of contrarians. New modelling procedures provide some optimism for dealing with uncertainty, and these should be pursued in the future.

References

Martin, T.E. and Maron, J.L. (2012). Climate impacts on bird and plant communities from altered animal-plant interactions. *Nature Climate Change*, 2, 195–200.

Møller, A.P., Fiedler, W., and Berthold, P. (2010). Conclusions. In: A.P. Møller, W. Fiedler, and P. Berthold (eds), *Effects of Climate Change on Birds*, pp. 311–13. Oxford University Press, Oxford, UK.

Møller, A.P., Milinski, M. and Slater, P.J.B. (1998). Stress and behavior. *Advances in the Study of Behaviour*, 27, 1–552.

Index

Abadi, F. 78
Abdu, S. 128
Abies balsamea 207
Abrams, P. 202, 206
Accipiter nisus 202
Acquarone, D. 210
Ådahl, E. 148
Advanced Very High Resolution Radiometer 30
Agapornis 228
age structure 70
Agosta, S. J. 193, 194
Agudo, R. 245
Aiello-Lammens, M. E. 156
Ainley, D. 83, 200, 208
Åkesson, S. 97
Alaemon alaudipes 125
Alauda arvensis 210
albatross, Amsterdam 156
albatross, black-browed 85, 148
albatross, wandering 86, 148
Albert, S. K. 39
Albright, T. P. 70, 122–123, 127
Alle alle 209
Allen, J. A. 126
Allen, J. R. M. 170
Altwegg, R. 141
Amarasekare, P. 258
Ambrosini, R. 96, 98
Amélineau, F. 209
Ammodytes marinus 209
AMO 6, 18
Anas platyrhynchos 156
Anser caerulescens 100
Anctil, A. 208
Andersson, M. B. 206
animal model 137
Anser brachyrhynchus 98
Anser caerulescens 150
antelope, saiga 171
Anthus berthelotii 190
Aptenodytes patagonicus 208
Aptenodytes forsteri 84–85, 148–151, 243, 246
Araújo, M. B. 67, 69, 128
Archilochus alexandri 260
Ardea alba 250
Arndt, D. S. 6
Arnold, J. M. 122, 129
Arnold, S. J. 135
arrival date 94
Ashcroft, M. B. 247
Athene cunicularia 126
Atkinson, C. T. 195, 258
Atlantic Multi-Decadal Oscillation 6, 18
atlases 38, 39
atmospheric circulation 13
attribution 173
auk, great 171
auk, little 209
auklet, Cassin's 151
Ault, T. R. 33
Auriparus flaviceps 122
Austin, G. T. 123, 200
Austin, M. P. 62
auto-correlation, environmental 79
AVHRR 30, 31
Avilés, J. 188
Aythya affinis 156
Aythya marila 156

babbler, southern pied 123, 124
Bagchi, R. 177
Bailey, L. 45, 47–49, 53, 55, 78, 142
Bakke, Ø. 148
Bakken, G. S. 120
Bale, J. 202–203
Balmer, D. E. 172–173
Barbet-Massin, M. 174, 223–224, 240
Barbraud, C. 75, 83, 148, 159, 208, 222
Barnard, P. 168, 174–176, 239
Barnes, I. 171
Barnston, A. G. 32
Barriocanal, C. 98
Bascompte, J. 199
Bastille-Rousseau, G. 201
Bateman, B. L. 257
Baudrot, V. 206
Bay, R. A. 140, 178
Beale, C. M. 249
bear, polar 212
Beaugrand, G. 199
Beissinger, S. R. 199, 225, 227
Berg, M. P. 251
Bellard, C. 199
Bennett, P. 208, 213
Benning, T. L. 228
Bergmann, C. 126, 136, 207
Berkeley Earth 31
Berry, P. M. 80
Berryman, A. 204
Berthold, P. 93, 221
Betts, M. G. 247
Bickford, D. 136, 149
Bijlsma, R. 206, 208
Bilharziella 189
Bird, J. P. 160
bird migration stations 38
bird observations 39
bird ringing 38, 30
bird survey data 38
Birkhead, T. R. 171
Birks, H. J. B. 170–171
Bischof, R. 258
Bitterlin, L. R. 94, 103
blackbird 205, 210
blackcap 195
Blair, M. J. 39, 174, 176
Blakers, M. 64
Blanckenhorn, W. 149
Blows, M. W. 138
Blunden, J. 6
Blunier, T. 170
Bocedi, G. 212
Boeckmann, M. 192
Boggs, C. L. 258
Böhning-Gaese, K. 200, 222, 227
Bolker, B. M. 77
Bolus, R. T. 104
Bonamour, S. 139
Bonhomme, R. 48

Bonnet, X. 236
Bonnot, T. W. 156
Bonsall, M. 206
booby, blue-footed 148
Borrelia burgdorferi 188, 192
Both, C. 93, 95, 98, 114, 200, 202, 205, 207, 209, 211, 243, 253
Botkin, D. B. 169
Bourgault, P. 110
Bowler, B. 206,
Bowler, D. E. 229
Boyce, M. 205
Bradshaw, A. D. 178
Brambilla, M. 168, 174
Brawn, J. D. 225
breeders' equation 137
breeding bird survey 38
Bretagnolle, V. 200, 207, 212
Brommer, J. E. 96, 138, 140, 172, 203, 211, 225
Brook, B. W. 86, 170, 228
Brotons, L. 221
Brown, J. 199
Brown, J. H. 221
bear, polar 207
bear, brown 203
Bell, J. R. 258
Bryant, D. 203, 207
Bubo scandiacus 170, 210, 258
Bubulcus ibis 172
Buchanan, K. L. 129
Buckley, L. B. 56, 128, 236
bunting, yellow-breasted 171
Burger, C. 251, 258
Burgess, M. D. 209
Burthe, S. 240
Bussière, E. M. S. 94
Bustamante, J. 108
Butchart, S. H. M. 160
Buteo buteo 210
Buteo lagopus 203
buzzard, common 210
buzzard, rough-legged 203
Byers, B. 240

Cadahia, L. 257
Cahill, A. E. 199
Cain, S. 247
Caizergues, A. 208
Calidris canutus 99, 136, 253
Callocephalon fimbriatum 126
Calonectris diomedea 151
Calyptorhynchus latirostris 121
Campbell-Tennant, D. J. E. 126
Campos, P. F. 171
Campylorhynchus brunneicapillus 122
Cantarero, A. 194
Canis lupus 204
Canis lupus dingo 207
Caprimulgus rufigena 122
Capuccino, N. 204
CaraDonna, P. J. 260
Cardellina pusilla 250
cardinal, northern 237
Cardone, J. S. 15
Cardinalis cardinalis 237
Cardinalis sinuatus 123
Carey, P. D. 47, 148
caribou 205
Carlson, A. D. 170, 194
Carmi, N. 129
Caro, S. P. 108–109, 111
Caron, J. M. 17
Carr, J. A. 240
carry-over effects and climate change 99
Case, M. 241
Castaño, F. 191
Caswell, H. 148, 156, 159
Catharacta skua 122
CDR 30
Cervus canadensis 228
Cervus elaphus 115
CH_4 8
Chamaille-Jammes, S. 211
Chambers, L. E. 94, 113, 199
Charmantier, A. 135, 137–140, 207
Chastel, O. 110, 111
Che-Castaldo, C. 84–85, 155
Chen, I. C. 172–173, 192
Cheng, L. 10
Chevin, L.-M. 142
Choat, B. 225
Chordeiles acutipennis 122–123
Cinclus cinclus 80, 156
citizen science data 3
Clamator glandarius 188
Clark, F. 78,
Clark, J. S. 147
Clark, R. G. 114
Clavero, M. 229, 230
Civitello, D. J. 193
climate change 5–25
climate refugia 247
Climate Research Unit 30
climate sensitivity 44–59
climate window analysis 47–48, 52
climatic sensitivity of individuals, populations and species 44
climatology network 29–33
Clobert, J. 200
CO_2 8
cockatoo, Carnaby's 121
cockatoo, gang-gang 126
Cockburn, A. 46–49, 51, 52
Coe, S. J. 241
Cohen, E. B. 98, 101, 103
Collins, M. 178
Combes, C. 193
community 221
community changes 225
communities and climate 222
communities and disturbance 230
communities and functional groups 227
communities and land-use 229
communities and species richness 223
conservation 236–256
constraints on migration 97
Cooch, E. G. 149
Coope, G. R. 170
Cooper, C. B. 230–231, 251
Cornelissen, J. 205
Cornulier, T. 204, 211
Cotton, P. A. 94, 222
Coulson, J. 204, 212
Courter, J. R. 257
Cramp, S. 171–172
Crampton, L. H. 258
Cresswell, K. A. 208–209
Crick, H. Q. P. 113
critical climatic predictors 45
crossbill 109
CRUTEM4 30
CRUTS 30
Cruz-McDonnell, K. K. 126
cuckoo, common 188
cuckoo, great spotted 188
Cuculus canorus 188
Cuervo, J. J. 252
Culp, L. A. 241, 245
Cunningham, S. J. 124–126, 173
Cyanistes caeruleus 110, 129, 136, 142, 150
Cyanomitra olivacea 190

Daly, C. 31
Dansgaard-Oeschger cycles 170
Dantas-Torres, F. 193
Darsie, Jr., R. F. 193
data 29–35
Daufresne, M. 125
Davies, S. 112
Dawson, T. P. 63, 108, 110, 177, 238, 244
Dawson, W. R. 121
de Valpine, P. 78
Deacy, W. 203
degree-days 95
DeGroote, L. W. 113

Delworth, T. L. 18
Dennis, B. 79
Dermanyssus gallinoides 191
Descamps, S. 111
Deser, C. 13, 15, 17, 20, 81
Deviche, P. 112
Devictor, V. 174, 199, 202, 222, 225–227, 259
Dey, C. J. 259
Dickinson, J. L. 251
Dicrostonyx groenlandicus 210
Dierschke, V. 99
Dietze, M. C. 147, 152, 156, 159–160
Diomedea amsterdamensis 156
Diomedea exulans 86, 148
dingo 207
dipper, white-throated 80–82, 156
dispersal 149
distribution 165–183
Divoky, G. J. 137
DNA collections 38
Doak, D. F. 79, 155, 159
Dobrowski, S. Z. 247
Dobson, F. S. 139
Doligez, B. 200
Dolman, P.M. 101
Donald, P. F. 149
Doswald, N. 174
dove, Eurasian collared 172
downscaling 34
Drent, R. H. 111
Drever, M. C. 114, 150, 156
drought 13
du Plessis, K. L. 123–124, 173
Dubuc-Messier, G. 137
Ducklow, H. 208, 209
Dudaniec, R. 191, 194
Dunn, P. O. 95, 108, 113–116, 135
Durant, J. M. 201, 208–209
Dybala, K. E. 147, 156
dynamical consequences of climate change 77–78

Easterling, D. R. 141
eBird 29
ecological niche 61
ecological niche modelling 60–73
egret, cattle 172
egret, great 250
Ehrlén, J. 151, 228
El Niño 11, 13, 16–17
Elith, J. 47, 60, 62, 68–69, 238
elk 115, 204, 228
Ellwood, E. 258
Elsen, P. 238
Elston, D. A. 47
Emberiza aureola 171
Emmerson, M. 212
Engen, S. 74, 77, 79–80, 86
Engler, J. O. 60, 70–71
Englund, G. 206
ENSO 5, 16, 18, 32
Erasmus, B. F. N. 127
Erlinge, S. 206
Estany-Tigerström, D. 229
Estes, J. A. 205
Estrada-Peña, A. 193
Etches, R. J. 120
Eudyptes chrysocome 209
Eudyptes chrysolophus 209
EURING 39
Evans, K. 204
Evans, M. I. 137, 176
evolution 134–146
extreme climatic events 141
Eyres, A. 70

fairywren, red-winged 126
fairywren, superb 137
Falconer, D. S. 137
FAPAR 30, 31
Fasullo, J. 17
FAUNMAP Working Group 170
Fay, R. 86
Feingold, G. 8
Ferrarini, A. 259
Fey, S. B. 142
Fick, S. E. 30
Ficedula albicollis 136
Ficedula hypoleuca 94, 114, 136, 194, 258
Fiedler, W. 39
Finch, T. 100
finch, zebra 129
Fink, D. 227
fir, balsam 207
Fisher, J. 171
Fisher, R. A. 79
Fishpool, L. D. C. 176
flycatcher, collared 136, 140
flycatcher, pied 94, 114, 136, 194, 258
Foden, W. B. 171, 241
Fontaine, J. B. 230
food abundance 111
Forchhammer, M. 200, 205
forcing 7–9, 69
Fordham, D. A. 70–71
Forrest, J. R. 258
Forsman, J. T. 221
Fourcade, Y. 67
Fowler, H. J. 34
Fraction of Absorbed Photosynthetically Active Radiation 30
Franklin, J. 60–61, 63, 65, 68, 238
Franks, S. E. 95, 115, 139, 203, 222
Fratercula arctica 209
Frederiksen, M. 45, 86, 149
Freeman, B. G. and Class Freeman 257
Frey, S. J. K. 247
Fuglei, E. 212
Fuller, T. 193
fulmar, northern 171
fulmar, southern 85–86, 149
Fulmarus glacialis 171
Fulmarus glacialoides 85–86, 149
functional response 206
fundamental niche 165
Fyfe, J. C. 18

Galaktionov, K. V. 259
Galbraith, H. 241
Gamelon, M. 80–82, 87, 160, 205
Garamszegi, L. Z. 190, 228
Garant, D. 138
Garcia, J. T. 170
Garcia, R. A. 200
Gardali, T. 241, 245
Gardner, J. L. 125–126, 149, 222
Garnett, S. 242
Gaston, A. J. 122
Gaston, K. J. 168, 249
Gätke, H. 39
Gauthier, G. 154, 159
Gaüzère, P. 174, 227, 249
generalist vs. specialist 210
genetic architecture 138
Germain, R. R. 138
Gerson, A. R. 120, 129
GHCN Daily 29
GHCN Monthly 29
Gibbons, D. W. 172–173
Gienapp, P. 46–48, 51, 135, 138, 140
Gigantobilharzia 189
Gilg, O. 210, 212
Gillings, S. 172–173
Gilman, S. E. 211
Gil-Tena, A. 229
Gillingham, P. 57, 249
Gillings, S. 225
Gilman, S. E. 199
Gilpin, M. E. 77
Gilroy, J. J. 96
Ginzburg, L. 202, 206
glaciers 12
Glad, A. 258
Glaucidium passerinum 210
Glick, P. 244
Global Biodiversity Information facility 39

Global Circulation Models 34
Global Climate Model 30
Global Climate Observing System Working Group on Surface Pressure 32
global mean surface temperature 7
global warming 3
GMST 7–8, 10
Godet, L. 226–227
godwit, black-tailed 267
godwit, Hudsonian 100
goldfinch, lesser 123
González-Quevedo, C. 190
goose, snow 100, 150
Gordo, O. 98, 99
Gordon, C. E. 207
Gosler, A. G. 40
Goward, S. N. 30
Goymann, W. 109
Grabherr, G. 172
Graham, C. H. 66
Graham, R. W. 170
Grant, G. S. 122
Grant, P. R. 142
Grant, R. B. 142
Green, R. E. 172–173, 210, 230
green wave hypothesis 257
greenhouse gases 5, 21
Gregory, R. D. 173
grosbeak, blue 250
Grosbois, V. 77, 171
Grøtan, V. 87
growth bars 40
Gu, L. 261
Guéry, L. 156
Guglielmo, C. G. 129
Guillemain, M. 189
Guillera-Arroita G. 39, 67
Guisan, A. 60, 61, 63, 65, 67–68, 80
gull, common 138
Gustafsson, L. 137
Gymnorhinus cyanocephalus 259

HadCRUTT4 30
Hadfield, J. D. 139
HadSST 30, 31
Haematopus ostralegus 80, 142
Haest, B. 46–47, 51–52
Haffer, J. 247
Hagemeier, W. J. M. 39, 174, 176
Hahn, T. P. 110–111
Hainsworth, F. R. 123
Hakkarainen, H. 206
Hallett, T. B. 151
Hamel, S. 156
Hamilton, C. D. 259
Hannah, L. 200, 249
Hansen, T. F. 138
Hanski, I. 61
Hare, S. R. 32
hare, snow-shoe 201, 259
Haridas, C. V. 79
Harrigan, R. J. 259
Harris, D. J. 156, 160
Harris, I. 30
Harris, J. B. C. 229
Harrison, J. A. 172, 176
Harrison, X. A. 99
Harvell, C. D. 187, 199
Hasle, G. 192
Hassell, M. 206
Hau, M. 109
Hawkins, E. 5, 9, 147–148, 151–153
Hays, J. D. 170
Hedenström, A. 97
Heino, M. 79
Heller, N. E. 229, 236
Henden, J. 204
Henderson, C. R. 137
Hendry, A. P. 139
Hennin, H. L. 110, 112, 116
Herakis gallinarum 189
Herfindal, I. 77, 86
heritability 137
Hickling, R. 167, 172, 173
Hieraaetus sp. 170
Hijmans, R. J. 30, 34, 67
Hiley, J. R. 249
Hilton, G. 209
Hindell, M. A. 113
Hirundo rustica 95, 96, 267
Hitch, A. T. 172, 257
Hjort, C. 39
Hochachka, W. M. 39
Hockey, P. A. R. 172–173, 229
Hof, C. 236, 242
Hoffmann, A. A. 139, 142
Holberton, R. L. 99
Hole, D. G. 174, 176, 177, 200
Holleman, L. J. M. 260
Holm, S. R. 221
Holloway, P. 46–49
Holloway, S. 172
honeyeater, white-plumed 126
Hõrak, P. 189
hormones and migration 109
hornbill, yellow-billed 125
Hornseth, M. L. 243
horseshoe crab 99
host 187–198, 258–259
Houle, D. 138
Hovinen, J. E. H. 209
Howard, C. 176, 223
Hoy, S. R. 206
Huang, Q. Y. 257
Huete, A. 31
Huey, R. B. 128
hummingbird, black-tailed 260
hummingbird, broad-tailed 259
Humphries, G. R. W. 160
Huntley, B. 39, 80, 168–175, 179, 200, 222, 223, 238
Hurlbert, A. H. 67
Hurrell, J. W. 13, 15–17, 19–20, 81
Husby, A. 47, 57, 139
Hutchinson, G. E. 165–166, 169
Hylander, K. 228
Hylocichla mustelina 102, 156
hyperthermia 121

Ikegami, K. 110
Iknayan, K. J. 199, 225
Iles, D. T. 56, 151, 155, 160
Illan, J. G. 80
Ilsoe, S. K. 259
immunity 192–194
Ims, R. A. 204, 211–212
indices of circulation variability 6
Inouye, D. W. 258–261
intensity of infection 189
Intergovernmental Panel on Climate Change 3, 5, 20–21
invasion 172
invasive species 228
IPBES 60, 68
IPCC 3, 5–9, 17, 122, 147, 244, 252
Irons, R. D. 111, 136
Iverson, S. A. 212
Ixodes scapularis 188, 192

Jablonski, D. 142
Jackson, S. T. 169–170
Jaksic, F. 206
Jarvis, A. 60
Jarzyna, M. A. 228
jay, piñon 259
Jenni, L. 95
Jenouvrier, S. 75, 83–87, 147–153, 155–156, 159, 208–209, 211, 222, 236, 243, 246
Jepsen, J. 211
Jetz, W. 67, 222–223, 238, 244
Jiguet, F. 221, 227–228
Johansson, J. 114, 116
Johnson, A. L. 110, 111
Johnson, C. N. 213, 259
Johnston, A. 249
Jones, G. M. 248–249
Jones, P. D. 30
Jonzén, N. 94

Joyner, T. A. 192
Justice, C. 31

Kahliq, I. 70, 128
Kalnay, E. 32
Kamp, J. 171
Kampichler, C. 229
Kanamitsu, M. 32
Kaňuščák, P. 126
Karell, P. 140, 207
Karter, A. J. 193
Kati, V. I. 224
Kausrud, K. 204
Kay, J. E. 81
Kearney, M. R. 52, 127
Kemp, A. C. 124
Kennedy, J. 30
Kéogan, K. 209
Keppel, G. 247
Kéry, M. 86, 95, 147
Kilpatrick, A. M. 193, 228
Kingsolver, J. G. 56, 236
Kirby, J. 199
Kissling, W. D. 224
kite, snail 210
kittiwake, black-legged 171, 209
Klassen, M. 129
Klausmeyer, K. R. 245
Klomp, H. 138
Kluen, E. 113, 222
Knape, J. 78
knot, red 136, 252
Knudsen, E. 236
Knutti, R. 154
Ko, C. Y. 225
Kolárová, E. 94, 257
Koop, J. A. H. 191
Korpimäki, E. 206
Krebs, C. 203
krill 208
Kristensen, N. P. 102, 113, 199
Krosby, M. 221
Kruuk, L. E. 46–47, 137
Kuczynski, L. 258
Kullberg, C. 257
Kuris, A. M. 194
Kutz, S. J. 187

La Niña 11, 18
La Sorte, F. A. 38, 222, 226, 227, 230, 251
Lack, D. 138, 172
Lafferty, K. D. 187, 188
Lagerholm, V. K. 70
Lagopus leucura 172
Lagopus muta 259
LAI 30–31
Lambrechts, M. M. 208
Lande, R. 74, 79–80, 87, 135, 155, 204
Lane, J. E. 258
Langham, G. M. 174–175, 239, 244
Lanius collaris 124
lark, hoopoe 125
Larus canus 138
Latimer, C. E. 247
Lattin, C. R. 109
Lambrechts, M. M. 150
Lameris, T. K. 258
Lanctot, R. B. 111
Laverty, C. 201
Lawler, J. J. 223, 228
Lawson, C. R. 87
laying date 112–116, 135–142
Leaf Area Index 30–31
Leathwick, J. R. 60, 62, 238
Leberg, P. L. 172, 257
Lee, A. T. K. 239
Legagneux, P. 100
Lehikoinen, A. 111, 134, 199, 203, 206, 210, 222, 225, 227
Leighton, P. A. 192
Leitner, S. 108
lemming 210
Lemoine, N. 200, 222, 227, 229
Lenoir, J. 231
Lennon, J. J. 39, 172, 257
Lensing, J. 207
Leonardsson, K. 206
Lepus americanus 201, 259
Leroux, S. J. 229
Letnic, M. 207
Leucocytozoon 190
Lewis, J. W. 187
Liao, W. 150, 160, 192, 258
Liebezeit, J. R. 111, 242
limits to species distributions 60–63
Limosa haemastica 100
Limosa limosa 267
Limulus polyphemus 99
Lindholm, C. G. 39
Lindström, A. 174, 227
Lindström, J. 125, 149
Ling, S. 213
Lister, A. M. 178
Littell, J. S. 248
Liukkonen-Anttila, T. 170
Livezey, R. E. 32
Logan, J. 202
Loiseau, C. 192, 259
Loison, A. 80
Lonicera korolkowii 32
Lonicera tatarica 32
long-term climate data 39
long-term data 37
López-Calderon, C. 98, 100
Lord, J. P. 207
lovebird 228
Lovejoy, T. 200
Loxia sp. 109
Ludwig, D. 86
Lurgi, M. 199
Lutscher, F. 204
Lymbery, A. J. 188
Lyme disease 192
Lynx 201
Lynx canadensis 201

Macauley Library 40
machine learning 47
Mackay, T. F. C. 137
MacLean, S. A. 199, 225, 227
MacLeod, C. J. 194
Magginni, R. 242
magpie 188
mallard 156
Mallord, J. 258, 260
Malurus cyaneus 137
Malurus elegans 126
mammoth, wooly 171
Mammuthus primigenius 171
Mann, M. E. 19
Mantua, N. J. 32
Marcogliese, D. J. 193
Mareca americana 156
Mariette, M. M. 129
Marini, M. A. 223
marmot, yellow-bellied 260
Marmota flaviventris 260
Maron, J. L. 116, 221, 228, 265
Marra, P. P. 95, 99, 252
Marrot, P. 135, 142
Martay, B. 75, 85
martin, crag 210
Martin, J. G. 56
Martin, T. 116, 200, 221, 228, 265
Martínez, J. 187, 193
Marzal, A. 188
Mas-Coma, S. 189
Mason, S. C. 167
Massa, C. 172
Massimino, D. 223
Mathewson, P. D. 128
Mawdsley, J. R. 236
MaxEnt 68
Mayhew, P. J. 221
McCann, K. 206
McCarthy, J. J. 236, 244
McCauley, L. A. 243
McClanahan, T. R. 193
McCleery, R. H. 40, 209

McCluney, K. E. 200
McClure, C. J. W. 213
McDermott, M. E. 113
McGrath, L. J. 99
McGuire, A. D. 260
McKechnie, A. E. 45, 121–123, 127–128
McKellar, A. E. 95, 98
McKinney, A. M. 258, 260
McLean, N. 53, 56–57, 113, 115, 142, 148
McMahon, S. M. 244
McNeilly, T. 178
McRae, B. H. 232
McWilliams, S. R. 97
Mech, L.H. 205
Meijer, T. 111
Meléndez, L. 190
Meller, K. 96, 222
melomys, Bramble Cay 171
Melomys rubicola 171
melting land ice 11
Merilä, J. 139
Merino, S. 187, 193–194
Merkle, J. A. 257
Mesquita, M. d. S. 46
metapopulation 61
methane 8
Methorst, J. 231
Michener, W. K. 230
Midgley, G. 229
migration 93
migration speed 97
migratory connectivity 93, 101
Millon, A. 203–206, 208, 210–212
Millar, C. I. 247
Miller, A. H. 129
Miller, C. R. 237
Miller, R. J. 138
Miller-Rushing, A. J. 260
Millien, V. 136
Mills, J. A. 42
Milne, R. 127
Mitchell, D. 128
Mitchell, G. W. 99
Mock, D. W. 125
model evaluation 64–65
model fitting 65
Moderate Resolution Imaging Spectroradiometer 31
MODIS 31
Møller, A. P. 37, 39–41, 46, 95, 101, 108, 111, 113, 115–116, 136–137, 142, 172, 188, 191, 193–194, 199–200, 202, 207–208, 210–211, 221, 237, 252, 259, 267
Mönkkönen, M. 221
Moore, F. R. 99
Morán-Ordoñez, A. 70
Morelli, T. L. 247
Moreno, J. 111, 142
Morice, C. P. 30
Moritz, C. 244
Morley, N. J. 187
Morris, T. 151, 203, 207
Morris, W. F. 79–80
Morrissey, M. B. 135
Mouquet, N. 74
Murdoch, W. 200, 202, 205
Murdock, C. C. 190
Murphy, E. 208
museum collections 38–40
Mustela vison 40
Myneni, R. B. 113
Mysterud, A. 78

Nakata, M. 94
NAO 6, 19–20, 30
National Audubon Society 174
National Ecological Observatory Network NEON 33
NCEP/NCAR Reanalysis I 32
NDVI 30, 99–100
Neelin, J. D. 246
NEON 33
Nereis diversicolor 83
Nerem, R. S. 11
nest records schemes 38
NestWatch programme 37
New, M. 67–68
Newbold, T. 211
Newman, E. A. 188
Newton, I. 60, 70, 94
Niche Mapper 127
Nicoll, M. A. C. 208
Niebuhr, C. N. 190
Niehaus, A. 209–210
Nielsen, J. T. 40, 200, 202, 206, 210–211
nighthawk, lesser 122
nightjar, rufous-cheeked 122–123
Nilsson, J.-Å 135
NO_2 8
NOAA Climate Data Records 30
NOAA's National Centers for Environmental Information 30
Noakes, M. J. 129
Nooker, J. K. 112
NPI 17
Normalized Difference Vegetation Index 30, 99–100
Norris, D. R. 99
Norris, K. 240
North Atlantic Oscillation 6, 18, 20, 30
North Pacific Index 6, 17
Northern Annular Mode 6
numerical response 206
Nussey, D. H. 138, 140

ocean expansion 11
ocean warming 11
Ockendon, N. 100, 150, 160, 228
O'Connor, C. M. 93, 99
O'Connor, J. A. 191
O'Connor, R. S. 122–123, 128
Ogilvie, J. E. 260
Olden, J. D. 193
Oliver, T. H. 70, 229
Olson, E. C. 138
Oppel, S. 136
Organization of Biological Field Stations 33
Orians, G. 208
Oro, D. 148, 209
Oswald, S. A. 122, 129, 173
Otranto, D. 193
Ottich, I. 99
Ovaskainen, O. 258
Overpeck, J. T. 170
Owens, I. P. F. 208, 213
owl, burrowing 126
owl, pygmy 210
owl, spotted 150, 248, 266
owl, snowy 170, 210, 258
owl, tawny 137, 140–141, 211
oystercatcher, Eurasian 80, 142
Ozgul, A. 260

Pacific Decadal Oscillation 16–17, 32
Pacific North American Index 6
Pacifici, M. 236
Pagel, M. 71
Paine, R. 199, 206
Palmer, G. 230
Palmer Drought Index 33
Pancerasa, M. 98
Paradis, E. 223
parakeet, ring-necked 228
parasite 187–198, 258–259
Pardo, D. 85, 147–148, 156
Paris Agreement of December 2015 5
Parmesan, C. 95, 103, 113, 172, 207, 236
parrot, mulga 126
parrot, night 128
parrot, red-rumped 126
Parus major 37, 86, 94, 110, 129, 135, 258
Pasanen-Mortensen, M. 205
Passer domesticus 110, 148
Passerculus sandwichensis 99
Passerina caerulea 250

Pattinson, N. B. 70
Pauli, J. N. 205
Paxton, K. L. 101
Peach, W. 208, 210
Pearce-Higgins, J. W. 57, 75, 86, 199, 203, 225, 242
Pearson, R. G. 63, 177, 208
Peckarsky, B. 207
Peers, M. J. L. 202, 212
Peery, M. Z. 150, 156, 248–249
Pellissier, L. 212
penguin, Adelie 84
penguin, emperor 84–85, 148–151, 243, 246
penguin, king 208
penguin, macaroni 209
penguin, rockhopper 209
Pérez-Rodríguez, A. 195
performance testing 50
Péron, C. 208
Petchey, O. L. 75
Peterson, A. T. 127, 174, 175, 201, 205, 239
Peterson, D. L. 239
Pettorelli, N. 212
Pezoporus occidentalis 128
phenology of parasites 188
phenotypic plasticity 134–146
Phillimore, A. B. 48, 51, 52, 54, 57, 142
Phillips, B. L. 187, 192
Phillips, R. A. 171
Philornis downsi 191, 194
Philornis torquans 191
photoperiod 108
Phylloscopus sibilatrix 260
Phylloscopus trochilus 115
PDO 16–17, 31
Pica pica 188
Picoides nuttallii 250
Piersma, T. 252
Pimm, S. L. 221
Pincebourde, S. 200
Pinguinus impennis 171
pipit, Berthelot's 190
Plasmodium 190
plasticity 138
Platycercus elegans 126
PNA 15
Pokallus, J. W. 205
Polis, G. 206
Polley, L. 187, 189
Pollock, L. J. 223
Pomacea paludosa 210
population consequences 148–164
population dynamics 74–90, 203
Population Prediction Interval 75
population structure 148
population studies 37–38
Porlier, M. 138
Porter, W. P. 52, 127, 200
Post, E. 200, 205, 207
Postma, E. 137, 138, 139
Potvin, D. A. 222
Poulin, R. 189, 193
Pounds, J. A. 225
Precheur, C. 208
precipitation 111
predator-prey interactions 199, 209
prevalence 189–190
prey 259
Prince, K. 222, 228
Price, J. 176
Price, P. 204
PRISM 31
Probst, J. C. 257
protected areas 249
Protocalliphora azurea 191
Przybylo, R. 140
Psephotus elegans 126
Psephotus haematonotus 126
Psittacula krameri 228
ptarmigan, rock 259
ptarmigan, white-tailed 172
Ptilotula penicillatus 126
Ptychoramphus aleuticus 151
Ptyonoprogne rupestris 210
puffin 208–209
Pulido, F. 96, 251
Pulliam, H. R. 61
Pygoscelis adeliae 84
Pyke, G. H. 260
pyrrhuloxia 123

Quadrelli, R. 16
Quaternary 169, 171
Qvarnström, A. 137
Quercus humilis 110
Quercus ilex 110

Rahel, F. J. 193
Rajasärkkä, A. 174
Ramakers, J. J. C. 138
Rand, T. 199, 206
Randolph, S. 193
Rangifer tarandus 205
range expansion, parasites 192
range shift of wintering area 95
Ranta, E. 201
Rattiste, K. 138
RCP 8
Réale, D. 260
realized niche 166
Redpath, S. 206
redstart, American 99, 100–101, 246
Reece, J. S 242
Reed, T. E. 75, 86, 115
Rempel, R. S. 243
Renner, S. S. 200
Representative Concentration Pathways 8
reproductive success 108
resilience 247
response to selection 139–140
Reudink, M. W. 101
Ricklefs, R. E. 123
Ripple, W. J. 200, 205–206
Rissa tridactyla 171, 209
Rivalan, P. 224
RNCEP 32
Roberts, A. M. I. 47–49
Robertson, C. J. R. 176
robin, American 250
Robinson, D. E. 120, 122
Robinson, R. A. 115, 173, 208
Robson, D. 98
rockjumper, Cape 127
Rockwell, S. M. 246
Rodionov, S. 206
Rodriguez, C. 108
Rodríguez, S. 129
Rohde, R. 31
Rohr, J. R. 187
Rolstad, J. 114
Root, T. 207, 236
rosella, crimson 126
Rosenblatt, A. E. 199
Rosenzweig, C. 5
Rostrhamus sociabilis 210
Rouse, Jr., J. W. 30
Rubolini, D. 39
Ruegg, K. 140
Ruffino, L. 111
Ruokonen, M. 170
Rushing, C. S. 102
Rutz, C. 206, 208

Saalfeld, S. T. 111
Saavedra, F. 260
Sæther, B.-E., 74, 75, 77, 80, 86, 114, 137, 207
Saiga tatarica 171
Saino, N. 93, 95, 97–100, 188
Saitoh, T. 170
Salido, L. 113
Santangeli, A. 225, 227, 229
Sakrejda, K. 135
Salewski, V. 40
Salvante, G. K. 111
SAM 21
Samplonius, J. M. 94
sand lance 209

Sanz, J. J. 222
Saraux, C. 111
Saunders, D. A. 121
Sauter, A. 41
Saxicola torquata axillaris 109
scaup, greater 156
scaup, lesser 156
Schaefer, H. C. 224
Schaper, S. V. 45, 110
Schaub, M. 86, 147
Scheffers, B. R. 199
Schindler, D. 200
Schleip, C. 46–48
Schmidt, N. M. 211, 258
Schmitz, O. 199, 205, 207, 212
Schnarr, E. 34
Schneider, A. 10
Schoech, S. J. 110
Scholander, P. F. 120
Schurr, F. M. 71
Schwagmeyer, P. 125
Schwartz, M. D. 32–33, 239
Scranton, K. 259
Scridel, D. 239
sea level pressure 13
sea surface temperatures 6
Seager, R. 18
seawater expansion 11
Sehgal, R. 190
Seibold, S. 199
Şekercioğlu, Ç. H. 222, 224, 239, 252
Selasphorus platycercus 259
selection 135
  on phenology 135–136
  on morphology 136
Senapathi, D. 136
Senner, N. R. 100, 258
Setophaga kirtlandi 169, 246, 267
Setophaga occidentalis 247–248
Setophaga petechia 140, 178
Setophaga ruticilla 99, 246
Sinclair, A. 203
Shakun, J. D. 170
Sharrock, J. T. R. 172
Shaw, M. R. 245
Shaw, P. 109
Shea, D. J. 19
shearwater, Scopoli's 151
Sheldon, B. C. 75, 140
Sheridan, J. A. 136, 149
Sheriff, M. J. 258
Shmueli, G. 50
shrike, common fiscal 124
Sibly, R. M. 148
Siegel, R. B. 243
Siepielski, A. M. 135
Silverin, B. 251
Simmons, K. E. L. 171
Simmons, R. E. 127
Sims, M. 48
Sinclair, A. 206
Sinelschikova, A. 97
Sirami, C. 228, 230
Siriwardena, G. M. 96
Sirkiä, P. M. 136
Skinner, W. R. 108
skua, great 122
skua, long-tailed 210, 258
skylark 210
Smallegange, I. 212
Smit, B. 45, 52, 70, 122, 127
Smith, A. D. 97
Smith, E. K. 122–123
Snow, D. W. 39
Snover, A. K. 151, 154, 160
SO_2 8
Soberón, J. 61
SOI 17
Solbu, E. B. 78
Solonen, T. 212
Solow, A. R. 53
Soulé, M. E. 205
sources of uncertainty 65–66
Southern Annular Mode 6, 21
Southern Oscillation Index 6
Sparks, T. 41, 47, 51
sparrow, Savannah 99
sparrow, house 110, 148
sparrowhawk, Eurasian 202
spatial heterogeneity 149
spatiotemporal predictions 64
species distribution models 238
species turnover 224
Spinus psaltria 123
Spooner, F. E. B. 203
Stearns, S. C. 149
Stempniewicz, L. 258
Stephens, P. A. 226
Stercorarius longicaudus 210, 258
Sterna paradisaea 237
Stevens, A. 201, 210
Stevens, B. 8
Stevenson, I. 203, 207
Stenseth, N. C. 5, 78, 201, 204–205, 207
Stocker, T. F. 172
stone-chat, African 109
Stott, I. 149
Stralberg, D. 222–224, 240
Streptopelia decaocto 172
Strix aluco 137, 140–141
Strix occidentalis 150, 248, 266
Stuart, A. J. 178
Studds, C. E. 95, 100, 251
Suggitt, A. J. 46–47, 53, 247
Sula nebouxii 148
sunbird, olive 190
Suraci, J. P. 207
Sutherland, W. J. 80, 101, 204
Sutherst, R. W. 188, 212
Suttle, K. 199
Sutton, R. 5, 9, 147–148, 151–153
Svenning, J. C. 221, 231
swallow, barn 95, 96, 100, 267
swallow, tree 114
Swanson, D. L. 128
Swanston, C. W. 238
Swart, N. C. 15
swift, Alpine 170
Sylvia atricapilla 195
Syringa chinensis 32
Szabo, J. K. 41
Szulkin, M. 114

Tabachnick, W. J. 193, 194
Tachycineta bicolor 114
Tachymarptis melba 170
Taeniopygia guttata 111, 129
Tarpley, J. 30
Tattersall, G. J. 126
Tavecchia, G. 151
Tayleur, C. M. 222, 228
Taylor, S. G. 258
teleconnections 15
Teller, B. J. 47–49, 54–55
Tempel, D. J. 248
temperature 10
  acute 121
  effects 110
  extreme 12
tern, Arctic 237
Teplitsky, C. 136, 138
Terraube, J. 200, 203, 207–208, 210–211
Thackeray, S. J. 47, 55, 112–113, 201, 252
Thalassarche melanophris 85, 148
thermal range and community shifts 225
thermo-neutrality models 128
thermoregulation 120
Thirgood, S. 206
Thomas, C. D. 39, 172, 221–222, 240, 249, 257
Thomas, D. W. 110, 114
Thompson, R. C. A. 187–188
Thompson, P. M. 171
Thompson, III, F. R. 96
Thorup, K. 99
thrasher, curve-billed 122–123
thrush, song 97, 205
thrush, wood 156

Thuiller, W. 60, 80
Tieleman, B. I. 125
timing of breeding 108
Tingley, M. W. 238, 257
tit, blue 110, 114, 129, 136, 142, 150
tit, great 37, 86, 94, 110, 129, 135, 258
Titeux, N. 230
Tockus leucomelas 123, 124–125
Toft, C. A. 193
Torti, V. M. 114
Tøttrup, A. P. 97–98
Toxostoma curvirostre 122–123
Trainor, A. M. 213
Tratham, P. 208
Travers, M. A. 193
Travis, J. M. J. 151
Trenberth, K. E. 14, 16–19, 32
Trichobilharzia 189
Trivedi, M. R. 57
trophic cascade 206
trophic networks 199
Tryjanowski, P. 51
Trzaska, S. 34
Tscharntke, T. 199, 206
Tuljapurkar, S. 79, 159
Tulloch, A. I. T. 41
Turchin, P. 204
Turdoides bicolor 123–124
Turdus migratorius 250
Turdus merula 205, 210
Turdus philomelos 97, 205
Turner, J. 12, 21
Tylianakis, J. 199, 200, 208
Tyrberg, T. 170
Tyson, R. 204

Ursus arctos 212
Ursus arctos middendorffi 203
U. S. A.-National Phenology
Network 33
U. S. Environmental Protection
Agency 33
U. S. Global Change Research
Program 33
Usui, T. 94, 98, 103

Valkema, J. 206
Vallecillo, S. 230
Valladares, V. 71
Valtonen, A. 258
Van Buskirk, J. 94, 103, 126
van Gils, J. A. 136, 252
van de Pol, M. 45–51, 53–57, 78–79,
83, 86–87, 141, 151, 156
van de Ven, T. M. F. N. 124, 126–127
van der Meer, J. 212
van der Putten, W. H. 251
van Oudenhove, L. 101
van Vuuren, D. P. 154
VanDerWal, J. 225
Vautard, R. 14, 111
Vedder, O. 142
Végvári, X. 94
verdin 122
Verhulst, S. 135
Veselý, L. 207
Villellas, J. 203
Virkkala, R. 134, 174, 199
Visser, M. 93, 98, 110–111, 114, 151,
199, 201, 222, 251, 258, 260
Voigt, W. 201
Vors, L. 205
Vucetich, J. 104, 205
vulnerability assessment 244

Wallace, J. M. 16
Walsberg, G. E. 122
Walsh, B. 138
Walther, B. A. 174, 221, 228–229
Wanless, S. 208–209
warbler, hermit 247–248
warbler, Kirtland's 169, 246, 267
warbler, willow 115
warbler, Wilson's 250
warbler, wood 260
warbler, yellow 140, 178
Ward, R. A. 193
Waters, H. 259
Watson, J. E. M. 249
Watts, H. E. 109
Wauchope, H. S. 252
weasel 40
Webster, P. J. 18, 101
Wegge, P. 114
Wegner, K. M. 193
Weimerskirch, H. 111, 148–149
Welbergen, J. A. 121
Wernham, C. 96
West Nile virus 192–193
Westerling, A. L. 248
Wetherington, M. T. 200
Wheatley, C. J. 246, 261
Whitfield, M. C. 122
Whittow, G. C. 120–121
wigeon, American 156
Wiens, J. J. 240
Wigley, T. M. L. 34
Wilby, R. L. 34
Williams, J. W. 165, 169, 170,
177, 200
Williams, T. S. 110–112
Willis, S. G. 231
Wilmers, C. 200, 205, 207
Wilson, S. 246
wind 14, 111
Winder, M. 200
Wingfield, J. C. 129
Winkler, D. W. 222
Winstanley, D. 39
Wise, D. 207
wolf 204
Wolf, B. O. 45, 122–123, 126–127
Wolf, J. B. 138, 151
Wolkovich, E. M. 56
woodpecker, Nuttall's 250
Woodward, F. I. 169
Woodworth, B. K. 129
WorldClim Version 1 30
WorldClim Version 2 30
Wright, J. 51
Wright, L. J. 173, 204
Wu, J. X. 250

Xie, S. 129

Yalcin, S. 229
Ydenberg, R. 209, 210
Yohe, G. 95, 103, 207, 236
Yom-Tov, Y. 125, 207, 222
Yong, W. 99
Young, B. 245
Young, I. 111
Younger, J. 170
Yoshimura, T. 110

Zack, S. 243
Zamora-Vilchis, I. 190, 192, 194
Zavaleta, E. S. 236
Zhao, Q. 156
Zimmermann, N. E. 60, 63
Zimova, M. 259
Zohner, C. M. 200
Zosterops lateralis 63–64
Zosterops lateralis chloronothus 64
Zuckerberg, B. 39, 172, 222, 228, 247
Zurell, D. 65, 69–71